The Origin of Mass and Strong Coupling Gauge Theories

Proceedings of the 2006 International Workshop

The Origin of Mass and Strong Coupling Gauge Theories

Proceedings of the 2006 International Workshop

Nagoya University, Nagoya, Japan 21 – 24 November 2006

Editors

M. Harada
M. Tanabashi
K. Yamawaki

Nagoya University, Japan

SCGT 06

World Scientific

NEW JERSEY • LONDON • SINGAPORE • BEIJING • SHANGHAI • HONG KONG • TAIPEI • CHENNAI

Published by

World Scientific Publishing Co. Pte. Ltd.
5 Toh Tuck Link, Singapore 596224
USA office: 27 Warren Street, Suite 401-402, Hackensack, NJ 07601
UK office: 57 Shelton Street, Covent Garden, London WC2H 9HE

British Library Cataloguing-in-Publication Data
A catalogue record for this book is available from the British Library.

THE ORIGIN OF MASS AND STRONG COUPLING GAUGE THEORIES (SCGT06)
Proceedings of the 2006 International Workshop

ISBN-13 978-981-270-641-6
ISBN-10 981-270-641-0

Printed in Singapore by B & JO Enterprise

PREFACE

Origin of Mass is the most urgent problem to be solved in particle theory and is the main target of the upcoming LHC experiments. Also, detailed mechanism of the origin of mass of hadrons should be tested through chiral symmetry restoration and deconfinement in QCD by the ongoing RHIC and upcoming LHC heavy ion collisions.

Based on the strong coupling gauge theories (SCGT), many Composite Models such as the Walking Technicolor, Top Quark Condensate, Little Higgs, Higgsless model, etc. have been proposed over decades to explain the dynamical origin of the mass of elementary particles, and are ready to be tested in the LHC. Many mechanisms proposed for the chiral phase transition in QCD should also be tested in the RHIC/LHC experiments.

The main obstacle of the strong coupling theories like QCD is the difficulty at obtaining the nonperturbative solution and making a definite prediction to be precisely compared with the experiments. Apart from the first-principle method of the lattice gauge theories, several alternative methods have been investigated to draw some, though not completely quantitative, predictions: Effective field theories (Chiral Perturbation Theory) with/without Hidden Local Symmetry (Moose or deconstructed extra dimensions), large N expansion, ladder Schwinger-Dyson Equation (gap equation) & Bethe-Salpeter equation, etc. as well as hints from some exact nonperturbative results of SUSY gauge theories.

Since the first meeting of the Nagoya SCGT Workshop held in 1988 (SCGT 88), which was motivated by the Walking Technicolor and some composite ideas like Hidden Local Symmetry, we have organized four SCGT workshops in 1988, 1990, 1996 and 2002 for discussing various developments of SCGT and new ideas. Physicists including many leading physicists from all over the world came together for the Workshops and created a new phase at each meeting.

From November 21-24, 2006, facing the start of the LHC experiment, we organized the fifth Nagoya SCGT workshop “International Workshop:

Origin of Mass and Strong Coupling Gauge Theories (SCGT 06)" in a spirit similar to the previous SCGT meetings. Among the 90 attendants induded Prof. G. 't Hooft and many eminent physicists. In addition to the traditional approaches, recent highlights include the holographic approach to the QCD and other SCGT, which shed new light on the related extra dimensions in terms of deconstruction or Moose/Hidden Local Symmetry. This volume contains 44 reports on the recent progress.

The workshop was financially supported by Daiko Foundation, The Mitsubishi Foundation, the Nagoya University Foundation, a JSPS Grant-in-Aid [(B) 18340059] and a JSPS Award for Eminent Scientists. On behalf of the Organizing Committee, we would like to express our sincere thanks to these organizations for their generous support. We also would like to acknowledge the Research Center for Materials Science, Nagoya University, for their generous offer of the Workshop site at Noyori Memorial Hall of the Center. Special thanks are due to young physicists at Nagoya University for their devoted assistance in preparing the workshop. Finally, we would like to thank Mrs. M. Kitajima for her patient assistance in administrative matters.

Editors
M. Harada
M. Tanabashi
K. Yamawaki

WORKSHOP ORGANIZATION

International Advisory Committee:

T.W. Appelquist (Yale)	M. Bando (Aichi)
W.A. Bardeen (FNAL)	S.J. Brodsky (SLAC)
G.E. Brown (Stony Brook)	R.S. Chivukula (Michigan State)
H. Georgi (Harvard)	C.T. Hill (FNAL)
M. Kobayashi (KEK)	J.B. Kogut (Illinois)
T. Maskawa (Kyoto)	V.A. Miransky (W. Ontario)
T. Muta (Hiroshima)	Y. Nambu (Chicago)
K. Nishijima (Tokyo)	M. Rho (Saclay)
G. 't Hooft (Utrecht)	

Organizing Committee:

Chairperson	K. Yamawaki (Nagoya)
Co-chairperson	M. Harada (Nagoya)
	M. Tanabashi (Tohoku)

K-I. Aoki (Kanazawa)	M. Hayakawa (Nagoya)
K. Higashijima (Osaka)	Y. Hosotani (Osaka)
K. Kanaya (Tsukuba)	Y. Kikukawa (Tokyo)
N. Kitazawa (Tokyo Metro.)	K.-I. Kondo (Chiba)
T. Kugo (Kyoto)	N. Maekawa (Nagoya)
S. Nojiri (Nagoya)	S. Sugimoto (Nagoya)
S. Uehara (Utsunomiya)	

Secretariat:

H. Fukano	K. Haba	M. Hashimoto
S. Matsuzaki	Y. Namekawa	Y. Nemoto
C. Nonaka	T. Yada	T. Yoshikawa

Sponsored by Nagoya University

1. S. Sawada
2. K. Kanaya
3. K. Yamawaki
4. G. 't Hooft
5. C. Nonaka
6. P. H. Frampton
7. Y. Hosotani
8. S. -Y. Wang
9. H. Fukaya
10. M. Tanabashi
11. M. Harada
12. H. Nishino
13. N. Watanabe
14. H. Nakkagawa
15. R. Wijewardhana
16. T. Inagaki
17. R. Casalbuoni
18. J. Hosek
19. K. Kondo
20. X. -H. Wu
21. N. Yokoi
22. K. Takenaga
23. N. Kitazawa
24. A. Matsuzaki
25. Y. Namekawa
26. G. W. Semenoff
27. S. Ichinose
28. H. -C. Cheng
29. T. Takeuchi
30. W. A. Bardeen
31. V. I. Zakharov
32. M. Hayakawa
33. J. M. Schechter
34. Y. Nemoto
35. Q. -S. Yan
36. H. Terao
37. T. Yoshikawa
38. J. -P. Lee
39. K. Aoki
40. N. J. Evans
41. R. Shrock
42. D. Anselmi
43. K. Konishi
44. N. Maru
45. D. -K. Hong
46. S. J. Brodsky
47. A. Shibata
48. Y. Furuhashi
49. T. Yada
50. J. Yasuda
51. M. Kurachi
52. T. Miyata
53. M. Kohda
54. K. Takei
55. S. Nojiri
56. M. Ishiduki
57. M. Kitajima
58. M. Hashimoto
59. M. Hayashi
60. I. Zahed
61. S. Nakamura
62. A. Tsuchiya
63. S. -G. Kim
64. S. Matsuzaki
65. S. Uno
66. N. Maekawa
67. T. Imoto
68. H. Fukano
69. P. Ko
70. D. K. Sinclair
71. S. Sugimoto
72. K. Haba
73. A. Ono
74. S. Yoshimoto
75. C. Kouvaris
76. K. Miyashita
77. T. Fujihara
78. J. W. Holt

CONTENTS

THE STRING IN AN EXCITED BARYON

Gerard 't Hooft

Institute for Theoretical Physics
Utrecht University, Leuvenlaan 4
3584 CE Utrecht, the Netherlands

and

Spinoza Institute
Postbox 80.195
3508 TD Utrecht, the Netherlands

e-mail: g.thooft@phys.uu.nl
internet: http://www.phys.uu.nl/~thooft/

A simple model is discussed in which baryons are represented as pieces of open string connected at one common point. There are two surprises: one is that, in the conformal gauge, the relative lengths of the three arms cannot be kept constant, but are dynamical variables of the theory. The second surprise is that, in the classical limit, the state with the three arms of length not equal to zero is unstable against collapse of one of the arms. After collapse, an arm cannot bounce back into existence. The implications of this finding, which agrees with an earlier report by Sharov, are briefly discussed.

An earlier version of this work was presented at Trento, Italy, Workshop on "Strings and QCD", July 2004.

1. Introduction

Mesons appear to be well-approximated by an effective string model in four dimensions, even if anomalies and lack of super-symmetry cause the spectrum of quantum states to violate unitarity to some extent. Presumably this is because there are no difficulties with classical (open or closed) strings in any number of dimensions, without any super-symmetry. String theory can simply be seen as a crude approximation for mesons, and as such it works reasonably well, although some observed bending of the Regge trajectories at low energies is difficult to reproduce in attempts at an improved treatment of the quantization in such a regime.

It would be very desirable to have an improved effective string model

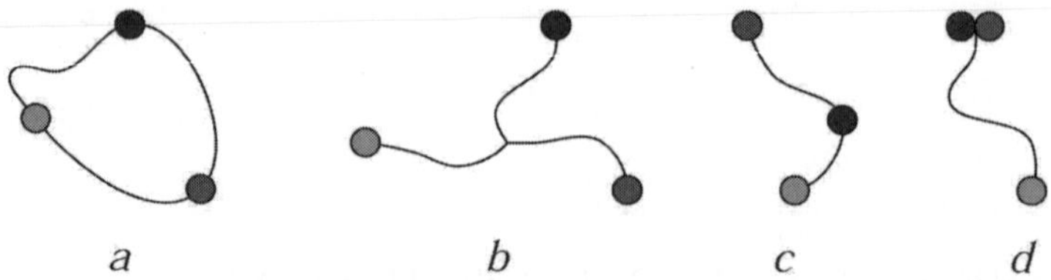

Fig. 1. Four possible string models for baryons. *a*) Three fermions on a closed string; *b*) Our starting point: quarks attached by a single string; *c*) After collapse of one arm; *d*) With two quarks in a $\overline{3}$ bound state.

for QCD even if no refuge can be taken into any supersymmetric deformations of the theory.[1] Such a string model should explain the approximately linear Regge trajectories, and if we can refine it, one might imagine using it as an alternative method to compute spectra and transition amplitudes in mesons and baryons. Ideally, a string model would not serve as a replacement of QCD, but as an elegant computational approach. Since QCD is not supersymmetric, lives in 4 rather than 10 or 26 dimensions, and has massive fermions at the end points of strings (as elementary representations of the color gauge group, rather than massless ones on the interior (which would be in the adjoint representation), one expects quite non-trivial 'corrections' before string theory can describe QCD accurately, but by inspecting the original Veneziano amplitudes for QCD, one nevertheless concludes that string theory, in its basic form, already works in an approximative sense.

The Regge trajectories for baryons appear to have the same slope as the mesonic ones. This can easily be explained if we assume baryons to consist of open strings that tend to have quarks at one end and diquarks at the other. The question asked in this short note is how a classical string model for baryons should be handled.[2,3] In the literature, there appears to be a preference for the Δ model,[4] consisting of a closed string with three fermion-like objects attached to it, see Fig. 1*a*. As closed strings can be handled using existing techniques, this is a natual thing to try.

However, from a physical point of view, the Y-configuration, sketched in Fig. 1*b* appears to be a better representation of the baryons. After all, given three quarks at fixed positions, the Y shape takes less energy. Furthermore, if two of these quarks stay close together, they behave as a diquark and the Regge spectrum (with the *same* slope as the mesons) would be readily explained. However, there turns out to be a good reason why the Y shape is dismissed;[6] we will discuss this, and then ask again whether "Δ" is to be preferred. As will be shown, our conclusion will not be that, but rather an open string with quarks at its end points. The third quark will be allowed to hover in between, attached on this single strand, or tend to stick at one of the ends, leading to a diquark. These will be the most stable configurations, but if completeness is asked for, one has to consider the set of all string configurations, containing open and/or closed pieces, with quarks at three end points.

We considered the exercise of solving the classical string equations for the Y-configuration. After we did this exercise, we found that it has been discussed already by Sharov.[5] The question will be whether we agree with his conclusions. An earlier version of this report appeared in Ref.[7]

2. The three arms

The three arms are described by the coordinate functions $X^{\mu,i}(\sigma,\,\tau)$, where the index i is a label for the arms: $i = 1, 2$, or 3. At the points $\sigma = 0$ (the Torricelli point), the three arms are connected:

$$X^{\mu,1}(0,\,\tau) = X^{\mu,2}(0,\,\tau) = X^{\mu,3}(0,\,\tau)\ . \tag{1}$$

σ is a coordinate along the three arms. It takes values on the segment $[\,0,\, L^i(\tau)\,]$, where as yet we keep the lengths $L^i(\tau)$ unspecified. Indeed, the relative lengths $L^i(\tau)$ of the three arms must be allowed to vary, and the reason for this is displayed in Fig. 2: signals running across the arms and back will have different arrival times, in general, see Fig. 2*a*. If we wish to fix the coordinates by choosing light cone coordinates, or equivalently, the conformal gauge, Eqs. (3) and (4), this will be important.

The Nambu action is

$$S = -\sum_{i=1}^{3}\int \mathrm{d}\tau \int_0^{L^i(\tau)} \mathrm{d}\sigma\,\sqrt{(\partial_\sigma X^{\mu,i}\cdot\partial_\tau X^{\mu,i})^2 - (\partial_\sigma X^{\mu,i})^2(\partial_\tau X^{\lambda,i})^2}\,\Big)\ . \tag{2}$$

Here, as in the expressions that follow, we assume the usual summation convention for the Lorentz indices $\mu,\ \lambda,\ \ldots$, but not for the branch indices i, where summation, if intended, will always be indicated explicitly. For simplicity, the masses of the quarks at the end points are neglected.

Assuming, in each of the three arms, the classical gauge condition

$$\partial_\sigma X^{\mu,i}\cdot\partial_\tau X^{\mu,i} = 0\ ; \tag{3}$$

$$(\partial_\sigma X^{\mu,i})^2 = -(\partial_\tau X^{\mu,i})^2\ , \tag{4}$$

Eq. (2) takes the form

$$S = \sum_{i=1}^{3} \int \mathrm{d}\tau \int_0^{L^i(\tau)} \mathrm{d}\sigma \left(\tfrac{1}{2}(\partial_\tau X^{\mu,i})^2 - \tfrac{1}{2}(\partial_\sigma X^{\mu,i})^2 \right) . \tag{5}$$

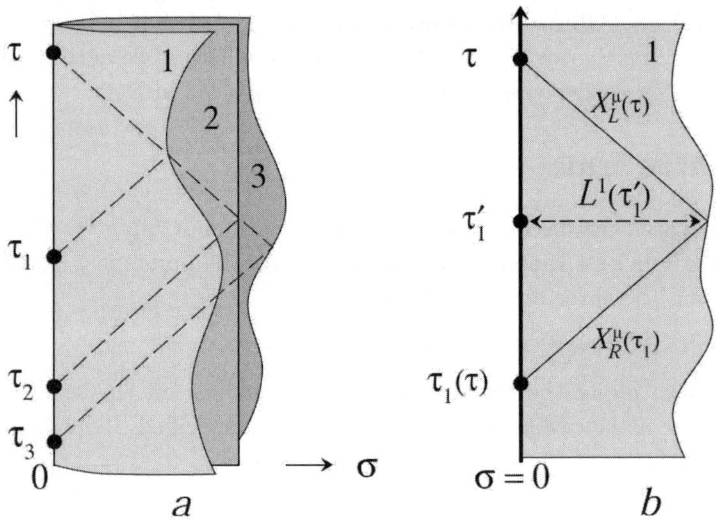

Fig. 2. *a*) The three world sheets. Their relative lengths are dynamical variables. One of the three may be kept fixed. Dashed lines: bouncing waves have differing arrival times. *b*) The variables $L^1(\tau)$ and $\tau_1(\tau_0)$ for branch #1, see text.

3. Boundary conditions

The boundary condition at the Torricelli point can be enforced by two Lagrange multipliers: we add to the action (5)

$$\int \mathrm{d}\tau (\lambda_1^\mu(\tau)(X^{\mu,1}(0,\tau) - X^{\mu,3}(0,\tau)) + \lambda_2^\mu(\tau)(X^{\mu,2}(0,\tau) - X^{\mu,3}(0,\tau)) . \tag{6}$$

Now, considering an infinitesimal variation of $X^{\mu,i}(\sigma,\tau)$, we find

$$\delta S = \sum_i \Bigg(\int \mathrm{d}\tau \mathrm{d}\sigma \, \delta X^{\mu,i} (\partial_\sigma^2 - \partial_\tau^2) X^{\mu,i} \; + $$

$$\int \mathrm{d}\tau \Big(\delta X^{\mu,i}(0,\tau)(\partial_\sigma X^{\mu,i}(0,\tau) + \lambda_i^\mu) - \delta X^{\mu,i}(L^i,\tau)\partial_\sigma X^{\mu,i}(L^i(\tau),\tau) \Big) \Bigg) , \tag{7}$$

where $\lambda_3 = -\lambda_1 - \lambda_2$. Thus, we see that the usual Neumann boundary condition holds at the three end points $\sigma = L^i$, while at the origin we have a Neumann boundary condition for the sum $\sum_{i=1}^3 X^{\mu,i}$ and a Dirichlet condition for the differences:

$$\begin{aligned} \sigma = 0 \; : \quad & \partial_\sigma (X^{\mu,1} + X^{\mu,2} + X^{\mu,3}) = 0 \; ; \\ & X^{\mu,1} - X^{\mu,3} = X^{\mu,2} - X^{\mu,3} = 0 \; , \\ \sigma = L^i(\tau) \; : \quad & \partial_\sigma X^{\mu,i} = 0 \; , \quad \forall i \; . \end{aligned} \tag{8}$$

4. Equations of motion

From Eq. (7), we read off the usual equations of motion implying that we have the sum of left- and right-going waves on all three branches[a]:

$$X^{\mu,i} = X_L^{\mu,i}(\sigma + \tau) + X_R^{\mu,i}(\tau - \sigma) \; . \tag{9}$$

The boundary conditions (8) now read

$$X_L^{\mu,i} = X_R^{\mu,i} \qquad \text{at } \sigma = L^i(\tau) \; , \quad \text{and} \tag{10}$$

$$\left. \begin{aligned} \partial_\tau X_R^{\mu,1} &= \partial_\tau (\tfrac{2}{3} X_L^{\mu,2} + \tfrac{2}{3} X_L^{\mu,3} - \tfrac{1}{3} X_L^{\mu,1}) \\ \partial_\tau X_R^{\mu,2} &= \partial_\tau (\tfrac{2}{3} X_L^{\mu,3} + \tfrac{2}{3} X_L^{\mu,1} - \tfrac{1}{3} X_L^{\mu,2}) \\ \partial_\tau X_R^{\mu,3} &= \partial_\tau (\tfrac{2}{3} X_L^{\mu,1} + \tfrac{2}{3} X_L^{\mu,2} - \tfrac{1}{3} X_L^{\mu,3}) \end{aligned} \right\} \text{ at } \sigma = 0 \; . \tag{11}$$

In addition to these, we have the constraints (3) and (4), now taking the form

$$\sum_\mu (\partial_\tau X_L^{\mu,i})^2 = \sum_\mu (\partial_\tau X_R^{\mu,i})^2 = 0 \; . \tag{12}$$

These constraints should be compatible with the boundary conditions. The conditions at the three end points, Eqs. (10), give nothing extra, but the equations at the origin, Eqs. (11), require that, at that point, also

$$\partial_\tau X_L^{\mu,1} \, \partial_\tau X_L^{\mu,2} \;=\; \partial_\tau X_L^{\mu,2} \, \partial_\tau X_L^{\mu,3} \;=\; \partial_\tau X_L^{\mu,3} \, \partial_\tau X_L^{\mu,1} \; , \tag{13}$$

---

[a] Our convention will be that *leftmoving* refers to movement towards the branch point, and *rightmoving* to movement in the direction of one of the end points.

and similarly for the right-going modes. From the point of view of causality, it might seem odd that there is a condition that the newly arriving modes should obey. However, it can easily be imposed by a relative rescaling of the coordinates $\sigma + \tau$ for the left-going modes in the three arms:

$$\partial_\tau X_L^{\mu,i}(\tau) = \frac{\mathrm{d}\tau_i}{\mathrm{d}\tau}\, \partial_{\tau_i} X_L^{\mu,i}(\tau_i) \,. \tag{14}$$

Indeed, as we shall now demonstrate, the requirement (13) fixes the functions $L^i(\tau)$.

Taking the fact that $X_L^{\mu,i}$ only depends on $\tau+\sigma$ and $X_R^{\mu,i}$ only on $\tau-\sigma$, the boundary condition (10) implies the existence of functions $\tau_i(\tau) < \tau$ such that

$$X_L^{\mu,i}(\tau) = X_R^{\mu,i}(\tau_i) \,, \tag{15}$$

where $\tau_i(\tau)$ are the solutions of

$$\tau - \tau_i = 2L^i(\tau_i') \,, \qquad \tau_i' \equiv \frac{\tau + \tau_i}{2} \,. \tag{16}$$

These are just the three departure times, from the center, of each of the three returning waves, see Fig. 2*b*. Now let

$$Q^{\mu,i}(\tau) \equiv \frac{\partial X_R^{\mu,i}(\tau)}{\partial \tau} \,; \tag{17}$$

$$R^i(\tau) \equiv -\, Q^{\mu,j}(\tau_j)\, Q^{\mu,k}(\tau_k) \,, \qquad i,j,k = 1,2,3 \textit{ cyclic}, \tag{18}$$

(i,j,k being any even permutation of 1,2,3). The minus sign in Eq. (18) is chosen because $Q^{\mu,i}\,Q^{\mu,j} = \vec{Q}^i \cdot \vec{Q}^j - |Q^i||Q^j| \le 0$, since $(Q^{\mu,i})^2 = 0$. Then, using (14), condition (13) implies

$$\frac{1}{R^1(\tau)}\frac{\mathrm{d}\tau_1}{\mathrm{d}\tau} = \frac{1}{R^2(\tau)}\frac{\mathrm{d}\tau_2}{\mathrm{d}\tau} = \frac{1}{R^3(\tau)}\frac{\mathrm{d}\tau_3}{\mathrm{d}\tau} \,. \tag{19}$$

Together with a gauge condition for the overall time parameter τ, for instance,

$$\tfrac{1}{3}\sum_i \tau_i = \tau - 1 \,; \qquad \sum_i \frac{\mathrm{d}\tau_i}{\mathrm{d}\tau} = 3 \,, \tag{20}$$

these form complete differential equations that determine the functions $\tau_i(\tau)$, and thus also the functions $L^i(\tau)$, through Eqs. (16):

$$\frac{\mathrm{d}\tau_i}{\mathrm{d}\tau} = \frac{3R^i}{R^1 + R^2 + R^3} \,. \tag{21}$$

The complete update of the functions $Q^{\mu,i}(\tau)$ at the central point then reads:

$$Q^{\mu,i}(\tau) \;=\; \frac{1}{\sum R}\Big(2R^j(\tau)Q^{\mu,j}(\tau_j) + 2R^k(\tau)Q^{\mu,k}(\tau_k) - R^i(\tau)Q^{\mu,i}(\tau_i)\Big)\;, \qquad i,j,k = 1,2,3 \; cyclic. \tag{22}$$

Using Eqs. (18), one verifies that indeed the constraint $(Q^{\mu,i}(\tau))^2 = 0$ continues to be respected.

5. An instability

In the special case when all R^i happen to be constant and equal, these equations can be solved. The three arms then have equal lengths, in the sense that waves running across the three arms bounce back in exactly the same time, which is constant. However, one readily convinces oneself that solutions with unequal arms should also exist. For instance, take an initial state described by functions $Q^{\mu,i}(\tau)$ consisting of superpositions of theta functions in τ, such that the condition

$$\sum_\mu (Q^{\mu,i})^2 = 0 \tag{23}$$

is always satisfied. One can then solve the equations stepwise for successive intervals in τ.

There is, however, a problem that needs to be addressed: what happens if one of the arms tends to vanish? This is for instance the case if, during some period, $\mathrm{d}\tau_1/\mathrm{d}\tau > 1$ until $\tau - \tau_1 \to 0$, see Fig. 3*a*. Do we need a new boundary condition addressing this situation? Indeed, it appears that, at least in the classical (unquantized) theory, an extra boundary condition will be needed. Our argument goes as follows.

If, at some value(s) of τ, one arm length parameter, say $L^1(\tau)$, tends to zero, then there will be waves running up and down this arm increasingly frequently, see Fig. 3*b*. The values for $Q^{\mu,i}$ for the two long arms, $i = 2,3$, will not vary rapidly during this small period (this is because these incoming waves were last updated at the times τ_2 and τ_3, both long before τ). Let us make an educated guess about what will happen, by keeping these Q vectors constant. In contrast, $Q^{\mu,1}(\tau)$ will be updated frequently, so, let us

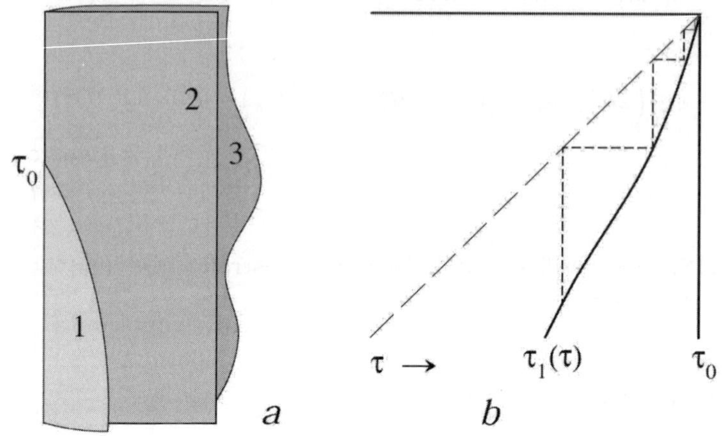

Fig. 3. *a*) A contracting arm. *b*) Indefinite series of 'updates' of the data along the contracting arm, due to bouncing waves.

consider this sequence. We have

$$Q^{\mu,1}(\tau) = \frac{1}{\sum_i R^i}\Big(2R^2(\tau)Q^{\mu,2}(\tau_2) + 2R^3(\tau)Q^{\mu,3}(\tau_3) - R^1(\tau)Q^{\mu,1}(\tau_1)\Big)\ , \tag{24}$$

$$Q^{\mu,2}(\tau_2) \quad \text{and} \quad Q^{\mu,3}(\tau_3) \quad \textit{constant.} \tag{25}$$

One finds that the parameters R^i are updated as follows:

$$\begin{aligned} R^1 &\rightarrow R^1 \equiv R \quad \text{(constant)}\ ; \\ R^2 &\rightarrow \frac{R^1\,R^2}{R^1+R^2+R^3}\ ; \\ R^3 &\rightarrow \frac{R^1\,R^3}{R^1+R^2+R^3}\ . \end{aligned} \tag{26}$$

This implies that R^2 and R^3 continue to decrease, until R^1 dominates. Thus,

$$\frac{d\tau_1}{d\tau} \rightarrow 3\ , \qquad \frac{d\tau_2}{d\tau} \rightarrow 0\ , \qquad \frac{d\tau_3}{d\tau} \rightarrow 0\ . \tag{27}$$

Since we started with $\tau_1 < \tau$, we find that $\tau_1 \rightarrow \tau$, rapidly. In fact, we found that, continuing along this line, assuming $Q^{\mu,2}$ and $Q^{\mu,3}$ to stay constant, the calculation can be done exactly. After n iterations, the vector $Q^{\mu,1,\,(n)}$

becomes

$$Q^{\mu,1\,(n)} = \frac{1}{n}\left(Q_0^\mu(-1)^n + \beta Q^{\mu,2} + (1-\beta)Q^{\mu,3}\right) , \qquad (28)$$

where Q_0^μ is a fixed vector, and β is a fixed coefficient, such that

$$Q_0^\mu\,Q^{\mu,2} = Q_0^\mu\,Q^{\mu,3} = 0\ ; \qquad (Q_0^\mu)^2 = R\beta(1-\beta)\ ; \qquad Q^{\mu,2}\,Q^{\mu,3} = -R\ . \qquad (29)$$

We also have

$$R^{2\,(n)} = \frac{\beta}{n}R\ ; \qquad R^{3\,(n)} = \frac{1-\beta}{n}R\ . \qquad (30)$$

In fact, Q_0^μ and β are only constant while following a bouncing wave, but they cannot be constant along the entire short arm. This is because of the alternating sign in Eq. (28). Since Q_0^μ is spacelike and orthogonal to $Q^{\mu,2}$ and $Q^{\mu,3}$, it must oscillate through zero. At such points, either β or $1-\beta$ must have quadratic zeros. Typically, one of these will oscillate like $\sin^2(\pi\sigma/2L^1)$. Thus, our solution consists of partly-periodic functions, in the sense that over periods lasting approximately $\ln 3$ in the parameter $\ln(\tau_0-\tau)$, the functions are updated using Eqs. (28), where the auxiliary functions (29) are exactly periodic.

Fig. 4. The Y shaped string just before one arm shrinks to zero. Ripples with increasing frequencies are generated on the two other arms. These disperse along the arms. Classicaly, the time-reversed configuration required to regenerate the third arm, occupies spots in phase space with vanishing probabilities.

We have no indication that small perturbances in the initial conditions will have drastic effects on these solutions, so that indeed generic solutions exist in which one arms rapidly shrinks to zero. What happens after such an event? At first sight, it seems reasonable to postulate some sort of bounce. By time-reversal symmetry, one might expect this arm to come back into existence. However, closer inspection makes such a 'solution' quite unlikely, or even impossible. Our observation is the following, see Fig. 4. Shortly before our shrinking event, the short arm has been violently oscillating, or rotating, around the central point. In doing so, the rapidly oscillating function $Q^{\mu,1}$ has been emitting high-frequency waves into the two long arms. These frequencies will always be much higher than the frequencies of oscillations entering from the long arms. The time-reversal of this configuration is hard to reconcile with causality requirements. Thus, the energy of the short arm has been dissipating in the form of high-frequency modes in the long arms, and the dissipated energy will be hard to recover.

6. Conclusion

We did not study possible quantization procedures for this model in any detail. Extreme non-linearity at the Torricelli point probably makes this impossible. Qualitatively, as is well-known, one expects Regge trajectories that are similar — and have the same slope — as the mesonic ones, since at any given energy, the highest angular momentum states will be achieved when one arm vanishes. Now, in our analysis, we find that, as soon as a classical string picture is adopted for baryonic states, at least one of the three arms will soon disappear, shedding its energy into the excitation modes of the two other arms, see Fig. 1*c*.

This is somewhat counter-intuitive. One might have thought that equipartition should take place: all corners of phase space should eventually be occupied with equal probability. The answer to this is, of course, that phase space is infinite, so that equipartition is impossible. It is the old problem of statistical physics before the advent of Quantum Mechanics: there is an infinite amount of phase space in the high frequency domain. Since physical baryons are quantum mechanical objects, we expect an effective cut-off at high frequencies, simply because of energy quantization. This does mean, however, that in the high energy domain, the majority of baryonic states will have these high-energy modes excited. If, due to some electro-weak interaction, or possibly due to a gluon hitting a quark, a baryonic state is created with three quarks energetically moving in different directions, we expect first the Y shape to form, but then the most likely baryonic excitation that is reached is one with a single open string connecting the three quarks.

We found that the classically stable configuration has a single open string with two quarks at the end points and one quark moving around on the string. However, because of an attraction between two quarks into a $\overline{3}$ bound state (as opposed to the 6), one expects quantum effects eventually to favor the configuration of one quark at one end and a di-quark at the other end of a single open string, see Fig. 1*d*. I still do not see a strong case in favor of the Δ configuration. The latter seems to be closer related to a baryon-glueball bound state that will probably have a rather low production cross section.

Finally, Y shaped string configurations have also emerged in string theory,[8] but only by viewing the Torricelli point as a *D*-brane, which as such must be handled in the classical approximation, so that most of the mass of the system is concentrated there. The dynamical properies of such configurations will again be different from what we studied here.

References

1. A.A. Polyakov, *A Few Projects in String Theory*, hep-th/9304146, Chapter 8.
2. X. Artru, *Nuc. Phys.* **B85** (1975) 442.
3. G.S. Sharov, *Four Various String Baryon Models and Regge Trajectories*, hep-ph/9809465.
4. G.S. Sharov, *String baryon model "triangle": hypocycloidal solutions*, hep-th/9808099.
5. G.S. Sharov, *Instability of Classic Rotational Motion for Three-String Baryon Model*, hep-ph/0001154.
6. G.S. Sharov, *Quasirotational Motions and Stability Problem in Dynamics of String Hadron Models*, hep-ph/0004003.
7. G. 't Hooft, *Minimal Strings for Baryons*, Presented at the Trento Workshop on "Hadrons and Strings", July 2004, ITP-UU-04/17; Spin-04/10; hep-th/0408148.
8. E. Witten, *Baryons and Branes in Anti de Sitter Space*, hep-th/9805112.

NON-ABELIAN DUALITY AND CONFINEMENT: MONOPOLES SEVENTY-FIVE YEARS LATER

K. KONISHI*

Dipartimento di Fisica, "E. Fermi", University of Pisa
Pisa, 56127, Italy
**E-mail: konishi@df.unipi.it*
http://www.df.unipi.it/~konishi/konishi.html

Recently, there has been a considerable improvement in our understanding of the quantum behavior of non-Abelian monopoles, a subject which remained obscure for so many years after the introduction of magnetic monopoles in quantum field theory by Dirac. In this talk, I will discuss how the dual gauge transformations among the non-Abelian monopoles can be understood through the study of non-Abelian *vortices* and how a (dual) picture of confinement naturally emerges from such a picture.

1. Introduction

Three quarters of a century have passed since the introduction of magnetic monopoles in quantum field theory by Dirac.[1] Our understanding of the soliton sector of spontaneously broken gauge theories[2] is still largely unsatisfactory. In particular, the development in our understanding of *non-Abelian* versions of monopoles[3–16] and vortices[17] have been very slow, in spite of many articles written on these subjects, and in spite of the important role these topological excitations are likely to play in various areas of physics. For instance, they might hold the key to the mystery of quark confinement in Quantum Chromodynamics (QCD).[18–20] Their quantum mechanical properties are gradually emerging, however, thanks to an ever improving grasp on the nonperturbative dynamics in the context of supersymmetric gauge theories. Some of the ingredients of this development include the Seiberg-Witten solution of $\mathcal{N} = 2$ supersymmetric gauge theories and exact instanton summations, better understanding of the properties of (super-) conformal field theories, exact results on the chiral condensates and symmetry breaking pattern in a wide class of $\mathcal{N} = 1$ supersymmetric gauge theories, and so on. Also, many new results on non-Abelian *vortices* and *domain walls* are now available, which are closely related to the problems concerning the monopoles.

We shall consider here concrete models to discuss our general ideas about non-Abelian monopoles: softly broken $\mathcal{N} = 2$ supersymmetric gauge theories with SU, SO and USp gauge groups with various matter multiplets in fundamental representation (quarks).

Our strategy is a two-way attack on the problem. On the one hand, we shall study the models in the fully quantum regime, by making use of the Seiberg-Witten curves and other exact properties of the theories.

On the other hand, the same models will be analyzed in the semi-classical region. The existence of certain parameters in the models allow one to move from the region where fully quantum mechanical results are applicable, back to the region where all VEVs are large and semi-classical analysis is justified. The idea to be discussed below in general terms, by using symmetry argument and homotopy maps, to define the dual group through the transformation properties of the low-energy vortices, is applied to the models in this regime. The properties of non-Abelian monopoles thus deduced match very non-trivially with what is found in the exact low-energy action.

2. Difficulties related to the GNOW duality

We start by reviewing briefly the standard construction of semiclassical "non-Abelian monopoles", and by recalling the difficulties associated with these. A system in which the gauge symmetry is spontaneously broken

$$G \stackrel{\langle\phi_1\rangle\neq 0}{\longrightarrow} H \tag{1}$$

where H is some non-Abelian subgroup of G, possesses a set of regular magnetic monopole solutions in the semi-classical approximation. They are natural generalizations of the Abelian 't Hooft-Polyakov monopoles,[2] found originally in the $G = SO(3)$ theory broken to $H = U(1)$ by a Higgs mechanism. When the "unbroken" gauge group is non-Abelian, the asymptotic gauge field can be written as

$$F_{ij} = \epsilon_{ijk} B_k = \epsilon_{ijk} \frac{r_k}{r^3} (\beta \cdot \mathbf{H}), \tag{2}$$

in an appropriate gauge, where $\mathbf{H}$ are the diagonal generators of H in the Cartan subalgebra. A straightforward generalization of the Dirac's quantization condition leads to

$$2\,\beta \cdot \alpha \in \mathbf{Z} \tag{3}$$

where α are the root vectors of H. The constant vectors β (with the number of components equal to the rank of the group H) label possible monopoles. It is easy to see that the solution of Eq. (3) is that β is any of the *weight vectors* of a group whose nonzero roots are given by

$$\alpha^* = \frac{\alpha}{\alpha \cdot \alpha}. \tag{4}$$

The group generated by Eq. (4) is known as the *dual* (we shall call it GNOW dual below) of H, let us call $\tilde{H}$. One is thus led to a set of semi-classical *degenerate* monopoles, with multiplicity apparently equal to that of a representation of $\tilde{H}$; this has led to the so-called GNOW conjecture, i.e., that they form a multiplet of the group $\tilde{H}$, dual of H^{4-}.[6] For simply-laced groups (with the same length of all nonzero roots) such as $SU(N)$, $SO(2N)$, the dual of H is basically the same group, except that the allowed representations tell us that

$$SU(N) \leftrightarrow \frac{SU(N)}{\mathbf{Z}_N}; \qquad SO(2N) \leftrightarrow SO(2N), \tag{5}$$

while

$$SO(2N+1) \leftrightarrow USp(2N), \tag{6}$$

There is no difficulty in explicitly constructing these degenerate set of monopoles.[6] The basic idea is to embed the 't Hooft-Polyakov monopoles in various broken $SU(2)$ subgroups. These set of monopoles constitute the prime candidates for the members of a multiplet of the dual group $\tilde{H}$.

There are however well-known difficulties in such an interpretation. The first concerns the topological obstruction:[11,15] in the presence of the classical monopole background, it is not possible to define a globally well-defined set of generators isomorphic to H. As a consequence, "no colored dyons exist". In a simplest case with the breaking

$$SU(3) \stackrel{\langle\phi_1\rangle\neq 0}{\longrightarrow} SU(2)\times U(1), \tag{7}$$

this means that

$$\textit{no monopoles with charges} \quad (\underline{2}, 1^*) \quad \textit{exist}, \tag{8}$$

where the asterisk indicates the dual, magnetic $U(1)$ charge.

The second can be regarded as an infinitesimal version of the same difficulty: certain bosonic zero modes around the monopole solution, corresponding to H gauge transformations, are non-normalizable (behaving as $r^{-1/2}$ asymptotically). Thus the standard procedure of quantization leading to H multiplets of monopoles, does not work. Some progress on the check of GNOW duality along this orthodox line of thought, has been reported nevertheless,[13] in the context of $\mathcal{N}=4$ supersymmetric gauge theories.

Both of these difficulties concern the transformation properties of the monopoles under the subgroup H, while the relevant question should be how they transform under the dual group, $\tilde{H}$. As field transformation groups, H and $\tilde{H}$ are relatively nonlocal, the latter should look like a nonlocal transformation group in the original, electric description.

The *multiciplicity* of the monopoles is, also, a subtle issue actually. Take again the case of the system with a breaking pattern, $SU(3) \stackrel{\langle\phi_1\rangle\neq 0}{\longrightarrow} SU(2)\times U(1)$. Naïvely one would argue that there are two monopoles, as they are supposed to belong to a doublet, according to the GNOW classification. Or, perhaps, one should conclude that there are infinitely many, continuously related solutions, as the two solutions obtained by embedding the 't Hooft solutions in $(1,3)$ and $(2,3)$ subspaces, are clearly part of the continuous set of (moduli of) solutions. Or perhaps one should conclude that there is only one monopole, as all the degenerate solutions are related by the unbroken *gauge* group $H = SU(2)$.[a] In brief, what is the multiplicity (#) of the monopoles:

$$\# = 1, \quad 2, \quad \text{or} \quad \infty\,? \tag{9}$$

We believe however that a proper formulation of the question is:

`What is the dual group?`
`How do the degenerate magnetic monopoles transform among themselves under that group?`

In the attempt to answer these questions, some general considerations are unavoidable. The first is the fact since H and $\tilde{H}$ groups are non-Abelian the dynamics of the system should enter the problem in essential way. For instance, the non-Abelian H interactions can become strongly-coupled at low energies and can break itself dynamically. This indeed occurs in pure $\mathcal{N}=2$ super Yang-Mills theories (*i.e.*, theories without quark hypermultiplets), where the exact quantum mechanical result is known in terms of the Seiberg-Witten curves.[21–23] Consider for instance, a pure $\mathcal{N}=2$, $SU(N+1)$ gauge theory. Even though partial breaking, *e.g.*, $SU(N+1) \to SU(N)\times U(1)$ looks perfectly possible semi-classically, in an appropriate region of classical degenerate vacua, no such vacua exist

[a]This interpretation however encounters the difficulties mentioned above. Also there are cases in which degenerate monopoles occur, which are not simply related by the group H.

quantum mechanically. In *all* vacua the light monopoles are abelian, the effective, magnetic gauge group being $U(1)^N$.

Generally speaking, the concept of a dual group multiplet is well-defined when $\tilde{H}$ interactions are weak (or at worst, conformal). This however means that one must study the original, electric theory in the regime of strong coupling, which would usually make the task exceedingly difficult. Fortunately, in $\mathcal{N}=2$ supersymmetric gauge theories, exact Seiberg-Witten curves describe the fully quantum mechanical consequences of the strong-interaction dynamics in terms of weakly-coupled dual magnetic variables.

This is how we know that the non-Abelian monopoles exist in fully quantum theories:[24] in the so-called r-vacua of softly broken $\mathcal{N}=2$, $SU(N)$ gauge theory, the light monopoles appear as the dominant infrared degrees of freedom and interact as pointlike particles having the charges of a fundamental multiplet $\underline{r}$ of an effective, dual $SU(r)$ gauge group. In an $SU(3)$ gauge theory broken to $SU(2) \times U(1)$ as in (7), with an appropriate number of quark multiplets ($N_f \geq 4$), for instance, light magnetic monopoles carrying the charges

$$(\underline{2}^*, 1^*) \tag{10}$$

under the dual $SU(2) \times U(1)$ appear in the low-energy effective action in one of the possible vacua. (Dual-) colored dyons do exist! The distinction between H and $\tilde{H}$ is crucial (*cfr.* Eq. (8)).

A closely related point concerns the phase of the system. If the dual group were in Higgs phase, the multiplet structure among the monopoles would get lost, generally. Therefore one must study the dual ($\tilde{H}$) system in confinement phase.[b] But then, according to the standard electromagnetic duality argument, *one must analyse the electric system in Higgs phase.* The monopoles will appear confined by the vortices of the H system, which can be naturally interpreted as confining string of the dual system $\tilde{H}$.

We are thus led to study the system with a hierarchical symmetry breaking,

$$G \xrightarrow{v_1} H \xrightarrow{v_2} \emptyset, \qquad v_1 \gg v_2, \tag{11}$$

instead of the original system (1). The smaller VEV breaks H completely. However, in order for the degeneracy among the monopoles not to be broken by the breaking at the scale v_2, we require that some global color-flavor

[b] Non-abelian monopoles in the *Coulomb phase* suffer from the difficulties already discussed.

diagnonal group

$$H_{C+F} \subset H_{color} \otimes G_F \tag{12}$$

remains unbroken (see below).

As we shall see, such a scenario is naturally realized in $\mathcal{N} = 2$ supersymmetric theories. An important lesson one learns from these considerations (and from the explicit models), is that the role of the massless flavor is fundamental. This manifests itself in more than one ways.

(i) H must be non-asymptotically free, this requires that there be sufficient number of massless flavors: otherwise, H interactions would become strong at low energies and H group can break itself dynamically;

(ii) The physics of the r vacua[25,27] indeed shows that the non-Abelian dual group $SU(r)$ appear only for $r \leq \frac{N_f}{2}$. This limit can be understood from the renormalization group: in order for a nontrivial r vacuum to exist, there must be at least $2r$ massless matter flavor in the original, electric theory;

(iii) Non-abelian vortices,[33,34] which as we shall see are closely related to the concept of non-Abelian monopoles, require also an exact flavor group. The non-Abelian flux moduli arise as a result of an exact color-flavor diagonal symmetry of the system, broken by individual soliton vortices.

3. General idea

It turns out that the properties of the monopoles induced by the breaking

$$G \to H \tag{13}$$

are closely related to the properties of the vortices, which develop when the low-energy H gauge theory is put in Higgs phase by a set of scalar VEVs, $H \to \emptyset$. The crucial instrument is the exact homotopy sequence,

$$\cdots \to \pi_2(G) \to \pi_2(G/H) \to \pi_1(H) \to \pi_1(G) \to \cdots \tag{14}$$

Assume for simplicity that $\pi_2(G)$ and $\pi_1(G)$ are both trivial. In this case it is clear that each element of $\pi_1(H)$ is an image of a corresponding element of $\pi_2(G/H)$: all monopoles are regular, 't Hooft-Polyakov monopoles. Consider instead the case $\pi_1(G)$ is nontrivial. Take for concreteness $G = SO(3)$, with $\pi_1(SO(3)) = \mathbf{Z}_2$, and $H = U(1)$, with $\pi_1(U(1)) = \mathbf{Z}$. For any compact Lie groups $\pi_2(G) = \emptyset$. The exact sequence illustrated in Fig. 1 in this case implies that the monopoles, classified by $\pi_1(U(1)) = \mathbf{Z}$ can further by divided into two classes, one belonging to the image of $\pi_2(SO(3)/U(1))$ – 't

Hooft-Polyakov monopoles! – and those which are not related to the breaking – the singular, Dirac monopoles. The correspondence is two-to-one: the monopoles of magnetic charges $2n$ times ($n = 1, 2, \ldots$) the Dirac unit are regular monopoles while those with charges $2n + 1$ are Dirac monopoles. In other words, the regular monopoles correspond to the kernel of the map $\pi_1(H) \to \pi_1(G)$ (Coleman[10]).

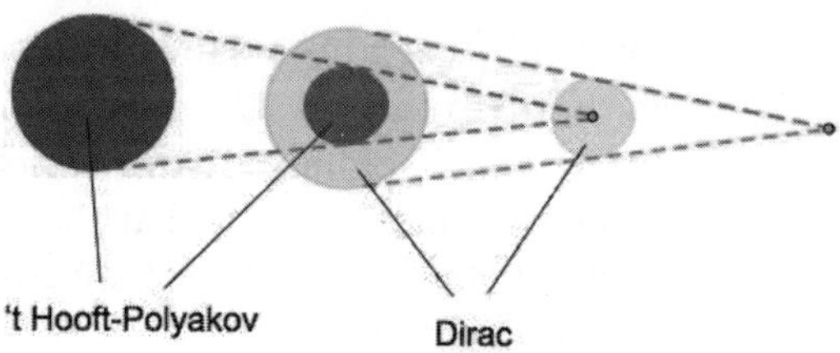

Fig. 1. A pictorial representation of the exact homotopy sequence, (14), with the leftmost figure corresponding to $\pi_2(G/H)$.

The exact sequence (14) assumes an important significance when we consider the system with a hierarchical symmetry breaking (11),

$$G \xrightarrow{v_1} H \xrightarrow{v_2} \emptyset.$$

As H is now completely broken the low-energy theory has vortices, classified by $\pi_1(H)$. If $\pi_1(G) = \emptyset$, however, the full theory cannot have vortices. This apparent paradox is solved when one realizes that there is another related paradox: monopoles representing $\pi_2(G/H)$ cannot be stable, because in the full theory the gauge group is completely broken, $G \to \emptyset$, and because for any Lie group, $\pi_2(G) = \emptyset$. These paradoxes solve themselves: the vortices of the low-energy theory end at the monopoles, which have large but finite masses. Or they are broken in the middle by (though suppressed) monopole-antimonopole pair production. Vice versa, the monopoles are not stable as its flux is carried away by the vortex. See Fig. 2

Applied to the case of $SO(3) \to U(1) \to \emptyset$, this was precisely the logic used by 't Hooft in his pioneering paper on the monopoles. As is seen from Fig. 1, the vortices ($\pi_1(U(1)) = \mathbf{Z}$) of the winding number two, corresponding to the trivial element of $\pi_1(SO(3)) = \mathbf{Z}_2$, should not be stable in the full theory: there must be a regular monopole-like configuration, having the magnetic charge twice the Dirac unit, $g_m = 4\pi/g$, where g is the the gauge

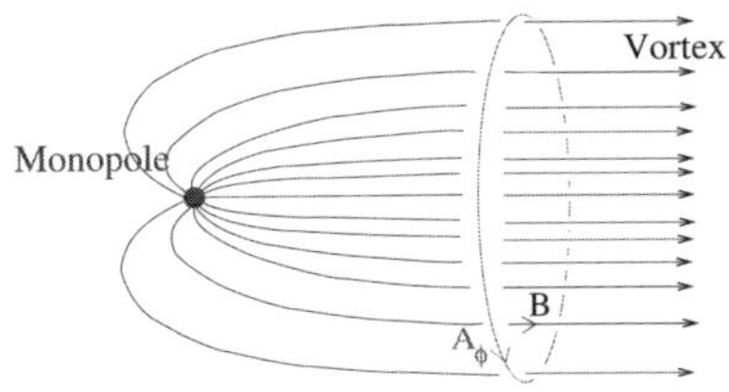

Fig. 2.

coupling constant of the $SO(3)$ theory, acting as a source or a sink of the magnetic flux (Fig. 2).

An important new aspect we have here, as compared to the case discussed by 't Hooft[2] is that now the unbroken group H is non-Abelian and that the low-energy vortices carry continuous, non-Abelian flux moduli.[33–35,40] As the color-flavor diagonal symmetry is an exact symmetry of the full theory, and the non-Abelian moduli of the low-energy vortices is a consequence of it, it follows that the monopoles appearing as the endpoints of such vortices carry the same continuous moduli and shares the same transformation properties of the vortices.[45] The idea is then:

`The dual group transformation among the monopoles can be defined through the (better understood) transformation properties of the vortices, which can be studied in the low-energy approximation.`

4. Concrete Model: Softly broken $\mathcal{N} = 2$ SU, SO, USp models

Physics of confining vacua and properties of light monopoles in these theories are studied by identifying all of the $\mathcal{N} = 1$ vacua (the points in the QMS – quantum moduli space, that is, the space of vacua – which survive the $\mathcal{N} = 1$ perturbation) and studying the low-energy action for each of them. The underlying $\mathcal{N} = 2$ theory, especially with $m_i = 0$ or with equal masses $m_i = m$, has a large continuous degeneracy of vacua (flat directions), which has been studied by using the Seiberg-Witten curves, non-renormalization of Higgs branch metrics, superconformal points and their universality, their moduli structure and symmetries, etc.[25,26] For the purpose of this section, however, we are most interested in the set of vacua which are picked up when the small generic bare quark masses m_i and a small nonzero adjoint mass μ are present. At the roots of these different branches of $\mathcal{N} = 2$

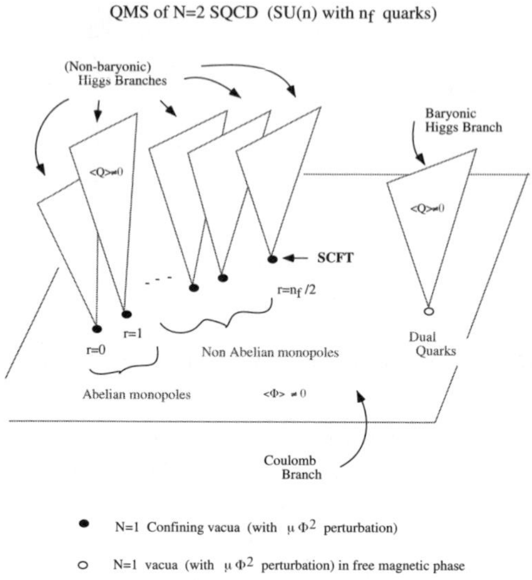

Fig. 3.

vacua where the Higgs branches meet the Coulomb branch, lie all these vacua, which survives the $\mathcal{N} = 1$ perturbation. In $SU(N_c)$ theories with N_f flavors with generic masses, all $\mathcal{N} = 1$ vacua arising this way have been completely classified.[27,28]

For nearly equal quark masses they fall into classes $r = 0, 1, \ldots, \frac{N_f}{2}$ groups of vacua near the "roots of non-baryonic Higgs branches", and for $N_f \geq N_c$, there are special vacua at the "root of baryonic Higgs branch". These names reflect the fact that in the respective Higgs branch non-baryonic or baryonic squark VEVS,

$$\langle Q_i^a \tilde{Q}_a^j \rangle, \qquad \langle \epsilon_{a_1 a_2 \ldots, a_{N_c}} Q_{i_1}^{a_1} Q_{i_2}^{a_2} \cdots Q_{i_{N_c}}^{a_{N_c}} \rangle, \tag{15}$$

are formed. See Fig. 3. Each group of vacua coalesce in a single vacuum where the gauge symmetry is enhanced into non-Abelian gauge groups in the equal mass ($m_i \to m$) limit, where the flavor symmetry $SU(N_f)$ becomes exact.

Our main interest is the first classes of the so-called "r-vacua", where the magnetic gauge group is

$$U(r) \times U(1)^{N_c - r},$$

and the massless matter multiplets consist of N_f monopoles in the fundamental representation of $U(r)$, and flavor-singlet Abelian monopoles carrying a single charge, each with respect to one of the $U(1)$ factors. Once the gauge group and the quantum numbers of the matter fields are all known, the $\mathcal{N}=2$ supersymmetry uniquely fixes the structure of the effective action. We find that:

- We see the non-Abelian monopoles in action, in the generic r ($2 \leq r \leq \frac{N_f}{2}$) vacua. They behave perfectly as point-like particles, albeit in a dual, magnetic gauge system. Upon $\mathcal{N}=1$ perturbation they condense (confinement phase) and induces flavor symmetry breaking $SU(N_f) \times U(1) \to U(r) \times U(N_f - r)$.
- The upper limit $r \leq \frac{N_f}{2}$ is a manifestation of monopole dynamics: only in this range of r the non-Abelian monopoles can appear as *recognizable infrared degrees of freedom.* We now see why in the $SU(2)$ Seiberg-Witten models, as well as in pure $\mathcal{N}=2$ Yang-Mills (i.e., $N_f=0$) models with different gauge groups, the low-energy monopoles were found always Abelian. In all these cases, non-Abelian monopoles would interact too strongly.
- Vice versa, there are homotopy and symmetry arguments[27,29] suggesting that non-Abelian monopoles appearing in the r-vacua are "baryonic constituents" of an Abelian ('t Hooft-Polyakov) monopople,

  $$\textit{Abelian monopole} \sim \epsilon^{a_1 \ldots a_r} q_{a_1}^{i_1} q_{a_2}^{i_2} \ldots q_{a_r}^{i_r},$$

  a_i being the dual color indices, and i_m the flavor indices. The $SU(r)$ gauge ineractions, being infrared-free, are unable to keep the Abelian monopole bound.
- That the effective degrees of freedom in the r vacua are non-Abelian rather than Abelian monopoles, is actually required also by symmetry of the system,[27,31] not only from the dynamics. If the Abelian monopoles of the r-th tensor flavor representation were the correct degrees of freedom, the low-energy effective theory would have too large an accidental symmetry – $SU(\binom{N_f}{r})$: 1 with a consequent paradox of too-many Nambu Goldstone bosons.
- An analogous argument might be given in the standard QCD, to exclude Abelian picture of confinement. We know from lattice simulations of $SU(3)$ theory that confinement and chiral symmetry breaking are closely related. If Abelian 't Hooft-Monopole-Mandelstam monopoles were the right degrees of freedom describing confinement, their condensation would somehow have to describe chiral symmetry breaking as well. We would then be led to assume that

they carry flavor quantum numbers of $SU(N_f)_L \times SU(N_f)_R$, *e.g.*,

$$Monopoles \sim M_i^j, \qquad \langle M_i^j \rangle \propto \delta_i^j \, \Lambda_{QCD},$$

where i, j are $SU(N_f)_L \times SU(N_f)_R$ indices. But such a system would have a far-too large accidental symmetry. Confinement would be accompanied by a large number of unexpected (and indeed unobserved) light Nambu-Goldstone bosons.

- The limiting case of r vacua, with $r = \frac{N_f}{2}$, as well as the massless limit of $USp(2N_c)$ and $SO(N_c)$ theories, are very interesting. The low-energy effective theory in these cases turn out to be conformally invariant (nontrivial infrared-fixed-point) theories. This is an analogue of an Abelian superconformal vacuum found first by Argyres and Douglas.[30] It can be explicitly checked that the low-energy degrees of freedom include relatively non-local monopoles and dyons.[27,31,32] There are no local effective Lagrangians describing the infrared dynamics: these are the most difficult cases to analyze, but are perhaps the most interesting ones, from the point of view of understanding QCD.

5. Vortices and Monopoles

The moral of the story is that the non-Abelian monopoles do exist in fully quantum mechanical systems. In typical confining vacua in supersymmetric gauge theories they are the relevant infrared degrees of freedom. Their condensation induces confinement and dynamical symmetry breaking. This brings us back to the problem of *understanding* these light, magnetic degrees of freedom as quantum solitons:

`What are their semi-classical counterparts?`
`Are they Goddard-Nuyts-Olive-Weinberg monopoles?`
`In which sense condensation of non-Abelian monopoles imply confinement?`
`How has the difficulty related to the dual group mentioned earlier been avoided?`

These are the questions we wish to answer. The idea is to take advantage of the fact that in supersymmetric theories there are parameters which can be varied, upon which the physical properties of the system depend in a holomorphic fashion. As m_i and μ are varied, there cannot be phase transition at some $|\mu|$ or at $|m_i|$: the number of Nambu-Goldstone bosons and hence the pattern of the symmetry breaking, must be invariant.

The model we study is the same one already discussed above, but now we analyze it in the region, $m_i \gg \mu \gg \Lambda$, so that the semiclassical reasoning makes sense.[45] For concreteness, we take as our model the standard $\mathcal{N}=2$ SQCD with N_f quark hypermultiplets, with a larger gauge symmetry, *e.g.*, $SU(N+1)$, which is broken at a much larger mass scale ($v_1 \sim |m_i|$) as

$$SU(N+1) \stackrel{v_1 \neq 0}{\longrightarrow} \frac{SU(N)\times U(1)}{\mathbf{Z}_N}. \tag{16}$$

The unbroken gauge symmetry is completely broken at a lower mass scale, $v_2 \sim |\sqrt{\mu m}|$, as in Eq. (23) below.

We are here back to our argument on the duality and non-Abelian monopoles, defined through a better-understood non-Abelian *vortices* presented in general terms in Section 3, but this time in the context of a concrete model, where the fully quantum mechanical answer is known.

The bosonic sector of this model is described, after elimination of the auxiliary fields, by

$$\mathcal{L} = \frac{1}{4g^2}F_{\mu\nu}^2 + \frac{1}{g^2}|\mathcal{D}_\mu \Phi|^2 + |\mathcal{D}_\mu Q|^2 + \left|\mathcal{D}_\mu \bar{\tilde{Q}}\right|^2 - V_1 - V_2, \tag{17}$$

where

$$V_1 = \frac{1}{8}\sum_A \left(t_{ij}^A \left[\frac{1}{g^2}(-2)\,[\Phi^\dagger,\Phi]_{ji} + Q_j^\dagger Q_i - \tilde{Q}_j \tilde{Q}_i^\dagger \right] \right)^2 ; \tag{18}$$

$$V_2 = g^2|\mu\,\Phi^A + \sqrt{2}\,\tilde{Q}\,t^A Q|^2 + \tilde{Q}\,[m+\sqrt{2}\Phi]\,[m+\sqrt{2}\Phi]^\dagger\,\tilde{Q}^\dagger + Q^\dagger\,[m+\sqrt{2}\Phi]^\dagger\,[m+\sqrt{2}\Phi]\,Q. \tag{19}$$

In the construction of the approximate monopole and vortex solutions we shall consider only the VEVs and fluctuations around them which satisfy

$$[\Phi^\dagger,\Phi] = 0, \qquad Q_i = \tilde{Q}_i^\dagger, \tag{20}$$

and hence the D-term potential V_1 can be set identically to zero throughout.

In order to keep the hierarchy of the gauge symmetry breaking scales, Eq. (11), we choose the masses such that

$$m_1 = \ldots = m_{N_f} = m, \qquad m \gg \mu \gg \Lambda. \tag{21}$$

Although the theory described by the above Lagrangian has many degenerate vacua, we are interested in the vacuum where[27]

$$\langle \Phi \rangle = -\frac{1}{\sqrt{2}} \begin{pmatrix} m & 0 & 0 & 0 \\ 0 & \ddots & \vdots & \vdots \\ 0 & \ldots & m & 0 \\ 0 & \ldots & 0 & -Nm \end{pmatrix} ; \tag{22}$$

$$Q = \tilde{Q}^\dagger = \begin{pmatrix} d & 0 & 0\, 0 \ldots \\ 0 & \ddots & 0 \vdots \ldots \\ 0 & 0 & d\, 0 \ldots \\ 0 & \ldots & 0\, 0 \ldots \end{pmatrix}, \qquad d = \sqrt{(N+1)\,\mu m}. \tag{23}$$

This is a particular case of the so-called r vacuum, with $r = N$. Although such a vacuum certainly exists classically, the existence of the quantum $r = N$ vacuum requires $N_f \geq 2\,N$, which we shall assume.

To start with, ignore the smaller squark VEV, Eq. (23). As $\pi_2(G/H) \sim \pi_1(H) = \pi_1(U(1)) = \mathbf{Z}$, the symmetry breaking Eq. (22) gives rise to regular magnetic monopoles with mass of order of $O(\frac{v_1}{g})$, whose continuous transformation property is our main concern here. At scales much lower than $v_1 = m$ but still neglecting the smaller squark VEV v_2, the theory reduces to an $SU(N) \times U(1)$ gauge theory with N_f light quarks $q_i, \tilde{q}^i$ (the first N components of the original quark multiplets $Q_i, \tilde{Q}^i$). By integrating out the massive fields, the effective Lagrangian valid between the two mass scales has the form,

$$\mathcal{L} = \frac{1}{4g_N^2}(F_{\mu\nu}^a)^2 + \frac{1}{4g_1^2}(F_{\mu\nu}^0)^2 + \frac{1}{g_N^2}|\mathcal{D}_\mu\phi^a|^2 + \frac{1}{g_1^2}|\mathcal{D}_\mu\phi^0|^2 + |\mathcal{D}_\mu q|^2 + |\mathcal{D}_\mu\bar{\tilde{q}}|^2$$
$$- g_1^2\left| - \mu m\sqrt{N(N+1)} + \frac{\tilde{q}\, q}{\sqrt{N(N+1)}}\right|^2 - g_N^2|\sqrt{2}\,\tilde{q}\, t^a q\,|^2 + \ldots \tag{24}$$

where $a = 1, 2, \ldots N^2 - 1$ labels the $SU(N)$ generators, t^a; the index 0 refers to the $U(1)$ generator $t^0 = \frac{1}{\sqrt{2N(N+1)}}\, diag(1, \ldots, 1, -N)$. We have taken into account the fact that the $SU(N)$ and $U(1)$ coupling constants (g_N and g_1) get renormalized differently towards the infrared.

The adjoint scalars are fixed to its VEV, Eq. (22), with small fluctuations around it,

$$\Phi = \langle\Phi\rangle(1 + \langle\Phi\rangle^{-1}\,\tilde{\Phi}), \qquad |\tilde{\Phi}| \ll m. \tag{25}$$

In the consideration of the vortices of the low-energy theory, they will be in fact replaced by the constant VEV. The presence of the small terms Eq. (25), however, makes the low-energy vortices not strictly BPS[44] (and this will be important in the consideration of their stability below). The quark fields are replaced, consistently with Eq. (20), as

$$\tilde{q} \equiv q^\dagger, \qquad q \to \frac{1}{\sqrt{2}}\, q, \tag{26}$$

where the second replacement brings back the kinetic term to the standard form.

We further replace the singlet coupling constant and the $U(1)$ gauge field as

$$e \equiv \frac{g_1}{\sqrt{2N(N+1)}}; \qquad \tilde{A}_\mu \equiv \frac{A_\mu}{\sqrt{2N(N+1)}}, \qquad \tilde{\phi}^0 \equiv \frac{\phi^0}{\sqrt{2N(N+1)}}. \tag{27}$$

The net effect is

$$\mathcal{L} = \frac{1}{4g_N^2}(F_{\mu\nu}^a)^2 + \frac{1}{4e^2}(\tilde{F}_{\mu\nu})^2 + |\mathcal{D}_\mu q|^2 - \frac{e^2}{2}\,|\,q^\dagger\,q - c\,\mathbf{1}\,|^2 - \frac{1}{2}\,g_N^2\,|\,q^\dagger\,t^a q\,|^2, \tag{28}$$

$$c = N(N+1)\sqrt{2\,\mu\,m}. \tag{29}$$

The transformation property of the vortices can be determined from the moduli matrix method, as was done in.[43] Indeed, the system possesses BPS saturated vortices described by the linearized equations

$$(\mathcal{D}_1 + i\mathcal{D}_2)\;q = 0, \tag{30}$$

$$F_{12}^{(0)} + \frac{e^2}{2}\left(c\,\mathbf{1}_N - q\,q^\dagger\right) = 0; \qquad F_{12}^{(a)} + \frac{g_N^2}{2}\,q_i^\dagger\,t^a\,q_i = 0. \tag{31}$$

The matter equation can be solved exactly as in[35–37] ($z = x^1 + ix^2$) by setting

$$q = S^{-1}(z,\bar{z})\,H_0(z), \quad A_1 + i\,A_2 = -2\,i\,S^{-1}(z,\bar{z})\,\bar{\partial}_z S(z,\bar{z}), \tag{32}$$

where S is an $N \times N$ invertible matrix over whole of the z plane, and H_0 is the moduli matrix, holomorphic in z.

The gauge field equations take a slightly more complicated form

$$\partial_z\,(\Omega^{-1}\partial_{\bar{z}}\,\Omega) = -\frac{g_N^2}{2}\,\mathrm{Tr}\,(\,t^a\,\Omega^{-1}\,q\,q^\dagger)\,t^a - \frac{e^2}{4N}\,\mathrm{Tr}\,(\,\Omega^{-1}q\,q^\dagger - \mathbf{1}), \quad \Omega = S\,S^\dagger. \tag{33}$$

than in the $U(N)$ model studied by other groups.[33,35–42] The last equation reduces to the master equation[37] of the $U(N)$ model in the limit, $g_N = e$.

5.1. *Dual gauge transformation from the vortex moduli*

The concepts such as the low-energy BPS vortices or the high-energy BPS monopole solutions are thus only approximate: their explicit forms are valid only in the lowest-order approximation, in the respective kinematical regions. Nevertheless, there is a property of the system which is exact and does not depend on any approximation: the full system has an exact, global $SU(N)_{C+F}$ symmetry, which is neither broken by the interactions nor by both sets of VEVs, v_1 and v_2. This symmetry is broken by individual soliton vortex, endowing the latter with non-Abelian orientational moduli, analogous to the translational zero-modes of a kink. Note that the vortex breaks the color-flavor symmetry as

$$SU(N)_{C+F} \to SU(N-1) \times U(1), \tag{34}$$

leading to the moduli space of the minimum vortices which is

$$\mathcal{M} \simeq \mathbf{C}P^{N-1} = \frac{SU(N)}{SU(N-1) \times U(1)}. \tag{35}$$

The fact that this moduli coincides with the moduli of the quantum states of an N-state quantum mechanical system, is a first hint that the monopoles appearing at the endpoint of a vortex, transform as a fundamental multiplet $\underline{N}$ of a group $SU(N)$.

The moduli space of the vortices is described by the moduli matrix (we consider here the vortices of minimal winding, $k = 1$)

$$H_0(z) \simeq \begin{pmatrix} 1 & 0 & 0 & -a_1 \\ 0 & \ddots & 0 & \vdots \\ 0 & 0 & 1 & -a_{N-1} \\ 0 & \dots & 0 & z \end{pmatrix}, \tag{36}$$

where the constants a_i, $i = 1, 2, \dots, N-1$ are the coordinates of $\mathbf{C}P^{N-1}$. Under $SU(N)_{C+F}$ transformation, the squark fields transform as

$$q \to U^{-1}\, q\, U, \tag{37}$$

but as the moduli matrix is defined *modulo* holomorphic redefinitions,[37] it is sufficient to consider

$$H_0(z) \to H_0(z)\, U. \tag{38}$$

Now, for an infinitesimal $SU(N)$ transformation acting on a matrix of the form Eq. (36), U can be taken in the form,

$$U = \mathbf{1} + X, \qquad X = \begin{pmatrix} \mathbf{0} & \vec{\xi} \\ -(\vec{\xi})^{\dagger} & 0 \end{pmatrix}, \tag{39}$$

where $\vec{\xi}$ is a small $N-1$ component constant vector. Computing $H_0\, X$ and making a V transformation from the left to bring back H_0 to the original form, we find

$$\delta a_i = -\xi_i - a_i\, (\vec{\xi})^{\dagger} \cdot \vec{a}, \tag{40}$$

which shows that a_i's indeed transform as the inhomogeneous coordinates of $\mathbf{C}P^{N-1}$.In fact, if a $\underline{N}$ vector $\vec{c}$ transforms as $\vec{c} \to (\mathbf{1} + X)\,\vec{c}$, the inhomogeneous coordinates $a_i = c_i/c_N$ transform as in Eq. (40). In other words, the vortex represented by the moduli matrix Eq. (36) transforms as a fundamental multiplet of $SU(N)$.

The fact that the vortices (seen as solitons of the low-energy approximation) transform as in the $\underline{N}$ representation of $SU(N)_{C+F}$, implies that there exist a set of monopoles which transform accordingly, as $\underline{N}$. The existence of such a set follows from the exact $SU(N)_{C+F}$ symmetry of the theory, broken by the individual monopole-vortex configuration.

This answers some of the questions formulated earlier (below Eq. (9)) unambiguously.[45] Note that in our derivation of continuous transformations of the monopoles, the explicit, semiclassical form of the latter is not used.

A subtle point is that in the high-energy approximation, and to lowest order of such an approximation, the semiclassical monopoles are just certain non-trivial field configurations involving $\phi(x)$ and $A_i(x)$ fields, and therefore apparently transform under the color part of $SU(N)_{C+F}$ only. When the full monopole-vortex configuration $\phi(x), A_i(x), q(x)$ (Fig. 2) are considered, however, only the combined color-flavor diagonal transformations keep the energy of the configuration invariant. In other words, the monopole transformations must be regarded as part of more complicated transformations involving flavor, when higher order effects in $O(\frac{v_1}{v_2})$ are taken into account. And this means that the transformations are among *physically distinct states*, as the vortex moduli describe obviously physically distinct vortices.[34]

This discussion highlights the crucial role played by the (massless) flavors in the underlying theory as has been already summarized at the end of Section 2.

Our construction has been generalized to the symmetry breaking $SO(2N+1) \to U(N) \to \emptyset$, $SO(2N+1) \to U(r) \times U(1)^{N-r} \to \emptyset$, in the concrete context of softly broken $\mathcal{N}=2$ models. There is an interesting difference in the quantum fate of the semiclassical monopoles in the case the unbroken SU factor has the maximum rank ($r = N$) and in the cases where $r \leq N-1$. The semiclassical (vortex-monopole complex) argument and the fully quantum mechanical results agree qualitatively, quite nontrivially.[45]

The fact that the vortices of the low-energy theory are BPS saturated, which allows us to analyze their moduli and transformation properties elegantly as discussed above, while in the full theory there are corrections which make them non BPS (and unstable), might cause some concern. Actually, the rigor of our argument is not affected by those terms which can be treated as perturbation. The attributes characterized by integers such as the transformation property of certain configurations as a multiplet of a non-Abelian group which is an exact symmetry group of the full theory, cannot receive renormalization. This is similar to the current algebra relations of Gell-Mann which are not renormalized. CVC of Feynman and Gell-Mann also hinges upon an analogous situation.

These basically answer the questions we asked earlier.

Ackowledgment

The author thanks the organizers of SCGT06, "Origin of Mass and Strong Coupling Gauge Theories" (21-24 Nov. 2006, Nagoya, Japan), for a stimulating occasion to discuss these and other interesting ideas with participants from different backgrounds.

References

1. P.A.M. Dirac, Proc. Roy. Soc. (1931) A **133**, 60; Phys. Rev. 74, 817 (1948).
2. G. 't Hooft, Nucl. Phys. B **79**, 817 (1974), A.M. Polyakov, JETP Lett. **20**, 194 (1974).
3. E. Lubkin, Ann. Phys. **23**, 233 (1963); E. Corrigan, D.I. Olive, D.B. Fairlie, J. Nuyts, Nucl. Phys. B **106**, 475 (1976).
4. P. Goddard, J. Nuyts, D. Olive, Nucl. Phys. B **125**, 1 (1977).
5. F.A. Bais, Phys. Rev. D **18**, 1206 (1978).
6. E.J. Weinberg, Nucl. Phys. B **167**, 500 (1980); Nucl. Phys. B **203**, 445 (1982); K. Lee, E. J. Weinberg, P. Yi, Phys. Rev. D **54** , 6351 (1996).
7. C.H. Taubes, Commun. Math. Phys. **80**, 343 (1980).
8. R.S. Ward, Commun. Math. Phys. **86**, 437 (1982).
9. N. Manton, Phys. Lett. B **154**, 397 (1985), Erratum-ibid. B **157**, 475 (1985).
10. S. Coleman, "The Magnetic Monopole Fifty Years Later", Lectures given at Int. Sch. of Subnuclear Phys., Erice, Italy (1981).
11. A. Abouelsaood, Nucl. Phys. B **226**, 309 (1983); P. Nelson, A. Manohar, Phys. Rev. Lett. **50**, 943 (1983); A. Balachandran, G. Marmo, M. Mukunda, J. Nilsson, E. Sudarshan, F. Zaccaria, Phys. Rev. Lett. **50**, 1553 (1983); P. Nelson, S. Coleman, Nucl. Phys. B **227**, 1 (1984); P.A. Horvathy, J.H. Rawnsley, Phys. Rev. D **32**, 968 (1985); Journ. Math. Phys. **27**, 982 (1986).
12. C. Rebbi, G. Soliani, "Soliton and particles", Singapore, World Scientific. 1984. Many earlier references on the solitons are collected in this book.
13. N. Dorey, C. Fraser, T.J. Hollowood, M.A.C. Kneipp, "NonAbelian duality in N=4 supersymmetric gauge theories," [arXiv: hep-th/9512116]; Phys.Lett. B **383**, 422 (1996).
14. C.J. Houghton, P.M. Sutcliffe, J. Math. Phys. **38**, 5576 (1997).
15. B.J. Schroers, F.A. Bais, Nucl. Phys. B **512**, 250 (1998); Nucl. Phys. B **535**, 197 (1998).

16. M. Strassler, Prog. Theor. Phys. Suppl. **131**, 439 (1998).
17. H.J. de Vega, Phys. Rev. D **18**, 2932 (1978); H.J. de Vega, F.A. Shaposnik, Phys. Rev. Lett. **56**, 2564 (1986); Phys. Rev. D34, 3206 (1986); J. Heo, T. Vachaspati, Phys. Rev. D **58**, 065011 (1998), P. Suranyi, hep-lat/9912023; F.A. Shaposnik, P. Suranyi, Phys. Rev. D **62**, 125002 (2000); J. Edelstein, W. Fuertes, J. Mas, J. Guilarte, Phys. Rev. D **62**, 065008 (2000); M. Kneipp, P. Brockill, Phys. Rev. D **64**, 125012 (2001).
18. G. 't Hooft, Nucl. Phys. B **190**, 455 (1981); S. Mandelstam, Phys. Lett. **53B**, 476 (1975); Phys. Rep. C **23**, 245 (1976).
19. Y.M. Cho, Phys. Rev. D **21**, 1080 (1980); L.D. Faddeev and A.J. Niemi, Phys. Rev. Lett. **82**, 1624 (1999); Phys. Lett. B **449**, 214 (1999).
20. T.T. Wu, C.N. Yang, in "Properties of Matter Under Unusual Conditions", Ed. H. Mark, S. Fernbach, Interscience, New York, 1969.
21. N. Seiberg, E. Witten, Nucl. Phys. B **426**, 19 (1994); Erratum *ibid.* B **430**, 485 (1994).
22. N. Seiberg, E. Witten, Nucl. Phys. B **431**, 484 (1994).
23. P. C. Argyres, A. F. Faraggi, Phys. Rev. Lett **74**, 3931 (1995); A. Klemm, W. Lerche, S. Theisen, S. Yankielowicz, Phys. Lett. B **344**, 169 (1995); Int. J. Mod. Phys. A **11**, 1929 (1996), A. Hanany, Y. Oz, Nucl. Phys. B **452**, 283 (1995) ; P. C. Argyres, M. R. Plesser, A. D. Shapere, Phys. Rev. Lett. **75**, 1699 (1995); P. C. Argyres, A. D. Shapere, Nucl. Phys. B **461**, 437 (1996); A. Hanany, Nucl.Phys. B **466**, 85 (1996).
24. S. Bolognesi, K. Konishi, Nucl. Phys. B **645**, 337 (2002).
25. P. C. Argyres, M. R. Plesser, N. Seiberg, Nucl. Phys. B **471**, 159 (1996); P.C. Argyres, M.R. Plesser, A.D. Shapere, Nucl. Phys. B **483**, 172 (1997); K. Hori, H. Ooguri, Y. Oz, Adv. Theor. Math. Phys. **1**, 1 (1998).
26. A. Hanany, Y. Oz, Nucl. Phys. B **466**, 85 (1996).
27. G. Carlino, K. Konishi, H. Murayama, JHEP **0002**, 004 (2000); Nucl. Phys. B **590**, 37 (2000).
28. G. Carlino, K. Konishi, S. P. Kumar, H. Murayama, Nucl. Phys. B **608**, 51 (2001).
29. R. Auzzi, S. Bolognesi, J. Evslin, K. Konishi, H. Murayama, Nucl. Phys. B **701**, 207 (2004).
30. P. C. Argyres, M. R. Douglas, Nucl. Phys. B **448**, 93 (1995); P. C. Argyres, M. R. Plesser, N. Seiberg, E. Witten Nucl. Phys. B **461**, 71 (1996); T. Eguchi, K. Hori, K. Ito, S.-K. Yang, Nucl. Phys. B **471**, 431 (1996).
31. G. Marmorini, K. Konishi, N. Yokoi, Nucl. Phys. B **741**, 180 (2006).
32. R. Auzzi, R. Grena, K. Konishi, Nucl. Phys. B **653**, 204 (2003).
33. A. Hanany, D. Tong, JHEP **0307**, 037 (2003); A. Hanany, D. Tong, JHEP **0404**, 066 (2004).
34. R. Auzzi, S. Bolognesi, J. Evslin, K. Konishi, A. Yung, Nucl. Phys. B **673**, 187 (2003).
35. Y. Isozumi, M. Nitta, K. Ohashi, N. Sakai, Phys. Rev. D **71**, 065018 (2005).
36. M. Eto, Y. Isozumi, M. Nitta, K. Ohashi, N. Sakai, Phys. Rev. Lett. **96**, 161601 (2006).
37. M. Eto, Y. Isozumi, M. Nitta, K. Ohashi, N. Sakai, J. Phys. A **39**, 315 (2006).

38. D. Tong, "TASI lectures on solitons: Instantons, monopoles, vortices and kinks" [arXiv: hep-th/0509216].
39. M. Shifman and A. Yung, Phys. Rev. D **66**, 045012 (2002); M. Shifman and A. Yung, Phys. Rev. D **70**, 045004 (2004).
40. A. Gorsky, M. Shifman, A. Yung, Phys. Rev. D **71**, 045010 (2005).
41. M. Shifman, A. Yung, Phys. Rev. D **73**, 125012 (2006).
42. M. Eto, Y. Isozumi, M. Nitta, K. Ohashi, N. Sakai, Phys. Rev. D **72**, 025011 (2005).
43. M. Eto, K. Konishi, G. Marmorini, M. Nitta, K. Ohashi, W. Vinci, N. Yokoi, Phys. Rev. D **74**, 065021 (2006).
44. E.B. Bogomolnyi, Sov. J. Nucl. Phys. **24**, 449 (1976); M.K. Prasad, C.M. Sommerfield, Phys. Rev. Lett. **35**, 760 (1975).
45. M. Eto, L. Ferretti, K. Konishi, G. Marmorini, M. Nitta, K. Ohashi, W. Vinci, N. Yokoi, "Non-abelian duality from vortex moduli: a dual model of color-confinement", [arXiv: hep-th/0611313].

ADS/CFT and QCD

Stanley J. Brodsky*

Stanford Linear Accelerator Center
Stanford University, Stanford, California 94309
**E-mail: sjbth@slac.stanford.edu*

Guy F. de Téramond*

Universidad de Costa Rica
San José, Costa Rica, and
Stanford Linear Accelerator Center
Stanford University, Stanford, California 94309
**E-mail: gdt@asterix.crnet.cr*

The AdS/CFT correspondence between string theory in AdS space and conformal field theories in physical space-time leads to an analytic, semi-classical model for strongly-coupled QCD which has scale invariance and dimensional counting at short distances and color confinement at large distances. Although QCD is not conformally invariant, one can nevertheless use the mathematical representation of the conformal group in five-dimensional anti-de Sitter space to construct a first approximation to the theory. The AdS/CFT correspondence also provides insights into the inherently non-perturbative aspects of QCD, such as the orbital and radial spectra of hadrons and the form of hadronic wavefunctions. In particular, we show that there is an exact correspondence between the fifth-dimensional coordinate of AdS space z and a specific impact variable ζ which measures the separation of the quark and gluonic constituents within the hadron in ordinary space-time. This connection allows one to compute the analytic form of the frame-independent light-front wavefunctions, the fundamental entities which encode hadron properties and allow the computation of decay constants, form factors, and other exclusive scattering amplitudes. New relativistic light-front equations in ordinary space-time are found which reproduce the results obtained using the 5-dimensional theory. The effective light-front equations possess remarkable algebraic structures and integrability properties. Since they are complete and orthonormal, the AdS/CFT model wavefunctions can also be used as a basis for the diagonalization of the full light-front QCD Hamiltonian, thus systematically improving the AdS/CFT approximation.

1. The Conformal Approximation to QCD

Quantum Chromodynamics, the Yang-Mills local gauge field theory of $SU(3)_C$ color symmetry provides a fundamental understanding of hadron and nuclear physics in terms of quark and gluon degrees of freedom. However, because of its strong-coupling nature, it is difficult to find analytic solutions to QCD or to make precise predictions outside of its perturbative domain. An important goal is thus to find an initial approximation to QCD which is both analytically tractable and which can be systematically improved. For example, in quantum electrodynamics, the Schrödinger and Dirac equations provide accurate first approximations to atomic bound state problems which can then be systematically improved by using the Bethe-Salpeter formalism and correcting for quantum fluctuations, such as the Lamb Shift and vacuum polarization.

One of the most significant theoretical advances in recent years has been the application of the AdS/CFT correspondence[1] between string states defined on the 5-dimensional Anti–de Sitter (AdS) space-time and conformal field theories in physical space-time. The essential principle underlying the AdS/CFT approach to conformal gauge theories is the isomorphism of the group of Poincare' and conformal transformations $SO(4,2)$ to the group of isometries of Anti-de Sitter space. The AdS metric is

$$ds^2 = \frac{R^2}{z^2}(\eta_{\mu\nu} dx^\mu dx^\nu - dz^2),$$

which is invariant under scale changes of the coordinate in the fifth dimension $z \to \lambda z$ and $x_\mu \to \lambda x_\mu$. Thus one can match scale transformations of the theory in $3+1$ physical space-time to scale transformations in the fifth dimension z.

QCD is not itself a conformal theory; however in the domain where the QCD coupling is approximately constant and quark masses can be neglected, QCD resembles a strongly-coupled conformal theory. As shown by Polchinski and Strassler,[2] one can simulate confinement by imposing boundary conditions in the holographic variable at $z = z_0 = 1/\Lambda_{QCD}$. Confinement can also be introduced by modifying the AdS metric to mimic a confining potential. The resulting models, although *ad hoc*, provide a simple semi-classical approximation to QCD which has both counting-rule behavior[3–6] at short distances and confinement at large distances. This simple approach, which has been described as a "bottom-up" approach, has been successful in obtaining general properties of scattering amplitudes of hadronic bound states[2,7–12] and the low-lying hadron spectra.[13–20] Studies of hadron couplings and chiral symmetry breaking,[16,21–24] quark potentials in confining backgrounds[25,26] and pomeron physics[27,28] has also been addressed within the bottom-up approach to holographic QCD, also known as AdS/QCD.

Recently, the behavior of the space-like form factors of the pion[29–31] and nucleons[32] has been discussed within the framework of AdS/QCD. Studies of geometric back-reaction controlling the infrared physics are given in refs.[33,34] It is also remarkable that the dynamical properties of the quark-gluon plasma observed at RHIC[35] can be computed within the AdS/CFT correspondence.[36] In contrast to the simple bottom-up approach described above, the introduction of additional higher dimensional branes to the $AdS_5 \times S^5$ background has been used to study chiral symmetry breaking,[37] and recently baryonic properties by using D4-D8 brane constructs.[38–40]

It was originally believed that the AdS/CFT mathematical tool would only be applicable to strictly conformal theories such as $\mathcal{N} = 4$ supersymmetry. In our approach, we will apply AdS/CFT to the low momentum, strong coupling regime of QCD where the coupling is approximately constant. Theoretical[41] and phenomenological[42] evidence is in fact accumulating that the QCD couplings defined from physical observables such as τ decay[43] become constant at small virtuality; *i.e.*, effective charges develop an infrared fixed point in contradiction to the usual assumption of singular growth in the infrared. Recent lattice gauge theory simulations[44] also indicate an infrared fixed point for QCD.

It is clear from a physical perspective that in a confining theory where gluons and quarks have an effective mass or maximal wavelength, all vacuum polarization corrections to the gluon self-energy must decouple at long wavelength; thus an infrared fixed point appears to be a natural consequence of confinement. Furthermore, if one considers a semi-classical approximation to QCD with massless quarks and without particle creation or absorption, then the resulting β function is zero, the coupling is constant, and the approximate theory is scale and conformal invariant.

In the case of hard exclusive reactions,[6] the virtuality of the gluons exchanged in the underlying QCD process is typically much less than the momentum transfer scale Q since typically several gluons share the total momentum transfer. Since the coupling is probed in the conformal window, this kinematic feature can explain why the measured proton Dirac form factor scales as $Q^4 F_1(Q^2) \simeq$ const up to $Q^2 < 35$ GeV^2[45] with little sign of the logarithmic running of the QCD coupling.

One can also use conformal symmetry as a *template*,[46] systematically correcting for its nonzero β function as well as higher-twist effects. For example, "commensurate scale relations"[47] which relate QCD observables to each other, such as the generalized Crewther relation,[48] have no renormalization scale or scheme ambiguity and retain a convergent perturbative structure which reflects the underlying conformal symmetry of the classical theory. In general, the scale is set such that one has the correct analytic behavior at the heavy particle thresholds.[49] The importance of using an analytic effective charge[50] such as the pinch scheme[51,52] for unifying the electroweak and strong couplings and forces is also important.[53] Thus conformal symmetry is a useful first approximant even for physical QCD.

In the AdS/CFT duality, the amplitude $\Phi(z)$ represents the extension of the hadron into the compact fifth dimension. The behavior of $\Phi(z) \to z^\Delta$ at $z \to 0$ must match the twist-dimension of the hadron at short distances $x^2 \to 0$. As we shall discuss, one can use holography to map the amplitude $\Phi(z)$ describing the hadronic state in the fifth dimension of Anti-de Sitter space AdS_5 to the light-front wavefunctions $\psi_{n/h}$ of hadrons in physical space-time,[19] thus providing a relativistic description of hadrons in QCD at the amplitude level. In fact, there is an exact correspondence between the fifth-dimensional coordinate of anti-de Sitter space z and a specific impact variable ζ in the light-front formalism which measures the separation of the constituents within the hadron in ordinary space-time. We derive this correspondence by noticing that the mapping of $z \to \zeta$ analytically transforms the expression for the form factors in AdS/CFT to the exact QCD Drell-Yan-West expression in terms of light-front wavefunctions.

Light-front wavefunctions are relativistic and frame-independent generalizations of the familiar Schrödinger wavefunctions of atomic physics, but they are determined at fixed light-cone time $\tau = t + z/c$—the "front form" advocated by Dirac—rather than at fixed ordinary time t. An important advantage of light-front quantization is the fact that it provides exact formulas to write down matrix elements as a sum of bilinear forms, which can be mapped into their AdS/CFT counterparts in the semi-classical approximation. One can thus obtain not only an accurate description of the hadron spectrum for light quarks, but also a remarkably simple but realistic model of the valence wavefunctions of mesons, baryons, and glueballs. The light-front wavefunctions predicted by AdS/QCD have many phenomenological applications ranging from exclusive B and D decays, deeply virtual Compton scattering and exclusive reactions such as form factors, two-photon processes, and two-body scattering. One thus obtains a connection between the theories and tools used in string theory and the fundamental phenomenology of hadrons.

2. Light-Front Wavefunctions in Impact Space

The light-front expansion is constructed by quantizing QCD at fixed light-cone time[54] $\tau = t + z/c$ and forming the invariant light-front Hamiltonian: $H_{LF}^{QCD} = P^+P^- - \vec{P}_\perp^2$ where $P^\pm = P^0 \pm P^z$.[55] The momentum generators P^+ and $\vec{P}_\perp$ are kinematical; *i.e.*, they are independent of the interactions. The generator $P^- = i\frac{d}{d\tau}$ generates light-cone time translations, and the eigen-spectrum of the Lorentz scalar H_{LF}^{QCD} gives the mass spectrum of the color-singlet hadron states in QCD; the projection of the eigensolution on the free Fock basis gives the hadronic light-front wavefunctions.

The holographic mapping of hadronic LFWFs to AdS string modes is most transparent when one uses the impact parameter space representation.[56] The total position coordinate of a hadron or its transverse center of momentum $\mathbf{R}_\perp$, is defined in terms of the energy momentum tensor $T^{\mu\nu}$

$$\mathbf{R}_\perp = \frac{1}{P^+}\int dx^- \int d^2\mathbf{x}_\perp\, T^{++}\,\mathbf{x}_\perp. \tag{1}$$

In terms of partonic transverse coordinates

$$x_i \mathbf{r}_{\perp i} = x_i \mathbf{R}_\perp + \mathbf{b}_{\perp i}, \tag{2}$$

where the $\mathbf{r}_{\perp i}$ are the physical transverse position coordinates and $\mathbf{b}_{\perp i}$ frame independent internal coordinates, conjugate to the relative coordinates $\mathbf{k}_{\perp i}$. Thus, $\sum_i \mathbf{b}_{\perp i} = 0$ and $\mathbf{R}_\perp = \sum_i x_i \mathbf{r}_{\perp i}$. The LFWF $\psi_n(x_j, \mathbf{k}_{\perp j})$ can be expanded in terms of the $n-1$ independent coordinates $\mathbf{b}_{\perp j}$, $j = 1, 2, \ldots, n-1$

$$\psi_n(x_j, \mathbf{k}_{\perp j}) = (4\pi)^{\frac{(n-1)}{2}} \prod_{j=1}^{n-1} \int d^2\mathbf{b}_{\perp j} \exp\Big(i \sum_{j=1}^{n-1} \mathbf{b}_{\perp j}\cdot\mathbf{k}_{\perp j}\Big) \widetilde{\psi}_n(x_j, \mathbf{b}_{\perp j}). \tag{3}$$

The normalization is defined by

$$\sum_n \prod_{j=1}^{n-1} \int dx_j d^2\mathbf{b}_{\perp j} \left|\widetilde{\psi}_n(x_j, \mathbf{b}_{\perp j})\right|^2 = 1. \tag{4}$$

One of the important advantages of the light-front formalism is that current matrix elements can be represented without approximation as overlaps of light-front wavefunctions. In the case of the elastic space-like form factors, the matrix element of the J^+ current only couples Fock states with the same number of constituents. If the charged parton n is the active constituent struck by the current, and the quarks $i = 1, 2, \ldots, n-1$ are spectators, then the Drell-Yan West formula[57–59] in impact space is

$$F(q^2) = \sum_n \prod_{j=1}^{n-1} \int dx_j d^2\mathbf{b}_{\perp j} \exp\Big(i\mathbf{q}_\perp \cdot \sum_{j=1}^{n-1} x_j \mathbf{b}_{\perp j}\Big) \left|\widetilde{\psi}_n(x_j, \mathbf{b}_{\perp j})\right|^2, \quad (5)$$

corresponding to a change of transverse momenta $x_j\mathbf{q}_\perp$ for each of the $n-1$ spectators. This is a convenient form for comparison with AdS results, since the form factor is expressed in terms of the product of light-front wave functions with identical variables.

We can now establish an explicit connection between the AdS/CFT and the LF formulae. It is useful to express (5) in terms of an effective single particle transverse distribution $\widetilde{\rho}$ [19]

$$F(q^2) = 2\pi \int_0^1 dx \frac{(1-x)}{x} \int \zeta d\zeta \, J_0\left(\zeta q \sqrt{\frac{1-x}{x}}\right) \tilde{\rho}(x, \zeta), \quad (6)$$

where we have introduced the variable

$$\zeta = \sqrt{\frac{x}{1-x}} \left| \sum_{j=1}^{n-1} x_j \mathbf{b}_{\perp j} \right|, \quad (7)$$

representing the x-weighted transverse impact coordinate of the spectator system. On the other hand, the expression for the form factor in AdS space is represented as the overlap in the fifth dimension coordinate z of the normalizable modes dual to the incoming and outgoing hadrons, Φ_P and $\Phi_{P'}$, with the non-normalizable mode, $J(Q, z) = zQK_1(zQ)$, dual to the external source[8]

$$F(Q^2) = R^3 \int \frac{dz}{z^3} \Phi_{P'}(z) J(Q, z) \Phi_P(z). \quad (8)$$

If we compare (6) in impact space with the expression for the form factor in AdS space (8) for arbitrary values of Q using the identity

$$\int_0^1 dx \, J_0\left(\zeta Q \sqrt{\frac{1-x}{x}}\right) = \zeta Q K_1(\zeta Q), \quad (9)$$

then we can identify the spectator density function appearing in the light-front formalism with the corresponding AdS density

$$\tilde{\rho}(x, \zeta) = \frac{R^3}{2\pi} \frac{x}{1-x} \frac{|\Phi(\zeta)|^2}{\zeta^4}. \quad (10)$$

Equation (10) gives a precise relation between string modes $\Phi(\zeta)$ in AdS_5 and the QCD transverse charge density $\tilde{\rho}(x, \zeta)$. The variable ζ represents a measure of the transverse separation between point-like constituents, and it is also the holographic variable z characterizing the string scale in AdS. Consequently the AdS string mode $\Phi(z)$ can be regarded as the propability amplitude to find n partons at transverse impact separation $\zeta = z$. Furthermore, its eigenmodes determine the hadronic spectrum.[19]

In the case of a two-parton constituent bound state, the correspondence between the string amplitude $\Phi(z)$ and the light-front wave function $\widetilde{\psi}(x, \mathbf{b})$ is expressed in closed form[19]

$$\left|\widetilde{\psi}(x,\zeta)\right|^2 = \frac{R^3}{2\pi}\, x(1-x)\, \frac{|\Phi(\zeta)|^2}{\zeta^4}, \tag{11}$$

where $\zeta^2 = x(1-x)\mathbf{b}_\perp^2$. Here $b_\perp$ is the impact separation and Fourier conjugate to $k_\perp$.

3. Holographic Light-Front Representation

The equations of motion in AdS space can be recast in the form of a light-front Hamiltonian equation[55]

$$H_{LC}\, |\psi_h\rangle = \mathcal{M}^2\, |\psi_h\rangle\,, \tag{12}$$

a remarkable result which allows the discussion of the AdS/CFT solutions in terms of light-front equations in physical 3+1 space-time. By substituting $\phi(\zeta) = \left(\frac{\zeta}{R}\right)^{-3/2} \Phi(\zeta)$, in the AdS wave equation describing the propagation of scalar modes in AdS space

$$\left[z^2\partial_z^2 - (d-1)z\,\partial_z + z^2\mathcal{M}^2 - (\mu R)^2\right]\Phi(z) = 0, \tag{13}$$

we find an effective Schrödinger equation as a function of the weighted impact variable ζ

$$\left[-\frac{d^2}{d\zeta^2} + V(\zeta)\right]\phi(\zeta) = \mathcal{M}^2\phi(\zeta), \tag{14}$$

with the effective potential $V(\zeta) \to -(1-4L^2)/4\zeta^2$ in the conformal limit, where we identity ζ with the fifth dimension z of AdS space: $\zeta = z$. We have written above $(\mu R)^2 = -4 + L^2$. The solution to (14) is $\phi(z) = z^{-\frac{3}{2}}\Phi(z) = Cz^{\frac{1}{2}}J_L(z\mathcal{M})$. This equation reproduces the AdS/CFT solutions for mesons with relative orbital angular momentum L. The holographic hadronic light-front wave functions $\phi(\zeta) = \langle\zeta|\psi_h\rangle$ are normalized according to

$$\langle\psi_h|\psi_h\rangle = \int d\zeta\, |\langle\zeta|\psi_h\rangle|^2 = 1, \tag{15}$$

and represent the probability amplitude to find n-partons at transverse impact separation $\zeta = z$. Its eigenmodes determine the hadronic mass spectrum.

The lowest stable state $L = 0$ is determined by the Breitenlohner-Freedman bound.[60] Its eigenvalues are set by the boundary conditions at $\phi(z = 1/\Lambda_{\rm QCD}) = 0$ and are given in terms of the roots of Bessel functions: $\mathcal{M}_{L,k} = \beta_{L,k}\Lambda_{\rm QCD}$. Normalized LFWFs $\widetilde{\psi}_{L,k}$ follow from (11)

$$\widetilde{\psi}_{L,k}(x,\zeta) = B_{L,k}\sqrt{x(1-x)}J_L\left(\zeta\beta_{L,k}\Lambda_{\rm QCD}\right)\theta\big(z \le \Lambda_{\rm QCD}^{-1}\big), \tag{16}$$

where $B_{L,k} = \Lambda_{\rm QCD}/\sqrt{\pi}J_{1+L}(\beta_{L,k})$. The resulting wavefunctions depicted in Fig. 1 display confinement at large interquark separation and conformal symmetry at short distances, reproducing dimensional counting rules for hard exclusive processes and the scaling and conformal properties of the LFWFs at high relative momenta in agreement with perturbative QCD.

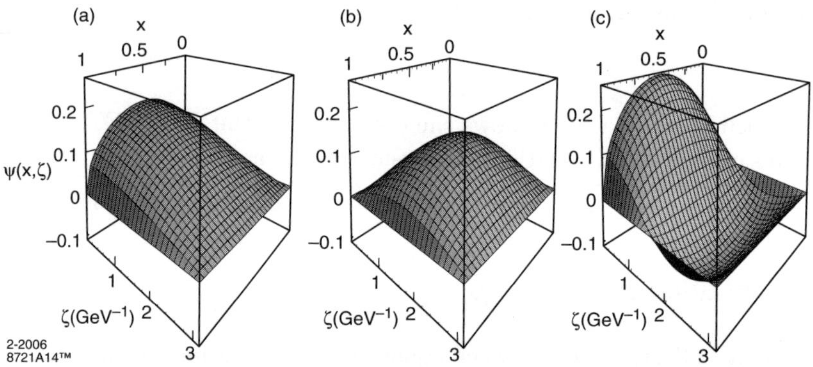

Fig. 1. AdS/QCD Predictions for the light-front wavefunctions of a meson.

Since they are complete and orthonormal, these AdS/CFT model wavefunctions can be used as an initial ansatz for a variational treatment or as a basis for the diagonalization of the light-front QCD Hamiltonian. We are now in fact investigating this possibility with J. Vary and A. Harinandrath. Alternatively one can introduce confinement by adding a harmonic oscillator potential $\kappa^4 z^2$ to the conformal kernel in Eq. (14). One can also introduce nonzero quark masses for the meson. The procedure is straightforward in the $\mathbf{k}_\perp$ representation by using the substitution $\frac{\mathbf{k}_\perp^2}{x(1-x)} \to \frac{\mathbf{k}_\perp^2 + m_2^2}{x} + \frac{\mathbf{k}_\perp^2 + m^2}{1-x}$.

4. Integrability of AdS/CFT Equations

The integrability methods of Ref. [61] find a remarkable application in the AdS/CFT correspondence. Integrability follows if the equations describing a physical model can be factorized in terms of linear operators. These ladder operators then generate all the eigenfunctions once the lowest mass eigenfunction is known. In holographic QCD, the conformally invariant 3 + 1 light-front differential equations can be expressed as ladder operators and their solutions can then be expressed in terms of analytical functions.

In the conformal limit the ladder algebra for bosonic (B) or fermionic (F) modes is given in terms of the operator ($\Gamma^B = 1, \quad \Gamma^F = \gamma_5$)

$$\Pi_\nu^{B,F}(\zeta) = -i\left(\frac{d}{d\zeta} - \frac{\nu+\frac{1}{2}}{\zeta}\,\Gamma^{B,F}\right), \tag{17}$$

and its adjoint

$$\Pi_\nu^{B,F}(\zeta)^\dagger = -i\left(\frac{d}{d\zeta} + \frac{\nu+\frac{1}{2}}{\zeta}\,\Gamma^{B,F}\right), \tag{18}$$

with commutation relations

$$\left[\Pi_\nu^{B,F}(\zeta), \Pi_\nu^{B,F}(\zeta)^\dagger\right] = \frac{2\nu+1}{\zeta^2}\,\Gamma^{B,F}. \tag{19}$$

For $\nu \geq 0$ the Hamiltonian is written as a bilinear form $H_{LC}^{B,F} = \Pi_\nu^{B,F\dagger}\Pi_\nu^{B,F}$. In the fermionic case the eigenmodes also satisfy a first order LF Dirac equation. For bosonic modes, the lowest stable state $\nu = 0$ corresponds to the Breitenlohner-Freedman bound. Higher orbital states are constructed from the L-th application of the raising operator $a^\dagger = -i\Pi^B$ on the ground state.

5. Hadronic Spectra in AdS/QCD

The holographic model based on truncated AdS space can be used to obtain the hadronic spectrum of light quark $q\bar{q}, qqq$ and gg bound states. Specific hadrons are identified by the correspondence of the amplitude in the fifth dimension with the twist dimension of the interpolating operator for the hadron's valence Fock state, including its orbital angular momentum excitations. Bosonic modes with conformal dimension $2 + L$ are dual to the interpolating operator $\mathcal{O}_{\tau+L}$ with $\tau = 2$. For fermionic modes $\tau = 3$. For example, the set of three-quark baryons with spin 1/2 and higher is described by the light-front Dirac equation

$$\left(\alpha\,\Pi^F(\zeta) - \mathcal{M}\right)\psi(\zeta) = 0, \tag{20}$$

where $i\alpha = \begin{pmatrix} 0 & I \\ -I & 0 \end{pmatrix}$ in the Weyl representation. The solution is

$$\psi(\zeta) = C\sqrt{\zeta}\left[J_{L+1}\left(\zeta\mathcal{M}\right) u_+ + J_{L+2}\left(z\mathcal{M}\right) u_-\right], \tag{21}$$

with $\gamma_5 u_\pm = u_\pm$. A discrete four-dimensional spectrum follows when we impose the boundary condition $\psi_\pm(\zeta = 1/\Lambda_{\rm QCD}) = 0$: $\mathcal{M}^+_{\alpha,k} = \beta_{\alpha,k}\Lambda_{\rm QCD}$, $\mathcal{M}^-_{\alpha,k} = \beta_{\alpha+1,k}\Lambda_{\rm QCD}$, with a scale-independent mass ratio.[15]

Figure 2(a) shows the predicted orbital spectrum of the nucleon states and Fig. 2(b) the Δ orbital resonances. The spin-3/2 trajectories are determined from the corresponding Rarita-Schwinger equation. The solution of the spin-3/2 for polarization along Minkowski coordinates, ψ_μ, is similar to the spin-1/2 solution. The data for the baryon spectra are from [64]. The internal parity of states is determined from the SU(6) spin-flavor symmetry. Since only one parameter, the QCD mass scale Λ_{QCD}, is introduced, the

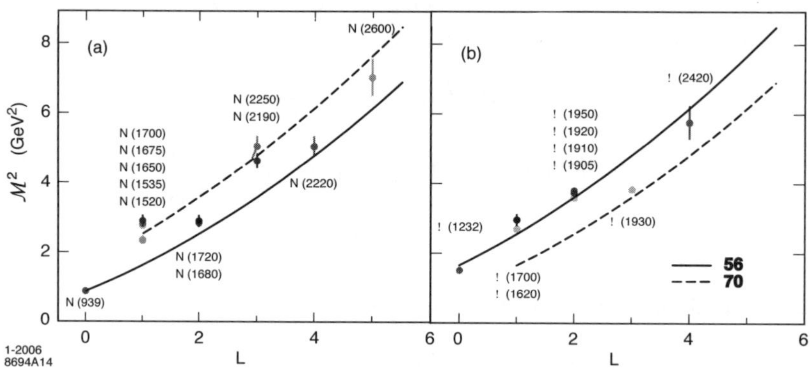

Fig. 2. Predictions for the light baryon orbital spectrum for $\Lambda_{QCD} = 0.25$ GeV. The **56** trajectory corresponds to L even $P = +$ states, and the **70** to L odd $P = -$ states.

agreement with the pattern of physical states is remarkable. In particular, the ratio of Δ to nucleon trajectories is determined by the ratio of zeros of Bessel functions. The predicted mass spectrum in the truncated space model is linear $M \propto L$ at high orbital angular momentum, in contrast to the quadratic dependence $M^2 \propto L$ in the usual Regge parameterization. One can obtain $M^2 \propto (L+n)$ dependence in the holographic model by the introduction of a harmonic potential $\kappa^2 z^2$ in the AdS wave equations.[18] This result can also be obtained by extending the conformal algebra written above. An account of the extended algebraic holographic model and a possible supersymmetric connection between the bosonic and fermionic operators used in the holographic construction will be described elsewhere.

6. Pion Form Factor

Hadron form factors can be predicted from the overlap integral representation in AdS space or equivalently by using the Drell-Yan West formula in physical space-time. For the pion string mode Φ in the harmonic oscillator model[18]

$$\Phi_\pi^{HO}(z) = \frac{\sqrt{2}\kappa}{R^{3/2}}\, z^2\, e^{-\kappa^2 z^2/2}, \tag{22}$$

the form factor has a closed form solution

$$F(Q^2) = 1 + \frac{Q^2}{4\kappa^2} \exp\left(\frac{Q^2}{4\kappa^2}\right) Ei\left(-\frac{Q^2}{4\kappa^2}\right), \tag{23}$$

where Ei is the exponential integral

$$Ei(-x) = \int_\infty^x e^{-t}\frac{dt}{t}. \tag{24}$$

Expanding the function $Ei(-x)$ for large arguments, we find for $-Q^2 \gg \kappa^2$

$$F(Q^2) \to \frac{4\kappa^2}{Q^2}, \tag{25}$$

and we recover the dimensional counting rule. The prediction for the pion form factor is shown in Fig. 3. The space-like behavior of the pion form factor in the harmonic oscillator (HO) model is almost indistinguishable from the truncated-space (TS) model result. The form factor at high Q^2 receives contributions from small ζ, corresponding to small $\vec{b}_\perp \sim \mathcal{O}(1/Q)$ (high relative $\vec{k}_\perp \sim \mathcal{O}(Q)$), as well as $x \to 1$. The AdS/CFT dynamics is thus distinct from endpoint models[29] in which the LFWF is evaluated solely at small transverse momentum or large impact separation.

The $x \to 1$ endpoint domain is often referred to as a "soft" Feynman contribution. In fact $x \to 1$ for the struck quark requires that all of the spectators have $x = k^+/P^+ = (k^0 + k^z)/P^+ \to 0$; this in turn requires high longitudinal momenta $k^z \to -\infty$ for all spectators – unless one has both massless spectator quarks $m \equiv 0$ with zero transverse momentum $k_\perp \equiv 0$, which is a regime of measure zero. If one uses a covariant formalism, such as the Bethe-Salpeter theory, then the virtuality of the struck quark becomes infinitely spacelike: $k_F^2 \sim -\frac{k_\perp^2 + m^2}{1-x}$ in the endpoint domain. Thus, actually, $x \to 1$ corresponds to high relative longitudinal momentum; it is as hard a domain in the hadron wavefunction as high transverse momentum.

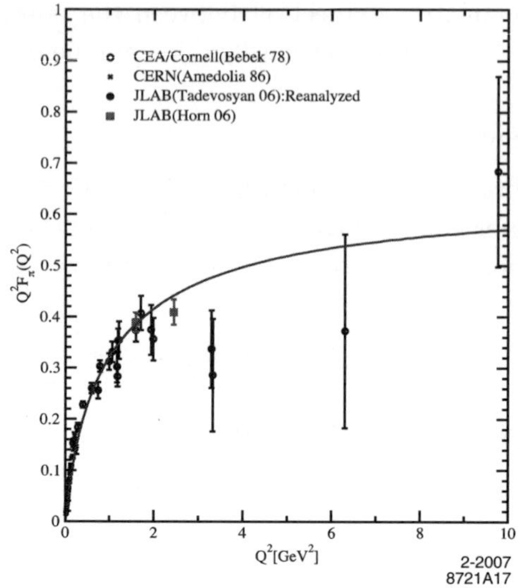

Fig. 3. $Q^2F_\pi(Q^2)$ in the harmonic oscillator model for $\kappa = 0.4$ GeV.

7. The Pion Decay Constant

The pion decay constant is given by the matrix element of the axial isospin current $J^{\mu 5a}$ between a physical pion and the vacuum state[62] $\langle 0|J_W^+(0)|\pi^-(P^+, \vec{P}_\perp)\rangle$, where J_W^+ is the flavor changing weak current. Only the valence state with $L_z = 0$, $S_z = 0$, contributes to the decay of the $\pi^\pm$. Expanding the hadronic initial state in the decay amplitude into its Fock components we find

$$f_\pi = 2\sqrt{N_C}\int_0^1 dx \int \frac{d^2\vec{k}_\perp}{16\pi^3}\, \psi_{\overline{q}q/\pi}(x, k_\perp). \tag{26}$$

This light-cone equation allows the exact computation of the pion decay constant in terms of the valence pion light-front wave function.[6]

The meson distribution amplitude $\phi(x, Q)$ is defined as[63]

$$\phi(x, Q) = \int^{Q^2} \frac{d^2\vec{k}_\perp}{16\pi^3}\, \psi(x, k_\perp). \tag{27}$$

It follows that

$$\phi_\pi(x, Q \to \infty) = \frac{4}{\sqrt{3}\pi} f_\pi \sqrt{x(1-x)}, \tag{28}$$

with

$$f_\pi = \frac{1}{8}\sqrt{\frac{3}{2}}\, R^{3/2} \lim_{\zeta\to 0} \frac{\Phi(\zeta)}{\zeta^2}, \tag{29}$$

since $\phi(x, Q \to \infty) \to \widetilde{\psi}(x, \vec{b}_\perp \to 0)/\sqrt{4\pi}$ and $\Phi_\pi \sim \zeta^2$ as $\zeta \to 0$. The pion decay constant depends only on the behavior of the AdS string mode near the asympototic boundary, $\zeta = z = 0$ and the mode normalization. For the truncated-space (TS) pion mode we find $f_\pi^{TS} = \frac{\sqrt{3}}{8J_1(\beta_{0,k})}\Lambda_{\rm QCD} = 83.4$ Mev, for $\Lambda_{QCD} = 0.2$ MeV. The corresponding result for the transverse harmonic oscillator (HO) pion mode (22) is $f_\pi^{HO} = \frac{\sqrt{3}}{8}\kappa = 86.6$ MeV, for $\kappa = 0.4$ GeV. The values of Λ_{QCD} and κ are determined from the space-like form factor data as discussed above. The experimental result for f_π is extracted form the rate of weak π decay and has the value $f_\pi = 92.4$ MeV.[64]

It is interesting to note that the pion distribution amplitude predicted by AdS/QCD (28) has a quite different x-behavior than the asymptotic distribution amplitude predicted from the PQCD evolution[63] of the pion distribution amplitude $\phi_\pi(x, Q \to \infty) = \sqrt{3} f_\pi x(1-x)$. The broader shape of the pion distribution increases the magnitude of the leading twist perturbative QCD prediction for the pion form factor by a factor of 16/9 compared to the prediction based on the asymptotic form, bringing the PQCD prediction close to the empirical pion form factor.[31]

Acknowledgments

This research was supported by the Department of Energy contract DE–AC02–76SF00515. We thank Alexander Gorsky, Chueng-Ryong Ji, and Mitat Unsal for helpful comments.

References

1. J. M. Maldacena, Adv. Theor. Math. Phys. **2**, 231 (1998) [Int. J. Theor. Phys. **38**, 1113 (1999)] [arXiv:hep-th/9711200].
2. J. Polchinski and M. J. Strassler, duality," Phys. Rev. Lett. **88**, 031601 (2002) [arXiv:hep-th/0109174].
3. S. J. Brodsky and G. R. Farrar, Phys. Rev. Lett. **31**, 1153 (1973).
4. S. J. Brodsky and G. R. Farrar, Phys. Rev. D **11**, 1309 (1975).
5. V. A. Matveev, R. M. Muradian and A. N. Tavkhelidze, Lett. Nuovo Cim. **7**, 719 (1973).
6. G. P. Lepage and S. J. Brodsky, Phys. Rev. D **22**, 2157 (1980).
7. R. A. Janik and R. Peschanski, Nucl. Phys. B **565**, 193 (2000) [arXiv:hep-th/9907177].
8. J. Polchinski and M. J. Strassler, JHEP **0305**, 012 (2003) [arXiv:hep-th/0209211].

9. R. C. Brower and C. I. Tan, Nucl. Phys. B **662**, 393 (2003) [arXiv:hep-th/0207144].
10. S. B. Giddings, Phys. Rev. D **67**, 126001 (2003) [arXiv:hep-th/0203004].
11. S. J. Brodsky and G. F. de Teramond, Phys. Lett. B **582**, 211 (2004) [arXiv:hep-th/0310227].
12. O. Andreev and W. Siegel, Phys. Rev. D **71**, 086001 (2005) [arXiv:hep-th/0410131].
13. H. Boschi-Filho and N. R. F. Braga, JHEP **0305**, 009 (2003) [arXiv:hep-th/0212207].
14. G. F. de Teramond and S. J. Brodsky, arXiv:hep-th/0409074.
15. G. F. de Teramond and S. J. Brodsky, Phys. Rev. Lett. **94**, 201601 (2005) [arXiv:hep-th/0501022].
16. J. Erlich, E. Katz, D. T. Son and M. A. Stephanov, Phys. Rev. Lett. **95**, 261602 (2005) [arXiv:hep-ph/0501128].
17. E. Katz, A. Lewandowski and M. D. Schwartz, Phys. Rev. D **74**, 086004 (2006) [arXiv:hep-ph/0510388].
18. A. Karch, E. Katz, D. T. Son and M. A. Stephanov, Phys. Rev. D **74**, 015005 (2006) [arXiv:hep-ph/0602229].
19. S. J. Brodsky, arXiv:hep-ph/0610115. 4th International Conference On Quarks And Nuclear Physics (QNP06), 5-10 June 2006, Madrid, Spain.
20. D. K. Hong, T. Inami and H. U. Yee, arXiv:hep-ph/0609270.
21. S. Hong, S. Yoon and M. J. Strassler, JHEP **0604**, 003 (2006) [arXiv:hep-th/0409118].
22. L. Da Rold and A. Pomarol, Nucl. Phys. B **721**, 79 (2005) [arXiv:hep-ph/0501218].
23. J. Hirn, N. Rius and V. Sanz, Phys. Rev. D **73**, 085005 (2006) [arXiv:hep-ph/0512240].
24. K. Ghoroku, N. Maru, M. Tachibana and M. Yahiro, Phys. Lett. B **633**, 602 (2006) [arXiv:hep-ph/0510334].
25. H. Boschi-Filho, N. R. F. Braga and C. N. Ferreira, Phys. Rev. D **73**, 106006 (2006) [Erratum-ibid. D **74**, 089903 (2006)] [arXiv:hep-th/0512295].
26. O. Andreev and V. I. Zakharov, Phys. Rev. D **74**, 025023 (2006) [arXiv:hep-ph/0604204].
27. H. Boschi-Filho, N. R. F. Braga and H. L. Carrion, Phys. Rev. D **73**, 047901 (2006) [arXiv:hep-th/0507063].
28. R. C. Brower, J. Polchinski, M. J. Strassler and C. I. Tan, arXiv:hep-th/0603115.
29. A. V. Radyushkin, Phys. Lett. B **642**, 459 (2006) [arXiv:hep-ph/0605116].
30. S. J. Brodsky, arXiv:hep-ph/0608005. Seventh Workshop on Continuous Advances in QCD, Minneapolis, Minnesota, 11-14 May 2006.
31. H. M. Choi and C. R. Ji, Phys. Rev. D **74**, 093010 (2006) [arXiv:hep-ph/0608148].
32. G. F. de Teramond, arXiv:hep-ph/0606143. Seventh Workshop on Continuous Advances in QCD, Minneapolis, May 11-14, 2006.
33. C. Csaki and M. Reece, arXiv:hep-ph/0608266.
34. J. P. Shock, F. Wu, Y. L. Wu and Z. F. Xie, arXiv:hep-ph/0611227.

35. B. Muller and J. L. Nagle, arXiv:nucl-th/0602029.
36. G. Policastro, D. T. Son and A. O. Starinets, Phys. Rev. Lett. **87**, 081601 (2001) [arXiv:hep-th/0104066].
37. J. Babington, J. Erdmenger, N. J. Evans, Z. Guralnik and I. Kirsch, Phys. Rev. D **69**, 066007 (2004) [arXiv:hep-th/0306018].
38. K. Nawa, H. Suganuma and T. Kojo, arXiv:hep-th/0612187.
39. D. K. Hong, M. Rho, H. U. Yee and P. Yi, arXiv:hep-th/0701276.
40. H. Hata, T. Sakai, S. Sugimoto and S. Yamato, arXiv:hep-th/0701280.
41. R. Alkofer, C. S. Fischer and F. J. Llanes-Estrada, Phys. Lett. B **611**, 279 (2005) [arXiv:hep-th/0412330].
42. S. J. Brodsky, S. Menke, C. Merino and J. Rathsman, Phys. Rev. D **67**, 055008 (2003) [arXiv:hep-ph/0212078].
43. S. J. Brodsky, J. R. Pelaez and N. Toumbas, Phys. Rev. D **60**, 037501 (1999) [arXiv:hep-ph/9810424].
44. S. Furui and H. Nakajima, Few Body Syst. **40**, 101 (2006) [arXiv:hep-lat/0612009].
45. M. Diehl, T. Feldmann, R. Jakob and P. Kroll, Eur. Phys. J. C **39**, 1 (2005) [arXiv:hep-ph/0408173].
46. S. J. Brodsky and J. Rathsman, arXiv:hep-ph/9906339.
47. S. J. Brodsky and H. J. Lu, Phys. Rev. D **51**, 3652 (1995) [arXiv:hep-ph/9405218].
48. S. J. Brodsky, G. T. Gabadadze, A. L. Kataev and H. J. Lu, Phys. Lett. B **372**, 133 (1996) [arXiv:hep-ph/9512367].
49. S. J. Brodsky, G. P. Lepage and P. B. Mackenzie, Phys. Rev. D **28**, 228 (1983).
50. S. J. Brodsky, M. S. Gill, M. Melles and J. Rathsman, Phys. Rev. D **58**, 116006 (1998) [arXiv:hep-ph/9801330].
51. M. Binger and S. J. Brodsky, Phys. Rev. D **74**, 054016 (2006) [arXiv:hep-ph/0602199].
52. J. M. Cornwall and J. Papavassiliou, Phys. Rev. D **40**, 3474 (1989).
53. M. Binger and S. J. Brodsky, Phys. Rev. D **69**, 095007 (2004) [arXiv:hep-ph/0310322].
54. P. A. M. Dirac, Rev. Mod. Phys. **21**, 392 (1949).
55. S. J. Brodsky, H. C. Pauli and S. S. Pinsky, Phys. Rept. **301**, 299 (1998) [arXiv:hep-ph/9705477].
56. D. E. Soper, Phys. Rev. D **15**, 1141 (1977).
57. S. D. Drell and T. M. Yan, Phys. Rev. Lett. **24**, 181 (1970).
58. G. B. West, Phys. Rev. Lett. **24**, 1206 (1970).
59. S. J. Brodsky and S. D. Drell, Phys. Rev. D **22**, 2236 (1980).
60. P. Breitenlohner and D. Z. Freedman, Annals Phys. **144**, 249 (1982).
61. L. Infeld, Phys. Rev. **59**, 737 (1941).
62. M. E. Peskin and D. V. Schroeder, "An Introduction To Quantum Field Theory, *Reading, USA: Addison-Wesley (1995) 842 p.*
63. G. P. Lepage and S. J. Brodsky, Phys. Lett. B **87**, 359 (1979).
64. S. Eidelman *et al.* [Particle Data Group], Phys. Lett. B **592**, 1 (2004).

HOLOGRAPHIC QCD & PERFECTION

NICK EVANS

Department of Physics and Astronomy, University of Southampton
Southampton, UK

A holographic description of chiral symmetry breaking in the pattern of QCD is reviewed. D7 brane probes are used to include quark fields in a simple non-supersymmetric deformation of the AdS/CFT Correspondence. The axial symmetry breaking is realized geometrically and the quark condensate and meson masses are computable. Surprisingly, treating the model as a description of QCD works quantitatively at the 15% level. Models of this AdS/QCD type typically have a strongly coupled, conformal UV regime that is far from QCD. To systematically move closer to QCD, we propose cutting out the large radius gravitational description and matching operators and couplings at a finite UV cut off in the spirit of a perfect lattice action. A simple example is discussed.

1. Introduction

Recently the first attempts have been made to bring the holographic techniques of the, string theory derived, AdS/CFT Correspondence[1] to bare on QCD. The hope is that there is some weakly coupled gravitational theory in five or more dimensions that describes the strong coupling regime of QCD. Here we will review a holographic description of chiral symmetry breaking[2] starting from the AdS/CFT Correspondence and discuss to what extent it can be used as a phenomenological tool for real QCD[a].

2. A Non-Supersymmetric Gravity Dual

The AdS/CFT Correspondence[1] is a duality between the conformal, large N_c, N=4 super Yang Mills theory and IIB strings (supergravity) on 5d Anti-de-Sitter space cross a five sphere. The field theory's global symmetries (an SO(2,4) superconformal symmetry and an $SU(4)_R$ symmetry) match to space-time symmetries of the *AdS* space and the five sphere respectively.

[a]I'm grateful to J Babington, J Erdmenger, Z Guralnik, I Kirsch, J Shock, A Tedder, and T Waterson for their contributions to the work reported here.

The supergravity fields enter the field theory in symmetry invariant ways and so appear as sources for field theory operators. The radial direction in AdS has the conformal symmetry properties of an energy scale and corresponds to the renormalization group scale. Thus the radial behaviour of the supergravity fields describes the RG flow of the field theory sources.

Let us consider a very simple example of AdS with a scalar field, the dilaton, switched on, due to Constable and Myers[3]

$$ds^2 = H^{-1/2}\left(\frac{u^4+b^4}{u^4-b^4}\right)^{\delta/4} dx_4^2 + H^{1/2}\left(\frac{u^4+b^4}{u^4-b^4}\right)^{(2-\delta)/4} \frac{u^4-b^4}{u^4} du_6^2 \quad (1)$$

$$H = \left(\frac{u^4+b^4}{u^4-b^4}\right)^{\delta} - 1, \qquad e^{\Phi} = \left(\frac{u^4+b^4}{u^4-b^4}\right)^{\Delta/2}, \qquad C_4 = H^{-1} \quad (2)$$

$$\delta = \frac{R^4}{2b^4}, \qquad \Delta^2 = 10 - \delta^2 \quad (3)$$

The x_4 directions correspond to the field theory's 4d space and u is the radial direction in the 6d transverse space. At large u the space becomes $AdS_5 \times S^5$ with radius R. Here b is a parameter that controls the size of the deformation from AdS - note it enters along with u and so has energy dimension one. The SO(6) isometry of the transverse plane survives at all u and thus the R symmetry of the field theory is not broken. From these facts we can deduce that b^4 corresponds in the field theory to a vacuum expectation value for the dimension four, R-chargeless operator TrF^2.

It is worth stressing that the vacuum of the $N = 4$ gauge theory has $TrF^2 = 0$ (this quantity is the D-term of a superfield and supersymmetry would be broken were it generated) and so the above geometry describes a non-vacuum state of the field theory. The geometry does though describe some non-supersymmetric, strongly coupled gauge configuration and is relatively simple - for these virtues we will use it below. A consequence of the supersymmetry breaking is that the dilaton (the gauge theory coupling) changes with u *ie* the gauge coupling runs with energy scale. It has a pole at $u = b$ which we interpret as playing the role of the pole in the QCD coupling at the scale Λ_{QCD}.

3. D7 Branes and Quarks

The $N = 4$ gauge theory only has adjoint matter fields - the original construction realized the gauge theory through open string modes with both ends tied to a D3 brane (they transform as $N_c, \bar{N}_c$). To generate fundamental representation quarks one must detach one of the string's ends from the D3 - it is useful to tie it to a D7 brane[4] as shown in Fig. 1.

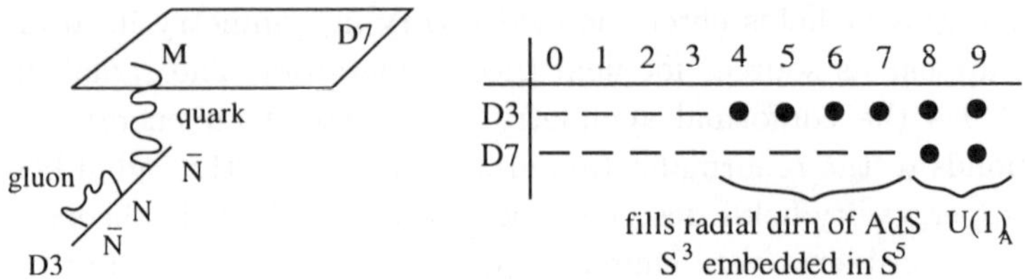

Fig. 1. D3/D7 configuration that introduces quarks into the AdS/CFT Correspondence.

The D3 and the D7 share the 0-3 directions, the D7 are in addition extended in the 4-8 directions (we will call the radial coordinate in this space ρ), and finally the D3 and D7 can be separated in the 8-9 directions (w_5 and w_6 below). This configuration which preserves $N = 2$ supersymmetry corresponds to the $N = 4$ gauge theory with an added fundamental representation quark hypermultiplet.

The minimum length D7-D3 string indicates (length $\times$ tension) the mass of the quark. If the D7 brane lies along the ρ axis then the quarks are massless and there is an SO(2) symmetry in the $w_5 - w_6$ plane. If the D7 lies off axis there is a non-zero quark mass and the SO(2) symmetry is explicitly broken. This indicates that the SO(2) symmetry is a geometric realization of the U(1) axial symmetry of the gauge theory (in the supersymmetric case that symmetry is part of a $\text{U}(1)_R$ symmetry). Note that at large N_c we neglect anomalies.

Using these techniques we can next include quarks into the dilaton deformed geometry above.[2] We will work in the approximation where the D7 brane is a probe (so there is no backreaction on the geometry) - this is the quenched limit where the number of flavours $N_f \ll N_c$. One simply embeds the D7 brane so as to minimize its world volume via it's Dirac Born Infeld action

$$S_{D7} = -T_7 \int d^8\xi \sqrt{P[G_{ab}]}, \qquad P[G_{ab}] = G_{MN} \frac{dx^M}{d\xi^a} \frac{dx^N}{d\xi^b} \tag{4}$$

where T_7 is the tension, ξ the coordinates on the D7, x^M are the spacetime coordinates and G_{MN} the background metric.

In the Constable Myers geometry one finds that the D7 brane is repelled by the singular core of the geometry and the regular embeddings of interest are those shown in Fig. 2. At large ρ the solutions become flat as the gauge theory returns to AdS. The solution is of the form $w_6 = m + c/\rho^2 + ...$ Here m corresponds to the quark mass and c to the $\bar{q}q$ condensate - we can read off the condensate as a function of the quark mass in this theory.

A more intuitive understanding of the embedding results from interpreting the separation of the D7 brane from the ρ axis as the effective quark mass. As one moves in ρ one is changing RG scale - at large ρ one sees a small bare quark mass but in the IR (small ρ) a dynamical mass is generated.

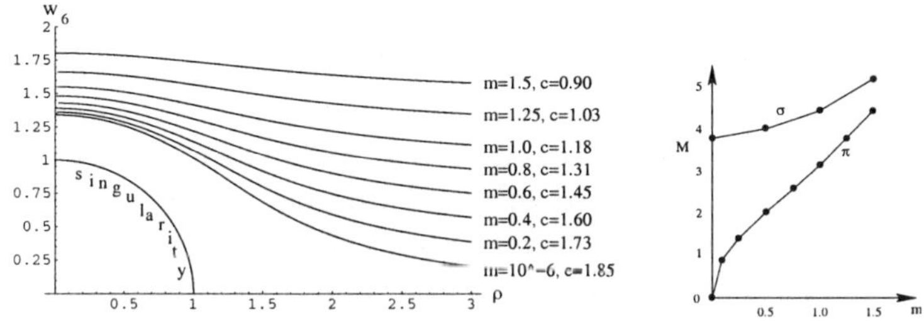

Fig. 2. Embedding solutions for a D7 probe in the Constable Myers geometry and a plot of the meson mass vs quark mass in that model.We chose $b = R$ here as an example.

In particular we can see that the solution exhibits chiral symmetry breaking. If we try to lie a D7 along the ρ axis, so $m = 0$, it is repelled from the origin and there is a non-zero value of the quark condensate. In fact the D7 may be deflected to any point on a circle in the $w_5 - w_6$ plane. We thus explicitly see the breaking of the SO(2) symmetry in that plane and the circle is the vacuum manifold. There should be a Goldstone boson associated with fluctuations of the D7 along the vacuum manifold. One can seek solutions to the equations of motion from the DBI action for those angular fluctuations of the form

$$\theta(\rho, x) = f(\rho) e^{-ikx}, \qquad k^2 = -M^2 \tag{5}$$

Only for particular values of M is $f(\rho)$ regular and hence the meson bound state masses are picked out. In Fig. 2 the meson masses as a function of quark mass are shown. There is a massless Goldstone at $m = 0$ and it's mass grows as $\sqrt{m}$ as in chiral perturbation theory. The mass of the meson associated with radial fluctuations is also shown - it always has a mass gap.

Note this simple model is sometimes criticized for the presence of a singularity in the metric. It is possible a source is present at $w = b$ that explains the singularity - one escapes addressing this problem because the D7 never penetrates the singularity. There are alternative D4-D6 descriptions of chiral symmetry breaking[5] that use completely smooth metrics and yet show the same generic structure. The UV of that theory is six dimensional though. See also other constructions in.[6]

4. AdS/QCD

The holographic description of chiral symmetry breaking above provides the pion spectrum. In addition a vector field on the D7 world volume describes the vector mesons. Solutions for these fields with non-trivial harmonics on the S^3 of the D7 brane also exist and describe R-charged mesons, reflecting the supersymmetric origin of the theory. There is no significant decoupling of these R-charged states since the theory is strongly coupled at the scale of the supersymmetry breaking parameter, b^4 or TrF^2.

In spite of the differences from QCD one can boldly move to a toy model of QCD in the spirit of the work in[7] (and[8]). In that work a five dimensional theory consisting of axial and vector gauge fields and $N_f^2 - 1$ pions in an AdS space with a hard IR (small r) cut off is studied as a model of the QCD pion, ρ and a mesons. We can now repeat that model but using the D7 world volume metric from the theory above.[9] This has the advantage that the conformal symmetry breaking is smoothly included in the metric which exists down to $r = 0$ rather than through an adhoc cut off. The condensate is also a prediction of the gauge dynamics in this model whereas it was included by hand in the pure AdS model.

The original AdS/CFT Correspondence was for a large N_c theory. N_c enters through the prediction for the relative coefficients of the scalar and vector fields' kinetic terms - in the phenomenological approach this is instead set by requiring that one reproduces the perturbative QCD result for the vector vector correlator.[7] Here one is hoping that the conformal nature of the UV asymptotics of AdS in someway mimics the conformal behaviour of weakly coupled QCD. The remaining parameters in the model are then the conformal symmetry breaking scale b (Λ_{QCD}) and the quark mass (position of the D7 at large r). Performing a global fit to meson data one finds the results below[9] - the fit is rather good (rms error 12.8%).

	holography	expt
m_π	139.0 MeV	139.6 MeV
m_ρ	742.7MeV	775.8 MeV
m_a	1337 MeV	1230 MeV

	holography	expt
f_π	83.9 MeV	92.4 MeV
f_ρ	297.0 MeV	345 MeV
f_a	491.4 MeV	433 MeV

5. Perfection

The success of the AdS/QCD approach is rather shocking - we used a quenched, large N_c gauge theory with superpartners present! Gauge gravity dualities are also a strong weak coupling duality and so by assuming the gravity dual is weakly coupled out to large radius we lost QCD's asymptotic freedom. Was the success of the fit just luck then? To answer this one must address systematic errors - this appears hard since the theory is a model and is not derived from QCD. Let us attempt to understand how, at least in principle, one could make a perfect holographic description of QCD.[10]

It is clear that a weakly coupled gravity description should only exist below the scale where QCD becomes strongly coupled. We should therefore impose a UV cut off, to represent where QCD undergoes this transition, and work in the gravity theory only at values of the radius below this. This is analogous to working in lattice QCD but with a rather coarse lattice. In fact it has been understood that one can simulate QCD on a coarse lattice and nevertheless precisely reproduce QCD.[11] The crucial point is that as one blocks from a fine lattice to a coarse lattice one must include higher dimension operator couplings. By analogy one should be careful to make sure all the couplings needed to reproduce QCD are present in the gravity dual with a UV cut off. One also needs to ensure all operators take their appropriate vacuum value and have the correct anomalous dimension.

In principle this is straight forward but there are an infinite number of possible operators and couplings and all could be large. One might worry about whether these couplings will be sufficiently small to keep the gravity theory perturbative - there is no guarantee but let us hope they will. In practice our only method to fix these values is phenomenological. One might pick on a small number of couplings and fix their values using a fit to measured hadron data. The hope is then that those are the significant changes needed and that the remainder of the physical spectrum will be predicted more accurately (here the analogy is to improving lattice actions).

As a toy example consider the AdS/QCD model in[12] of the ρ meson and it's excited states. The theory is just a gauge field in AdS_5 with a non-zero dilaton that blows up in the IR, $\Phi \sim r^{-2}$. In the usual approach one would fix the large r behaviour of the gauge field to enforce the $\bar{q}\gamma^\mu q$ operator to be dimension 3 in the UV. We now though impose that boundary condition not at infinity but at some finite UV cut off.[10] In other words we ensure the scaling dimension of the operator is three down to the scale where QCD becomes non-perturbative. In Fig. 3 we plot the ρ meson masses for the low excitation numbers and compare to the experimental values. Lowering the cut off improves the fit (rms error of 2% for the best fit). Thus changing the anomalous dimension of the quark bilinear operators seems to be an example of an improvement of the holographic dual. One should caution that the importance of many other operators and couplings should be checked although it is far from clear how to include some of these in the gravity dual. Hopefully though one has understood how to be more systematic in the approach.

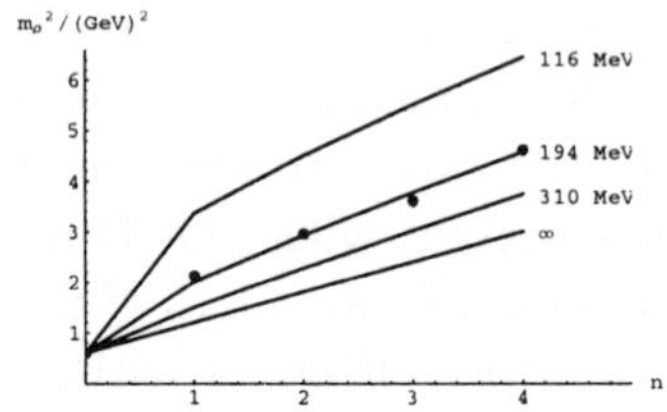

Fig. 3. The ρ meson and its excited states' masses with varying UV cut off in the model in.[12] The dots are the QCD data.

References

1. J. Maldacena, Adv. Theor. Math. Phys. **2** 231 (1998); S.S. Gubser, I.R. Klebanov and A.M. Polyakov, Phys. Lett. **B428** 105 (1998); E. Witten, Adv. Theor. Math. Phys. **2** 253 (1998).
2. J. Babington, J. Erdmenger, N. J. Evans, Z. Guralnik and I. Kirsch, Phys. Rev. D **69** (2004) 066007; N. J. Evans and J. P. Shock, Phys. Rev. D **70** (2004) 046002.
3. N. R. Constable and R. C. Myers, JHEP **9911** (1999) 020.
4. A. Karch and E. Katz, JHEP **0206** (2002) 043; M. Bertolini, P. Di Vecchia, M. Frau, A. Lerda and R. Marotta, Nucl. Phys. **B621**, 157; M. Graña and J. Polchinski, Phys. Rev. **D65**:126005 (2002); M. Kruczenski, D. Mateos, R. C. Myers and D. J. Winters, JHEP **0307**, 049 (2003) .
5. M. Kruczenski, D. Mateos, R.C. Myers, D.J. Winters, JHEP **0405**, 041 (2004).
6. K. Ghoroku and M. Yahiro, Phys. Lett. B **604** (2004) 235; T. Sakai and S. Sugimoto, Prog. Theor. Phys. **113** (2005) 843; E. Antonyan, J. A. Harvey, S. Jensen and D. Kutasov, hep-th/0604017; V. G. Filev, C. V. Johnson, R. C. Rashkov and K. S. Viswanathan, hep-th/0701001.
7. J. Erlich, E. Katz, D. T. Son and M. A. Stephanov, Phys. Rev. Lett. **95** (2005) 261602; L. Da Rold and A. Pomarol, Nucl. Phys. B **721** (2005) 79.
8. S. J. Brodsky and G. F. de Teramond, Phys. Rev. Lett. **96** (2006) 201601; Phys. Rev. Lett. **94** (2005) 201601; H. Boschi-Filho and N. R. F. Braga, Eur. Phys. J. C **32** (2004) 529; JHEP **0305** (2003) 009.
9. N. Evans, A. Tedder and T. Waterson, hep-ph/0603249.
10. N. Evans and A. Tedder, Phys. Lett. B **642** (2006) 546; N. Evans, J. P. Shock and T. Waterson, Phys. Lett. B **622** (2005) 165.
11. P. Hasenfratz and F. Niedermayer, Nucl. Phys. B **414** (1994) 785; M. Luscher and P. Weisz, Phys. Lett. B **158** (1985) 250.
12. A. Karch, E. Katz, D. T. Son, M. A. Stephanov, Phys. Rev. D **74** (2006) 015005.

MESONS AND BARYONS FROM STRING THEORY*

S. SUGIMOTO

Department of Physics, Nagoya University,
Nagoya, 464-8502, Japan
E-mail: sugimoto@eken.phys.nagoya-u.ac.jp

Recently, we proposed a holographic description of 4 dim QCD by using a D4/D8-brane system in type IIA string theory. (hep-th/0412141, hep-th/0507073 with T. Sakai) It was shown that the model nicely catches various features in QCD and hadron physics. Here we give an overview of the model and present some new results about construction of baryons in this framework.

Keywords: QCD; D-brane; Supergravity.

1. Introduction

Since Maldacena proposed the duality between string theory in AdS space and conformal field theory,[3] there have been various attempts to extend this idea to more general situations. (See for example Ref. 4 for a review.) A natural question one would ask is whether we can analyze the realistic QCD using this technique.

A key step toward this direction was given by Witten in Ref. 5, in which a supergravity dual of the 4 dimensional pure Yang-Mills theory was proposed. The basic idea is to use D4-branes compactified on a supersymmetry breaking circle to realize 4 dimensional Yang-Mills theory and replace the D4-branes with the corresponding supergravity solution to obtain a holographic dual description. Though it is in general quite difficult to obtain a rigorous result in the duality of such non-supersymmetric theory, a lot of non-trivial evidence for the duality has been found and it encourages us to further investigate along this line toward the analysis of QCD via supergravity or superstring theory.

---

*Talk given at the international workshop SCGT06 "Origin of Mass and Strong Coupling Gauge Theories" (Nov.21-24, 2006, Nagoya, Japan) based on works done with T.Sakai (Ref. 1) and with H.Hata, T.Sakai and S.Yamato (Ref. 2).

In Ref. 1, we proposed a way to extend this model to obtain a holographic description of QCD with fundamental quarks. Our strategy is to use the above D4-brane background that represents pure Yang-Mills theory and add probe D8-branes to add fundamental quarks to the system.[a] One of the advantages of our model is that the chiral $U(N_f)_R \times U(N_f)_L$ symmetry[b] in QCD with N_f massless flavors is manifestly realized. Thanks to this property, we can argue spontaneous chiral symmetry breaking and the appearance of the associated Nambu-Goldstone bosons which are interpreted as pions. Moreover, massive (axial-)vector mesons are also found in the meson spectrum. Some of the masses and couplings of these mesons are calculated and they are compared with experimental data. We should, however, keep in mind that the classical analysis in the supergravity side can only be trusted if the number of color N_c and the 't Hooft coupling $\lambda = g_{\rm YM}^2 N_c$ are large enough, which might sound still a bit far from the realistic situation. Nevertheless, as we will see, our model reproduces various results expected in realistic QCD. The aim of this article is to explain the basic idea of the model and summarize the main results given in Ref. 1 and present some new results for the baryons in Ref. 2.

2. Construction of QCD

We construct $U(N_c)$ QCD with N_f massless flavors by using N_c D4-banes and N_f D8-$\overline{\rm D8}$ pairs in type IIA string theory. D4-branes are extended along x^μ ($\mu = 0, \ldots, 3$) directions and wrapped on an S^1 parametrized by $x^4 = \tau$. The period of the parameter τ is written as $2\pi M_{\rm KK}^{-1}$, where $M_{\rm KK}$ gives the mass scale of massive Kaluza-Klein modes. Following Ref. 5, we impose the anti-periodic boundary condition along this S^1 to all the fermions in the system. The D8-branes and $\overline{\rm D8}$-branes are placed at antipodal points on the S^1 and extended along all the other directions as depicted in left side of Fig. 1. Since we are going to consider the near horizon limit of the supergravity solution corresponding to the D4-branes in the next section, we are interested in the strings attached on the D4-branes, that is, 4-4 strings, 4-8 strings and 4-$\overline{8}$ strings. As it is well-known, a $U(N_c)$ gauge field is created by the 4-4 strings. Since the supersymmetry is completely broken by the anti-periodic boundary condition for fermions along the S^1, the other

[a]The same idea was used in another interesting model proposed by Kruczenski et al. in Ref. 6, in which they used probe D6-branes to add quarks.

[b]The $U(1)_A$ subgroup of this chiral symmetry is actually anomalous. However, the effect of this anomaly is negligible in the large N_c limit.

unwanted adjoint fields created by the 4-4 strings are expected to become massive. On the other hand, the massless modes in the 4-8 strings and 4-$\overline{8}$ strings turn out to be N_f flavors of fermions that belong to the fundamental representation of the $U(N_c)$ gauge group. Furthermore, as discussed in Ref. 8, the chirality of the fermions created by 4-8 strings are opposite to those created by the 4-$\overline{8}$ strings. Therefore the $U(N_f)_{\mathrm{D8}} \times U(N_f)_{\overline{\mathrm{D8}}}$ gauge symmetry of the N_f D8-$\overline{\mathrm{D8}}$ pairs is interpreted as the $U(N_f)_L \times U(N_f)_R$ chiral symmetry of QCD. After all, we obtain 4 dimensional $U(N_c)$ QCD with N_f massless flavors with manifest $U(N_f)_L \times U(N_f)_R$ chiral symmetry realized on the D4-brane world-volume.

3. Holographic description of QCD

Let us move to the supergravity description of the D4/D8/$\overline{\mathrm{D8}}$ system considered in the previous section. Here we use so-called probe approximation[7] and replace the D4-branes with the corresponding supergravity solution given in Ref. 5. The D8 and $\overline{\mathrm{D8}}$-branes are treated as probe branes embedded in this curved background and all the backreaction to the background caused by these probe branes are neglected. This approximation can be justified when $N_c \gg N_f \sim \mathcal{O}(1)$. The metric of this background is given as

$$ds^2 \propto \frac{4}{9} M_{\mathrm{KK}}^2\, u^{3/2}(\eta_{\mu\nu} dx^\mu dx^\nu + f(u) d\tau^2) + u^{-3/2}\left(\frac{du^2}{f(u)} + u^2 d\Omega_4^2\right) \quad (1)$$

where u (≥ 1) represents the radial direction transverse to the D4-brane and $f(u) \equiv 1 - u^{-3}$. $d\Omega_4^2$ is the line element of a unit S^4 surrounding the D4-brane. From this metric, we see that the radius of the S^1 parameterized by τ shrinks to zero at $u \to 1$. Though the metric looks singular at $u \to 1$, the geometry is actually everywhere smooth and the topology of the space-time is $\mathbb{R}^{1,3} \times \mathbb{R}^2 \times S^4$, where $\mathbb{R}^{1,3}$ is the Minkowski space parametrized by x^μ ($\mu = 0, \ldots, 3$) and $\mathbb{R}^2$ is the (u, τ) plane.

This geometry implies that the D8-brane and $\overline{\mathrm{D8}}$-brane must be connected at $u = 1$ as depicted in the right side of Fig. 1. Therefore, we have only one connected component of the D8-brane in this background and the gauge group $U(N_f)_{\mathrm{D8}} \times U(N_f)_{\overline{\mathrm{D8}}}$ is now broken to the diagonal subgroup $U(N_f)_V$. This phenomena is interpreted as the chiral symmetry breaking in QCD.

In this model, the closed strings are interpreted as glueballs and the D4-branes wrapped on the S^4 are interpreted as baryons. These objects are studied by many people around 1998 (See Ref. 4 and references therein.). In the following, we mainly consider the open strings attached on the probe

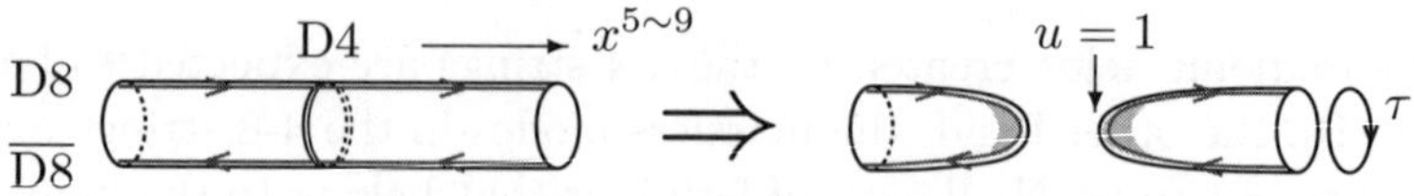

Fig. 1. Chiral symmetry breaking.

D8-branes. Since they carry flavor indices associated with the end points of the open strings, they are interpreted as mesons.

The effective theory of the open strings on the D8-branes is a 9 dimensional $U(N_f)$ gauge theory. In this article, we only consider fields that are invariant under $SO(5)$ symmetry that acts as rotation of the S^4, and ignore all the higher Kaluza-Klein modes. Then, the effective theory is reduced to a 5 dimensional gauge theory. By inserting the supergravity background to the effective action of the D8-brane and integrating over the S^4, we obtain the 5 dimensional effective action [c]

$$S_{\rm D8} \simeq \kappa \int d^4x\, dz\, \mathrm{Tr}\left(\frac{1}{2}K^{-1/3}F_{\mu\nu}^2 + M_{\rm KK}^2 K F_{\mu z}^2\right) + \frac{N_c}{24\pi^2}\int_5 \omega_5(A)\ , \quad (2)$$

where $\kappa = \frac{\lambda N_c}{216\pi^3}$ is a constant of $\mathcal{O}(N_c)$, $\omega_5(A)$ is the Chern-Simons 5-form, z is the fifth coordinate related to u by $u^3 = 1 + z^2$ and $K(z) = 1 + z^2$. The claim is that this 5 dimensional $U(N_f)$ Yang-Mills - Chern-Simons theory is considered as the effective theory of mesons. It is quite interesting to note that this 5 dimensional description of mesons is closely related to 5 dimensional phenomenological models proposed in Ref. 9.

4. Mesons

Let us next explain how to extract 4 dimensional physics from the 5 dimensional action (2). First we expand the gauge field (A_μ, A_z) using some complete sets $\{\psi_n(z)\}_{n\geq 1}$ and $\{\phi_n(z)\}_{n\geq 0}$ as

$$A_\mu(x^\mu, z) = \sum_{n\geq 1} B_\mu^{(n)}(x^\mu)\psi_n(z)\ , \quad A_z(x^\mu, z) = \sum_{n\geq 0}\varphi^{(n)}(x^\mu)\phi_n(z)\ . \quad (3)$$

These complete sets are chosen so that the kinetic and mass terms for the 4 dimensional fields $B_\mu^{(n)}$ and $\varphi^{(n)}$ become diagonal. We choose $\{\psi_n\}_{n\geq 1}$ as eigenfunctions satisfying

$$-K^{1/3}\partial_z(K\partial_z\psi_n) = \lambda_n\psi_n\ , \quad \kappa\int dz\, K^{-1/3}\psi_n\psi_m = \delta_{nm}\ . \quad (4)$$

[c]There is also a scalar field on the D8-brane, but here we will omit it for simplicity.

Here λ_n are the eigenvalues. Then we choose $\{\phi_n\}_{n\geq 1}$ as $\phi_n = \partial_z \psi_n$ which satisfy the ortho-normal condition

$$\kappa \int dz\, K \phi_n \phi_m = \lambda_n \delta_{nm} \,. \tag{5}$$

An important point here is that there is one more normalizable mode

$$\phi_0(z) = \frac{c}{K(z)} \,, \tag{6}$$

where the normalization constant c is chosen as $c = M_{\rm KK}^{-1}(\kappa\pi)^{-1/2}$.

Inserting the expansion (3) into the action (2) we obtain

$$S_{D8} \sim \int d^4x \, {\rm Tr} \left[\partial_\mu \varphi^{(0)2} + \sum_{n\geq 1} \left(\frac{1}{2} F_{\mu\nu}^{(n)2} + \lambda_n M_{\rm KK}^2 \left(B_\mu^{(n)} - \partial_\mu \varphi^{(n)} \right)^2 \right) \right] + (\text{interaction terms}) \,, \tag{7}$$

where $F_{\mu\nu}^{(n)} = \partial_\mu B_\nu^{(n)} - \partial_\nu B_\mu^{(n)}$. From this we see that $\varphi^{(n)}$ with $n \geq 1$ are eaten by $B_\mu^{(n)}$, which become massive vector fields. On the other hand, $\varphi^{(0)}$ does not have a partner vector field '$B_\mu^{(0)}$' in the expansion (3) and remain as a massless scalar field. We interpret $\varphi^{(0)}$ as the pion field, the lightest vector meson $B_\mu^{(1)}$ as the ρ meson, the second lightest vector meson $B_\mu^{(2)}$ as the a_1 meson and so on. The spin, parity and charge conjugation parity of π, ρ, a_1 mesons are consistent with this interpretation. Actually, from the fact that ϕ_0 and ψ_{2k-1} are even functions while ψ_{2k} are odd function, it can be shown that $\varphi^{(0)}$ is a pseudo-scalar meson, $B_\mu^{(2k-1)}$ are vector mesons and $B_\mu^{(2k)}$ are axial-vector mesons for $k = 1, 2, \cdots$. In contrast to usual construction of effective theory of mesons, in which each meson field is introduced independently, various mesons π, ρ, a_1, $\cdots$ are now elegantly unified in the 5 dimensional gauge field (A_μ, A_z) in our description.

As we have seen in (7), the mass of the n-th vector meson is given by $m_n^2 = \lambda_n M_{\rm KK}^2$ where λ_n is the eigenvalue of the eigenequation (4). It is tempting to compute the eigenvalues numerically and compare the results with the observed meson table.[10] Of course, we should not be too serious about the agreement or disagreement since the approximation is still very crude. For example, we have assumed that $N_c \gg N_f \sim \mathcal{O}(1)$, for which the realistic values are $N_c = 3$, $N_f = 2$.[d] We also know that our model deviates from QCD above the energy scale around $M_{\rm KK}$. It is unfortunately difficult to take a limit $M_{\rm KK} \to \infty$, since the asymptotic freedom of QCD requires

[d] Here we only consider up and down quarks whose masses are reasonably small compared to $\Lambda_{\rm QCD}$.

$\lambda \to 0$ in this limit and supergravity approximation breaks down. Therefore, our numerical results may only be useful to see if our model is too bad or not.

Here is the result:

	ρ	a_1	ρ'	(a_1')	ρ''
exp.(MeV)	776	1230	1459	(1647)	1720
our model	[776]	1189	1607	2023	2435

where we have fixed the value of $M_{\rm KK}$ to fit the rho meson mass.

We can also extract the interaction terms in (7). Here we skip all the details and show the results.

coupling	our model	fitting m_ρ and f_π	experiment
f_π	$1.13 \cdot \kappa^{1/2} M_{\rm KK}$	[92.4 MeV]	92.4 MeV
L_1	$0.0785 \cdot \kappa$	0.584×10^{-3}	$(0.1 \sim 0.7) \times 10^{-3}$
L_2	$0.157 \cdot \kappa$	1.17×10^{-3}	$(1.1 \sim 1.7) \times 10^{-3}$
L_3	$-0.471 \cdot \kappa$	-3.51×10^{-3}	$-(2.4 \sim 4.6) \times 10^{-3}$
L_9	$1.17 \cdot \kappa$	8.74×10^{-3}	$(6.2 \sim 7.6) \times 10^{-3}$
L_{10}	$-1.17 \cdot \kappa$	-8.74×10^{-3}	$-(4.8 \sim 6.3) \times 10^{-3}$
$g_{\rho\pi\pi}$	$0.415 \cdot \kappa^{-1/2}$	4.81	5.99
g_ρ	$2.11 \cdot \kappa^{1/2} M_{\rm KK}^2$	0.164 GeV^2	0.121 GeV^2
$g_{a_1\rho\pi}$	$0.421 \cdot \kappa^{-1/2} M_{\rm KK}$	4.63 GeV	$2.8 \sim 4.2$ GeV

The middle column of this table is the values obtained by fixing $M_{\rm KK}$ and κ as

$$M_{\rm KK} \simeq 949 \text{ MeV} \ , \quad \kappa \simeq 0.00745 \tag{8}$$

to fit the rho meson mass and pion decay constant f_π. Though the agreement of our numerical results with the experimental data is not extremely good, we think it is much better than expected.

We close this section by listing some other topics in our paper[1] that are omitted here.

Skyrme model: If we define a $U(N_f)$ valued field

$$U(x^\mu) = P\exp\left\{-\int_{-\infty}^{\infty} dz' A_z(x^\mu, z')\right\} \ , \tag{9}$$

it transforms as

$$U(x^\mu) \to g_+ U(x^\mu) g_-^{-1} \tag{10}$$

under the 5 dimensional gauge symmetry. Here $g_\pm \equiv \lim_{z\to\pm\infty} g(x^\mu, z)$ is the asymptotic value of the gauge function $g(x^\mu, z)$ in the limit $z \longrightarrow \pm\infty$ and it is interpreted as an element of the chiral symmetry $(g_+, g_-) \in U(N_f)_L \times U(N_f)_R$. (10) is nothing but the transformation property for the pion field in the chiral lagrangian. Using this, we can write down the effective action with manifest the chiral symmetry. The result is

$$S_{\rm D8} \simeq \int d^4x \,{\rm Tr}\left[\frac{f_\pi^2}{4}(U^{-1}\partial_\mu U)^2 + \frac{1}{32e_S^2}[U^{-1}\partial_\mu U, U^{-1}\partial_\nu U]^2 \right] + \mathcal{O}(B_\mu^{(n)2}) \tag{11}$$

where

$$f_\pi^2 = \frac{4\kappa}{\pi}M_{\rm KK}^2 \ , \quad e_S^{-2} \simeq 2.51 \cdot \kappa \ . \tag{12}$$

The effective action (11) coincides with that for the Skyrme model.

Vector meson dominance: We can show, in a certain gauge choice, the external photon field interacts with mesons only through vector meson exchange. This shows that the complete vector meson dominance[11] is realized.

Chiral anomaly and WZW term: The chiral anomaly and the WZW term in QCD are elegantly reproduced from the CS-term in (2).

$U(1)_A$ anomaly and η' meson mass: The $U(1)_A$ anomaly can also be understood in the supergravity description. Taking this anomaly into account, the mass of the η' meson is estimated and shown to satisfy the Witten-Veneziano formula[12]

$$m_{\eta'}^2 = \frac{2N_f}{f_\pi^2}\chi_g \ , \tag{13}$$

where χ_g is the topological susceptibility.

ω meson decay: The Feynman diagrams relevant to the omega meson decay, $\omega \to \pi^0\gamma$ and $\omega \to \pi^0\pi^+\pi^-$, turn out to be those depicted in Fig. 2. There are no direct ω-π-γ or ω-π-π-π couplings and these decays are induced by vector meson exchange. This structure is exactly the same as that proposed in Gell-Mann-Sharp-Wagner model.[13] Furthermore, the decay width of $\omega \to \pi^0\gamma$ is written by using ρ-π-π coupling $g_{\rho\pi\pi}$ as

$$\Gamma(\omega \to \pi^0\gamma) = \frac{N_c^2}{3}\frac{\alpha}{64\pi^4 f_\pi^2}\, g_{\rho\pi\pi}^2 |\vec{p}_\pi|^3 \ . \tag{14}$$

This expression is nothing but that proposed by Fujiwara et al.[14] based on hidden local symmetry approach.[15]

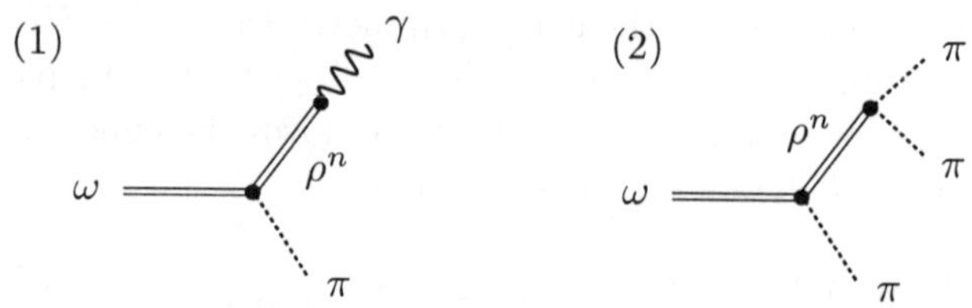

Fig. 2. The relevant diagrams for (1) $\omega \to \pi\gamma$ and (2) $\omega \to \pi\pi\pi$.

5. Baryons

As we saw in the previous section, we obtained the Skyrme model (11) as the effective action for the pion field (9). In the early 60's, Skyrme proposed that the baryons appear as solitons (called Skyrmion) in the Skyrme model.[16] The baryon number is identified as the winding number

$$N_B = \frac{1}{24\pi^2} \int \mathrm{Tr}(UdU^{-1} \wedge UdU^{-1} \wedge UdU^{-1}) \tag{15}$$

carried by the soliton solution.

On the other hand, baryons in the AdS/CFT context are constructed by D-branes wrapped on non-trivial cycles.[17] In our case, there is a non-trivial S^4 in the background. A D4-brane wrapped on the S^4 behaves as a point particle in the 4 dimensional space-time which is interpreted as a baryon. This wrapped D4-brane can be embedded in the world-volume of the D8-branes. It is well-known that a Dp-brane within D$(p+4)$-brane is equivalent to an instanton configuration in the world-volume gauge theory on the D$(p+4)$-branes.[18,19] In terms of the 5 dimensional gauge theory (2), the baryon is realized as an instanton in the 4 dimensional space parameterized by (x^1, x^2, x^3, z). It is actually possible to show that the baryon number expressed in three different ways, i.e. winding number of the Skyrmion (15), the instanton number

$$N_B = \frac{1}{8\pi^2} \int \mathrm{Tr}(F \wedge F) \tag{16}$$

and the number of the D4-branes wrapped on the S^4, are all equal.

The baryon expressed as the instanton in the 5 dimensional gauge theory is quite analogous to the Skyrmion in the Skyrme model. We can apply various techniques developed in the Skyrme model to analyze baryons from the instanton description. In particular, we obtain baryon spectrum by considering a quantum mechanics on the instanton moduli space, just as Adkins-Nappi-Witten (ANW) did in the Skyrme model.[20] The important point here is that the contributions from (axial-)vector mesons are included in the 5 dimensional gauge theory. Here we only present some of the results. See Ref. 2 for the details. We obtain baryons with $I = J$, where I and J are isospin and spin, respectively, just as ANW found in the Skyrme model. The baryon mass is given by

$$M \simeq M_0 + \left(\sqrt{\frac{(l+1)^2}{6} + \frac{2}{15} N_c^2} + \sqrt{\frac{2}{3}} (n_\rho + n_z) \right) M_{\rm KK} \, . \tag{17}$$

Here $l = 1, 3, 5, \cdots$ is related to isospin I and spin J as $I = J = l/2$. The quantum numbers n_z and n_ρ are non-negative integers associated with the fluctuations of position in the z-direction and size of the instanton, respectively. It is interesting to note that the parity of the baryons is given by $(-1)^{n_z}$. M_0 is a constant of order $\mathcal{O}(N_c)$ given by

$$M_0 = 8\pi^2 \kappa \, M_{\rm KK} + \mathcal{O}(N_c^0) \, . \tag{18}$$

However, since we are not able to calculate the $\mathcal{O}(N_c^0)$ contribution in M_0, we only consider the mass differences among the baryons and treat M_0 as a free parameter.

For comparison with our mass formula (17) to be made below, we list baryons with $I = J$ in the PDG baryon summary table[10] and a possible interpretation of the quantum numbers (n_ρ, n_z). In the table, superscripts $\pm$ represent the parity. The subscript $*$ indicates that the evidence of existence is poor.

(n_ρ, n_z)	$(0,0)$	$(1,0)$	$(0,1)$	$(1,1)$	$(2,0)/(0,2)$	$(2,1)/(0,3)$	$(1,2)/(3,0)$
$N\ (l=1)$	940^+	1440^+	1535^-	1655^-	1710^+, ?	2090_*^-, ?	2100_*^+, ?
$\Delta\ (l=3)$	1232^+	1600^+	1700^-	1940_*^-	1920^+, ?	?, ?	?, ?

(19)

If we use the value of $M_{\rm KK}$ in (8), the mass difference between the $l = 3$ states and the $l = 1$ states in the formula (17) is

$$M_{l=3} - M_{l=1} = \sqrt{\frac{8}{3} + \frac{6}{5}} M_{\rm KK} - \sqrt{\frac{2}{3} + \frac{6}{5}} M_{\rm KK} \simeq 569 \, {\rm MeV} \, . \tag{20}$$

The mass difference between the $(n_\rho, n_z) = (1,0)$ or $(0,1)$ states and the $(0,0)$ states with a common l is

$$M_{(1,0)/(0,1)} - M_{(0,0)} = \frac{2}{\sqrt{6}} M_{\rm KK} \simeq 774 \, {\rm MeV} \, . \tag{21}$$

Unfortunately, these values are a bit too large compared with the experimental values. If $M_{\rm KK}$ were 500 MeV, the predicted values using (17) would become very close to those listed in (19):[e]

(n_ρ, n_z)	$(0,0)$	$(1,0)$	$(0,1)$	$(1,1)$	$(2,0)/(0,2)$	$(2,1)/(0,3)$	$(1,2)/(3,0)$
$N\,(l=1)$	$[940]^+$	1348^+	1348^-	1756^-	$1756^+ \times 2$	$2164^- \times 2$	$2164^+ \times 2$
$\Delta\,(l=3)$	1240^+	1648^+	1648^-	2056^-	$2056^+ \times 2$	$2464^- \times 2$	$2464^+ \times 2$

(22)

Here we have to make some negative comments. Although this mass spectrum (22) looks very nice, there are a lot of reasons that we should not trust these numerical values. Note that we have assumed N_c and the 't Hooft coupling λ are large in our analysis. We have to keep in mind that there may be $1/N_c$ and $1/\lambda$ corrections that will become important especially for large quantum numbers l, n_z, n_ρ in the mass formula (17). We have also ignored the higher derivative corrections to the action (2) that may contribute to the analysis. Furthermore, we know that the model deviates from realistic QCD at high energies, roughly above the Kaluza-Klein scale $M_{\rm KK}$. Because the baryons are heavier than the value of $M_{\rm KK}$ used in (8) or (22), our quantitative analysis cannot be justified anyway. In fact, as we have observed, the value of $M_{\rm KK}$ in (8) does not work well for the baryon spectrum. Clearly, more investigation is needed for sensible quantitative tests.

Finally, we make a few comments on the N_c dependence of the baryon mass formula (17). For $N_c \gg l$, (17) is approximated as

$$M \simeq \widetilde{M}_0 + \left(\frac{1}{4}\sqrt{\frac{5}{6}}\,\frac{l(l+2)}{N_c} + \sqrt{\frac{2}{3}}(n_\rho + n_z) \right) M_{\rm KK}\ , \tag{23}$$

where $\widetilde{M}_0$ is a constant of order $\mathcal{O}(N_c)$. This mass formula (23) is consistent with the expected N_c dependence in large N_c QCD.[20,21] It has been known that the main part of the baryon mass is of order $\mathcal{O}(N_c)$ and the mass splittings among the low-lying baryons with different spins are of order $\mathcal{O}(1/N_c)$, while those among excited baryons are of order $\mathcal{O}(N_c^0)$. This is exactly what we observe in (23). The states considered in Ref. 20 correspond to the states with $n_\rho = n_z = 0$. Their mass formula is

$$M \simeq M_0 + \frac{l(l+2)}{8L}\ , \tag{24}$$

where M_0 and L are parameters of order $\mathcal{O}(N_c)$, which is also reproduced in (23).

---

[e]The nucleon mass 940 MeV is used as an input to fix M_0.

6. Summary and discussion

A holographic dual of QCD with N_f massless quarks is proposed on the basis of D4/D8-brane configuration in type IIA string theory. The effective theory of open strings is reduced to the 5 dimensional Yang-Mills - Chern-Simons theory in a curved background. Various mesons are unified in the 5 dimensional gauge field. Baryons are described as instantons in the spatial 4 dimensional space. Though the approximation made in our analysis is still very crude, our model catches various features of realistic QCD and provides new insights in hadron physics. Actually, the agreement with experimental data is much better than expected and it is quite unnatural to think it is just coincidence.

The fact that open strings represent mesons reminds us of the old idea of string theory around late 60's when it was born as a theory of hadron. One of the problem at that time was that string theory requires unrealistic 10 dimensional space-time. Our model seems to suggest that this idea was essentially correct. New ingredients added here are D-branes, curved background and holography. Using D-branes and a curved background, we have seen that 4 dimensional QCD can be described by 10 dimensional string theory. Having 10 dimensional space-time is no longer a problem once we accept the idea of holography.

Nowadays everyone believes that QCD is the fundamental theory of hadrons and the hadronic string is considered as an effective object. However, gauge/string correspondence states that QCD and its dual string theory (if it really exists) are actually equivalent, suggesting that both QCD and string theory could be fundamental theory of hadrons at the same time. In this way, it may provide a new perspective to the concept of "elementary particle" in the real world.

Acknowledgments

We would like to thank T. Sakai, H. Hata, S. Yamato for collaboration. This work was partly supported by the Grant-in-Aid for Young Scientists (B)No. 17740143 from the Ministry of Education, Culture, Sports, Science and Technology, Japan.

References

1. T. Sakai and S. Sugimoto, Prog. Theor. Phys. **113**, 843 (2005); Prog. Theor. Phys. **114**, 1083 (2005).
2. H. Hata, T. Sakai, S. Sugimoto and S. Yamato, arXiv:hep-th/0701280.

3. J. M. Maldacena, Adv. Theor. Math. Phys. **2**, 231 (1998) [Int. J. Theor. Phys. **38**, 1113 (1999)].
4. O. Aharony, S. S. Gubser, J. M. Maldacena, H. Ooguri and Y. Oz, Phys. Rept. **323**, 183 (2000).
5. E. Witten, Adv. Theor. Math. Phys. **2**, 505 (1998).
6. M. Kruczenski, D. Mateos, R. C. Myers and D. J. Winters, JHEP **0405**, 041 (2004). J. L. F. Barbon, C. Hoyos, D. Mateos and R. C. Myers, JHEP **0410**, 029 (2004).
7. A. Karch and E. Katz, JHEP **0206**, 043 (2002).
8. S. Sugimoto and K. Takahashi, JHEP **0404**, 051 (2004).
9. D. T. Son and M. A. Stephanov, Phys. Rev. D **69**, 065020 (2004); J. Erlich, E. Katz, D. T. Son and M. A. Stephanov, Phys. Rev. Lett. **95** (2005) 261602; L. Da Rold and A. Pomarol, Nucl. Phys. B **721** (2005) 79. J. Hirn and V. Sanz, JHEP **0512**, 030 (2005).
10. W. M. Yao *et al.* [Particle Data Group], J. Phys. G **33**, 1 (2006).
11. M. Gell-Mann and F. Zachariasen, Phys. Rev. **124**, 953 (1961); J. J. Sakurai, *Currents and Mesons,* (University of Chicago Press, Chicago, 1969).
12. E. Witten, Nucl. Phys. B **156**, 269 (1979); G. Veneziano, Nucl. Phys. B **159**, 213 (1979).
13. M. Gell-Mann, D. Sharp and W. G. Wagner, Phys. Rev. Lett. **8**, 261 (1962).
14. T. Fujiwara, T. Kugo, H. Terao, S. Uehara and K. Yamawaki, Prog. Theor. Phys. **73**, 926 (1985).
15. M. Bando, T. Kugo, S. Uehara, K. Yamawaki and T. Yanagida, Phys. Rev. Lett. **54**, 1215 (1985); M. Bando, T. Kugo and K. Yamawaki, Phys. Rep. **164**, 217 (1988).
16. T. H. R. Skyrme, Proc. R. Soc. London A **260** (1961), 127; Proc. R. Soc. London A **262** (1961), 237; Nucl. Phys. **31** (1962), 556.
17. E. Witten, JHEP **9807**, 006 (1998).
18. E. Witten, Nucl. Phys. B **460**, 541 (1996).
19. M. R. Douglas, "Branes within branes," arXiv:hep-th/9512077.
20. G. S. Adkins, C. R. Nappi and E. Witten, Nucl. Phys. B **228**, 552 (1983).
21. E. Witten, Nucl. Phys. B **160**, 57 (1979).

Giant loops and the AdS/CFT Correspondence

Gordon W. Semenoff

Department of Physics and Astronomy, University of British Columbia
Vancouver, British Columbia, Canada V6T 1Z1

A review of recent results from the study of 1/2 BPS Wilson loops in N=4 supersymmetric Yang-Mills theory is presented. Some speculations about the behavior of the Wilson loop operator restricted to the space of the 1/2 BPS chiral primary operators are presented.

1. Matrix models for Wilson loops

The study of highly symmetric states has provided some insight into the AdS/CFT correspondence. For the case of 1/2 BPS chiral primary operators in $\mathcal{N}=4$ supersymmetric Yang-Mills theory, there is a beautiful picture of their string theory dual in a family of 1/2-BPS geometries.[1] The gauge theory states are well understood. Their quantum numbers are protected from quantum corrections and are thus given exactly at the classical level. The problem of finding the multiplicity of states with given quantum numbers is nicely encoded in a quantum mechanical matrix oscillator.[2,3] Moreover, the states of the oscillator and their classical limit have an interpretation in the boundary data which specify 1/2-BPS geometries.[1,4]

Recently, an analogous picture of 1/2-BPS Wilson loops has emerged. The loop operator which is relevant to the AdS/CFT correspondence is[5]

$$W_R[C] = \frac{1}{N}\mathrm{Tr}_R \mathcal{P}\exp\left(\oint_C d\tau\left(i\dot{x}^\mu(\tau)A_\mu(x(\tau)) + |\dot{x}(\tau)|\theta^I\Phi^I(x(\tau))\right)\right) \quad (1)$$

where R is an irreducible representation of GL(N). This loop measures the holonomy of the wave-function of a heavy W-boson which is created when $SU(N+1)$ gauge symmetry is Higgsed to $SU(N)\times U(1)$ and θ^I is the orientation of the condensate.

In the two special cases when the contour C is a circle or an infinite straight line, $W_R[C]$ commutes with half of the 16 supercharges and 16 conformal supercharges of $\mathcal{N}=4$ Yang-Mills theory and is 1/2 BPS. The circle and line are related by a conformal transformation. However, it has been found that the circle and straight line have different expectation values[6,7] (for a review see Ref.[8]). The straight line is believed to be completely protected from quantum corrections, so that $< W_R[\text{straight line}] >= 1$, whereas the circle is thought to be given by the Hermitian matrix integral[6]

$$< W_R[\text{circle}] >= \frac{\int [dM]\ \text{Tr}_R e^M\ e^{-\frac{2N}{\lambda}\text{Tr}M^2}}{\int [dM]\ e^{-\frac{2N}{\lambda}\text{Tr}M^2}} \tag{2}$$

The difference between the two is attributed to the fact that the straight line commutes with half of the supercharges and therefore has a supersymmetric regularization where it can be computed whereas the circle commutes with linear combinations which always contain conformal supercharges and is supersymmetric only in the conformal field theory which is obtained once regulators are removed.

It is important to remember that (2) is a *conjecture* which has not been proven yet. It is known to be the sum of rainbow ladder Feynman diagrams which contain no internal vertices.[6] Further evidence for it is the fact that the leading corrections, diagrams with one or more internal vertices, cancel identically.[6] The conjecture is that all such diagrams cancel, leaving nonzero only the sum over ladders (2). However, even if it is the sum of all Feynman diagrams, there could also be non-perturbative contributions, for example from instantons.[9,10] It may be plausible that the latter are suppressed in the 't Hooft large N limit. Thus we might consider (2) more reliable when this limit is taken. In the fundamental representation, $F = \square$, explicit evaluation of the matrix integral gives

$$< W_\square[\text{circle}] >= \frac{1}{N} L^1_{N-1}(-\lambda/4N) e^{\lambda/8N} = \frac{2}{\sqrt{\lambda}} I_1(\sqrt{\lambda}) + \frac{\lambda}{48N^2} I_2(\sqrt{\lambda}) + \ldots \tag{3}$$

with the Laguerre polynomial $L^m_n(x) = \frac{1}{n!} e^x x^{-m} \frac{d^n}{dx^n}\left(e^{-x} x^{m+n}\right)$ and $I_k(x)$ are Bessel functions. The first term on the right-hand-side is the 'tHooft limit. Its extrapolation to large λ gives

$$< W_\square[\text{circle}] > \rightarrow \sqrt{\frac{2}{\pi}} \frac{1}{\lambda^{3/4}} e^{\sqrt{\lambda}} \tag{4}$$

The exponent agrees with the string dual that is computed using Maldacena's presctiption[5] of finding the extremal area of a world-sheet embedded in $AdS_5 \times S^5$ with boundary C, itself residing on the boundary of AdS_5. The pre-factor has never been computed from string theory, but it has been observed that its λ-dependence, as well as that of higher-order terms in $1/N^2$ in the large-λ limit of (3) are what would be expected from string sigma model path integrals.[7] Agreement of the exponent is already considered a stunning confirmation of the AdS/CFT picture.

There is a *further conjecture* that a matrix model computes the leading asymptotic of a correlation function of the circle Wilson loop and the chiral primary operators[11]

$$\frac{< W_R[\text{circle}]\text{Tr}_{R'}Z(x) >}{< W_R[\text{circle}] >} \sim \left(\frac{2\pi r}{4\pi^2 x^2}\right)^{J'} \frac{\int d^2z\ \text{Tr}_R e^{\frac{1}{2}(z+\bar{z})}\text{Tr}_{R'} z\ e^{-\frac{2N}{\lambda}\text{Tr}\bar{z}z}}{\int d^2z\ \text{Tr}_R e^{\frac{1}{2}(z+\bar{z})}\ e^{-\frac{2N}{\lambda}\text{Tr}\bar{z}z}} \tag{5}$$

J' is the number of boxes in the R' Young tableau. The matrices Z are complex and cannot be diagonalized by conjugation with a unitary matrix, so generally this is not an eigenvalue model. However, in some cases it can be rewritten as a normal matrix model,[12] which is a similar integral but restricted to matrices whose real and imaginary parts commute with each other and can be simultaneously diagonalized. In that case, practical computations can be done. Solving the matrix model in that picture has a nice analog to a two dimensional electrostatics problem and a number of examples have been worked in Ref.[12]. The matrix model picture was further used to compute the correlators of chiral primary operators[13] where

$$< \text{Tr}_R Z(x)\text{Tr}_{R'}\bar{Z}(y) >= N_R\delta_{RR'}\left(\frac{1}{4\pi^2(x-y)^2}\right)^J \tag{6}$$

and the normalization at large N is $N_R = J\lambda^J$.

2. Giant loops

The large N limit of the matrix model representations of the Wilson loop expectation value with a large purely anti-symmetric[14,15] or symmetric[16–18] representations can be computed:

$$\left\langle W_{\tiny\begin{array}{|c|}\hline\\\hline\\\hline\\\hline\end{array}\}_{J\ \text{boxes}}}[\text{circle}]\right\rangle = \frac{N!}{J!(N-J)!}e^{N\frac{2\sqrt{\lambda}}{3\pi}\sin^3\theta}\ ,\ \theta - \frac{1}{2}\sin 2\theta = \pi\frac{J}{N} \tag{7}$$

$$\langle W_{\square\square\ldots\square}[\text{circle}]\rangle = e^{N\left(2\kappa\sqrt{\kappa^2+1}+2\sinh^{-1}\kappa\right)}\quad \kappa = \frac{J}{N}\frac{\sqrt{\lambda}}{4} \tag{8}$$

In the limit where $N \to \infty$ and J finite, both Eqns. (7) and (8) reduce to the appropriate generalization of the strong coupling limit (4) $< W >\sim e^{J\sqrt{\lambda}}$. However, if we let J become large, of order N, the right-hand-sides of both Eqs. (7) and (8) are larger than $e^{J\sqrt{\lambda}}$. This means that there is a saddle point of the matrix model which has lower energy than the semi-circle eigenvalue distribution which is used to compute (3). This modification of the matrix model has been studied in various contexts.[12,16,17,19]

This is a hint at a similar phenomenon on the string side of AdS/CFT duality – existence of a state of string theory obeying the boundary condition but with lower energy than J world-sheets wrapped on a minimal surface AdS_2. The idea is that a large number of stacked fundamental strings can polarize into a D-brane with J units of fundamental string charge dissolved in it. Consider the symmetry of the circle loop. The loop breaks the $PSU(2,2|4)$ supersymmetry which has bosonic subgroup $SO(2,4) \times SO(6)$ to $Osp(4^*|4)$ with bosonic group $SL(2,R) \times SO(3) \times SO(5)$. This means that the string solution is allowed to lie uniformly on a subspace $(AdS_2 \times S^2 \subset AdS_5) \times (S^4 \subset S^5)$. It can thus contain a D3-brane wrapping $AdS_2 \times S^2$ or a D5-brane wrapping $AdS_2 \times S^4$. Given this hint, it has been shown that the exponents in (7) and (8) can be gotten by extremizing the Born-Infeld action for the D5-brane and D3-branes, respectively, with J units of electric flux subject to the condition that the objects pinch down to the circle contour at the boundary of AdS_5. The pre-factor in (7) is not apparent in the classical string sigma model. It comes from quantum fluctuations and it is an important contribution to the expectation value.

These blown up, or giant loops are analogs of giant gravitons. A giant graviton is a 1/2 BPS state whose gauge theory dual are the operators $\mathrm{Tr}_R Z(x)$ where R corresponds to a Young tableau with J boxes. When $\frac{J}{N} \to 0$ at large N, these operators correspond to 1/2 BPS gravitons propagating on the $AdS_5 \times S^5$ space-time. They have angular momentum J on an equator of the S^5 and have a static world-line in AdS_5. They have a residual symmetry group with bosonic $R^1 \times SO(4) \times SO(4)$ which suggests that this string dual could spread out on its world-line plus two S^3's. In fact, when J is large, with $\frac{J}{N}$ finite, they become giant gravitons, blowing up into a D3-branes with radius $\sim \frac{J}{N}$ wrapping either the $S^3 \subset AdS_5$ for an antisymmetric representation $R = \left.\begin{array}{c}\square\\ \vdots\\ \square\end{array}\right\} J$ boxes or $S^3 \subset S^5$ for a symmetric representation $R = \square\square\ldots\square$. For representations with multiple rows of varying lengths, the string dual wraps both S^3's. These giant gravitons were originally found as classical solutions of the Born-Infeld action for the D3-brane where the angular momentum J is large.[20]

When, as we have assumed in the above discussion, J grows no faster than N in the large N limit, the giant loop or giant graviton are probe branes which are too light to back-react on the $AdS_5 \times S^5$ geometry. When J grows faster than N in the large N limit, the back-reaction of the graviton or loop changes the geometry of space-time. The 1/2 BPS solutions of IIB supergravity in this case are well understood in the case of the giant graviton where there is an almost explicit solution[1] and less understood in the case of the giant loop.[17,21]

3. Correlation function of giant loop and giant graviton

The correlator of the fundamental representation with the single trace operator, for example, in the 'tHooft large N limit, is[a]

$$\frac{< W_{\square} \frac{1}{\sqrt{J\lambda^J}} \mathrm{Tr} Z^J(x) >}{< W_{\square} >} \rightarrow \left(\frac{2\pi r}{4\pi^2 x^2}\right)^J \frac{1}{N} \frac{\sqrt{J\lambda}}{2} \frac{I_J(\sqrt{\lambda})}{I_1(\sqrt{\lambda})}$$
$$\sim \left(\frac{2\pi r}{4\pi^2 x^2}\right)^J \left[\frac{1}{N}\frac{\sqrt{J\lambda}}{2} + \mathcal{O}(1)\right] \text{ at } \lambda \rightarrow \infty \quad (9)$$

This strong coupling limit also agrees with a string theory computation.[22] In that case, the asymptotic of this matrix element are obtained by computing the connected correlator of two classical world-sheets. The leading contribution is exchange of the lightest supergraviton. The asymptotic form of its propagator, plus the integral of its vertex operator over the world-sheets gives the result.

For large representations, this should give a correlator between giant loops and giant gravitons. This correlator has been computed recently for the case of a giant loop and a point graviton,[23]

$$\frac{< W_{\square\square\ldots\square} \frac{1}{\sqrt{J\lambda^J}} \mathrm{Tr} Z^J(x) >}{< W_{\square\square\ldots\square} >} \rightarrow \left(\frac{2\pi r}{4\pi^2 x^2}\right)^J \frac{2}{\sqrt{J}} \sinh\left(J \sinh^{-1}\kappa\right) \quad (10)$$

$$\frac{< W_{\begin{smallmatrix}\square\\\square\\\vdots\\\square\end{smallmatrix}} \frac{1}{\sqrt{J\lambda^J}} \mathrm{Tr} Z^J(x) >}{< W_{\begin{smallmatrix}\square\\\square\\\vdots\\\square\end{smallmatrix}} >} \rightarrow \left(\frac{2\pi r}{4\pi^2 x^2}\right)^J \frac{2(J-2)!}{\pi(J+1)!} C^{(2)}_{J-2}(\cos\theta) \sin^3\theta \quad (11)$$

where κ and θ are given in (7) and (8) and $C^{(2)}_k(x)$ is a Gegenbauer polynomial.

They also computed the strong coupling limit using the string dual. There, they extracted the large distance asymptotic of the correlator by computing the 2-point function of giant loops. They used the D3-brane or the D5-brane geometry for the loop and coupled to gravitons with the appropriate vertex operator. and compared it with the extrapolation of the matrix model result to large λ is matched by a computation on the string theory side.

---

[a]For brevity, we denote $W_R[\text{circle}] \equiv W_R$.

4. 1/2 BPS matrix model

In Ref.[12] a parallel was drawn between the matrix model representation of the circle Wilson loop and the matrix quantum mechanics representation of the 1/2 BPS states corresponding to $\mathrm{Tr}_R Z(x)$. The Wilson loop can be written as a normal matrix model

$$< W_{\Box} >= \frac{\int d^2 z_i |\Delta(z_i)|^2 e^{-\frac{2N}{\lambda}\sum \bar{z}_i z_i} e^{\frac{1}{\sqrt{2}}(z+\bar{z})}}{\int dz_i |\Delta(z_i)|^2 e^{-\frac{2N}{\lambda}\sum \bar{z}_i z_i}} = \frac{< \Psi_0|\frac{1}{N}\sum_i e^{x_i}|\Psi_0 >}{< \Psi_0|\Psi_0 >} \tag{12}$$

where $|\Psi_0>$ is the ground state of a system of N fermions in a harmonic potential. This is precisely the ground atate of the matrix quantum mechanical model with action

$$S = \int dt\, \frac{1}{2}\mathrm{Tr}\left\{(\dot{X} + i[A, X])^2 - X^2\right\}$$

where both A and X are hermitian $N \times N$ matrices. It is well known that this model is solved by a gas of free fermions in the harmonic well. If we imagine that the ground state of this gas is analogous to the vacuum of $\mathcal{N} = 4$ Yang-Mills, A chiral primary operator is simply an operator which creates fermions in excited states. For a Young tableau with rows of lengths $\ell_1 \geq \ell_2 \geq \ldots \geq \ell_N$, the excited state has levels $\ell_1 + N, \ell_2 + N - 1, ..., \ell_N$ occupied. There is a 1–1 correspondence between Young tableau and excited states of the oscillator. We could restrict our attention to such states. They are simply the 1/2 BPS states that are created from the Yang-Mills vacuum by chiral primary operators. Then, the results for the expectation value of the Wilson loop, as well as for the asymptotics of the correlators of the Wilson loop and chiral primaries would imply that, on this restricted space, the Wilson loop operator simply acts as the operator $W_R = \mathrm{Tr}_R e^X$.

Acknowledgments

GS acknowledges the Institut des Hautes Etudes Scientifiques (IHES), Bures-sur-Yvette where parts of this work were completed. This research is supported by NSERC of Canada.

References

1. H. Lin, O. Lunin and J. M. Maldacena, "Bubbling AdS space and 1/2 BPS geometries," JHEP **0410**, 025 (2004) [arXiv:hep-th/0409174].
2. S. Corley, A. Jevicki and S. Ramgoolam, "Exact correlators of giant gravitons from dual N = 4 SYM theory," Adv. Theor. Math. Phys. **5**, 809 (2002) [arXiv:hep-th/0111222].
3. D. Berenstein, "A toy model for the AdS/CFT correspondence," JHEP **0407**, 018 (2004) [arXiv:hep-th/0403110].

4. L. Grant, L. Maoz, J. Marsano, K. Papadodimas and V. S. Rychkov, "Minisuperspace quantization of 'bubbling AdS' and free fermion droplets," JHEP **0508**, 025 (2005) [arXiv:hep-th/0505079].
5. J. M. Maldacena, "Wilson loops in large N field theories," Phys. Rev. Lett. **80**, 4859 (1998) [arXiv:hep-th/9803002].
6. J. K. Erickson, G. W. Semenoff and K. Zarembo, "Wilson loops in N = 4 supersymmetric Yang-Mills theory," Nucl. Phys. B **582**, 155 (2000) [arXiv:hep-th/0003055].
7. N. Drukker and D. J. Gross, "An exact prediction of N = 4 SUSYM theory for string theory," J. Math. Phys. **42**, 2896 (2001) [arXiv:hep-th/0010274].
8. G. W. Semenoff and K. Zarembo, "Wilson loops in SYM theory: From weak to strong coupling," Nucl. Phys. Proc. Suppl. **108**, 106 (2002) [arXiv:hep-th/0202156].
9. M. Bianchi, M. B. Green and S. Kovacs, "Instantons and BPS Wilson loops," arXiv:hep-th/0107028.
10. M. Bianchi, M. B. Green and S. Kovacs, "Instanton corrections to circular Wilson loops in N = 4 supersymmetric Yang-Mills," JHEP **0204**, 040 (2002) [arXiv:hep-th/0202003].
11. G. W. Semenoff and K. Zarembo, "More exact predictions of SUSYM for string theory," Nucl. Phys. B **616**, 34 (2001) [arXiv:hep-th/0106015].
12. K. Okuyama and G. W. Semenoff, "Wilson loops in N = 4 SYM and fermion droplets," JHEP **0606**, 057 (2006) [arXiv:hep-th/0604209].
13. C. Kristjansen, J. Plefka, G. W. Semenoff and M. Staudacher, Nucl. Phys. B **643**, 3 (2002) [arXiv:hep-th/0205033].
14. S. Yamaguchi, "Wilson loops of anti-symmetric representation and D5-branes," JHEP **0605**, 037 (2006) [arXiv:hep-th/0603208].
15. J. Gomis and F. Passerini, "Holographic Wilson loops," JHEP **0608**, 074 (2006) [arXiv:hep-th/0604007].
16. N. Drukker and B. Fiol, "All-genus calculation of Wilson loops using D-branes," JHEP **0502**, 010 (2005) [arXiv:hep-th/0501109].
17. S. Yamaguchi, "Bubbling geometries for half BPS Wilson lines," arXiv:hep-th/0601089.
18. J. Gomis and F. Passerini, "Wilson loops as D3-branes," JHEP **0701**, 097 (2007) [arXiv:hep-th/0612022].
19. S. A. Hartnoll and S. P. Kumar, "Higher rank Wilson loops from a matrix model," JHEP **0608**, 026 (2006) [arXiv:hep-th/0605027].
20. J. McGreevy, L. Susskind and N. Toumbas, "Invasion of the giant gravitons from anti-de Sitter space," JHEP **0006**, 008 (2000) [arXiv:hep-th/0003075].
21. O. Lunin, "On gravitational description of Wilson lines," JHEP **0606**, 026 (2006) [arXiv:hep-th/0604133].
22. D. Berenstein, R. Corrado, W. Fischler and J. M. Maldacena, "The operator product expansion for Wilson loops and surfaces in the large N limit," Phys. Rev. D **59**, 105023 (1999) [arXiv:hep-th/9809188].
23. S. Giombi, R. Ricci and D. Trancanelli, "Operator product expansion of higher rank Wilson loops from D-branes and matrix models," JHEP **0610**, 045 (2006) [arXiv:hep-th/0608077].

GAUGED NONLINEAR SIGMA MODEL IN AdS_5 SPACE AND HADRON PHYSICS

Xiao-Hong Wu

School of Physics, Korea Institute for Advanced Study,
207-43, Cheongryangri 2-dong, Dongdaemun-gu, Seoul 130-722, Korea
E-mail: wuxh@kias.re.kr

We study the low energy pion, ρ and a_1 phenomenology in the framework of a five-dimensional gauged linear sigma model in AdS_5 space with two dimension-6 operators. we calculate the particle spectra and decay constants, and work out the phenomenology of $a_1 \to \rho\pi$, $a_1 \to \pi\gamma$, and $\rho \to \pi\pi$ channels. As a noval feature, we get non-vanishing $\text{Br}(a_1 \to \pi\gamma)$, and calculate the pion electromagnetic charge radius, which agrees with the experimental results.

Keywords: QCD; AdS/CFT correspondence; holography.

Recent progress in string theory, the AdS/CFT correspondence, shed new light on the strongly coupled gauge theory.[1] There are several successful attempts to understand QCD and hadron physics with the construction of D-brane models.[2–5] In Refs. 6 and 7, a gauged linear sigma model in AdS_5, "AdS/QCD", has been constructed as a bottom-up approach to describe low energy hadron physics, under the inspiration of AdS/CFT correspondence. The model can reproduce the pion, ρ, a_1 mass spectra and decay constants very well. Some phenomenology of $\rho \to \pi\pi$ and $a_1 \to \pi\gamma$ is also considered. However the $\rho\pi\pi$ coupling is a little small, and $\text{Br}(a_1 \to \pi\gamma)$ is vanishing. It has been realized long time ago that dimension-6 operators in the framework of gauged linear sigma model in 4D can improve the phenomenology of low energy hadron physics, and therefore generate non-vanishing contribution to $\text{Br}(a_1 \to \pi\gamma)$.[8,9] We incorporate higher dimensional operators, especially two dim-6 terms, into the AdS/QCD model. In this work, we only consider the vector, axial-vector and pseudoscalar sectors, as a first step of our study. Naively, these new operators will have nontrivial effects on the interaction vertex, as well as mass spectra and decay constants. We expect they may contribute to the $\text{Br}(a_1 \to \pi\gamma)$, which is vanishing in the original AdS/QCD model.[6,7] We also study the phenomenology of $\rho \to \pi\pi$ and $a_1 \to \rho\pi$, the branching ratios and D/S wave amplitude ratio in the latter channel. By introducing the photon as external field, we study the pion electromagnetic form factor, and calculate the pion electromagnetic charge radius, which agrees with the experimental results in our numerical study.[10]

The Lagrangian of the AdS/QCD model[6,7] defined in a slice of AdS_5 is given by

$$\mathcal{L}_5^{\mathrm{dim}-4} = \sqrt{g} M_5 \ \mathrm{Tr}\left[-\frac{1}{4} L_{MN} L^{MN} - \frac{1}{4} R_{MN} R^{MN} \right.$$
$$\left. + \frac{1}{2} D_M \Phi^\dagger D^M \Phi - \frac{1}{2} M_\Phi^2 \Phi^\dagger \Phi \right] , \tag{1}$$

where $M_\Phi^2 = -3/L^2$ from AdS/CFT correspondence,[1] $D_M \Phi = \partial_M \Phi + iL_M \Phi - i\Phi R_M$, $L_M = L_M^a \tau^a/2$ with τ^a being the Pauli matrix, and $M, N = 0, 1, 2, 3, 5(\text{or}\, z\,)$. We define $\Phi = S e^{iP/v(z)}$ with $\langle S \rangle = v(z)$. Under $\mathrm{SU}(2)_V$, S and P transform as singlet and triplet, respectively. The AdS_5 space is characterized in the conformally flat metric, $ds^2 = (L/z)^2 (dx^\mu dx_\mu - dz^2)$. L is the curvature of the 5-dimensional AdS space. In this model, the AdS_5 space is compactified such that $L_0 < z < L_1$, where $L_0 \to 0$ is a ultraviolet (UV) cutoff and L_1 is an infrared (IR) cutoff. Solving the equation of motion for S, we obtain

$$\langle S \rangle \equiv v(z) = c_1 z + c_2 z^3,$$

with the integration constants $c_{1,2}$ determined by the following boundary conditions

$$M_q = \frac{L}{L_0} v \bigg|_{L_0}, \qquad \xi = L v \bigg|_{L_1},$$

where M_q is the current quark mass matrix, which breaks chiral symmetry explicitly, and ξ is related to $\langle \bar{q} q \rangle$, which breaks chiral symmetry spontaneously.

In this work, we consider corrections to the model from higher dimensional operators, and for simplicity, we only consider dim-6 operators in the chiral limit. We note that part of large N_c corrections through meson-loop contributions is discussed in Ref. 11. The following mass dimensions for scalar field Φ and vector fields L_M and R_M, $\dim(\Phi) = \dim(L_M) = \dim(R_M) = 1$, are adopted. The Lagrangian with dimension-6 operators reads

$$\mathcal{L}_5^{\mathrm{dim}-6} = \sqrt{g}M_5\ \mathrm{Tr}\Big[- i\frac{\kappa}{M_5^2}\Big(L_{MN}D^M\Phi D^N\Phi^\dagger + R_{MN}D^M\Phi^\dagger D^N\Phi\Big) + \frac{\zeta}{M_5^2}L_{MN}\Phi R^{MN}\Phi^\dagger\Big], \tag{2}$$

where κ and ζ are constants that will be fixed later.

We work in the chiral limit. Then $v(z)$ is proportional to $\mathbf{1}$, $v(z) \simeq \xi\frac{z^3}{L_1^3}\mathbf{1}$. The vector and axial gauge bosons are defined by

$$V_M = \frac{1}{\sqrt{2}}(L_M + R_M)$$
$$A_M = \frac{1}{\sqrt{2}}(L_M - R_M)\,.$$

It is straightforward to derive the quadratic terms for vector, axial-vector and pseudoscalar, and the VAP, VPP and four-pion interaction vertices.[10]

We calculate the two-point correlation functions for vector and axial-vector with respect to the UV boundary external source fields v_μ and a_μ, which couple to the vector and axial-vector currents operators, respectively,

$$V_\mu|_{z=L_0} = v_\mu, \qquad A_\mu|_{z=L_0} = a_\mu.$$

From the AdS/CFT correspondence, in order to calculate the current-current correlation function in the strongly coupled CFT side, we can do it in the weakly interacting AdS side instead. Then the effective Lagrangian in momentum space in term of the correlators is

$$\mathcal{L}_{\mathrm{eff}} = v_\mu \Pi_V^{\mu\nu}(p^2)v_\nu + a_\mu \Pi_A^{\mu\nu}(p^2)a_\nu \tag{3}$$

with $\Pi_{V,A}^{\mu\nu}(p^2) = (g^{\mu\nu} - p^\mu p^\nu/p^2)\,\Pi_{V,A}(p^2)$.

In the large N_c limit, the correlators can be written as the sum in terms of the resonance masses and decay constants,

$$\Pi_A(p^2) = p^2\sum_n \frac{f_{A_n}^2}{p^2 - M_{A_n}^2} + f_\pi^2 \tag{4}$$

$$\Pi_V(p^2) = p^2\sum_n \frac{f_{V_n}^2}{p^2 - M_{V_n}^2} \tag{5}$$

Then the vector and axial meson masses are determined as the poles of their corresponding correlators, and the decay constants are related with the residue.

Now we study hadronic observables such as decay widths and form factors using our model given in Eq. (1) and Eq. (2). Primarily we investigate how those dimension-6 operators in Eq. (2) affect the results obtained with only interactions in Eq. (1). We first Kaluza-Klein (KK) decompose the vector field as $V_\mu(x,z) = \frac{1}{\sqrt{M_5 L}}\sum_{n=1}^{\infty}\tilde{V}_\mu^{(n)}(x) f_V^{(n)}(z)$ and also for the axial-vector and pseudoscalar fields. The first resonances of the vector, axial-vector and pseudoscalar fields are associated with ρ, a_1 and π respectively.

We first consider the process $a_1 \to \rho\pi$ and obtain the following structure

$$\begin{aligned}\mathcal{L}_{a_1\rho\pi} &= ig_{1a_1\rho\pi}\mathrm{Tr}(\tilde{A}_\mu[\tilde{V}_\mu, \tilde{A}_z]) + ig_{2a_1\rho\pi}\mathrm{Tr}(\tilde{A}_\mu[\tilde{V}_{\mu\nu}, \partial_\nu\tilde{A}_z]) \\ &\quad + ig_{3a_1\rho\pi}\mathrm{Tr}(\tilde{A}_{\mu\nu}[\tilde{V}_{\mu\nu}, \tilde{A}_z]) \end{aligned} \quad (6)$$

where the coefficients $g_{ia_1\rho\pi}$ $(i = 1,2,3)$ are defined in.[10] With the interaction vertex above, it is straightforward to derive the amplitude of the process, which can be written as

$$\mathcal{A}(a_1 \to \rho\pi) = -i\epsilon^\mu(s_{a_1})\epsilon^\nu(s_\rho)\left[f_{a_1\rho\pi}g_{\mu\nu} + g_{a_1\rho\pi}p_{\pi\,\mu}p_{\pi\,\nu}\right].$$

The S/D wave amplitudes are defined as in Ref.[12]

The $\rho\pi\pi$ vertex can be expressed as

$$\mathcal{L}_{\rho\pi\pi} = \frac{i}{\sqrt{2}}g_{\rho\pi\pi}\mathrm{Tr}(\tilde{V}^\mu[\tilde{A}_5, \partial_\mu\tilde{A}_5]) + \frac{i}{\sqrt{2}}f_{\rho\pi\pi}\mathrm{Tr}(\tilde{V}^{\mu\nu}[\partial_\mu\tilde{A}_5, \partial_\nu\tilde{A}_5]) \quad (7)$$

with the couplings defined in.[10] We also calculate the decay width $\Gamma(\rho \to \pi\pi)$, which includes the non-minimal coupling $f_{\rho\pi\pi}$, even though its contribution is numerically small.

Before we study the process $a_1 \to \pi\gamma$, we introduce the photon as an external gauge field and rewrite the bulk vector field decomposition as

$$V_\mu(x,z) = e\tilde{F}_\mu(x)\tau_3 + \frac{1}{\sqrt{M_5 L}}\sum_{n=1}^{\infty}\tilde{V}_\mu^{(n)}(x) f_V^{(n)}(z)\,, \quad (8)$$

with $\tau_3 = \sigma_3/\sqrt{2}$, where σ is the Pauli matric, and e is identified with the physical electron charge at chiral symmetry breaking scale. The advantage of our treatment of photon as external field, compared with the treatment of photon as the electromagnetic subgroup of $SU(3)_V$,[7] is that we don't need to worry about the photon KK excitations, as well as the mixing between photon KK excitations and ρ^0 KK excitations. With the help of vector KK decomposition Eq.(8), the $a_1 \to \pi\gamma$ channel has similiar results as $a_1 \to \rho\pi$. We have verified that the gauge non-invariant term of structure $\mathrm{Tr}(\tilde{A}^\mu[F_\mu, \tilde{A}_z])$ is cancelled out.

We also calculate the electromagnetic form factor of pions. In additional to the usual structure of contact $\gamma\pi\pi$ interaction $\mathrm{Tr}(F^\mu[\tilde{A}_5, \partial_\mu\tilde{A}_5])$, we have non-minimal structure $\mathrm{Tr}(F^{\mu\nu}[\partial_\mu\tilde{A}_5, \partial_\nu\tilde{A}_5])$, which comes from the dim-6

κ term. And we also find $g_{\gamma\pi\pi} = e$. From the kinetic term and the vector KK decomposition (8), we can derive the kinetic mixing of γ and ρ,

$$\mathcal{L}_{\gamma\rho} = -\frac{1}{2} e g_{\gamma\rho} F^{\mu\nu} \tilde{V}_{\mu\nu} \,, \tag{9}$$

with

$$g_{\gamma\rho} = \frac{M_5}{\sqrt{M_5 L}} \int_{L_0}^{L_1} dz a Z_v f_V(z) \,. \tag{10}$$

In small momentum limit, the electromagnetic form factor of pions can be expressed as

$$F(q^2) = 1 + \frac{1}{6} r_\pi^2 q^2 + \mathcal{O}(q^4) \,, \tag{11}$$

with the pion charge radius r_π calculated as

$$r_\pi^2 = 6\left[-\frac{f_{\gamma\pi\pi}}{g_{\gamma\pi\pi}} + \frac{g_{\gamma\rho} g_{\rho\pi\pi}}{m_\rho^2} \right] . \tag{12}$$

Our vector meson dominance (VMD) is different from the usual VMDs as discussed in ref.,[13] where we have an additional non-minimal $\gamma\pi\pi$ contact interaction, $\mathrm{Tr}(\tilde{F}^{\mu\nu}[\partial_\mu \tilde{A}_5, \partial_\nu \tilde{A}_5])$.

Before we discuss the $\mathcal{O}(p^4)$ chiral Lagrangian, we consider the vector field ρ effective Lagrangian,[14]

$$\begin{aligned} \mathcal{L}_V = & -\frac{1}{4} \mathrm{Tr}[V^{\mu\nu} V_{\mu\nu}] + \frac{1}{2} m_\rho^2 \mathrm{Tr}[V_\mu - \frac{i}{g} \Gamma_\mu]^2 \\ & - \frac{1}{2\sqrt{2}} e g_{\gamma\rho} \mathrm{Tr}[V_{\mu\nu} f_+^{\mu\nu}] + \frac{i}{\sqrt{2}} f_{\rho\pi\pi} f_\pi^2 \mathrm{Tr}[V_{\mu\nu} u^\mu u^\nu] \end{aligned} \tag{13}$$

with ρ transforming as gauge field of $SU(2)_V$ and the notation of Γ_μ, $f_+^{\mu\nu}$, u^μ the same as in ref.[14] The coefficients in the effective Lagrangian are determined by matching with our theory with dim-6 operators. The $\mathcal{O}(p^4)$ chiral Lagrangian for the pions is given in ref,[15] In this work, we do not discuss scalar and pseudoscalar resonances contribution to $L_{3,4,5,6,7,8}$, and only study the vector and axial resonances contribution to $L_{1,2,3,9,10}$. After Integrating out the vector rho meson, we obtain the following chiral coefficients,

$$\begin{aligned} L_1 &= \frac{f_\pi^4}{8 m_\rho^4} g_{\rho\pi\pi}^2 - \frac{f_\pi^4}{4 m_\rho^4} g_{\rho\pi\pi} f_{\rho\pi\pi} \,, \qquad L_2 = 2 L_1 \,, \qquad L_3 = -6 L_1 \,, \\ L_9 &= \frac{f_\pi^4}{m_\rho^4} g_{\rho\pi\pi}^2 + \frac{f_\pi^2}{2 m_\rho^2} e g_{\rho\pi\pi} f_{\rho\pi\pi} - \frac{2 f_\pi^4}{m_\rho^2} g_{\gamma\rho} g_{\rho\pi\pi} . \end{aligned} \tag{14}$$

L_{10} can be calculated from the two-point correlators of vector and axial, $\Pi_{V,A}$,

$$L_{10} = \frac{1}{4}[\Pi'_A(0) - \Pi'_V(0)], \tag{15}$$

where the derivative is over p^2.

We also calculate the electromagnetic mass difference of pion from the operator of $\mathrm{Tr}[Q_R U Q_L U^\dagger]$,

$$\delta m_\pi \equiv m_{\pi^+} - m_{\pi^0} \simeq \frac{3\alpha_{\mathrm{em}}}{8\pi m_\pi f_\pi^2} \int_0^\infty dp^2(\Pi_A - \Pi_V). \tag{16}$$

We will present our numerical results of various hadronic observables and chiral coefficients bellow. We use χ^2 to fit the four parameters L_1, ξ, κ, ζ from m_ρ, m_{a_1}, D/S ratio, $\Gamma(\rho \to \pi\pi)$ in case B and m_ρ, m_{a_1}, $\Gamma(\rho \to \pi\pi)$, f_π in case C. Our results are summarized in Table **1**, **2** and **3**. As a comparison, we also give Da Rold and Pomarol's results[7] in case A.

In both cases B and C, $\Gamma(a_1 \to \pi\gamma)$ is non-vanishing, but small (less than 100KeV), while $\Gamma(a_1 \to \rho\pi)$ is a little small in case B, but consistent with experimental measurement in case C. We have checked that the dominant contribution to $\Gamma(a_1 \to \rho\pi)$ comes from the leading order structure $\mathrm{Tr}(\tilde{A}^\mu[\tilde{V}_\mu, \tilde{A}_z])$. However, $\mathrm{Tr}(\tilde{A}^\mu[\tilde{F}_\mu, \tilde{A}_z])$ term is cancelled out for $a_1 \to \pi\gamma$ channel. Only dim-6 κ and ζ terms contribute to the above process. This is different from usual 4D models with large $\Gamma(a_1 \to \pi\gamma)$, where the ratio between $\Gamma(a_1 \to \pi\gamma)$ and $\Gamma(a_1 \to \rho\pi)$ is roughly $e^2/g^2_{\rho\pi\pi}$, and only a single type of operator $\mathrm{Tr}(\tilde{A}^{\mu\nu}[\tilde{V}_{\mu\nu}, \pi])$ contributes to both channels.[8]

Table 1. Mass spectra and decay constants. The unit of masses, decay constants is MeV.

case f_π	L_1 ξ	κ (10^{-6}) ζ (10^{-6})	m_ρ f_ρ	m_{a_1} f_{a_1}
exp 86.4 ± 9.7			775.8 ± 0.5	1230 ± 40
A 85.0	3.125 4.0	0. 0.	769.6 138	1253 163
B 71.9	2.836 2.56	-5.930 -39.72	775.8 144	1230 182
C 78.7	3.102 4.010	-16.03 0.09188	775.8 140	1246 172

The pion charge radius r_π agrees with the experiment in both case B and C. Pion decay constant f_π is a little small in case B, while the D/S ratio of $a_1 \to \rho\pi$ is small in case C. As in other 5D models, the KSRF

Table 2. Various hadronic observables. The unit of decay widths is MeV.

case	$\Gamma(\rho \to \pi\pi)$ $g_{\rho\pi\pi}$	$\Gamma(a_1 \to \pi\gamma)$ r_π(fm)	$\Gamma(a_1 \to \rho\pi)$ D/S ratio
exp	146.4 ± 1.5	0.640 ± 0.246 0.672 ± 0.008	$250 \sim 600$ -0.108 ± 0.016
A	95.4 4.8	0. 0.585	295.5 -0.055
B	146.5 5.8	0.088 0.654	165.3 -0.094
C	146.4 5.6	0.042 0.640	409.8 -0.027

relation $g_{\rho\pi\pi}^2/m_\rho^2 = c/f_\pi^2$ with $c = 1/2^{16}$ is not satisfied very well. In both case B and C, c is roughly 0.3, which means nearly complete vector meson dominance of order $\mathcal{O}(p^2)$ four-pion interaction. The higher ρ resonace and scalar exchange, and contact four-pion interaction contribution are less than $\sim 10\%$.

Table 3. The chiral coefficients L_i in unit 10^{-3}.

case	L_1	L_2	L_3	L_9	L_{10}	δm_π
exp	0.4 ± 0.3	1.4 ± 0.3	-3.5 ± 1.1	6.9 ± 0.7	-5.5 ± 0.7	4.6
A	0.43	0.86	-2.6	5.1	-5.5	3.4
B	0.32	0.65	-1.9	4.0	-5.0	1.5
C	0.46	0.93	-2.8	5.3	-5.1	2.9

The chiral coefficients of relevance and pion mass difference are given in Table **3**. The results do not significantly change, compared with Da Rold and Pomarol's case.

In summary, we have considered a AdS/QCD model with two dim-6 dimension operators. We study the mass spectra and decay constants of vector, axial and pseudoscalar sectors, and phenomenology of $a_1 \to \rho\pi$, $\rho \to \pi\pi$ and $a_1 \to \pi\gamma$ channel. We obtain non-vanishing branching ratio of $a_1 \to \pi\gamma$, which is a new feature of our model. The photon is introduced as an external field. We also study the pion electromagnetic form factor, and calculate the pion charge radius, which agrees with the experimental results.

Acknowledgments

The author would like to acknowledge the invitation of the SCGT 06 organizers and giving me the opportunity to present our work. This work is in collaboration with Youngman Kim and Pyungwon Ko.

References

1. J. Maldacena, Adv. Theor. Math. Phys. **2** (1998) 231; S. S. Gubser, I. R. Klebanov, A. M. Polyakov, Phys. Lett. **B428** (1998) 105 [hep-th/9802109]; E. Witten, Adv. Theor. Math. Phys. 2 (1998) 253 [hep-th/9802150].
2. A. Karch and E. Katz, JHEP **0206**, 043 (2002) [arXiv:hep-th/0205236].
3. S. J. Brodsky, this proceeding.
4. N. Evans, this proceeding.
5. S. Sugimoto, this proceeding.
6. J. Erlich, E. Katz, D. T. Son and M. A. Stephanov, Phys. Rev. Lett. **95**, 261602 (2005) [hep-ph/0501128].
7. L. Da Rold and A. Pomarol, Nucl. Phys. B **721**, 79 (2005) [arXiv:hep-ph/0501218].
8. U. G. Meissner, Phys. Rept. **161**, 213 (1988).
9. P. Ko and S. Rudaz, Phys. Rev. **D 50**, 6877 (1994).
10. Y. Kim, P. Ko and X.-H. Wu, in preparation.
11. M. Harada, S. Matsuzaki and K. Yamawaki, Phys. Rev. D **74**, 076004 (2006) [arXiv:hep-ph/0603248].
12. N. Isgur, C. Morningstar and C. Reader, Phys. Rev. D **39**, 1357 (1989).
13. H. B. O'Connell, B. C. Pearce, A. W. Thomas and A. G. Williams, Prog. Part. Nucl. Phys. **39**, 201 (1997) [arXiv:hep-ph/9501251].
14. G. Ecker, J. Gasser, H. Leutwyler, A. Pich and E. de Rafael, Phys. Lett. B **223**, 425 (1989).
15. G. Ecker, J. Gasser, A. Pich and E. de Rafael, Nucl. Phys. B **321**, 311 (1989).
16. K. Kawarabayashi and M. Suzuki, Phys. Rev. Lett. **16** (1966) 255; Riazuddin and Fayyazuddin, Phys. Rev. **147** (1966) 1071.

MAGNETIC STRINGS AS PART OF YANG–MILLS PLASMA

M. N. CHERNODUB

ITEP, B. Cheremushkinskaya 25, 117118 Moscow, Russia
RIISE, Hiroshima University, Higashi-Hiroshima, 739-8527, Japan

V. I. ZAKHAROV

INFN - Sezione di Pisa, Largo Pontecorvo, 3, 56127 Pisa, Italy
Max-Planck Institut für Physik, Föhringer Ring 6, 80805, München, Germany

Magnetic strings are defined as infinitely thin surfaces which are closed in the vacuum and can be open on an external monopole trajectory (that is, defined by 't Hooft loop). We review briefly lattice data on the magnetic strings which refer mostly to SU(2) and SU(3) pure Yang-Mills theories and concentrate on implications of the strings for the Yang-Mills plasma. We argue that magnetic strings might be a liquid component of the Yang-Mills plasma and suggest tests of this hypothesis.

Keywords: Confinement; Yang-Mills plasma; AdS/QCD correspondence.

1. Generalities

1.1. *Definition of magnetic strings*

There are two fundamental probes of Yang-Mills vacuum, Wilson and 't Hooft loops, related to the heavy-quark and heavy-monopole potentials:

$$\langle W\rangle \sim \exp\left(-V_{Q\bar{Q}}(r)\,T\right)\,, \qquad \langle H\rangle \sim \exp\left(-V_{M\bar{M}}(r)\,T\right). \tag{1}$$

In the string picture, one postulates that the Wilson and 't Hooft loops are given by sums over all surfaces which can be spanned on the loops,

$$\begin{aligned}\langle W\rangle &\sim \sum \exp(\ -f(...)A_C)\,, \\ \langle H\rangle &\sim \sum \exp(\ -\tilde{f}(...)A_H)\,,\end{aligned} \tag{2}$$

where A_C and A_H are the area of the surfaces spanned on the Wilson contour C and the 't Hooft loop, respectively, and $f(...)$, $\tilde{f}(...)$ are certain weight functions. We do not specify the arguments of the weight functions here. In models with extra dimensions, for example, the weight functions depend on the background metrics.

Equations (2) imply that there exist strings which can be open on the Wilson and 't Hooft loops. We call these strings as electric and magnetic strings, respectively. Explicit realization of (2) is achieved in dual formulations of highly symmetrical versions of Yang-Mills theories, for a review see Ref. 1.

1.2. *Percolation of strings*

At zero temperature, heavy quarks are confined while heavy monopoles are not confined. In terms of the string tensions this means:

$$\sigma_{\text{electric}}(T=0) \sim \Lambda^2_{\text{QCD}}\,, \qquad (3)$$
$$\sigma_{\text{magnetic}}(T=0) = 0\,.$$

These relations are realized explicitly in case of large number of colors N_c and supersymmetry.[1] A reservation is that the tension depends actually on the length of the strings and Eqs. (3) hold only if the strings are long enough.

Connection between the weight factors which enter (2) and the string tensions (3) is not so straightforward. Consider first the case when the weight factors $f, \tilde{f}$ are constants. Then they still do not coincide with the string tensions. The point is that the weight factors specify suppression due to the action associated with the surface. However, there is a hidden entropy factor in Eq. (2) which is the number of various surfaces of the same area. Thus, the tension (3) reflects the balance between suppression due to the action and enhancement due to the entropy:

$$(\text{tension})\cdot(\text{area}) \quad \sim \quad (\text{action}) \;-\; (\text{entropy})\,. \qquad (4)$$

One can estimate readily that the entropy factor results in exponential enhancement of contribution of (large) areas.

One of central points is that vanishing string tension does not mean that the strings are not relevant and uninteresting. In a way, just to the contrary. Vanishing tension means percolation of strings, or that they 'are everywhere'. For the first time this observation was made in connection with deconfinement transition in Yang–Mills theories at finite temperature.[2] Namely, it was suggested that at the critical temperature the electric strings become tensionless,

$$\sigma_{\text{electric}}(T=T_c) = 0\,, \qquad (5)$$

and percolate through the vacuum.

It is worth emphasizing that percolation assumes cancellation between energy and entropy for long strings.[a] In particular, strong cancellation between energy and entropy of strings at temperatures above the phase transition was observed in Ref. 3 by studying the Wilson loop:

$$\langle W\rangle \sim \exp(-F)\,, \qquad F \;=\; U \;+\; T(\partial F/\partial T)\,, \tag{6}$$

where F is the free energy, U is the potential energy. Measured values of energy and entropy are about 10 and cancel each other.

Let us go back to discuss the $T = 0$ case. Vanishing tension of magnetic strings, see (3) implies percolation of magnetic strings at zero temperature. And, indeed, percolation of magnetic strings has been observed on the lattice. In lattice terminology, magnetic strings are known as center vortices whose role in confinement has been extensively studied.[4]

We are more interested in the structure of the strings themselves. In particular, if the strings are infinitely thin indeed then both energy and entropy should be divergent at short distances and cancel each other in free energy, or tension.[b] This remarkable phenomenon was indeed observed.[5] In particular the non-Abelian action associated with the surfaces equals to:

$$S_{\text{strings}} \;\approx\; 0.54\,(\text{Area})\,a^{-2}\;\;, \tag{7}$$

where a is the lattice spacing, $a \to 0$ in the continuum limit, and $(Area)$ is the total area which is in physical units. The data refers to $a \geq 0.06\,\text{fm}$.

1.3. *Supercritical phase*

The tachyonic nature of the magnetic strings is manifested in existence of an infinite cluster of surfaces in the vacuum.[c] Indeed, the spatial extension of cluster is obviously related to correlation length,

$$\langle R_{\text{finite cluster}}\rangle \;\sim\; l_{\text{corr}} \;\sim\; m^{-1}\,, \tag{8}$$

where m is the excitation mass. With a stretch of imagination, one can conclude that existence of an infinite cluster corresponds then either to zero or tachyonic mass.[d]

---

[a] Short strings which quantum mechanically correspond to glueballs of low spin can well have finite tension and, respectively, glueballs remain massive at the phase transition.
[b] For further comments and explanation see Ref. 6.
[c] For introduction to the percolation theory and further references see, e.g., Ref. 7.
[d] The absolute value of the tachyonic mass controls in fact density of the infinite cluster.

The most non-trivial ingredient of the percolation theory is that there is continuity of description across the threshold of emergence of a tachyonic mode. In other words, one can still use the language of the false vacuum even when there exists a negative mode. An example of such a description is an expression for the density of the infinite cluster as proportional to ζ^{α} where $\zeta = 0$ at the point of the phase transition, where ζ is a non-negative function of couplings of the model and α is a positive number.

The density of the percolating cluster of the magnetic strings does satisfy such a relation:

$$\theta_{\text{plaq}} \approx \left(\frac{a}{0.5\,\text{fm}}\right)^2 , \qquad (9)$$

where θ_{plaq} is the probability for a given plaquette on the lattice to belong to the infinite, or percolating cluster of surfaces. Note that in the continuum limit of $a \to 0$ the infinite cluster occupies a vanishing part of the total volume. The value of $(a \cdot \Lambda_{\text{QCD}})$ specifies closeness to the point of the phase transition.

Apart from the probability θ_{plaq} one can introduce many other observables and corresponding critical exponents. Theory of percolation reduces then to relations among the critical exponents.

1.4. *Magnetic monopoles*

Theory of percolation of surfaces is much less developed than theory of percolation of trajectories (respectively, of strings and of particles). Thus, it is a great simplification that magnetic strings can in fact be substituted for our purposes by particles living on the strings. In lattice terminology[e] the particles are magnetic monopoles, for further references see Ref. 6. Geometrically the substitution is possible since the monopole trajectories (which are closed loops) cover densely the surfaces associated with the strings. Physicswise, the relation is that monopoles correspond to a tachyonic mode in terms of the string excitations.

The probability of a given link to belong to a monopole trajectory is[8]

$$\theta_{\text{link}} \approx \left(\frac{a}{0.5\,\text{fm}}\right)^3 + \left(\frac{a}{0.8\,\text{fm}}\right)^2 \left(1 - \frac{a}{0.25\,\text{fm}}\right), \qquad (10)$$

[e]The terminology is somewhat misleading since the non-Abelian field of the monopole-like objects considered here is aligned with the surfaces and in general is not quantized. The field of conventional "monopoles" in a common sense of this word would be, instead, spherically symmetric and quantized.

where $a < 0.13\,\text{fm}$ for the theory to converge to the continuum limit. The first term here corresponds to the infinite cluster, or the tachyonic mode. The second term corresponds to finite clusters, or excitations and dominates in the limit of $a \to 0$. Moreover, numerically about (90-95)% of monopole trajectories fall on the magnetic-strings surfaces, see Ref. 5 and references therein. This is just clarification what we meant above by "particles living on the strings": the evidence is pure numerical but convincing.

2. Magnetic component of Yang-Mills plasma

2.1. *From tachyonic mode to thermal plasma*

Percolating magnetic strings is a fascinating phenomenon. There is a word of caution, however: magnetic strings describe vacuum structure, or the structure of the ground state. What is actually observable are excitations of the ground state. The percolating vacuum magnetic strings represent a tachyonic mode with respect to the perturbative vacuum and cannot be observed directly in an experimental set-up. Things may change at finite temperature, as we will explain now.

Indeed, it is a general rule that degrees of freedom which are virtual at zero temperature become real particles at non-zero temperature. Consider, for example, theory of free photons. At zero temperature the lowest state, vacuum is realized as zero-point fluctuations. If we evaluate the energy density it is ultraviolet divergent:

$$\epsilon_{\text{vac}} \sim \sum \frac{\omega(\mathbf{k})}{2} \sim \Lambda_{\text{UV}}^4,$$

where $\mathbf{k}$ is the momentum and Λ_{UV} is an ultraviolet cut off. At finite temperature, the energy density of the photon gas is given by the Stefan-Boltzmann law

$$\epsilon(T) \sim T^4,$$

and, obviously, the coefficients in front of the ultraviolet divergence at zero temperature and in front of T^4 are related to each other.

An example closer to our case: if there is Higgs mechanism of spontaneous symmetry breaking at zero temperature, at finite temperature the symmetry is restored and the number of degrees of freedom of scalar particles in plasma is the same as in the Lagrangian (while at zero temperature the tachyonic mode is unobservable).

Thus we can expect, that at deconfinement phase transition magnetic strings, or monopoles are released into the thermal plasma.[9]

2.2. *Percolation at finite temperature*

Since the magnetic monopoles are directly observable on the lattice one can try to check these expectation against the lattice data. First, however, we should clarify how to distinguish between real and virtual particles in the Euclidean formulation of the theory. The problem is that in Euclidean space the wave functions of real particles exponentially decay with time and the difference between virtual and real particles is not obvious.

Consider first free particles. The answer to our question can be found[9] then in closed form within so called polymer representation of field theory, see, e.g. Ref. 10. One starts with classical action of a particle of mass M,

$$S_{cl} \;=\; M \cdot L\,, \tag{11}$$

where L is the length of trajectory. The Feynman propagator is given then by the sum over all the paths $P_{x,y}$ connecting the points x, y:

$$G(x-y) \;\sim\; \sum_{P_{x,y}} \exp(-S_{cl}\{P_{x,y}\})\,. \tag{12}$$

To enumerate all the paths one uses latticized space and the sum (12) with classical action (11) can be evaluated exactly.[10] The result is that $G(x-y)$ is indeed proportional to the propagator of a scalar particle with mass m related to the original mass parameter M (see (11)) in the following way:

$$m^2 \;\approx\; \frac{\text{const}}{a} \cdot \Big(M(a) \;-\; \frac{\text{const}'}{a}\Big)\,, \tag{13}$$

where the constants are known for a particular lattice regularization. Note that for the physical mass, m^2 be independent of a the original mass parameter M(a) is to be tuned to const'/a.

Finite temperature T is introduced through compactification of the time direction into a circle of length $1/T$ so that the points

$$x \;=\; (\mathbf{x}, x_4 + s/T)$$

where s is an integer number, are identified. The physical meaning of s is that it counts the number of wrapping in time direction. Intuitively, it is quite clear that the wrapped trajectories in the Euclidean space encode information on real particles at $T \neq 0$ in Minkowski space. Indeed, compactness of the time direction means that the wrapped trajectories correspond to particles which exist 'for ever', that is real particles.

Formally, the proof runs as follows.[9] One introduces Fourier transform of trajectories wrapped s times:

$$G_s(\mathbf{p}) \;=\; \int d^3x\, G(\mathbf{x}, t = s/T)\, \exp(-i\mathbf{p}\mathbf{x})\,.$$

The $s = 0$ case corresponds to unwrapped trajectories which are independent of temperature, $G_0 \equiv G_{\rm vac}$. Then one can derive equation:

$$\frac{1}{e^{\omega_{\mathbf{p}}/T} - 1} = \frac{\sum_s G_s}{2\, G_0}\,, \tag{14}$$

which explicitly realizes our intuition by relating the sum over wrapped trajectories to the Bose-Einstein distribution.

Further refinement on (14) are possible. First, normalization on the propagator at $T = 0$,

$$G_0 \;=\; 4/(a^2 \omega_{\mathbf{p}})$$

is awkward since G_0 depends explicitly on the lattice spacing a. Also, we expect that the properties of the monopoles are actually constrained by environment. To at least partly account for the interaction with the environment we introduce non-zero chemical potential μ and effective number of degrees of freedom of the monopoles, N_{df}. One can show then that the density of real particles,

$$\rho(T) = \int \frac{d^3\mathbf{p}}{(2\pi)^3} \frac{N_{df}}{e^{(\omega_{\mathbf{p}}+\mu)/T} - 1}\,, \tag{15}$$

equals to the average number of wrapping per unit 3d volume,

$$\rho(T) \;=\; n_{wr} \;=\; \langle |s| \rangle / V_{3d}\,. \tag{16}$$

Let us mention the simplest case of static trajectories. The static trajectories look the same in Euclidean and Minkowski spaces and obviously correspond to real particles. Thus, Eq. (16) is true in case of static trajectories, also if the particles interact strongly. At first sight, the static case is quite an academic limit. In fact, it is not. The point is that one can argue on general grounds that the whole of non-perturbative physics at high temperatures reduces to magnetostatics[11] with strong interaction. Moreover, detailed lattice studies[12] reveal that the approximation of static trajectories works well for monopoles already around $T = 2.4T_c$ where T_c is the temperature of the deconfinement phase transition.

We will accept (16) as definition of density of real particles for lattice studies of the monopole trajectories.

2.3. *Transition chain: condensate → liquid → gas*

Lattice data on the wrapped trajectories turn in fact exciting. First of all, the change of character of the trajectories near the phase transition has been noticed in many papers. Namely, the infinite, or percolating cluster disappears. Thus, no tachyonic mode at $T > T_c$, as is expected.

Even more remarkable, the long monopole trajectories tend to become time-oriented. As we have just discussed, time-oriented trajectories correspond to real particles in thermal plasma. Moreover, remember that monopoles are just marking magnetic strings, which is the primary object.[f] We just follow monopoles because the equations, like (16) can be derived explicitly in case of particles. However, as far as the qualitative behavior is concerned, direct observations on strings (that is, surfaces) reveal the same qualitative change: above T_c the strings percolate only in space–, not time–directions, for discussion and references see Ref. 4.

One can conclude that, qualitatively, there is no doubt that around $T = T_c$ there is transition of part of the tachyonic mode into physical degrees of freedom of the Yang-Mills plasma. The reality of the magnetic component of the plasma seems to be confirmed. However, it is only quantitative analysis that cannot tell, how significant this component is.

Quantitative analysis[9] of the lattice data on wrapped trajectories[14] reveals that dependence of the density of monopoles in plasma is different in two ranges of temperatures. First, at temperatures

$$T < T < 2T_c$$

the density of thermal monopoles is approximately constant:

$$\rho(T) \approx T_c^3 . \tag{17}$$

It is interesting to compare this number with density of ultrarelativistic non-interacting particles at such temperatures. At $T = T_c$ the density (10) is about 10 times higher than the density of the ideal gas (with one degree of freedom).

Thus, at temperatures just above the phase transition the monopoles are quite dense and, moreover, their density does not depend on the temperature, similarly to a liquid.

---

[f]Note that presence of monopoles in thermal plasma was speculated also in recent papers.[13] Monopoles considered in these papers are essentially Dirac monopoles and it is not clear how to reconcile such models with the well known observation that there is no local field theory which could accommodate both magnetic and electric charges. There is no such no-go theorem for the 'monopoles' considered here since they are rather building blocks of the magnetic strings than particles, see discussion above. Moreover, there is no relation between the monopoles introduced in Ref. 13 at non-zero temperature and condensate at zero temperature while such a relation is a guiding principle in paper 9.

Starting from $T \approx 2T_c$ the density of thermal monopoles begins to grow and at large T the density can be approximated as

$$\rho(T) \approx (0.25T)^3 \qquad (T > 2T_c)\,, \tag{18}$$

where we keep only the term dominating at high temperatures. If we compare now (18) to the ideal-gas case we find that at very high temperatures the monopoles correspond to much less than one degree of freedom.

Thus, there is a chain of transformation of the magnetic degrees of freedom, as function of temperature. First, there is quantum condensate. Then, it melts into a liquid. And at higher temperatures the liquid is evaporated into a gas.

Note an amusing analogy with physics of superfluidity.[15] In case of liquid helium there exist a superfluid and ordinary components of liquid. With increasing temperature the superfluid component undergoes transition into ordinary liquid. This is analogy of the deconfinement phase transition which is associated with vanishing of the quantum condensate and release of magnetic degrees of freedom into plasma. In terms of trajectories, infinite cluster existing at zero temperature is transformed into time oriented trajectories (or, respectively, surfaces in the string language). At higher temperatures, $T > 2T_c$ the liquid is evaporated into gluon gas. One can expect that the properties of plasma are described by perturbative physics with better and better accuracy.

2.4. *Resolving mysteries of phenomenology?*

The observation (17) that the density of the magnetic component of plasma is approximately a constant,

$$\rho_{\text{magnetic}} \approx \text{const}\,, \qquad T_c < T < 2T_c\,,$$

might have remarkable phenomenological implications. The point is that the properties of the plasma, as are observed at RHIC,[16] indicate that near the phase transition the plasma is rather a liquid. Such a conclusion follows, in particular, from evaluation of the viscosity which turns to be small.[17] However, up to now there has been no key to answer the question, what is dynamics behind these plasma properties. Lattice measurements which imply (17) might provide a key to identify particular component of the plasma which looks as a liquid. Of course much more should be done to be sure of this conclusion.

3. Phase transition as change of dimensionality

3.1. *Two types of strings, in 4d and 3d*

In Sec. 2.2. we considered percolation at finite temperature in some detail. In particular, we argued that wrapped trajectories in Euclidean space correspond to real (thermal) particles in the Minkowski space. The case we considered is in fact Higgs theory of a single complex field. Now we would like to emphasize that there is deep difference between the Yang-Mills case and the ϕ^4 case. Namely, in the Yang-Mills case the deconfinement phase transition is in some sense a change from four- to three-dimensional physics. The difference between the two theories can be readily understood in simple geometrical terms if we use the strings language for the Yang-Mills case.

To begin with, we argued above that in the deconfinement phase both electric and magnetic strings have vanishing tension:

$$\sigma_{\text{electric}} = \sigma_{\text{magnetic}} = 0\,, \tag{19}$$

which implies percolation of two types of strings. In four dimensions, however, such percolation is hardly possible since each string is a 2d object and two types of percolating strings would occupy the whole space:

$$2d + 2d = 4d\,.$$

True, the situation is rather marginal since strings are expected to intersect along a submanifold of dimension $d = 0$ and whether it is possible to avoid intersections depends on details of the problem.

From the classical paper[18] we learn that under feasible assumptions (like existence of mass gap) the resolution of the uncertainty is that percolation of both types of strings in 4d is not possible. Indeed in the language of the strings the result of Ref. 18 is that either

$$\sigma_{\text{electric}} \neq 0\,, \qquad \sigma_{\text{magnetic}} = 0\,,$$

or vice verse.

How it is possible then to realize (19) at all? The answer is that the price is violation of O(4) rotational invariance. The strings should percolate differently in time and in space directions. At finite temperature this does not contradict any general principle since it is only the time coordinate that is compactified. Thus, at $T \neq 0$ one can consider separately temporal and spatial Wilson loops,[g] and temporal and spatial 't Hooft loops.

[g] In fact, instead of the temporal Wilson loop one should introduce correlator of two Polyakov lines at large spatial distances. For recent discussion see, e.g., Ref. 19.

Here we come to one of central observations concerning strings in 4d and 3d spaces. Namely, strings in 3d (or in a time slice of 4d space) are loops (trajectories),[h] and for two types of strings

$$1d \; + \; 1d \; = 2d \; < 3d \,,$$

so that the strings can percolate simultaneously.

For the string tension associated with the temporal loops we get from our basic relation (19):

$$\sigma^{\rm temporal}_{\rm electric} \;=\; \sigma^{\rm temporal}_{\rm magnetic} \;=\; 0\,, \qquad T \;>\; T_c\,. \tag{20}$$

Thus, zero tension have strings which can be open on the lines parallel to the time direction. As we explained above, vanishing tension means percolation. To ensure (20) we need both electric and magnetic strings percolate in directions perpendicular to the time axis, i.e. in 3d space.[i]

Next, one can readily understand that at $T > T_c$ both spatial Wilson and 't Hooft loops exhibit area laws:

$$\sigma^{\rm spatial}_{\rm electric} \neq 0\,, \qquad \sigma^{\rm spatial}_{\rm magnetic} \neq 0\,, \qquad T > T_c\,. \tag{21}$$

Unlike the 4d case, Eq. (21) is consistent with percolation of strings in 3d because in 3d there is no representation like (2). On the other hand, percolation of magnetic strings (1d objects) still disorders the spatial Wilson line and percolation of electric strings disorders the 't Hooft line. Hence, Eqs. (21) are valid.[j] In other words, at $T > T_c$ we expect percolation of both types of strings in 3d. From the 4d perspective these strings are 2d surfaces nearly parallel to the x_4 direction. In the 3d projection they are percolating lines.

3.2. *Three mechanisms under same disguise*

In Sec. 2.2 we emphasized connection between wrapped trajectories and real particles in thermal plasma. Here, we would like to emphasize that we discussed in fact different mechanisms which can be revealed by counting the wrapped trajectories.

Percolation of trajectories in 4d. Consider theory of a free complex scalar field in 4d and at temperature $T = 0$. In the polymer representation[10,22]

---

[h] We do not discuss here finite thickness of strings.

[i] For discussion of geometry of strings, their percolation at finite temperature and corresponding lattice data, in particular, Ref. 20.

[j] For a related discussion see.[21]

the theory is encoded in properties of particle trajectories. In particular, if there is no tachyonic mode, $m^2 > 0$, then there exist only finite clusters of trajectories. The distribution in length L of the clusters is

$$N(L) \sim L^{-3} \exp(-m^2 L a) . \tag{22}$$

Moreover, the average radius of the cluster is given by

$$R^2 \sim L \cdot a . \tag{23}$$

Let us now switch on finite temperature. Extension of the cluster in the time direction cannot then exceed $R \sim 1/T$. Which means that some of finite clusters become wrapped trajectories. Counting these trajectories, one can reproduce the Planck distribution, see (14).

String percolation: from 4d to 3d. Another mechanism was discussed in Sec. 3.1. Here we argued that at $T > T_c$ both magnetic and electric strings percolate only in spatial directions. In other words, the long trajectories become time-oriented. Let us emphasize that this is absolutely different mechanism than simple percolation. In case of percolation, the properties of trajectories are not changed by switching on temperature. Just some long trajectories are 'caught' by the periodicity condition at $x_4 = s/T$ and become wrapped trajectories.

In case of the "change of dimensionality" of string percolation, the trajectories become time-oriented. The effect is not simply periodicity imposed on the boundary. How one can understand such a non-local effect? An explanation is suggested by AdS/QCD correspondence (for references see, e.g., Ref. 6). The basic idea is that the tension of strings (both electric and magnetic) depends on their length. In particular, $\sigma_{\text{magnetic}} = 0$ only if the string is long enough, $l \sim \Lambda_{\text{QCD}}^{-1}$. And only tensionless strings percolate. If one imposes periodic boundary condition at $x_4 = s/T$ then at the critical temperature the percolating string is no longer long enough, and there can be no percolation in the time direction. There is a non-local change of properties of strings because the condition for percolation is non-local: string is to be long enough. There is no such condition in case of the standard percolation. Since percolation in the time direction is no longer allowed time-oriented strings become energetically favorable.

At $T = T_c$ not only the magnetic string changes its tension. For example, the string tension associated with the spatial Wilson loop in the model[23] depends on temperature in the following way:[19]

$$\sigma_{\text{electric}}^{\text{spatial}} \approx \frac{T^2}{T_c^2} \exp\left(\frac{T_c^2}{T^2} - 1\right) , \qquad T > T_c \tag{24}$$

Probably, the very position of the phase transition, T_c could be found from condition of self-consistency of predictions for various strings. But this issue has not been addressed in literature so far.

Dimensional reduction. Non-perturbative nature of magnetostatics in the Yang-Mills case was discovered first within the scheme of dimensional reduction which should work at high temperatures, when the running coupling $g^2(2\pi T)$ is small.[11] The picture is quite similar to the case just discussed with even more asymmetry between the spatial and time directions. For example, instead of time-oriented oriented trajectories one would discuss strictly static trajectories.

3.3. *Identifying the mechanism*

Using lattice data on wrapped trajectories and various string tensions one can in fact distinguish between various mechanisms summarized in Sec. 3.2.

In particular, in case of standard, or particle percolation the length of the wrapped trajectories would be divergent in the continuum limit, $a \to 0$:

$$\langle L_{\text{wrapped}} \rangle \sim T^{-2} a^{-1} . \tag{25}$$

Such dependence on the lattice spacing is typical of the random walk.

On the other hand, the string model with running string tension is non-local and it allows for a change of character of trajectories through the whole lattice. The relation (25) is not valid any longer. Also, there is natural scale for T_c which is position of the horizon in the AdS/QCD schemes, see, e.g., Ref. 19.

The lattice data on the wrapped trajectories which we discussed in Sec. 2 clearly favor the non-local model, or string percolation.

Next, let us discuss how to distinguish a general scheme of "change of dimensionality" from the standard dimensional reduction. A key point here is that the dimensional reduction is derived within perturbation theory. The perturbation theory, in turn, can be applied to evaluate $\sigma^{\text{spatial}}_{\text{electric}}$ and $\sigma^{\text{spatial}}_{\text{magnetic}}$, see, e.g., Ref. 24 and references therein. There is ample evidence that at temperatures near the phase transition, $T < 2T_c$ perturbation theory does not work. Thus, dimensional reduction seems not to apply at temperatures crucial for the "magnetic liquid" discussed in Sec. 2.

As a further potential check of the string mechanism let us mention possible role of strings in the anomaly. The point is that strings naturally contribute to the temperature-dependent conformal anomaly:

$$\langle G^2(T) \rangle \sim \Lambda^2_{\text{QCD}}\, T^2 , \tag{26}$$

where $\langle G^2(T)\rangle$ is the temperature dependent part of the gluon condensate which can be studied directly on the lattice, for review see, e.g., Ref. 25.

Indeed, in time-slice (or in 3d) strings percolate as trajectories, or loops. For the QCD-related strings which we are discussing the total length of trajectories is of order

$$L_{\text{total}} \sim \Lambda^2_{\text{QCD}} V_{3d} . \tag{27}$$

The strings are time-oriented and extend in the time direction at distance of order T^{-1}. The tension for such strings (modulo possible 'threshold effects' near T_c) is of order $\sigma \sim T^2$, see, e.g., (24). In this way we come to the estimate (26) for the action density associated with the strings.

4. Conclusions

There is amusing possibility that magnetic degrees of freedom which condense and tachyonic (unphysical) at zero temperature become a component of thermal Yang-Mills plasma. There are first indications from lattice simulations that this is indeed the case but much more is to be done to really confirm the picture.

Acknowledgments

We are thankful to O. Andreev, V. Bornyakov, A. Nakamura, M. I. Polikarpov, T. Suzuki for useful discussions and to B. Shklovskii and A. I. Vainshtein for communications. The work of M.N.Ch. was supported by the JSPS grants No. L-06514. V.I.Z. is thankful to the organizers of the SCGT06 workshop, and especially to Prof. K. Yamawaki for the invitation and hospitality.

References

1. I. R. Klebanov, *Int. J. Mod. Phys. A* **21**, 1831 (2006), [arXiv:hep-ph/0509087];
   O. Aharony et al. *Phys. Rept.* **323**, 183 2000, [arXiv:hep-th/9905111].
2. A. M. Polyakov, *Phys. Lett. B* **72**, 477 (1978).
3. O. Kaczmarek et al, *Prog. Theor. Phys. Suppl.* **153**, 287 (2004), [arXiv:hep-lat/0312015].
4. J. Greensite, *Prog. Part. Nucl. Phys.* **51**, 1 (2003), [arXiv:hep-lat/0301023].
5. F. V. Gubarev et al, *Phys. Lett. B* **574**, 136 (2003), [arXiv:hep-lat/0212003];
   A. V. Kovalenko, M. I. Polikarpov, S. N. Syritsyn and V. I. Zakharov, *Nucl. Phys. Proc. Suppl.* **129**, 665 (2004), [arXiv:hep-lat/0309032];
   V. G. Bornyakov et al, *Phys. Lett. B* **537**, 291 (2002), [arXiv:hep-lat/0103032].

6. V. I. Zakharov, *"From confining fields on the lattice to higher dimensions in the continuum"*, [arXiv:hep-ph/0612342]; *AIP Conf. Proc.* **756**, 182 (2005), [arXiv:hep-ph/0501011].
7. G. Grimmelt, *"Percolation"*, Springer, Berlin, (1999).
8. V. G. Bornyakov, P. Y. Boyko, M. I. Polikarpov, V. I. Zakharov, *Nucl. Phys. B* **672**, 222 (2003), [arXiv:hep-lat/0305021].
9. M.N. Chernodub, V.I. Zakharov, *Phys. Rev. Lett.* **98** (2007), [arXiv:hep-ph/0611228].
10. A.M. Polyakov, *"Gauge Fields and Strings"* Harvard Academic Publishers, Chur, Switzerland, (1987);
J. Ambjorn, *"Quantization of geometry"*, [arXiv:hep-th/9411179].
11. P. H. Ginsparg, *Nucl. Phys. B* **170**, 388 (1980);
T. Appelquist and R. D. Pisarski, *Phys. Rev. D* **23**, 2305 (1981).
12. M. N. Chernodub, K. Ishiguro and T. Suzuki, *JHEP* **0309**, 027 (2003), [arXiv:hep-lat/0204003].
13. C. P. Korthals Altes, *"Magnetic monopoles in hot QCD"*, [arXiv:hep-ph/0607154];
J. Liao and E. Shuryak, *"Strongly coupled plasma with electric and magnetic charges"*, [arXiv:hep-ph/0611131].
14. S. Ejiri, *Phys. Lett. B* **376**, 163 (1996), [arXiv:hep-lat/9510027].
15. E. M. Lifshitz and L. P. Pitaevskii, *"Statistical Physics, part 2"*, Volume 9 of the Landau and Lifshitz Course of Theoretical Physics, chapter III, Pergamon Press, New York, (1980).
16. E. V. Shuryak, *"Strongly coupled quark-gluon plasma: The status report"*, [arXiv:hep-ph/0608177].
17. D. Teaney, *Phys. Rev. C* **68**, 034913 (2003), [arXiv:nucl-th/0301099];
A. Nakamura and S. Sakai, *Phys. Rev. Lett.* **94**, 072305 (2005), [arXiv:hep-lat/0406009].
18. G. 't Hooft, *Nucl. Phys. B* **138**, 1 (1978).
19. O. Andreev and V. I. Zakharov, *Phys. Lett. B* **645**, 437 (2007), [arXiv:hep-ph/0607026].
20. P. de Forcrand, D. Noth, *Phys. Rev. D* **72**, 114501 (2005), [arXiv:hep-lat/0506005];
G. S. Bali et al, *Phys. Rev. Lett.* **71**, 3059 (1993), [arXiv:hep-lat/9306024].
21. M.N. Chernodub, R. Feldmann, E. M. Ilgenfritz, A. Schiller, *Phys. Lett. B* **605**, 161 (2005), [arXiv:hep-lat/0406015].
22. M. N. Chernodub, V. I. Zakharov, *Nucl. Phys. B* **669**, 233 (2003).
23. O. Andreev, *Phys. Rev. D* **73**, 107901 (2006), [arXiv:hep-th/0603170];
O.Andreev, V.I.Zakharov, *Phys. Rev. D* **74**, 025023 (2006), [arXiv:hep-ph/0604204].
24. F. Karsch, E. Laermann and M. Lutgemeier, *Phys. Lett. B* **346**, 94 (1995), [arXiv:hep-lat/9411020];
P. Giovannangeli and C. P. Korthals Altes, *Nucl. Phys. B* **721**, 25 (2005), [arXiv:hep-ph/0412322].
25. R. D. Pisarski, *"Fuzzy bags and Wilson lines"*, [arXiv:hep-ph/0612191].

A GAUGE-INVARIANT MECHANISM FOR QUARK CONFINEMENT AND A NEW APPROACH TO THE MASS GAP PROBLEM

KEI-ICHI KONDO

Department of Physics, Faculty of Science, Chiba University, Chiba 263-8522, Japan
E-mail: kondok@faculty.chiba-u.jp

We give a gauge-invariant description of the dual superconductivity for deriving quark confinement and mass gap in Yang-Mills theory.

Keywords: Quark confinement; vacuum condensate; mass gap; glueball mass.

1. Introduction

The fundamental degrees of freedom of QCD, i.e., quarks and gluons, have never been observed in experiments. Only the color singlet combinations, hadrons (mesons and baryons) and glueballs, are expected to be observed. We wish to answer the question why color confinement occurs. In particular, we wish to clarify what the mechanism for quark confinement is.

Dual superconductivity picture proposed by Nambu, 't Hooft and Mandelstam[1] in 1970s is based on the electric–magnetic duality of the ordinary superconductivity. In order to apply this idea to describe the dual superconductivity in Yang-Mills theory, however, we must answer the questions:

(1) How to extract the "Abelian" part responsible for quark confinement from the non-Abelian gauge theory in the gauge-invariant way without losing characteristic features of non-Abelian gauge theory, e.g., asymptotic freedom.[7]
(2) How to define the magnetic monopole to be condensed in Yang-Mills theory in the gauge-invariant way even in absence of any fundamental scalar field, in sharp contrast to the Georgi-Glashow model.

In this direction, a crucial idea called the Abelian projection was proposed by 't Hooft.[2] Recall that the Wilson criterion of quark confinement is the area decay law of the Wilson loop average:

$$\left\langle \mathrm{tr}\left[\mathscr{P}\exp\left\{ig\oint_C dx^\mu \mathscr{A}_\mu(x)\right\}\right]\right\rangle_{YM} \sim e^{-\sigma_{NA}|S|}. \tag{1}$$

Recent investigations have shown that quark confinement based on the dual superconductor picture is realized in the maximal Abelian gauge (MAG):[4,8] the continuum form of MAG for SU(2) is (a background gauge) given by

$$[\partial_\mu \delta^{ab} - g\epsilon^{ab3}A_\mu^3(x)]A_\mu^b(x) = 0 \ (a,b=1,2), \tag{2}$$

for the Cartan decomposition: $\mathscr{A}_\mu = A_\mu^a \frac{\sigma^a}{2} + A_\mu^3 \frac{\sigma^3}{2}$ $(a=1,2)$. Numerical simulations on a lattice in MAG yield surprisingly the area decay law for the Abelian(-projected) Wilson loop in Yang-Mills theory:

$$\left\langle \exp ig\oint_C dx^\mu A_\mu^3(x)\right\rangle_{YM}^{MAG} \sim e^{-\sigma_{Abel}|S|} \ ! \tag{3}$$

Remarkable results are the saturation of the string tension:[3] Abelian dominance[5] $\Leftrightarrow \sigma_{NA} \sim \sigma_{Abel}$ and the Monopole dominance[6] $\Leftrightarrow \sigma_{Abel} \sim \sigma_{monopole}$ for the decomposition: $A_\mu^3 =$ Photon part + Monopole part.

However, we have still problems:

- The Abelian projection and MAG break SU(2) color symmetry explicitly.
- Abelian dominance has never been observed in gauge fixings other than MAG.

Then one raises the question: the dual superconductivity might not be a gauge-invariant concept? We wish to discuss how to cure these shortcomings. In this talk, we argue that

(1) [gauge-invariant "Abelian" projection] The "Abelian" part $\mathbb{V}_\mu$ responsible for quark confinement can be extracted from the non-Abelian gauge theory by using a nonlinear change of variables in the gauge-invariant way without breaking color symmetry.

(2) [infrared "Abelian" dominance] The remaining part $\mathbb{X}_\mu$ acquires the mass to decouple in the low-energy region, leading to infrared "Abelian" $\mathbb{V}_\mu$ dominance. The dynamical mass originates from the existence of gauge-invariant dimension-2 vacuum condensate $\langle \mathbb{X}_\mu^2 \rangle \neq 0$.

2. Non-Abelian Stokes theorem for the Wilson loop operator

The Wilson loop operator for non-Abelian gauge field

$$W_C[\mathscr{A}] := \mathrm{tr}\left[\mathscr{P}\exp\left\{ig\oint_C dx^\mu \mathscr{A}_\mu(x)\right\}\right]/\mathrm{tr}(\mathbf{1}), \tag{4}$$

is rewritten into an Abelian form called the Diakonov-Petrov (DP) version of the non-Abelian Stokes theorem which was for the first time derived in Ref. 10 for SU(2). This was rederived in Refs. 11 and 13 to be extented to SU(N) in Refs. 12 and 13. Once a unit vector field $\boldsymbol{n}(x)$ is introduced, the SU(2) version is

$$\begin{aligned} W_C[\mathscr{A}] &= \int d\mu_S(\boldsymbol{n}) \exp\left\{ ig\frac{J}{2}\int_{S:\partial S=C} dS^{\mu\nu}G_{\mu\nu}\right\}, \\ G_{\mu\nu}(x) &= \partial_\mu[\boldsymbol{n}(x)\cdot\mathscr{A}_\nu(x)] - \partial_\nu[\boldsymbol{n}(x)\cdot\mathscr{A}_\mu(x)] - g^{-1}\boldsymbol{n}(x)\cdot[\partial_\mu\boldsymbol{n}(x)\times\partial_\nu\boldsymbol{n}(x) \\ n^A(x)\sigma^A &= U^\dagger(x)\sigma^3U(x), \quad \mathscr{A}_\mu(x) = \mathscr{A}_\mu^A(x)\sigma^A/2, \end{aligned} \tag{5}$$

and $d\mu_S(\boldsymbol{n})$ is the product of the invariant measure on SU(2)/U(1) over S:

$$d\mu_S(\boldsymbol{n}) := \prod_{x\in S} d\mu(\boldsymbol{n}(x)), \quad d\mu(\boldsymbol{n}(x)) = \frac{2J+1}{4\pi}\delta(\boldsymbol{n}(x)\cdot\boldsymbol{n}(x)-1)d^3\boldsymbol{n}(x). \tag{6}$$

The "Abelian" field strength $G_{\mu\nu}$ is SU(2) gauge invariant, since it is cast into the manifestly SU(2) invariant form:

$$\begin{aligned} G_{\mu\nu}(x) &= \boldsymbol{n}(x)\cdot\mathscr{F}_{\mu\nu}(x) - g^{-1}\boldsymbol{n}(x)\cdot(D_\mu\boldsymbol{n}(x)\times D_\nu\boldsymbol{n}(x)) \\ &= 2\mathrm{tr}\left\{\boldsymbol{n}(x)\mathscr{F}_{\mu\nu}(x) + ig^{-1}\boldsymbol{n}(x)[D_\mu\boldsymbol{n}(x), D_\nu\boldsymbol{n}(x)]\right\}, \end{aligned} \tag{7}$$

where the gauge transformation is given by $\boldsymbol{n}(x) \to U^\dagger(x)\boldsymbol{n}(x)U(x)$, $D_\mu\boldsymbol{n}(x) \to U^\dagger(x)D_\mu\boldsymbol{n}(x)U(x)$, $\mathscr{F}_{\mu\nu}(x) \to U^\dagger(x)\mathscr{F}_{\mu\nu}(x)U(x)$. Note that $G_{\mu\nu}$ has the same form as the 't Hooft–Polyakov tensor under the identification $n^A(x) \leftrightarrow \hat\phi^A(x) := \phi^A(x)/|\phi(x)|$. Therefore, the magnetic current k_μ obeying the topological conservation law $\partial^\mu k_\mu = 0$ is defined by

$$k^\mu(x) := \partial_\nu {}^*G^{\mu\nu}(x) = (1/2)\epsilon^{\mu\nu\rho\sigma}\partial_\nu G_{\rho\sigma}(x). \tag{8}$$

3. Reformulation of Yang-Mills theory based on the non-linear change of variables

Using the Diakonov-Petrov NAST, we arrive at the expression for the Wilson loop average $W(C) := \langle W_C[\mathscr{A}]\rangle_{YM}$:

$$W(C) = \tilde Z_{\rm YM}^{-1}\int \mathcal{D}\mu[\boldsymbol{n}]\int \mathcal{D}\mathscr{A}_\mu e^{iS_{\rm YM}}\tilde W_{\mathscr{A}}(C) = \frac{\int \mathcal{D}\mu[\boldsymbol{n}]\int \mathcal{D}\mathscr{A}_\mu e^{iS_{\rm YM}}\tilde W_{\mathscr{A}}(C)}{\int \mathcal{D}\mu[\boldsymbol{n}]\int \mathcal{D}\mathscr{A}_\mu e^{iS_{\rm YM}}}, \tag{9}$$

where we have defined the *reduced* Wilson loop operator

$$\tilde W_{\mathscr{A}}(C) = \exp\left\{ ig\frac{J}{2}\int_{S:\partial S=C} d^2S^{\mu\nu}G_{\mu\nu}\right\}, \tag{10}$$

and the new partition function $\tilde{Z}_{\rm YM} = \int \mathcal{D}\mu[\boldsymbol{n}] \int \mathcal{D}\mathscr{A}_\mu \exp(iS_{\rm YM}[\mathscr{A}])$. Here we have inserted the unity into the functional integration, $1 = \int \mathcal{D}\mu[\boldsymbol{n}] \equiv \int \mathcal{D}\boldsymbol{n}\delta(\boldsymbol{n}\cdot\boldsymbol{n}-1) := \prod_{x\in\mathbb{R}^D} \int [d\boldsymbol{n}(x)]\delta(\boldsymbol{n}(x)\cdot\boldsymbol{n}(x)-1)$.

Suppose a unit vector field $\boldsymbol{n}(x)$ is given as a functional of $\mathscr{A}_\mu(x)$

$$\boldsymbol{n}(x) = \boldsymbol{n}_{\mathscr{A}}(x). \tag{11}$$

Then the "decomposition" (once known as Cho-Faddeev-Niemi (CFN) decomposition.[14–17])

$$\mathscr{A}_\mu(x) := \overbrace{\underbrace{c_\mu(x)\boldsymbol{n}(x)}_{\mathbb{C}_\mu:\text{restricted potential}} + \underbrace{g^{-1}\partial_\mu \boldsymbol{n}(x)\times\boldsymbol{n}(x)}_{\mathbb{B}_\mu:\text{magnetic potential}}}^{\mathbb{V}_\mu} + \underbrace{\mathbb{X}_\mu(x)}_{\text{covariant potential}}, \tag{12}$$

with

$$c_\mu(x) = \boldsymbol{n}(x)\cdot\mathscr{A}_\mu(x), \quad \mathbb{X}_\mu(x) = g^{-1}\boldsymbol{n}(x)\times D_\mu[\mathscr{A}]\boldsymbol{n}(x), \tag{13}$$

is regarded as a non-linear change of variables (NLCV):

$$(\mathscr{A}_\mu^A(x)\to) \quad \boldsymbol{n}^A(x), \mathscr{A}_\mu^A(x) \to \boldsymbol{n}^A(x), c_\mu(x), \mathbb{X}_\mu^A(x). \tag{14}$$

A remarkable property of NLCV is that the curvature tensor $\mathscr{F}_{\mu\nu}[\mathbb{V}]$ obtained from the connection $\mathbb{V}_\mu$ is parallel to $\boldsymbol{n}$ and its magnitude $G_{\mu\nu}$ coincides exactly with $G_{\mu\nu}$ appearing in NAST of the Wilson loop operator:

$$\begin{aligned} \mathscr{F}_{\mu\nu}[\mathbb{V}](x) :=& \partial_\mu \mathbb{V}_\nu(x) - \partial_\nu \mathbb{V}_\mu(x) + g\mathbb{V}_\mu(x)\times\mathbb{V}_\nu(x) = \boldsymbol{n}(x)G_{\mu\nu}(x), \\ \to G_{\mu\nu}(x) =& \boldsymbol{n}(x)\cdot\mathscr{F}_{\mu\nu}[\mathbb{V}](x) = \partial_\mu[\boldsymbol{n}(x)\cdot\mathscr{A}_\nu(x)] - \partial_\nu[\boldsymbol{n}(x)\cdot\mathscr{A}_\mu(x)] \\ & - g^{-1}\boldsymbol{n}(x)\cdot[\partial_\mu\boldsymbol{n}(x)\times\partial_\nu\boldsymbol{n}(x)]. \end{aligned} \tag{15}$$

The remaining issue is to answer how to define and obtain the color unit vector field $\boldsymbol{n}(x)$ from the original Yang-Mills theory written in terms of $\mathscr{A}_\mu(x)$ alone. A procedure has been given in our reformulation of Yang-Mills theory in the continuum spacetime.[18,19] By introducing $\boldsymbol{n}(x)$ field in addition to $\mathscr{A}_\mu(x)$, we have a gauge theory with the enlarged gauge symmetry $\tilde{G}$, called the master Yang-Mills theory. We propose a constraint[18](called the new Maximal Abelian gauge, nMAG) by which $\tilde{G} := SU(2)_\omega \times [SU(2)/U(1)]_\theta$ in the master Yang-Mills theory is broken down to the diagonal subgroup: $G' = SU(2)$: Minimize the functional $\int d^Dx \frac{1}{2}g^2\mathbb{X}_\mu^2$ w.r.t. the enlarged gauge transformations:

$$0 = \delta_{\omega,\theta}\int d^Dx\frac{1}{2}g^2\mathbb{X}_\mu^2 = \delta_{\omega,\theta}\int d^Dx(D_\mu[\mathscr{A}]\boldsymbol{n})^2. \tag{16}$$

It has been shown that nMAG determines the color field $\boldsymbol{n}(x)$ as a functional of a given configuration of $\mathscr{A}_\mu(x)$. Therefore, if we impose the new MAG

(16) to the master-Yang–Mills theory, we have a gauge theory (called the Yang–Mills theory II) with the local gauge symmetry $G' := SU(2)^{\omega'}_{local}$ with $\boldsymbol{\omega}'(x) = (\boldsymbol{\omega}_{\parallel}(x), \boldsymbol{\omega}_{\perp}(x) = \boldsymbol{\theta}_{\perp}(x))$, which is a diagonal SU(2) part of the original $\tilde{G} := SU(2)^{\omega}_{local} \times [SU(2)/U(1)]^{\theta}_{local}$. The local gauge symmetry G' of the Yang–Mills theory II for new variables is

$$\delta_{\omega'}\boldsymbol{n} = g\boldsymbol{n} \times \boldsymbol{\omega}', \tag{17}$$

$$\delta_{\omega'}c_{\mu} = \boldsymbol{n} \cdot \partial_{\mu}\boldsymbol{\omega}', \tag{18}$$

$$\delta_{\omega'}\mathbb{X}_{\mu} = g\mathbb{X}_{\mu} \times \boldsymbol{\omega}', \tag{19}$$

$$\Longrightarrow \delta_{\omega'}\mathbb{V}_{\mu} = D_{\mu}[\mathbb{V}]\boldsymbol{\omega}', \quad \delta_{\omega'}\mathscr{A}_{\mu} = D_{\mu}[\mathscr{A}]\boldsymbol{\omega}'. \tag{20}$$

This should be compared with the conventional MAG. The old MAG leaves local $U(1)$ and global U(1) unbroken, but breaks global SU(2). The new MAG leaves local G'=SU(2) and global SU(2) unbroken (color rotation invariant). This is an advantage of this reformulation.

4. Gauge-invariant gluon mass and infrared Abelian dominance

Note that the mass dimension-2 operator $\mathbb{X}_{\mu}^2$ is invariant under the local gauge transformation II:

$$\delta'_{\omega}\mathbb{X}_{\mu}^2(x) = 0. \tag{21}$$

We claim the existence of gauge-invariant dimension-two condensate[20]

$$\langle 0|\mathbb{X}_{\mu}(x) \cdot \mathbb{X}_{\mu}(x)|0\rangle_{\rm YM} \neq 0. \tag{22}$$

How does such a gauge-invariant dimension-two condensates occur? It can be induced from the self-interactions among gluons:

$$-\frac{1}{4}(g\mathbb{X}_{\mu} \times \mathbb{X}_{\nu})^2, \tag{23}$$

since a rough observation yields

$$\begin{aligned} &-\frac{1}{4}(g\mathbb{X}_{\mu} \times \mathbb{X}_{\nu}) \cdot (g\mathbb{X}^{\mu} \times \mathbb{X}^{\nu}) \\ &\rightarrow \frac{1}{2}g^2\mathbb{X}_{\mu}^A \left[\langle -\mathbb{X}_{\rho}^2\rangle \delta^{AB} - \langle -\mathbb{X}_{\rho}^A\mathbb{X}_{\rho}^B\rangle\right] \mathbb{X}_{\mu}^B = \frac{1}{2}M_X^2\mathbb{X}_{\mu} \cdot \mathbb{X}_{\mu}, \quad M_X^2 = \frac{2}{3}g^2\langle -\mathbb{X}_{\rho}^2\rangle \end{aligned} \tag{24}$$

This idea can be made precise. See Ref. 20 for details. Consequently, the gauge-invariant mass term for "off-diagonal" $\mathbb{X}_{\mu}$ gluons is generated. This mass term is invariant under the local SU(2) gauge transformation II. Then the $\mathbb{X}_{\mu}$ gluons to be decoupled in the low-energy (or long-distance) region.

This leads to the infrared Abelian dominance: The Wilson loop average is entirely estimated by the field $\mathbb{V}_\mu$ alone.

5. Lattice formulation and numerical simulations

Our reformulation of Yang-Mills theory in the continuum spacetime[18,19] has been implemented on a lattice to perform numerical simulations as follows.
• Non-compact formulation[21] has lead to
  · generation of color field configuration $\boldsymbol{n}(x)$
  · restoration of color symmetry (global gauge symmetry)
  · gauge-invariant definition of magnetic monopole charge
• Compact formulation I[22] has succeeded to show
  · magnetic charge quantization subject to the Dirac quantization condition $gg_m/(4\pi) \in \mathbb{Z}$
  · magnetic monopole dominance in the string tension
• Compact formulation II[23] has confirmed
  · the non-vanishing gluon mass $M_X = 1.2\text{GeV}$ (cf. MAG result[9])

6. Conclusion and discussion

Using a nonlinear change of variables, we have succeeded to separate the original SU(2) gluon field variables $\mathscr{A}_\mu$ into "Abelian" part $\mathbb{V}_\mu$ and the "remaining" part $\mathbb{X}_\mu$ *without breaking color symmetry*:

$$\mathscr{A}_\mu = \mathbb{V}_\mu + \mathbb{X}_\mu,$$

in the following sense.
• $\mathbb{V}_\mu$ are responsible for quark confinement: the DP version of the non-Abelian Stokes theorem tells us that the non-Abelian Wilson loop operator is entirely rewritten in terms of the SU(2) invariant *Abelian* field strength $G_{\mu\nu}$ defined from the variable $\mathbb{V}_\mu$. This specification leads to a definition of gauge-invariant magnetic monopoles with the magnetic charge subject to Dirac quantization condition (which is confirmed by analytical and numerical methods) and magnetic monopole dominance in the string tension (confirmed by numerical method).
• $\mathbb{X}_\mu$ could decouple in the low-energy regime: This is because the $\mathbb{X}_\mu$ gluon acquires the gauge-invariant mass dynamically through the non-vanishing vacuum condensation of mass dimension–two $\langle \mathbb{X}_\mu^2 \rangle \neq 0$. This leads to the infrared "Abelian" dominance. We have examined a possibility of dimension–two condensate $\langle \mathbb{X}_\mu^2 \rangle \neq 0$ by analytical[20] and numerical[24] methods.

In Ref. 25, we have discussed how the Faddeev model can be regarded as a low-energy effective theory of Yang-Mills theory to see the mass gap.

References

1. Y. Nambu, Phys. Rev. D **10**, 4262–4268 (1974).
   G. 't Hooft, in: High Energy Physics, edited by A. Zichichi (Editorice Compositori, Bologna, 1975).
   S. Mandelstam, Phys. Report **23**, 245–249 (1976).
2. G. 't Hooft, Nucl.Phys. B **190** [FS3], 455–478 (1981).
3. Z.F. Ezawa and A. Iwazaki, Phys. Rev. D **25**, 2681–2689 (1982).
4. A. Kronfeld, M. Laursen, G. Schierholz and U.-J. Wiese, Phys.Lett. B **198**, 516–520 (1987).
5. T. Suzuki and I. Yotsuyanagi, Phys. Rev. D **42**, 4257–4260 (1990).
6. J.D. Stack, S.D. Neiman and R. Wensley, [hep-lat/9404014], Phys. Rev. D**50**, 3399–3405 (1994). H. Shiba and T. Suzuki, [hep-lat/9404015], Phys.Lett.B**333**, 461–466 (1994).
7. K.-I. Kondo, [hep-th/9709109], Phys. Rev. D **57**, 7467–7487 (1998).
8. K.-I. Kondo, [hep-th/9801024], Phys. Rev. D **58**, 105019 (1998).
9. K. Amemiya and H. Suganuma, [hep-lat/9811035], Phys. Rev. D**60**, 114509 (1999).
   V.G. Bornyakov, M.N. Chernodub, F.V. Gubarev, S.M. Morozov and M.I. Polikarpov, [hep-lat/0302002], Phys. Lett. B**559**, 214–222 (2003).
10. D.I. Diakonov and V.Yu. Petrov, Phys. Lett. B **224**, 131–135 (1989).
11. K.-I. Kondo, [hep-th/9805153], Phys. Rev. D **58**, 105016 (1998).
12. K.-I. Kondo and Y. Taira, [hep-th/9906129], Mod. Phys. Lett. A **15**, 367–377 (2000); [hep-th/9911242], Prog. Theor. Phys. **104**, 1189–1265 (2000).
13. M. Hirayama and M. Ueno, [hep-th/9907063], Prog. Theor. Phys. **103**, 151–159 (2000).
14. Y.S. Duan and M.L. Ge, Sinica Sci., **11**, 1072 (1979).
15. Y.M. Cho, Phys. Rev. D **21**, 1080–1088 (1980). D **23**, 2415–2426 (1981).
16. L. Faddeev and A.J. Niemi, [hep-th/9807069], Phys. Rev. Lett. **82**, 1624–1627 (1999). [hep-th/0608111], Nucl. Phys. B (2007), to appear.
17. S.V. Shabanov, [hep-th/9903223], Phys. Lett. B **458**, 322–330 (1999).
    S.V. Shabanov, [hep-th/9907182], Phys. Lett. B **463**, 263–272 (1999).
18. K.-I. Kondo, T. Murakami and T. Shinohara, [hep-th/0504107], Prog. Theor. Phys. **115**, 201–216 (2006).
19. K.-I. Kondo, T. Murakami and T. Shinohara, [hep-th/0504198], Eur. Phys. J. C **42**, 475–481 (2005).
20. K.-I. Kondo, [hep-th/0609166], Phys. Rev. D**74**, 125003 (2006).
21. S. Kato, K.-I. Kondo, T. Murakami, A. Shibata, T. Shinohara and S. Ito, [hep-lat/0509069], Phys. Lett. B **632**, 326–332 (2006).
22. S. Ito, S. Kato, K.-I. Kondo, T. Murakami, A. Shibata and T. Shinohara, [hep-lat/0604016], Phys. Lett. B **645**, 67–74 (2007).
23. A. Shibata, S. Ito, S. Kato, K.-I. Kondo, T. Murakami, and T. Shinohara, hep-lat/0610023, A talk in Lattice2006.
24. S. Kato, K.-I. Kondo, T. Murakami, A. Shibata and T. Shinohara, hep-ph/0504054.
25. K.-I. Kondo, A. Ono, A. Shibata, T. Shinohara and T. Murakami, [hep-th/0604006], J. Phys. A: Math. Gen. **39**, 13767–13782 (2006).

TOY MODEL FOR MIXING OF TWO CHIRAL NONETS

A.H. FARIBORZ[A], R. JORA[B] and J. SCHECHTER[C*]

(A) Department of Mathematics/Science,
SUNY Institute of Technology,
Utica, NY 13504-3050.
E-mail: fariboa@sunyit.edu

(B) Physics Department,
Syracuse University,
Syracuse, NY 13244-1130, USA.
E-mail: cjora@phy.syr.edu

(C) Physics Department,
Syracuse University,
Syracuse, NY 13244-1130, USA.
E-mail: schechte@phy.syr.edu

We propose a systematic procedure to study a generalized linear sigma model which can give a physical picture of possible mixing between $q\bar{q}$ and $qq\bar{q}\bar{q}$ low lying spin zero states. In the limit of zero quark masses, we derive the model independent results for the properties of the Nambu Goldstone pseudoscalar particles. For getting information on the scalars it is necessary to make a specific choice of terms. We impose two plausible physical criteria - the modeling of the axial anomaly and the suppression of effective vertices representing too many fermion lines - for limiting the large number of terms which are allowed on general grounds. We calculate the tree-level spectrum based on the leading terms in our approach and find that it prominently exhibits a very low mass isosinglet scalar state.

The fields of our "toy" model[1] consist of a 3×3 matrix chiral nonet field M, which represents $q\bar{q}$ type states as well as a 3×3 matrix chiral nonet field M', which represents four quark type states. They have the decompositions into scalar and pseudoscalar pieces: $M = S + i\phi$, $M' = S' + i\phi'$ and behave under $\mathrm{SU(3)_L} \times \mathrm{SU(3)_R}$ transformations as $M \rightarrow U_L M U_R^\dagger$ and $M' \rightarrow U_L M' U_R^\dagger$. However, the $\mathrm{U(1)_A}$ transformation which acts at the

---

*Speaker

quark level as $q_{aL} \rightarrow e^{i\nu} q_{aL}$, $q_{aR} \rightarrow e^{-i\nu} q_{aR}$ distinguishes the two fields according to

$$M \rightarrow e^{2i\nu} M, \qquad M' \rightarrow e^{-4i\nu} M'. \tag{1}$$

Note that our treatment is based only on the symmetry structure and hence applies when M' is any linear combination of qq-$\bar{q}\bar{q}$ and $q\bar{q}$-$q\bar{q}$ type fields. We will be interested in the situation where non-zero vacuum values of S and S' may exist: $\langle S_a^b \rangle = \alpha \delta_a^b$, $\langle S_a'^b \rangle = \beta \delta_a^b$, corresponding to an assumed SU(3)$_\mathrm{V}$ invariant vacuum. The Lagrangian density which defines our model is

$$\begin{aligned} \mathcal{L} = &- \frac{1}{2} \mathrm{Tr} \left(\partial_\mu M \partial_\mu M^\dagger \right) - \frac{1}{2} \mathrm{Tr} \left(\partial_\mu M' \partial_\mu M'^\dagger \right) \\ &- V_0 \left(M, M' \right) - V_{SB}, \end{aligned} \tag{2}$$

where $V_0(M, M')$ stands for a general function made from SU(3)$_\mathrm{L}$ $\times$ SU(3)$_\mathrm{R}$ (but not necessarily U(1)$_\mathrm{A}$) invariants formed out of M and M'. The quantity V_{SB} which represents the effective chiral symmetry breaking light quark mass terms will be set to zero here. It is accepted that this is a reasonable qualitative approximation since the largest parts of the masses of all particles made of light quarks, other than the lightest 0^- octet are expected to arise from spontaneous breakdown of chiral symmetry. The U(1)$_\mathrm{A}$ transformation, which plays a special role in this model, suggests another useful simplification. In QCD there is a special instanton induced term- the "'t Hooft determinant" - which breaks the U(1)$_\mathrm{A}$ symmetry and can be modeled as $\det(M) + \det(M^\dagger)$. It thus may be natural to require all the terms to satisfy U(1)$_\mathrm{A}$ invariance except for a particular subset which could model the U(1)$_\mathrm{A}$ anomaly. If one demands that, as in QCD, the U(1)$_\mathrm{A}$ variation of the effective Lagrangian be proportional to the gluonic axial anomaly then a similar effective term $\mathcal{L}_\eta = -c_3[\ln(\det(M)) - \ln(\det(M^\dagger))]^2$, where c_3 is a numerical parameter, seems appropriate.

A lot of information concerning especially the pseudoscalar particles in the model may be obtained in general without even specifying the terms in the potential. This may be achieved by studying "generating" equations which arise from the demand that the infinitesimal symmetry transformations in the model hold. For the nine axial transformations one finds:

$$\begin{aligned} &[\phi, \frac{\partial V_0}{\partial S}]_+ - [S, \frac{\partial V_0}{\partial \phi}]_+ + (\phi, S) \rightarrow (\phi', S') = \\ &1[2\mathrm{Tr}(\phi' \frac{\partial V_0}{\partial S'} - S' \frac{\partial V_0}{\partial \phi'}) - 8c_3 i \ln(\frac{\det M}{\det M^\dagger})]. \end{aligned} \tag{3}$$

To get constraints on the particle masses we will differentiate these equations once with respect to each of the two matrix fields ϕ and ϕ' and evaluate the equations in the ground state, taking into account the "minimum" conditions, $\langle \frac{\partial V_0}{\partial S} \rangle = 0$ and $\langle \frac{\partial V_0}{\partial S'} \rangle = 0$. Further differentiations with respect to all four matrix fields will similarly yield "model independent" information on 3 and 4 point vertices. We also require the Noether currents, $(J_\mu^{axial})_a^b = \alpha \partial_\mu \phi_a^b + \beta \partial_\mu {\phi'}_a^b + \cdots$, where the dots stand for terms bilinear in the fields. Using Eq.(3) the squared mass matrix which mixes the degenerate two quark and degenerate four quark pseudoscalar octets is:

$$(M_\pi^2) = y_\pi \begin{bmatrix} \beta^2/\alpha^2 & -\beta/\alpha \\ -\beta/\alpha & 1 \end{bmatrix},$$

where $y_\pi = \langle \frac{\partial^2 V_0}{\partial {\phi'}_1^2 \partial {\phi'}_2^1} \rangle$. Clearly, $\det(M_\pi^2) = 0$ and the zero mass pion octet is a mixture of two quark and four quark fields. The transformation between the diagonal fields π^+ and π'^+ and the original pion fields is defined as:

$$\begin{bmatrix} \pi^+ \\ \pi'^+ \end{bmatrix} = R_\pi^{-1} \begin{bmatrix} \phi_1^2 \\ {\phi'}_1^2 \end{bmatrix} = \begin{bmatrix} \cos\theta_\pi & -\sin\theta_\pi \\ \sin\theta_\pi & \cos\theta_\pi \end{bmatrix} \begin{bmatrix} \phi_1^2 \\ {\phi'}_1^2 \end{bmatrix},$$

which also defines the transformation matrix, R_π. The explicit diagonalization yields:

$$\tan\theta_\pi = -\frac{\beta}{\alpha}, \tag{4}$$

which may be interpreted as the ratio of the four quark condensate to the two quark condensate in the underlying QCD. We see that the mixing between the two quark pion and the four quark pion would vanish if the four quark condensate were to vanish in this model. Rewriting the Noether current as $(J_\mu^{axial})_1^2 = F_\pi \partial_\mu \pi^+ + F_{\pi'} \partial_\mu \pi'^+ + \cdots$ shows that

$$F_\pi = 2\sqrt{\alpha^2 + \beta^2}, \qquad F_{\pi'} = 0. \tag{5}$$

Note that the physical higher mass pion state decouples from the axial current. Altogether the (eight) zero mass pseudoscalars are characterized by the three parameters α, β and y_π. On the other hand, there are only two experimental inputs: $F_\pi = 131$ MeV and the mass of $\pi(1300)$, the presumed higher mass pion candidate. Thus, the interesting question of the four quark content of the pion has an inevitably model dependent answer.

Next let us consider the "model independent" information available for the two pseudoscalar SU(3) singlet states. This sector is related to the QCD axial anomaly. In the single M model, the anomaly can be modeled by the term $\mathcal{L}_\eta$ mentioned above. In the M-M' model under consideration this form is no longer unique and it is natural to consider a generalization in which $\ln(\det(M)/\det(M^\dagger))$ is replaced by $\gamma_1[\ln(\det(M)/\det(M^\dagger))] + (1-\gamma_1)[\ln(\mathrm{Tr}(MM'^\dagger)/\mathrm{Tr}(M'M^\dagger))]$, where γ_1 is a dimensionless parameter. Then the squared mass matrix which mixes the two SU(3) pseudoscalar singlet states is obtained as:

$$(M_0^2) = \begin{bmatrix} -\frac{8c_3(2\gamma_1+1)^2}{3\alpha^2} + z_0^2 y_0 & -z_0 y_0 + \frac{8c_3(1-\gamma_1)(2\gamma_1+1)}{3\alpha\beta} \\ -z_0 y_0 + \frac{8c_3(1-\gamma_1)(2\gamma_1+1)}{3\alpha\beta} & y_0 - \frac{8c_3(1-\gamma_1)^2}{3\beta^2} \end{bmatrix}.$$

Here $z_0 = -2\beta/\alpha$ and $y_0 = \langle \frac{\partial^2 V}{\partial \phi_0' \partial {\phi'}_0} \rangle$. Note that when c_3 is set to zero, making the entire Lagrangian $\mathrm{U(1)_A}$ invariant, $\det(M_0^2) = 0$. Then one of the singlet pseudoscalars becomes, as well known, a Nambu Goldstone particle. This occurs in the large number of colors limit but we will not make that approximation here.

In order to get information about the scalar meson masses and mixings as well as to complete the description of the pseudoscalars it is necessary to make a specific choice of interaction terms. To proceed in a systematic way we define the following quantity for each term,

$$N = 2n + 4n', \tag{6}$$

where n and n' are respectively the number of M fields and the number of M' fields contained in that term. We shall restrict our choice to the lowest non-trivial value of N, which corresponds physically to the total number of quark and antiquark lines at each vertex. In addition to the two special terms which saturate the $\mathrm{U(1)_A}$ anomaly already mentioned, this gives the leading (N=8) potential

$$\begin{aligned} V_0 = &- c_2 \,\mathrm{Tr}(MM^\dagger) + c_4^a \,\mathrm{Tr}(MM^\dagger MM^\dagger) \\ &+ d_2 \,\mathrm{Tr}(M'M'^\dagger) + e_3^a(\epsilon_{abc}\epsilon^{def} M_d^a M_e^b M_f'^c + h.c.) \\ &+ \cdots , \end{aligned} \tag{7}$$

where the dots stand for the $\mathrm{U(1)_A}$ non-invariant terms. For simplicity, we have neglected the N=8 term, $c_4^b[\mathrm{Tr}(MM^\dagger)]^2$ which is suppressed, in the single M model by the quark line rule. It may be noted that the quantities $\det(M)$ and $\mathrm{Tr}(MM'^\dagger)$ which enter into those two terms which saturate the $\mathrm{U(1)_A}$ anomaly have N=6. In this counting scheme, $\mathrm{U(1)_A}$ invariant terms with N=12 (and higher) might be successively added to improve the approximation. The minimum equations for this potential are:

$$\left\langle \frac{\partial V_0}{\partial S_a^a} \right\rangle = 2\,\alpha\,\left[-c_2 + 2\,c_4^a\,\alpha^2 + 4\,e_3^a\,\beta\right] = 0, \tag{8}$$

$$\left\langle \frac{\partial V_0}{\partial S'^a_a} \right\rangle = 2\left[d_2\,\beta + 2\,e_3^a\,\alpha^2\right] = 0. \tag{9}$$

The U(1)$_A$ violating c_3 terms do not contribute to these equations. Notice that α is an overall factor in Eq. (8) so that, in addition to the physical spontaneous breakdown solution where $\alpha \neq 0$ there is a solution with $\alpha = 0$. On the other hand, β is not an overall factor of Eq. (9) and it is easy to see that β, which measures the "4 quark condensate", is necessarily non-zero in the physical situation where α is non-zero. From the specific form in Eq.(7) we find the mixing squared mass matrices for the degenerate octet scalars as well as the SU(3) singlet scalars:

$$(X_a^2) = \begin{bmatrix} 2\left[-c_2 + 6\,c_4^a\,\alpha^2 - 2\,e_3^a\,\beta\right] & -4\alpha e_3^a \\ -4\alpha e_3^a & 2d_2 \end{bmatrix} \tag{10}$$

$$(X_0^2) = \begin{bmatrix} 2\left[-c_2 + 6\,c_4^a\,\alpha^2 + 4\,e_3^a\,\beta\right] & 8\alpha e_3^a \\ 8\alpha e_3^a & 2d_2 \end{bmatrix} \tag{11}$$

Now let us consider the comparison of the model with experiment. To start with there are 8 parameters (α, β, c_2, d_2, c_4^a, e_3^a, c_3 and γ_1). The last two parameters appear only in the mass matrices of the pseudoscalar SU(3) singlets and are conveniently discussed separately. The other six are effectively reduced to four by using the two minimum equations (8) and (9). As the corresponding four experimental inputs we take the non-strange quantities:

$$\begin{aligned} m(0^+\text{octet}) &= m[a_0(980)] = 984.7 \pm 1.2\,\text{MeV} \\ m(0^+\text{octet}') &= m[a_0(1450)] = 1474 \pm 19\,\text{MeV} \\ m(0^-\text{octet}') &= m[\pi(1300)] = 1300 \pm 100\,\text{MeV} \\ F_\pi &= 131\,\text{MeV} \end{aligned} \tag{12}$$

Evidently, the largest experimental uncertainty appears in the mass of $\pi(1300)$; we shall consider the other masses as fixed at their central values and vary this mass in the indicated range. From studying the scalar SU(3) singlet states we find the consistency condition for positivity of the eigenvalues of their squared mass matrix, Eq.(11):

$$m[\pi(1300)] < 1302\,\text{MeV}. \tag{13}$$

The model predicts, as $m[\pi(1300)]$ varies from 1200 to 1300 MeV,

$$\begin{aligned} m(0^+\text{singlet}) &= 510 \rightarrow 28\ \text{MeV}, \\ m(0^+\text{singlet}') &= 1506 \rightarrow 1555\ \text{MeV}. \end{aligned} \tag{14}$$

Clearly, the most dramatic feature is the very low mass of the lighter SU(3) singlet scalar meson. Of course, one expects the addition of quark mass type terms to modify the details somewhat.

To calculate the masses of the SU(3) singlet pseudoscalars we must diagonalize Eq.(6) with the specific choices of parameters $y_0 = 2d_2$ and $z_0 = 4e_3^a\alpha/d_2$ corresponding to the potential of Eq.(7). This enables us to fit in principle, for any choice of $m[\pi(1300)]$, the two parameters c_3 and γ_1 in terms of the experimental masses of $\eta(958)$ and one of the candidates $\eta(1295)$, $\eta(1405)$, $\eta(1475)$ and $\eta(1760)$. However, it turns out that the positivity of the eigenvalues of the matrix (M_0^2) imposes additional constraints on the choice of $m[\pi(1300)]$ in Eq.(13). Furthermore the first two candidates for the heavier η are also ruled out on grounds of this positivity. For $\eta(1475)$ the allowed range of $m[\pi(1300)]$ is restricted to 1200 to 1230 MeV. On the other hand, there is no additional restriction if $\eta(1760)$ is chosen. If the choice of $\eta(1475)$ is made, the predicted range of $m(0^+\text{singlet})$ is narrowed from that given in Eq.(14) to 510 $\rightarrow$ 410 MeV.

It is very interesting to see what the model has to say about the four quark percentages of the particles it describes. These percentages are displayed in Fig.1 as functions of the precise value of the input parameter $m[\pi(1300)]$. The pion four quark content (equal to 100 $\sin^2\theta_\pi$) is seen to be about 17 percent. Of course the heavier pion would have about an 83 percent four quark content. On the other hand, the octet scalar states present a reversed picture: the $a_0(980)$ has a large four quark content while the $a_0(1450)$ has a smaller four quark content. The very light and the rather heavy 0^+ singlets are about maximally mixed, having roughly equal contributions from the 4 quark and 2 quark components.

The perhaps more plausible scenario in the case of the 0^- singlets takes $\eta(1475)$ as the heavy 0^- singlet state. Fig. 1 shows that there are two solutions for each value of $m[\pi(1300)]$; the dotted line gives a mainly $q\bar{q}$ content while the solid line gives a mainly four quark content. Note that this scenario does not allow $m[\pi(1300)]$ to be higher than about 1230 MeV. The choice (not shown) of $\eta(1760)$ as the partner of $\eta(958)$ also leads to two solutions with small and large two quark content.

There are two reasons for next briefly discussing the pi-pi scattering in this model. First, since the iso-singlet scalar resonances above are being considered at tree level, one expects, as can be seen in the single M model, both with two and three flavors, that unitarity corrections for the scattering amplitudes will alter their masses and widths. Second, since the pion looks unconventional in this model (having a non-negligible four quark compo-

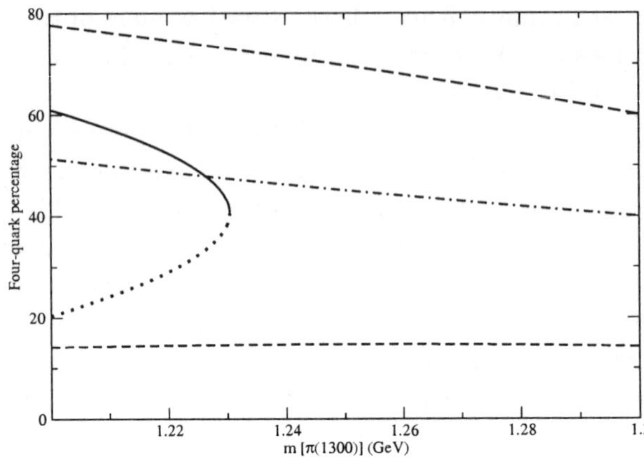

Fig. 1. Four quark percentages of the pion (dashed line), the $a_0(980)$ (top long-dashed line), the very light 0^+singlet (dotted-dashed line) and the $\eta(958)$ in the scenario where the higher state is identified as the $\eta(1475)$ (curve containing both solid and dotted pieces) as functions of the undetermined input parameter, $m[\pi(1300)]$. Note that there are two solutions for the $\eta(958)$: the dotted curve choice gives it a predominant two quark structure and the solid curve choice, a larger four quark content.

nent) one might worry that the fairly precise "current algebra" formula for the near to threshold scattering amplitude might acquire unacceptably large corrections. We have verified in detail[1] that the current algebra theorem will "tolerate" any amount of four quark component in the massless pion. The present model clearly shows that while the (lighter) pion is mainly two quark, the lighter scalars have very large four quark components. This is perhaps the opposite of what one might initially think and is related to the characteristic mixing pattern emerging in a transparent form here.

Acknowledgments

We would like to thank the organizers of SCGT06 for the opportunity to present our work at this very stimulating workshop and for the excellent job they have done.

References

1. For acknowledgments, references and further details please see arXiv:hep-ph/0612200 and the Syracuse report SU-4252-845 (in preparation) by the present authors.

DROPPING ρ AND A_1 MESON MASSES AT THE CHIRAL PHASE TRANSITION IN THE GENERALIZED HIDDEN LOCAL SYMMETRY*

Masayasu Harada

Department of Physics, Nagoya University, Nagoya, 464-8602, Japan
E-mail: harada@hken.phys.nagoya-u.ac.jp

Chihiro Sasaki

Gesellschaft für Schwerionenforschung (GSI), 64291 Darmstadt, Germany
E-mail: c.sasaki@gsi.de

In this contribution to the proceedings, we present our recent analysis on the chiral symmetry restoration in the generalized hidden local symmetry (GHLS) which incorporates the rho meson, A_1 meson and the pion consistently with the chiral symmetry of QCD. It is shown that a set of parameter relations, which ensures the first and second Weinberg sum rules, is invariant under the renormalization group evolution. Then, it is found that the Weinberg sum rules together with the matching of the vector and axial-vector current correlators to the OPE inevitably leads to *the dropping masses of both rho and A_1 mesons* at the symmetry restoration point, and that the mass ratio flows into one of three fixed points.

1. Introduction

Changes of the hadron masses are indications of the chiral symmetry restoration occurring in hot and/or dense QCD.[2] Dropping masses of hadrons following the Brown-Rho (BR) scaling[3] can be one of the most prominent candidates of the strong signal of the melting quark condensate $\langle \bar{q}q \rangle$ which is the order parameter of the spontaneous chiral symmetry breaking.

The vector manifestation (VM)[4] is the Wigner realization in which the ρ meson becomes massless degenerate with the pion at the chiral phase transition point. The VM is formulated[5–7] in the effective field theory (EFT)

*Talk presented by M. Harada at 2006 International Workshop on "Origin of Mass and Strong Coupling Gauge Theories" (SCGT06) based on the work done in Ref. 1.

based on the hidden local symmetry (HLS).[8,9] The VM gives a theoretical description of the dropping ρ mass, which is protected by the existence of the fixed point (VM fixed point).

The dropping mass is supported by the enhancement of dielectron mass spectra below the ρ/ω resonance observed at CERN SPS,[10,11] the mass shift of the ω meson in nuclei measured by the KEK-PS E325 Experiment,[12] the CBELSA/TAPS Collaboration[13] and also that of the ρ meson observed in the STAR experiment.[14] Recently NA60 Collaboration has provided data for the dimuon spectrum[15] and it seems difficult to explain the data by a naive dropping ρ.[16] However, there are still several ambiguities which were not considered.[17–19] Especially, the strong violation of the vector dominance (VD), which is one of the significant predictions of the VM,[20] plays an important role[17,21] to explain the data.

In the VM, it was assumed that the axial-vector and scalar mesons are decoupled from the theory near the phase transition point. However, the masses of these mesons may decrease following the BR scaling.[22] There were several analyses with models including axial-vector mesons such as in Ref. 23. These analyses are not based on the fixed point structure and found no significant reduction of the masses of axial-vector meson. Then, it is desirable to construct an EFT which includes the axial-vector meson as a dynamical degree of freedom, and to study whether a fixed point structure exists and it can realize the light axial-vector meson.

In Ref. 1, we picked up the model based on the generalized hidden local symmetry (GHLS),[9,24–26] and developed the chiral perturbation theory (ChPT) with GHLS. We showed that a set of the parameter relations, which satisfies the pole saturated forms of the first and second Weinberg sum rules, is stable against the renormalization group evolution. Then, we found that the Weinberg sum rules together with the matching to the operator product expansion necessarily leads to the dropping masses of both vector and axial-vector mesons. Interestingly, the ratio of masses of vector and axial-vector mesons as well as the mixing between the pseudoscalar and axial-vector mesons flows into one of three fixed points: They exhibit the VM–like, Ginzburg-Landau–like and Hybrid–like patterns of the chiral symmetry restoration.

In the following, we briefly summarize the work done in Ref. 1.

2. Generalized Hidden Local Symmetry

The Lagrangian of the generalized hidden local symmetry (GHLS)[9,24,25] is based on the $G_{\rm global} \times G_{\rm local}$ symmetry, where $G_{\rm global} = [SU(N_f)_L \times$

$SU(N_f)_R]_{\rm global}$ is the chiral symmetry and $G_{\rm local} = [SU(N_f)_L \times SU(N_f)_R]_{\rm local}$ is the GHLS. The whole symmetry $G_{\rm global} \times G_{\rm local}$ is spontaneously broken to a diagonal $SU(N_f)_V$. The basic quantities are the GHLS gauge bosons L_μ and R_μ, which are identified with the vector and axial-vector mesons as $V_\mu = (R_\mu + L_\mu)/2$ and $A_\mu = (R_\mu - L_\mu)/2$, and three matrix valued variables ξ_L, ξ_R and ξ_M, which are introduced as $U = \xi_L^\dagger \xi_M \xi_R$. Here $N_f \times N_f$ special-unitary matrix U is a basic ingredient of the chiral perturbation theory (ChPT).[27]

The fundamental objects are the Maurer-Cartan 1-forms defined by

$$\hat{\alpha}^\mu_{L,R} = D^\mu \xi_{L,R} \cdot \xi^\dagger_{L,R}/i\,, \quad \hat{\alpha}^\mu_M = D^\mu \xi_M \cdot \xi^\dagger_M/(2i)\,, \tag{1}$$

where the covariant derivatives of $\xi_{L,R,M}$ are given by

$$\begin{aligned} D_\mu \xi_L &= \partial_\mu \xi_L - i L_\mu \xi_L + i \xi_L \mathcal{L}_\mu\,, \quad D_\mu \xi_R = \partial_\mu \xi_R - i R_\mu \xi_R + i \xi_R \mathcal{R}_\mu\,, \\ D_\mu \xi_M &= \partial_\mu \xi_M - i L_\mu \xi_M + i \xi_M R_\mu\,, \end{aligned} \tag{2}$$

with $\mathcal{L}_\mu$ and $\mathcal{R}_\mu$ being the external gauge fields introduced by gauging $G_{\rm global}$. There are four independent terms with the lowest derivatives:

$$\begin{aligned} \mathcal{L}_V &= F^2 {\rm tr}\big[\hat{\alpha}_{\parallel\mu}\hat{\alpha}^\mu_\parallel\big]\,, \quad \mathcal{L}_A = F^2 {\rm tr}\big[\hat{\alpha}_{\perp\mu}\hat{\alpha}^\mu_\perp\big]\,, \quad \mathcal{L}_M = F^2 {\rm tr}\big[\hat{\alpha}_{M\mu}\hat{\alpha}^\mu_M\big]\,, \\ \mathcal{L}_\pi &= F^2 {\rm tr}\big[\big(\hat{\alpha}_{\perp\mu} + \hat{\alpha}_{M\mu}\big)\big(\hat{\alpha}^\mu_\perp + \hat{\alpha}^\mu_M\big)\big]\,, \end{aligned} \tag{3}$$

where F is the parameter carrying the mass dimension 1 and $\hat{\alpha}^\mu_{\parallel,\perp} = \big(\xi_M \hat{\alpha}^\mu_R \xi^\dagger_M \pm \hat{\alpha}^\mu_L\big)/2$. The kinetic term of the gauge bosons is given by

$$\mathcal{L}_{\rm kin}(L_\mu, R_\mu) = -\frac{1}{4g^2}{\rm tr}\big[L_{\mu\nu}L^{\mu\nu} + R_{\mu\nu}R^{\mu\nu}\big]\,, \tag{4}$$

where g is the GHLS gauge coupling and the field strengths are defined by $L_{\mu\nu} = \partial_\mu L_\nu - \partial_\nu L_\mu - i\big[L_\mu, L_\nu\big]$ and $R_{\mu\nu} = \partial_\mu R_\nu - \partial_\nu R_\mu - i\big[R_\mu, R_\nu\big]$.

By combining the above terms, the GHLS Lagrangian is given by

$$\mathcal{L} = a\mathcal{L}_V + b\mathcal{L}_A + c\mathcal{L}_M + d\mathcal{L}_\pi + \mathcal{L}_{\rm kin}(L_\mu, R_\mu)\,, \tag{5}$$

where a, b, c and d are dimensionless parameters. From this we find the following expressions for the masses of vector and axial-vector mesons M_{ρ,A_1}, the ρ-γ mixing strength g_ρ and strength of the coupling of the A_1 meson to the axial-vector current g_{A_1}:

$$M_\rho = g\sqrt{a}F\,, \quad M_{A_1} = g\sqrt{b+c}F\,, \quad g_\rho = gaF^2\,, \quad g_{A_1} = gbF^2\,. \tag{6}$$

3. Weinberg's Sum Rules

Let us start with the axial-vector and vector current correlators defined by

$$G_A(Q^2)(q^\mu q^\nu - q^2 g^{\mu\nu})\delta_{ab} = \int d^4x\, e^{iqx}\, \langle 0|\, T\, J^\mu_{5a}(x) J^\nu_{5b}(0)|0\rangle\ ,$$

$$G_V(Q^2)(q^\mu q^\nu - q^2 g^{\mu\nu})\delta_{ab} = \int d^4x\, e^{iqx}\, \langle 0|\, T\, J^\mu_{a}(x) J^\nu_{b}(0)|0\rangle\ , \qquad (7)$$

where $Q^2 = -q^2$ is the space-like momentum, J^μ_{5a} and J^μ_a are the axial-vector and vector currents and $(a,b) = 1, \ldots, N_f^2 - 1$. At the leading order of the GHLS the current correlators $G_{A,V}$ are expressed as

$$G_A(Q^2) = \frac{F_\pi^2}{Q^2} + \frac{F_{A_1}^2}{M_{A_1}^2 + Q^2}\,, \quad G_V(Q^2) = \frac{F_\rho^2}{M_\rho^2 + Q^2}\,, \qquad (8)$$

where the A_1 and ρ decay constants are defined by

$$F_{A_1}^2 = \left(\frac{g_{A_1}}{M_{A_1}}\right)^2 = \frac{b^2}{b+c} F^2\,, \quad F_\rho^2 = \left(\frac{g_\rho}{M_\rho}\right)^2 = aF^2\,. \qquad (9)$$

The same correlators can be evaluated by the OPE,[28] which shows that the difference between two correlators scales as $1/Q^6$:

$$G_A^{(\mathrm{OPE})}(Q^2) - G_V^{(\mathrm{OPE})}(Q^2) = \frac{32\pi}{9} \frac{\alpha_s \langle \bar{q}q\rangle^2}{Q^6}\,. \qquad (10)$$

We require that the high energy behavior of the difference between two correlators in the GHLS agrees with that in the OPE: $G_A(Q^2) - G_V(Q^2)$ in the GHLS scales as $1/Q^6$. This requirement can be satisfied only if the following relations are satisfied:

$$F_\pi^2 + F_{A_1}^2 = F_\rho^2\,, \quad F_{A_1}^2 M_{A_1}^2 = F_\rho^2 M_\rho^2\,, \qquad (11)$$

which are nothing but the pole saturated forms of the Weinberg first and second sum rules.[29] In terms of the parameters of the GHLS Lagrangian, the above relations can be satisfied if we take

$$a = b\,, \quad d = 0\,. \qquad (12)$$

In Ref. 1, we calculated the RGEs for the parameters a, b, c, d and the gauge coupling g. It was shown that the parameter relations $a = b$ and $d = 0$ are stable against the renormalization group evolution, i.e., *the non-renormalization of the Weinberg sum rules expressed in terms of the leading order parameters in the GHLS.*

4. Chiral Symmetry Restoration

The maching condition together with the Weinberg sum rules (11) in the high-energy region is obtained as

$$F_{A_1}^2 M_{A_1}^4 - F_\rho^2 M_\rho^4 = F_\rho^2 M_\rho^2 \left(M_{A_1}^2 - M_\rho^2\right) = \frac{32\pi\alpha_s}{9} \langle \bar{q}q \rangle^2 , \tag{13}$$

which is a measure of the spontaneous chiral symmetry breaking. When the chiral restoration point is approached, the quark condensate approaches zero: $\langle \bar{q}q \rangle \rightarrow 0$. Then, the parameters of the GHLS Lagrangian scales as $a^2 \cdot c \cdot g^4 \propto \langle \bar{q}q \rangle^2 \rightarrow 0$ near the chiral symmetry restoration point. Since $a = 0$ and $c = 0$ are unstable against the RGEs and $g = 0$ is a fixed point, the symmetry restoration in the GHLS can be realized only if

$$g \rightarrow g^* = 0 . \tag{14}$$

This implies the massless ρ and A_1 mesons, since both masses are proportional to the gauge coupling g. Thus we conclude that, *when we require the first and second Weinberg sum rules to be satisfied, the chiral symmetry restoration in the GHLS required through the matching to QCD can be realized with masses of ρ and A_1 mesons vanishing at the restoration point*:

$$M_\rho \rightarrow 0 , \quad M_{A_1} \rightarrow 0 . \tag{15}$$

To study the phase structure of the GHLS through the RGEs for a, c and g, it is convenient to introduce the following dimensionless parameters associated with a, c and g:

$$W(\mu) = \frac{N_f}{2(4\pi)^2} \frac{a+c}{ac} \frac{\mu^2}{F^2} = \frac{N_f}{2(4\pi)^2} \frac{\mu^2}{F_\pi^2(\mu)} , \quad \zeta(\mu) = \frac{a}{a+c} = \frac{M_\rho^2}{M_{A_1}^2} . \tag{16}$$

The phase of the GHLS is determined by the on-shell pion decay constant $F_\pi(\mu = 0)$, or equivalently W defined in Eq. (16), as

$W(\mu = 0) = 0$: broken phase , $\qquad W(\mu = 0) \neq 0$: symmetric phase .

The flow diagram shown in Fig. 1 has three fixed points:

$$\begin{aligned} &\text{GL-type}: \quad (\zeta^*, W^*) = (1, 1/2) , \\ &\text{VM-type}: \quad (\zeta^*, W^*) = (0, 1) , \\ &\text{Hybrid-type}: \quad (\zeta^*, W^*) = (1/3, 3/2) . \end{aligned} \tag{17}$$

From this we can distinguish three patterns of the chiral symmetry restoration characterized by three fixed points by the values of the ratio of ρ and A_1 meson masses expressed by ζ as in Eq. (16) as follows: At the fixed point "GL-type", ζ goes to 1, which implies that the ρ meson mass degenerates

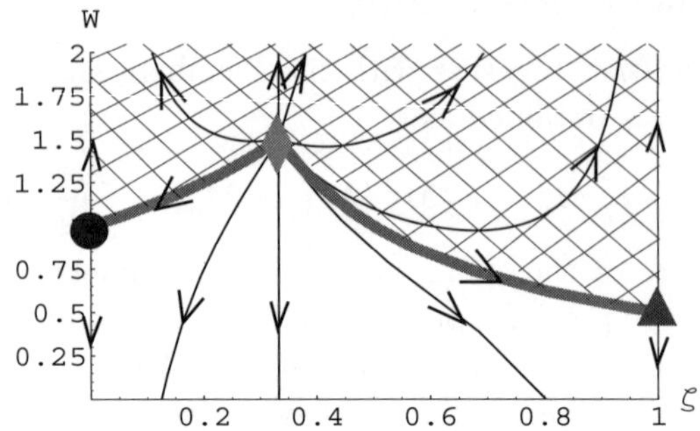

Fig. 1. Phase diagram on ζ-W plane. Arrows on the flows are written from the ultraviolet to the infrared. Gray lines divide the broken phase (lower side) and the symmetric phase (upper side; cross-hatched area). Points denoted by ▲, • and ♦ express the fixed point $(\zeta, W) = (1, 1/2)$, $(0, 1)$ and $(1/3, 3/2)$ respectively.

into the A_1 meson mass. At the fixed point "VM-type", on the other hand, the ρ meson becomes massless faster than the A_1 meson since ζ goes to zero. The fixed point "Hybrid-type" is the ultraviolet fixed point in any direction, so that it is not so stable as to "GL-type" and "VM-type".

To summarize, we found that the chiral symmetry restoration in the GHLS required through the matching to QCD can be realized only if the masses of ρ and A_1 mesons vanish at the restoration point:

$$M_\rho \to 0, \quad M_{A_1} \to 0, \tag{18}$$

and that the mass ratio flows into one of the following three fixed points:

$$\begin{aligned} \text{GL-type} &: M_\rho^2/M_{A_1}^2 \to 1, \\ \text{VM-type} &: M_\rho^2/M_{A_1}^2 \to 0, \\ \text{Hybrid-type} &: M_\rho^2/M_{A_1}^2 \to 1/3. \end{aligned} \tag{19}$$

In Ref. 1 we studied the strength of the direct γ-π-π coupling, which measures the validity of the vector dominace (VD). We have shown that, $g_{\gamma\pi\pi} \to 0$ for the GL-type, $g_{\gamma\pi\pi} \to 1/2$ for the VM-type and $g_{\gamma\pi\pi} \to 1/3$ for the Hybrid-type. This strongly affects to the understanding of the experimental data on dilepton productions based on the dropping ρ.

Acknowledgment

The work of M.H. is supported in part by the Daiko Foundation #9099, the 21st Century COE Program of Nagoya University provided by Japan Society for the Promotion of Science (15COEG01), and the JSPS Grant-in-Aid for Scientific Research (c) (2) 16540241. The work of C.S. is supported in part by the Virtual Institute of the Helmholtz Association under the grant No. VH-VI-041.

References

1. M. Harada and C. Sasaki, Phys. Rev. D **73**, 036001 (2006).
2. See, e.g., V. Bernard and U. G. Meissner, Nucl. Phys. A **489**, 647 (1988); T. Hatsuda and T. Kunihiro, Phys. Rept. **247**, 221 (1994) [hep-ph/9401310]; R. D. Pisarski, hep-ph/9503330; R. Rapp and J. Wambach, Adv. Nucl. Phys. **25**, 1 (2000); F. Wilczek, hep-ph/0003183; G. E. Brown and M. Rho, Phys. Rept. **363**, 85 (2002).
3. G. E. Brown and M. Rho, Phys. Rev. Lett. **66**, 2720 (1991).
4. M. Harada and K. Yamawaki, Phys. Rev. Lett. **86**, 757 (2001).
5. M. Harada and K. Yamawaki, Phys. Rept. **381**, 1 (2003).
6. M. Harada and C. Sasaki, Phys. Lett. B **537**, 280 (2002).
7. M. Harada, Y. Kim and M. Rho, Phys. Rev. D **66**, 016003 (2002).
8. M. Bando, T. Kugo, S. Uehara, K. Yamawaki and T. Yanagida, Phys. Rev. Lett. **54**, 1215 (1985).
9. M. Bando, T. Kugo and K. Yamawaki, Phys. Rept. **164**, 217 (1988).
10. G. Agakishiev *et al.* [CERES Collaboration], Phys. Rev. Lett. **75**, 1272 (1995).
11. G. Q. Li, C. M. Ko and G. E. Brown, Phys. Rev. Lett. **75**, 4007 (1995).
12. K. Ozawa *et al.* [E325 Collaboration], Phys. Rev. Lett. **86**, 5019 (2001); M. Naruki *et al.*, Phys. Rev. Lett. **96**, 092301 (2006).
13. D. Trnka *et al.* [CBELSA/TAPS Collaboration], Phys. Rev. Lett. **94**, 192303 (2005).
14. E. V. Shuryak and G. E. Brown, Nucl. Phys. A **717**, 322 (2003).
15. S. Damjanovic *et al.* [NA60 Collaboration], J. Phys. G **31**, S903 (2005).
16. S. Damjanovic *et al.* [NA60 Collaboration], Nucl. Phys. A **774**, 715 (2006).
17. G. E. Brown and M. Rho, arXiv:nucl-th/0509001; arXiv:nucl-th/0509002.
18. H. van Hees and R. Rapp, arXiv:hep-ph/0604269.
19. B. Schenke and C. Greiner, Phys. Rev. Lett. **98**, 022301 (2007).
20. M. Harada and C. Sasaki, Nucl. Phys. A **736**, 300 (2004).
21. M. Harada and C. Sasaki, Phys. Rev. D **74**, 114006 (2006).
22. G. E. Brown, C. H. Lee and M. Rho, arXiv:nucl-th/0507073.
23. See, e.g., C. Song and V. Koch, Phys. Lett. B **404**, 1 (1997); M. Urban, M. Buballa and J. Wambach, Phys. Rev. Lett. **88**, 042002 (2002).
24. M. Bando, T. Kugo and K. Yamawaki, Nucl. Phys. B **259**, 493 (1985).
25. M. Bando, T. Fujiwara and K. Yamawaki, Prog. Theor. Phys. **79**, 1140 (1988).
26. N. Kaiser and U. G. Meissner, Nucl. Phys. A **519**, 671 (1990).
27. S. Weinberg, Physica A **96**, 327 (1979); J. Gasser and H. Leutwyler, Annals Phys. **158**, 142 (1984); Nucl. Phys. B **250**, 465 (1985).
28. M. A. Shifman, A. I. Vainshtein and V. I. Zakharov, Nucl. Phys. B **147**, 385 (1979); Nucl. Phys. B **147**, 448 (1979).
29. S. Weinberg, Phys. Rev. Lett. **18**, 507 (1967).

On the Ground State of QCD inside a Compact Stellar Object

Roberto Casalbuoni
Department of Physics, University of Florence,
Florence, Italy and
Sezione INFN, Florence, Italy
E-mail: casalbuoni@fi.infn.it
www.theory.fi.infn.it/casalbuoni/

We describe the effects of the strange quark mass and of the color and electric neutrality on the superconducing phases of QCD.

Keywords: QCD; Color Superconductivity; Finite Density.

1. Introduction

It is now a well-established fact that at zero temperature and sufficiently high densities quark matter is a color superconductor.[1,2] The study starting from first principles was done in Refs. 3–5. At baryon chemical potentials much higher than the masses of the quarks u, d and s, the favored state is the so-called Color-Flavor-Locking (CFL) state, whereas at lower values, when the strange quark decouples, the relevant phase is called two-flavor color superconducting (2SC).

An interesting possibility is that in the interior of compact stellar objects (CSO) some color superconducting phase may exist. In fact the central densities for these stars could be up to 10^{15} g/cm^3, whereas the temperature is of the order of tens of keV. However the usual assumptions leading to prove that for three flavors the favored state is CFL should now be reviewed. Matter inside a CSO should be electrically neutral and should not carry color. Also conditions for β-equilibrium should be fulfilled. As far as color is concerned, it is possible to impose a simpler condition, that is color neutrality, since in Ref. 6 it has been shown that there is no free energy cost in projecting color singlet states out of color neutral ones. Furthermore one has to take into account that at the interesting density the mass of the strange quark is a relevant parameter. All these effects, the mass of the strange quark, β-equilibrium and color and electric neutrality, imply that the radii of the Fermi spheres of quarks that would pair are not the same. This difference in radius, as we shall see, is going to create a problem with the usual BCS pairing. Let us start from the mass effects. Suppose to have two fermions of masses $m_1 = M$ and $m_2 = 0$ at the same chemical potential μ. The corresponding Fermi momenta are $p_{F_1} = \sqrt{\mu^2 - M^2}$ and $p_{F_2} = \mu$. We see that the radius of the Fermi sphere of the massive fermion is smaller than the one of the massless particle. If we assume $M \ll \mu$ the massive particle has an effective chemical potential $\mu_{\rm eff} = \sqrt{\mu^2 - M^2} \approx \mu - M^2/2\mu$ and the mismatch between the two Fermi spheres is given by

$$\delta\mu \approx \frac{M^2}{2\mu} \tag{1}$$

This shows that the quantity $M^2/(2\mu)$ behaves as a chemical potential. Therefore for $M \ll \mu$ the mass effects can be taken into account through the introduction of the mismatch between the chemical potentials of the two fermions given by eq. (1). This is the way that we will follow in our study.

Now let us discuss β-equilibrium. If electrons are present (as generally required by electrical neutrality) chemical potentials of quarks of different electric charge are different. In fact, when at the equilibrium for $d \to ue\bar{\nu}$, we have

$$\mu_d - \mu_u = \mu_e \tag{2}$$

From this condition it follows that for a quark of charge Q_i the chemical potential μ_i is given by

$$\mu_i = \mu + Q_i\mu_Q \tag{3}$$

where μ_Q is the chemical potential associated to the electric charge. Therefore

$$\mu_e = -\mu_Q \tag{4}$$

Notice also that μ_e is not a free parameter since it is determined by the neutrality condition

$$Q = -\frac{\partial\Omega}{\partial\mu_e} = 0 \tag{5}$$

At the same time the chemical potentials associated to the color generators T_3 and T_8 are determined by the color neutrality conditions

$$\frac{\partial\Omega}{\partial\mu_3} = \frac{\partial\Omega}{\partial\mu_8} = 0 \tag{6}$$

We see that in general there is a mismatch between the quarks that should pair according to the BCS mechanism at $\delta\mu = 0$. Increasing the mismatch has the effect of destroying the BCS phase and the system either goes into the normal phase or to some different phase. In the next Sections we will explore some of these possible alternatives.

2. Neutrality and β-equilibrium

Just as a very simple example of the effect of the neutrality and β-equilibrium conditions, let us consider three non interacting quarks, u, d and s. The β-equilibrium requires

$$\mu_{d,s} = \mu_u + \mu_e \tag{7}$$

The chemical potentials of the single species in term of the baryon chemical potential, $\bar{\mu}$, and of the charge chemical potential, $\mu_Q = -\mu_e$, are therefore

$$\mu_u = \bar{\mu} - \frac{2}{3}\mu_e, \quad \mu_d = \mu_s = \bar{\mu} + \frac{1}{3}\mu_e \tag{8}$$

The numerical densities of different quarks are given by

$$N_{u,d} = \frac{\mu_{u,d}^3}{\pi^2}, \quad N_s = \frac{(\mu_s^2 - M_s^2)^{3/2}}{\pi^2}, \quad N_e = \frac{\mu_e^3}{3\pi^2} \tag{9}$$

On the other hand the neutrality condition requires

$$\frac{2}{3}N_u - \frac{1}{3}N_d - \frac{1}{3}N_s - N_e = 0 \tag{10}$$

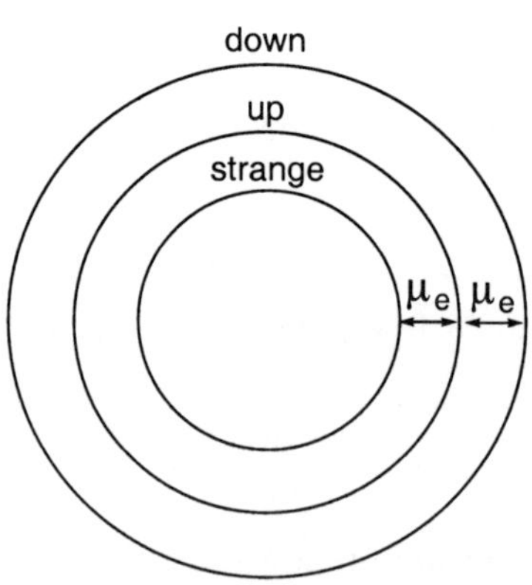

Fig. 1. The Fermi spheres for three non interacting quarks, u, d and s by taking into account the mass of the strange quark (see text).

If the strange quark mass is neglected the previous equation has the simple solution

$$N_u = N_d = N_s, \quad N_e = 0 \tag{11}$$

In this case the Fermi spheres of the three quarks have the same radius (remember that for a single fermion the numerical density is given by $N = p_F^3/(3\pi^2)$). However if we take into account $M_s \neq 0$ at the lowest order in M_s/μ we get

$$\mu_e \approx \frac{M_s^2}{4\mu}, \quad p_F^d - p_F^u \approx p_F^u - p_F^s \approx \mu_e \tag{12}$$

The result is shown in Fig. 1. It can also be shown that in the normal phase the chemical potentials associated to the color charges T_3 and T_8 vanish. We will make use of these results when we will discuss the LOFF phase.

3. Gapless quasi-fermions

When a mismatch is present, the spectrum of the quasi-particles is modified as follows

$$E_{\delta\mu=0} = \sqrt{(p-\mu)^2 + \Delta^2} \to E_{\delta\mu} = \left|\delta\mu \pm \sqrt{(p-\mu)^2 + \Delta^2}\right| \tag{13}$$

Therefore for $|\delta\mu| < \Delta$ we have gapped quasi-particles with gaps $\Delta \pm \delta\mu$. However, for $|\delta\mu| = \Delta$ a gapless mode appears and from this point on there are regions of the phase space which do not contribute to the gap equation (blocking region).

The gapless modes are characterized by

$$E(p) = 0 \Rightarrow p = \mu \pm \sqrt{\delta\mu^2 - \Delta^2} \tag{14}$$

Since the energy cost for pairing two fermions belonging to Fermi spheres with mismatch $\delta\mu$ is $2\delta\mu$ and the energy gained in pairing is 2Δ, we see that the fermions begin to unpair for $2\delta\mu \geq 2\Delta$. These considerations will be relevant for the study of the gapless phases when neutrality is required.

4. The gCFL phase

The gCFL phase is a generalization of the CFL phase which has been studied both at $T = 0$[7,8] and $T \neq 0$.[9] The condensate has now the form

$$\langle 0|\psi_{aL}^{\alpha}\psi_{bL}^{\beta}|0\rangle = \Delta_1\epsilon^{\alpha\beta 1}\epsilon_{ab1} + \Delta_2\epsilon^{\alpha\beta 2}\epsilon_{ab2} + \Delta_3\epsilon^{\alpha\beta 3}\epsilon_{ab3} \tag{15}$$

The CFL phase corresponds to all the three gaps Δ_i being equal. Varying the gaps one gets many different phases. In particular we will be interested to the CFL, to the g2SC characterized by $\Delta_3 \neq 0$ and $\Delta_1 = \Delta_2 = 0$ and to the gCFL phase with $\Delta_3 > \Delta_2 > \Delta_1$. Notice that in the g2SC phase defined here the strange quark is present but unpaired.

In flavor space the gaps Δ_i correspond to the following pairings in flavor

$$\Delta_1 \Rightarrow ds, \quad \Delta_2 \Rightarrow us, \quad \Delta_3 \Rightarrow ud \tag{16}$$

The mass of the strange quark is taken into account by shifting all the chemical potentials involving the strange quark as follows: $\mu_{\alpha s} \to \mu_{\alpha s} - M_s^2/2\mu$. It has also been shown in Ref. 10 that color and electric neutrality in CFL require

$$\mu_8 = -\frac{M_s^2}{2\mu}, \quad \mu_e = \mu_3 = 0 \tag{17}$$

At the same time the various mismatches are given by

$$\delta\mu_{bd-gs} = \frac{M_s^2}{2\mu}, \quad \delta\mu_{rd-gu} = \mu_e = 0, \quad \delta\mu_{rs-bu} = \mu_e - \frac{M_s^2}{2\mu} \tag{18}$$

It turns out that in the gCFL the electron density is different from zero and, as a consequence, the mismatch between the quarks d and s is the first one to give rise to the unpairing of the corresponding quarks. This unpairing is expected to occur for

$$2\frac{M_s^2}{2\mu} > 2\Delta \quad \Rightarrow \quad \frac{M_s^2}{\mu} > 2\Delta \tag{19}$$

This has been substantiated in Ref. 8 by a calculation in the NJL model based on one gluon-exchange. The authors assume for their calculation a chemical potential, $\mu = 500\ MeV$ and a CFL gap given by $\Delta = 25\ MeV$. The transition from the CFL phase, where all gaps are equal, to the gapless phase occurs roughly at $M_s^2/\mu = 2\Delta$. In Fig. 2 we show the free energy of the various phases with reference to the normal phase. The CFL phase is the stable one up to $M_s^2/\mu \approx 2\Delta$. Then the gCFL phase takes over up to about 130 MeV, where the system goes to the normal phase. Notice that except in a very tiny region around this point, the CFL and gCFL phases dominate over the corresponding 2SC and g2SC ones. The thin short-dashed line represents the free energy of the CFL phase up to the point where it becomes equal to the free-energy of the normal phase. This happens for $M_s^2/\mu \approx 4\Delta$.

Although the gCFL phase appears to be energetically favored it cannot be the real ground state. In fact, it has been shown in Refs. 11 and 12 that in this phase there is a chromomagnetic instability. This instability manifests itself in the masses of the gluons 1, 2, 3, 8 becoming pure imaginary at the

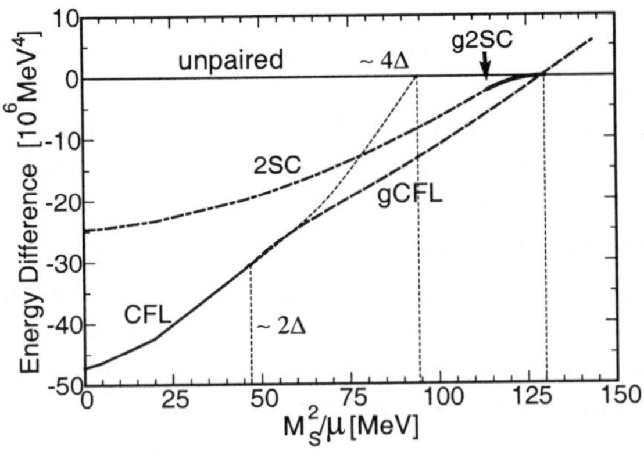

Fig. 2. Free energy of the various phases discussed in the text with reference to the normal phase, named unpaired in the figure.

transition CFL-gCFL. An analogous instability (relative to the gluons 4, 5, 6, 7, 8) occurs in the g2SC phase[13–16] and it seems to be related to the gapless modes present in homogeneous phases, as conjectured in Ref. 17.

5. Possible solutions of the problem of the chromagnetic instability

There have been various proposals to solve the problem of the chromomagnetic instability. We will shortly review these attempts before discussing the proposal that at the moment seems to be the favored one, that is the one corresponding to the LOFF phase (see next Section):

- **Gluon condensation**. If one assumes artificially that the expectation values of $A_{\mu 3}$ and $A_{\mu 8}$ are not zero, and of the order of $10\ MeV$, the instability goes away.[11] This argument has been done more accurate for the g2SC phase in Refs. 18–21, where it has been considered a model exhibiting chromo-magnetic condensation. It turns out that the rotational symmetry is broken and this makes some connection with the LOFF phase. At the moment these models have not been extended to the three flavor case.
- **CFL-K^0 phase**. If the mismatch is not too large (meaning $\delta\mu/\mu \ll 1$) the CFL pattern can be modified by a flavor rotation of the condensate. This is equivalent to have a condensate of kaons.[22] The transition to this phase occurs roughly for a strange quark mass

satisfying $M_s > m^{1/3}\Delta^{2/3}$, with m the light quark mass and Δ the CFL condensate. Also this phase exhibits gapless modes and the gluon instability occurs.[23–25] Allowing for a space dependent condensate, a current is generated which eliminates the instability.[26] Also in this case, a space dependent condensation brings a relation to the LOFF phase.

- **Single flavor pairing.** If the stress caused by the mismatch is too big, single flavor pairing could occur. However the gap appears to be too small. It could be important at low chemical potential before the nuclear phase (see, for instance Ref. 27).
- **Secondary pairing.** The gapless modes could form a secondary gap, but here too the gap is far too small.[28,29]
- **Mixed phases.** The possibility of mixed phases both of nuclear and quark matter[30] as well as mixed phases of different Color Superconducting[31,32] phases has been considered. However all these possibilities are either unstable or energetically disfavored.
- LOFF phase. In Ref. 33 it has been shown that the chromagnetic instability of the g2SC phase is just what is needed in order to make the crystalline, or LOFF phase, energetically favored. Also it turns out that in the LOFF phase there is no chromomagnetic instability although gapless modes are present.[34]

The previous considerations make the LOFF phase worth to be considered and this is what we will do in the next Section.

6. The LOFF Phase

According to the authors of Refs. 35 and 36 when fermions belong to different Fermi spheres, they might prefer to pair staying as much as possible close to their own Fermi surface. The total momentum of the pair is not zero, $\vec{p}_1 + \vec{p}_2 = 2\vec{q}$ and, as we shall show, $|\vec{q}|$ is fixed variationally whereas the direction of $\vec{q}$ is chosen spontaneously. Since the total momentum of the pair is not zero the condensate breaks rotational and translational invariance. The simplest form of the condensate compatible with this breaking is just a simple plane wave (more complicated possibilities will be discussed later)

$$\langle\psi(x)\psi(x)\rangle \approx \Delta\, e^{2i\vec{q}\cdot\vec{x}} \tag{20}$$

It should also be noticed that the pairs use much less of the Fermi surface than they do in the BCS case. For instance, if both fermions are sitting at their own Fermi surface, they can pair only if they belong to circles fixed by $\vec{q}$. More generally there is a quite large region in momentum space (the so called blocking region) which is excluded from pairing. This leads to a condensate generally smaller than the BCS one.

Let us now consider in more detail the LOFF phase (for reviews of this phase see Refs. 37–40). For two fermions at different densities we have an extra term in the hamiltonian which can be written as

$$H_I = -\delta\mu\sigma_3 \tag{21}$$

where, in the original LOFF papers[35,36], $\delta\mu$ is proportional to the magnetic field due to the impurities, whereas in the actual case $\delta\mu = (\mu_1 - \mu_2)/2$ and σ_3 is a Pauli matrix acting on the two fermion space. According to Refs. 35 and 36 this favors the formation of pairs with momenta

$$\vec{p}_1 = \vec{k} + \vec{q}, \quad \vec{p}_2 = -\vec{k} + \vec{q} \tag{22}$$

We will discuss in detail the case of a single plane wave (see eq. (20)). The interaction term of eq. (21) gives rise to a shift in the quasi-particles energy due both to the non-zero momentum of the pair and to the different chemical potentials

$$E(\vec{p}) - \mu \to E(\pm\vec{k} + \vec{q}) - \mu \mp \delta\mu \approx E(\vec{p}) \mp \bar{\mu} \tag{23}$$

with

$$\bar{\mu} = \delta\mu - \vec{v}_F \cdot \vec{q} \tag{24}$$

Notice that the previous dispersion relations show the presence of gapless modes at momenta depending on the angle of $\vec{v}_F$ with $\vec{q}$. Here we have assumed $\delta\mu \ll \mu$ (with $\mu = (\mu_1 + \mu_2)/2$) allowing us to expand E at the first order in $\vec{q}/\mu$.

The study of the gap equation shows that increasing $\delta\mu$ from zero we get first the BCS phase. Then at $\delta\mu = \delta\mu_1$ there is a first order transition to the LOFF phase,[35,37] and at $\delta\mu = \delta\mu_2 > \delta\mu_1$ there is a second order phase transition to the normal phase.[35,37] We start comparing the grand potential in the BCS phase to the one in the normal phase. Their difference is given by (see for example Ref. 39)

$$\Omega_{\rm BCS} - \Omega_{\rm normal} = -\frac{p_F^2}{4\pi^2 v_F}\left(\Delta_0^2 - 2\delta\mu^2\right) \tag{25}$$

where the first term comes from the energy necessary to the BCS condensation, whereas the last term arises from the grand potential of two free fermions with different chemical potential. We recall also that for massless fermions $p_F = \mu$ and $v_F = 1$. We have again assumed $\delta\mu \ll \mu$. This implies that there should be a first order phase transition from the BCS to the normal phase at $\delta\mu = \Delta_0/\sqrt{2}$,[41] since the BCS gap does not depend on $\delta\mu$. In order to compare with the LOFF phase one can expand the gap equation around the point $\Delta = 0$ (Ginzburg-Landau expansion) to explore the possibility of a second order phase transition.[35] The result for the free energy is

$$\Omega_{\rm LOFF} - \Omega_{\rm normal} \approx -0.44\,\rho(\delta\mu - \delta\mu_2)^2 \tag{26}$$

At the same time, looking at the minimum in q of the free energy one finds

$$q v_F \approx 1.2\,\delta\mu \tag{27}$$

Since we are expanding in Δ, in order to get this result it is enough to minimize the coefficient of Δ^2 in the free-energy (the first term in the Ginzburg-Landau expansion).

We see that in the window between the intersection of the BCS curve and the LOFF curve and $\delta\mu_2$, the LOFF phase is favored. Also at the intersection there is a first order transition between the LOFF and the BCS phase. Furthermore, since $\delta\mu_2$ is very close to $\delta\mu_1$ the intersection point is practically given by $\delta\mu_1$. The window of existence of the LOFF phase $(\delta\mu_1, \delta\mu_2) \simeq (0.707, 0.754)\Delta_0$ is rather narrow, but there are indications that considering the realistic case of QCD[42] the window opens up. Such opening occurs also for different crystalline structures than the single plane wave.[38,43]

7. The LOFF phase with three flavors

In the last Section we would like to illustrate some preliminary result about the LOFF phase with three flavors. This problem has been considered in Ref. 44 under various simplifying hypothesis:

- The study has been made in the Ginzburg-Landau approximation.
- Only electrical neutrality has been required and the chemical potentials for the color charges T_3 and T_8 have been put equal to zero (see later).
- The mass of the strange quark has been introduced as it was done previously for the gCFL phase.
- The study has been restricted to plane waves, assuming the follow-

ing generalization of the gCFL case:

$$\langle \psi^{\alpha}_{aL} \psi^{\beta}_{bL} \rangle = \sum_{I=1}^{3} \Delta_I(\vec{x}) \epsilon^{\alpha\beta I} \epsilon_{abI}, \quad \Delta_I(\vec{x}) = \Delta_I e^{2i\vec{q}_I \cdot \vec{x}} \tag{28}$$

- The condensate depends on three momenta, meaning three lengths of the momenta q_i and three angles. In Ref. 44 only four particular geometries have been considered: 1) all the momenta parallel, 2) $\vec{q}_1$ antiparallel to $\vec{q}_2$ and $\vec{q}_3$, 3) $\vec{q}_2$ antiparallel to $\vec{q}_1$ and $\vec{q}_3$, 4) $\vec{q}_3$ antiparallel to $\vec{q}_1$ and $\vec{q}_2$.

The minimization of the free energy with respect to the $|\vec{q}_I|$'s leads to the same result as in eq. (27), $|\vec{q}_I| = 1.2\delta\mu_I$. Let us notice that consistently with the Ginzburg-landau approximation requiring to be close to the normal phase, we assume $\mu_3 = \mu_8 = 0$ as discussed in Section 2. We remember also that close to the normal phase the Fermi surfaces are given in Fig. 1 and as a consequence at the same order of approximation we expect $\Delta_2 = \Delta_3$ (since ud and us mismatches are equal) and $\Delta_1 = 0$, due to the sd mismatch being the double of the other two. Once we assume $\Delta_1 = 0$ only the two configurations with q_2 and q_3 parallel or antiparallel remain. However the antiparallel is unlike. In fact, as it can be seen from Fig. 3, in the antiparallel configuration we have two u quarks in the same ring reducing the phase space, and correspondingly the gap, due to the Fermi statistics. This observation is indeed verified by numerical calculations. Then, one has

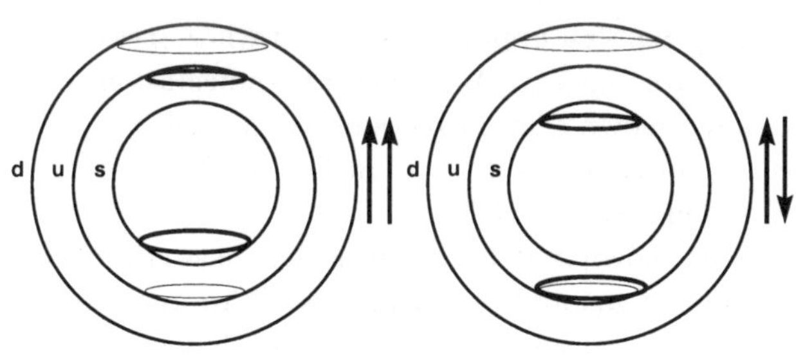

Fig. 3. The two Fermi spheres corresponding to q_2 (left arrow) and q_3 (right arrow) respectively parallel and antiparallel. The pairing rings du and us are shown by thin and thick lines respectively.

to minimize with respect to the gap and μ_e in order to require electrical neutrality. The results are given in Fig. 4 using the same input parameters as in Section 4 for the gCFL case. We see that below 150 MeV the LOFF

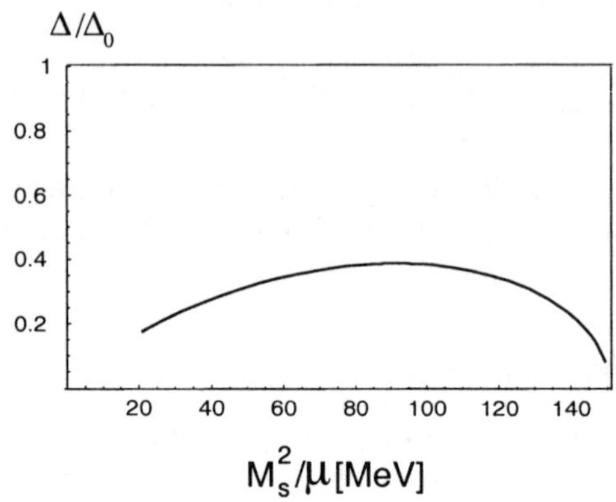

Fig. 4. The ratio of the gap Δ/Δ_0 for LOFF with three flavors vs. M_s^2/μ. Here Δ_0 is the CFL gap and $\Delta = \Delta_2 = \Delta_3$.

phase is favored over the normal phase with a gap arriving at almost 0.4 the CFL gap. Of course, it is interesting to compare this result with the gCFL result given in Fig. 2. The comparison is made in Fig. 5. We see that the LOFF phase dominates over gCFL in the interval between 128 MeV and 150 MeV where the transition to the normal phase is located. These results have been confirmed by an exact calculation with respect to the gap (but always at the leading order in the chemical potential), done in Ref. 45. The result found by these authors show that in the range of M_s considered here the Ginzburg-Landau approximation is rather accurate and if any it overestimates the free energy. As a further confirmation of these results, in Ref. 46 we have shown that corrections at the order $1/\mu$ do not modify qualitatively the previous results but rather tend to enlarge the window where LOFF dominates over gCFL.

It has also been shown in Ref. 47 that in the phase studied in this Section the chromo-magnetic instability disappears. Here one has to distinguish the longitudinal and transverse masses of the gluons with respect to the direction of the total momentum of the pair. It results that all these masses are real.

More recently an extension of the simple ansatz of a single plane wave for each gap, as considered in this Section, has been made in Ref. 48. The simple ansatz of eq. (28) has been generalized in the following way

$$\langle ud\rangle \approx \Delta \sum_a e^{2i\vec{q}_3^a\cdot\vec{r}}, \quad \langle us\rangle \approx \Delta \sum_a e^{2i\vec{q}_2^a\cdot\vec{r}} \quad \langle ds\rangle \approx 0 \tag{29}$$

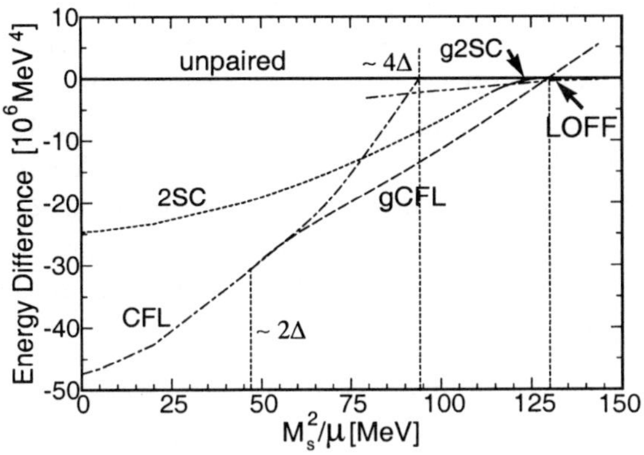

Fig. 5. Comparison of the free energy of the various phases already considered in Fig. 2 (same notations as here) with the LOFF phase with three flavors.

with the index a running from 1 up to a maximum value of 8. In practice

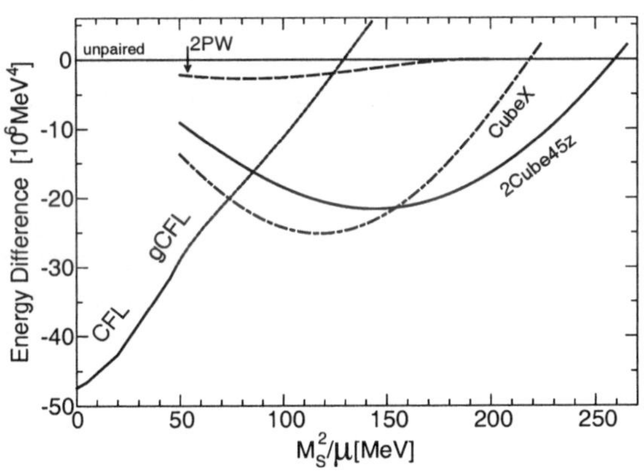

Fig. 6. Comparison of the free energy of the various phases already considered in Figs. 2 and 5 (same notations as here) with various crystalline structures in the three flavor case.

for any choice of the range of the index a one gets a particular crystalline structure defined by the vectors $\vec{q}^{\,a}$ pointing at the vertices of the crystal. In Ref. 48 the study has been extended to 11 crystals. The favored structures are the so called CubeX and 2Cube45z. The CubeX is a cube characterized by 4 vectors $\vec{q}_2^{\,a}$ and 4 $\vec{q}_3^{\,a}$. Each set of vectors lies in a plane and the two planes cut at 90 degrees forming a cube. In the 2Cube45z, there are 8 vectors in each set defining two cubes which are rotated one with respect to the other of 45 degrees along the z axis. The free energies for these two crystals are compared with the case of a single plane wave for each pairing (called in this context 2PW) in Fig. 6. We see that the CubeX and the 2Cube45z take over the gCFL phase in almost all the relevant range of M_s^2/μ. Taking into account that this calculation has been made in the Ginzburg-Landau approximation it looks plausible that these two phases are the favorite ones up to the CFL phase.

8. Conclusions

As we have seen there have been numerous attempts in trying to determine the fundamental state of QCD under realistic conditions existing inside a compact stellar objects, that is to say, neutrality in color and electric charge, β-equilibrium and a non vanishing strange quark mass. Many competing phases have been found. Most of them have fermionic gapless modes. However, gapless modes in presence of a homogeneous condensate seem to lead unavoidably to a chromo-magnetic instability and it seems necessary to consider space dependent condensates. In this respect the LOFF phase, where the space dependence comes about in relation to the non zero total momentum of the pair, seems to be a natural candidate. This phase in the presence of three flavors has been recently considered.[44,45,48] It has been found that there are no chromo-magnetic instabilities[47] and that energetically it is favored almost up to the CFL phase. However, considering the approximations involved in these calculations, before to draw sounded conclusions one should attend for more careful investigations.

References

1. B. Barrois, *Nuclear Physics* **B129**, 390 (1977); S. Frautschi, *Proceedings of workshop on hadronic matter at extreme density*, Erice 1978; D. Bailin and A. Love, *Physics Report* **107** (1984) 325 .
2. M. Alford, K. Rajagopal, and F. Wilczek, *Phys. Lett.* **B422**(1998) 247 [hep-ph/9711395]; R. Rapp, T. Schafer, E. V. Shuryak and M. Velkovsky, *Phys. Rev. Lett.* **81**, 53 (1998) [hep-ph/9711396].
3. D.T. Son, *Phys. Rev.* **D59** (1999) 094019 [hep-ph/9812287]; T. Schäfer and F. Wilczek, *Phys. Rev.* **D60** (1999) 114033 [hep-ph/9906512]; D.K. Hong, V.A. Miransky, I.A. Shovkovy, and L.C.R. Wijewardhana, *Phys. Rev.* **D61** (2000) 056001 [hep-ph/9906478]; S.D.H. Hsu and M. Schwetz, *Nucl. Phys.* **B572** (2000) 211 [hep-ph/9908310]; W.E. Brown, J.T. Liu, and H.-C. Ren, *Phys. Rev.* **D61** (2000) 114012 [hep-ph/9908248].

4. R.D. Pisarski and D.H. Rischke, *Phys. Rev.* **D61** (2000) 051501 [nucl-th/9907041].
5. I.A. Shovkovy and L.C.R. Wijewardhana, *Phys. Lett.* **B470** (1999) 189 [hep-ph/9910225]; T. Schäfer, *Nucl. Phys.* **B575** (2000) 269 [hep-ph/9909574].
6. P. Amore, M. C. Birse, J. A. McGovern and N. R. Walet, *Phys. Rev.* **D65** (2002) 074005 [hep-ph/0110267].
7. M. Alford, C. Kouvaris and K. Rajagopal, *Phys. Rev. Lett.* **92** (2004) 222001 [hep-ph/0311286].
8. M. Alford, C. Kouvaris and K. Rajagopal, *Phys. Rev.* **D71** (2005) 054009 [hep-ph/0406137].
9. M. Alford, P. Jotwani, C. Kouvaris, J. Kundu and K. Rajagopal, *Phys. Rev.* **D71** (2005) 114011 [astro-ph/0411560].
10. M. Alford and K. Rajagopal, *JHEP* **06** (2002) 031 [hep-ph/0204001].
11. R. Casalbuoni, R. Gatto, M. Mannarelli, G. Nardulli and M. Ruggieri, *Phys. Lett.* **B605** (2005) 362 [hep-ph/0410401].
12. K. Fukushima, *Phys. Rev.* **D70** (2005) 07002 [hep-ph/0506080].
13. M. Huang and I. A. Shovkovy, *Phys. Rev.* **D70** (2004) 051501 [hep-ph/0407049]; *ibidem Phys. Rev.* **D70** (2004) 094030 [hep-ph/0408268].
14. M. Hashimoto, *Phys. Lett.* **B642** (2006) 93 [arXiv:hep-ph/0605323].
15. O. Kiriyama, *Phys. Rev.* **D74** (2006) 074019 [arXiv:hep-ph/0608109].
16. O. Kiriyama, arXiv:hep-ph/0609185.
17. M. Alford and Q. Wang, *J. Phys.* **G31** (2005) 719 [hep-ph/0501078].
18. E. V. Gorbar, M. Hashimoto and V. A. Miransky, *Phys. Lett.* **B632** (2006) 305 [arXiv:hep-ph/0507303].
19. E. V. Gorbar, M. Hashimoto and V. A. Miransky, *Phys. Rev. Lett.* **96** (2006) 022005 [arXiv:hep-ph/0509334].
20. O. Kiriyama, D. H. Rischke and I. A. Shovkovy, *Phys. Lett.* **B643** (2006) 331 [arXiv:hep-ph/0606030].
21. L. He, M. Jin and P. Zhuang, arXiv:hep-ph/0610121.
22. P. F. Bedaque and T. Schafer, *Nucl. Phys.* **A697** (2002) 802 [arXiv:hep-ph/0105150].
23. A. Kryjevski and T. Schafer, *Phys. Lett.* **B606** (2005) 52 [arXiv:hep-ph/0407329].
24. A. Kryjevski and D. Yamada, *Phys. Rev.* **D71** (2005) 014011 [arXiv:hep-ph/0407350].
25. A. Kryjevski, arXiv:hep-ph/0508180.
26. A. Gerhold and T. Schafer, *Phys. Rev.* **D73** (2006) 125022 [arXiv:hep-ph/0603257].
27. M. G. Alford, arXiv:hep-lat/0610046.
28. I. Shovkovy and M. Huang, *Prepared for NATO Advanced Study Institute: Structure and Dynamics of Elementary Matter, Kemer, Turkey, 22 Sep - 2 Oct 2003.*
29. M. Alford and Q. h. Wang, *J. Phys.* G **32** (2006) 63 [arXiv:hep-ph/0507269].
30. M. G. Alford, K. Rajagopal, S. Reddy and F. Wilczek, *Phys. Rev.* **D64** (2001) 074017 [arXiv:hep-ph/0105009].
31. M. Buballa, F. Neumann and M. Oertel, *AIP Conf. Proc.* **660** (2003) 196.

32. M. Alford, C. Kouvaris and K. Rajagopal, arXiv:hep-ph/0407257.
33. I. Giannakis and H.C. Ren, *Phys. Lett.* **B611** (2005) 137 [hep-ph/0412015].
34. I. Giannakis and H.C. Ren, *Nucl. Phys.* **B723** (2005) 255 [hep-th/0504053].
35. A. I. Larkin and Yu. N. Ovchinnikov, *Sov. Phys. JETP* **20** (1965) 762.
36. P. Fulde and R. A. Ferrell, *Phys. Rev.* **135** (1964) A550.
37. M. G. Alford, J. A. Bowers and K. Rajagopal, *Phys. Rev.* **D63** (2001) 074016 [hep-ph/0008208].
38. J. A. Bowers and K. Rajagopal, *Phys. Rev.* **D66** (2002) 065002 [hep-ph/0204079].
39. R. Casalbuoni and G. Nardulli, *Rev. Mod. Phys.* **76** (2004) 263 [hep-ph/0305069].
40. J. A. Bowers, hep-ph/0305301.
41. B. S. Chandrasekhar, *App. Phys. Lett.* **1** (1962) 7.
42. A. K. Leibovich, K. Rajagopal and E. Shuster, *Phys. Rev.* **D64** (2001) 094005 [hep-ph/0104073]; see also I. Giannakis, J. T. Liu and H. C. Ren, *Phys. Rev.* **D66** (2002) 031501 [hep-ph/0202138].
43. R. Casalbuoni, M. Ciminale, M. Mannarelli, G. Nardulli, M. Ruggieri and R. Gatto, *Phys. Rev.* **D70** (2004) 054004 [hep-ph/0404090].
44. R. Casalbuoni, R. Gatto, N. Ippolito, G. Nardulli and M. Ruggieri, *Phys. Lett.* **B627** (2005) 89 [Erratum-ibid. **B634** (2006) 565] [arXiv:hep-ph/0507247].
45. M. Mannarelli, K. Rajagopal and R. Sharma, *Phys. Rev.* **D73** (2006) 114012 [arXiv:hep-ph/0603076].
46. R. Casalbuoni, M. Ciminale, R. Gatto, G. Nardulli and M. Ruggieri, *Phys. Lett.* **B642** (2006) 350 [arXiv:hep-ph/0606242].
47. M. Ciminale, G. Nardulli, M. Ruggieri and R. Gatto, *Phys. Lett.* **B636** (2006) 317 [arXiv:hep-ph/0602180].
48. K. Rajagopal and R. Sharma, *Phys. Rev.* **D74** (2006) 094019 [arXiv:hep-ph/0605316].

DENSE HADRONIC MATTER IN HOLOGRAPHIC QCD

Keun-Young Kim[a], Sang-Jin Sin[b] and Ismail Zahed[a]

[a] *Department of Physics and Astronomy, SUNY Stony-Brook, NY 11794*
[b] *Department of Physics, Hanyang University, Seoul 133-791, Korea*

We provide a method to study hadronic matter at finite density in the context of the Sakai-Sugimoto model. We introduce the baryon chemical potential through the external $U(1)_v$ in the induced (DBI plus CS) action on the D8-probe-brane, where baryons are skyrmions. Vector dominance is manifest at finite density. We derive the baryon density effect on the dispersion relations of pion and vector mesons at large N_c. At large density the pion velocity drops to zero. Holographic dense matter enforces exactly the tenets of vector dominance, and screens efficiently vector mesons. At the freezing point the $\rho - \pi\pi$ coupling vanishes with a finite rho mass of about 20% its vacuum value.

1. Introduction

Recently there has been much interest[1,2] in the AdS/CFT approach[3] to study the non-perturbative aspects of gauge theories at large N_c and strong coupling with even applications to heavy ion collision experiments such as jet quenching.[4] On the other hand, dense QCD is of much relevance to compact stars and current heavy ion collisions. This subject has been intensely studied in the past decade following the observation that at asymptotic densities QCD matter may turn to a color superconductor due to asymptotic freedom.[5] At very large N_c there is evidence that the Overhauser effect takes over with the formation of density waves.[6] Indeed, at weak coupling the high degeneracy of the Fermi surface causes quark-quark (antiquark-antiquark) pairing of the BCS kind with different color-flavor arrangements, while at large number of colors quark-antiquark pairing of the Overhauser kind is favored. In strong coupling, first principle calculations are elusive and the two pairings may compete.

To address the issue of strongly coupled QCD at large N_c in dense matter we explore in this paper the holographic principle. In recent years, the gauge/string duality has provided a framework for addressing a number of problems in strong coupling where little is known from first principles in the continuum. To do so, we consider a model for holographic QCD recently discussed by Sakai and Sugimoto. In the limit $N_f \ll N_c$, chiral symmetry in QCD is generated by immersing N_f D8-$\overline{\mathrm{D8}}$ into a D4 background in 5 dimensions where supersymmetry is broken by compactification (Kaluza-Klein mechanism with M_{KK} scale). The induced DBI action on D8 yields a 4 dimensional effective theory of pions and infinitly many vector mesons where the M_{KK} scale plays the role of an upper cutoff (the analogue of the chiral scale $4\pi f_\pi$). At large N_c baryons are Skyrmions.

In this paper, we extend the analysis by Sakai and Sugimoto[2] to finite baryon density. Although there have been many papers[7] on finite chemical potential in the holographic approach, all involve the R-charge or isospin instead of fermion number. The difficulty for the fermion number is that the $U(1)$ charge for R-symmetry is already dual to the $U(1)$ charge of AdS Reisner-Nordstrom Black hole charge. There is no additional $U(1)$ in bulk to modify the geometry of the AdS black hole. In this paper we follow the more traditional route of introducing the chemical potential via the nonperturbative induced DBI plus CS action through the external $U(1)_V$ source in Sakai and Sugimoto chiral model. Specifically, we introduce the baryon chemical potential as $\mathcal{V}_0 = -i\mu_B/N_c$ where $\mathcal{V}_\mu$ is the external vector field in the induced DBI plus CS action. While $\mathcal{V}_0$ drops in the DBI part it contributes to the CS part through the WZW as it should. A redefinition of the vector fields in the Sakai-Sugimoto model necessary to enforce vector dominance, causes a reshuffling of $\mathcal{V}_0$ from the CS to the DBI action as we detail below, with important phenomenological consequences.

In section 2 we review the emergence of the induced effective action on the probe D8 brane from the 5 dimensional D4-brane background, and we detail the contributions of the DBI action to third order in the vector fields, as well as the Chern-Simons part. The expanded effective action enforces total vector dominance. In section 3 we introduce the baryon chemical potential and details the nature of the effective expansion at finite density. In section 4, we discuss the pion and leading vector meson parameters and interactions in holographic dense matter. Our conclusions and suggestions are summarized in section 5.

2. Duality and Branes

2.1. *D4/D8 Branes*

Doubly compactified D=6 is the boundary of AdS_7 in which the M-theory reduces to type IIA string theory. As a result M5 branes wrapping around the first S^1 with radius R^1 transmute to D4-branes. N_c copies of these branes yield near extremal black-hole background.

The metric, dilaton ϕ and the RR three-form field C_3 in a D4-brane background are

$$ds^2 = \left(\frac{U}{R}\right)^{3/2} (\eta_{\mu\nu}dx^\mu dx^\nu + f(U)d\tau^2) + \left(\frac{R}{U}\right)^{3/2} \left(\frac{dU^2}{f(U)} + U^2 d\Omega_4^2\right),$$
$$e^\phi = g_s \left(\frac{U}{R}\right)^{3/4}, \quad F_4 \equiv dC_3 = \frac{2\pi N_c}{V_4}\epsilon_4 , \quad f(U) \equiv 1 - \frac{U_{\rm KK}^3}{U^3} . \tag{1}$$

Here $\mu = 0, 1, 2, 3$ and τ is the compact variable on S^1. $U > U_{KK}$ is the radial coordinate along 56789, ϵ_4 the 4-form and $V_4 = 8\pi^2/3$ the volume of a unit S^4 surrounding the D-4 brane. $R^3 \equiv \pi g_s N_c l_s^3$, where g_s and l_s are the string coupling and length respectively. Again, this background represents N_c D-4 branes wrapped on S^1.

Let us consider N_f $D8$-probe-branes in this background, which may be described by $\tau(U)$. We choose a specific configuration such that τ is constant($\tau = \frac{\pi}{4M_{KK}}$). It corresponds to the maximal asymptotic separation between $D8$ and $\bar{D}8$. with a new variable z, instead of U, defined by the relation,

$$U = (U_{KK}^3 + U_{KK}z^2)^{\frac{1}{3}} \equiv U_z \tag{2}$$

the induced metric on D8 is

$$\begin{aligned} ds_{D8}^2 &= g_{MN}dx^M dx^N \\ &= \left(\frac{U_z}{R}\right)^{3/2} \eta_{\mu\nu}dx^\mu dx^\nu + \frac{4}{9}\left(\frac{R}{U_z}\right)^{3/2} \frac{U_{KK}}{U_z}dz^2 + \left(\frac{R}{U_z}\right)^{3/2} U_z^2 d\Omega_4^2 , \end{aligned} \tag{3}$$

where $M(N) = \{\mu(\nu)(0,1,2,3), z(4), \alpha(5,6,7,8)\}$.

2.2. *DBI and CS Action on D8*

Consider the $U(N_f)$ gauge field A_M on the probe D8-brane configuration. The effective action is 9-dimensional and composed of the DBI action and the Chern-Simons action.

$$S_{DBI} = -T \int d^9x \,\mathrm{tr}\left(e^{-\phi}\sqrt{-\det(g_{MN} + 2\pi\alpha' F_{MN})}\right) , \tag{4}$$
$$S_{CS} = \frac{1}{48\pi^3}\int_{D8} C_3 \,\mathrm{tr}\, F^3 , \tag{5}$$

where $e^{-\phi} = g_s\left(\frac{U_z}{R}\right)^{-\frac{3}{4}}$, $F_4 = dC_3$ is the RR 4-form field strength.

Assuming that $A_\alpha = 0$ and A_μ and A_z are independent of the coordinates on the S^4, the induced DBI action becomes 5-dimensional

$$S_{DBI} = -\widetilde{T} \int d^4x dz\, U_z^2$$
$$\mathrm{tr}\sqrt{1 + (2\pi\alpha')^2 \frac{R^3}{2U_z^3}\eta^{\mu\nu}\eta^{\rho\sigma}F_{\mu\rho}F_{\nu\sigma} + (2\pi\alpha')^2 \frac{9}{4}\frac{U_z}{U_{\rm KK}}\eta^{\mu\nu}F_{\mu z}F_{\nu z} + \mathcal{O}(F^3)}\ , \tag{6}$$

where $\widetilde{T} = \frac{N_c}{216\pi^5}\frac{M_{KK}}{\alpha'^3}$.

2.3. *Effective Action in 4-dimension*

In this part we review how the 5-dimendional DBI action yields the 4-dimensional effective action. Essentially, the 5-dimensional induced action with compact S^1 is 4-dimensional for all excitations with wavelengths larger than the compactification radius of the order of $1/M_{KK}$. Throughout and for zero baryon density, our discussion parallels the original discussion by Sakai and Sugimoto.[2] We quote it for notation and completeness, and refer to their work for further details.

The leading terms in the $1/\lambda \approx 1/M_{KK}$ expansion of the DBI action is

$$S_{\rm D8}^{\rm DBI} = \kappa \int d^4x dZ\ \mathrm{tr}\left[\frac{1}{2}K^{-1/3}F_{\mu\nu}^2 + KF_{\mu z}^2\right]\ , \tag{7}$$

where

$$\kappa \equiv \widetilde{T}(2\pi\alpha')^2 R^3 = \frac{\lambda N_c}{108\pi^3}\ ,\quad Z \equiv \frac{z}{U_{KK}}\ ,\quad K \equiv 1 + Z^2\ . \tag{8}$$

In order to extract four-dimensional meson fields out of the five dimensional gauge field, we expand the gauge field as

$$A_\mu(x^\mu, z) = \sum_{n=1}^{\infty} B_\mu^{(n)}(x^\mu)\psi_n(z)\ , \tag{9}$$

$$A_z(x^\mu, z) = \varphi^{(0)}(x^\mu)\phi_0(z) + \sum_{n=1}^{\infty}\varphi^{(n)}(x^\mu)\phi_n(z)\ , \tag{10}$$

We choose the functions $\psi_n(z)$ to be the eigenfunctions satisfying the self-adjoint differential equation

$$-K^{1/3}\partial_Z\left(K\,\partial_Z\psi_n\right) = \lambda_n\psi_n \tag{11}$$

where λ_n is the eigenvalue and $\{\psi_n\}$ is a complete set. the normalization is fixed by the Laplacian in (11)

$$\kappa \int dZ\, K^{-1/3} \psi_n \psi_m = \delta_{nm} \,. \tag{12}$$

This implies

$$\kappa \int dZ\, K \partial_Z \psi_n \partial_Z \psi_m = \lambda_n \delta_{nm} \tag{13}$$

The $\phi_n(Z)$ are chosen such that

$$\begin{aligned} \phi_n(Z) &= (m_n U_{KK})^{-1} \partial_Z \psi_n(Z) \quad (n \geq 1) \\ \phi_0(Z) &= \frac{1}{\sqrt{\pi\kappa} M_{kk} U_{KK} K} \end{aligned} \tag{14}$$

satisfying the normalization condition:

$$(\phi_m, \phi_n) \equiv \frac{9}{4} \widetilde{T} (2\pi\alpha')^2 U_{\mathrm{KK}}^3 \int dZ\, K\, \phi_m \phi_n = \delta_{mn} \,, \tag{15}$$

which is compatible with (11) and (12).

Inserting (9) and (10) into (7) and using the orthonomality of ψ_n and ϕ_n, yield

$$\begin{aligned} S_{DBI} = \int d^4x \ \mathrm{tr} \Bigg[(\partial_\mu \varphi^{(0)})^2 + \sum_{n=1}^{\infty} \Bigg(\frac{1}{2} (\partial_\mu B_\nu^{(n)} - \partial_\nu B_\mu^{(n)})^2 \\ + \lambda_n M_{KK}^2 (B_\mu^{(n)} - \lambda_n^{-1/2} \partial_\mu \varphi^{(n)})^2 \Bigg) \Bigg] \\ + (\text{interaction terms}) \,. \end{aligned} \tag{16}$$

In the expansion (9) and (10), we have implicitly assumed that the gauge fields are zero asymptotically, i.e. $A_M(x^\mu, z) \to 0$ as $z \to \pm\infty$. The residual gauge transformation that does not break this condition is obtained by a gauge function $g(x^\mu, z)$ that asymptotes a constant $g(x^\mu, z) \to g_\pm$ at $z \pm \infty$. We interpret (g_+, g_-) as elements of the chiral symmetry group $U(N_f)_L \times U(N_f)_R$ in QCD with N_f massless flavors.

2.3.1. *External photons*

By weakly gauging the $U(N_f)_L \times U(N_f)_R$ chiral symmetry, we may introduce the external gauge fields $(A_{L\mu}, A_{R\mu})$. For the interaction between mesons and photon A_μ^{em}, we may choose

$$A_{L\mu} = A_{R\mu} = eQA_\mu^{\mathrm{em}} \,, \tag{17}$$

where e is the electromagnetic coupling constant and Q is the electric matrix-valued charge

$$Q = \frac{1}{3}\begin{pmatrix} 2 & & \\ & -1 & \\ & & -1 \end{pmatrix} , \tag{18}$$

for $N_f = 3$. To insert the external-source gauge fields, we impose the asymptotic values of the gauge field A_μ on the D8-probe-brane as

$$\lim_{z\to+\infty} A_\mu(x^\mu, z) = A_{L\mu}(x^\mu) , \qquad \lim_{z\to-\infty} A_\mu(x^\mu, z) = A_{R\mu}(x^\mu) . \tag{19}$$

This is implemented by modifying the mode expansion (9) as

$$A_\mu(x^\mu, z) = A_{L\mu}(x^\mu)\psi_+(z) + A_{R\mu}(x^\mu)\psi_-(z) + \sum_{n=1}^{\infty} B_\mu^{(n)}(x^\mu)\psi_n(z) , \tag{20}$$

where $\psi_\pm(z)$ are defined as

$$\psi_\pm(z) \equiv \frac{1}{2}(1 \pm \psi_0(z)) , \quad \psi_0(z) \equiv \frac{2}{\pi}\arctan z , \tag{21}$$

which are the non-normalizable zero modes of (11) satisfying $\partial_z\psi_\pm(z) \propto \phi_0(z)$.

2.3.2. *Vector meson dominance*

The induced DBI action (7) carries exact vector dominance. Indeed, by diagonalizing the kinetic terms of the vector meson fields, vector meson dominance emerges from the underlying DBI action naturally. To see that, we recall that the gauge fields are

$$A_\mu = \mathcal{V}_\mu + \mathcal{A}_\mu\psi_0 + \sum_{n=1}^{\infty} v_\mu^n\,\psi_{2n-1} + \sum_{n=1}^{\infty} a_\mu^n\,\psi_{2n} , \tag{22}$$

$$A_z = -i\,\Pi\,\phi_0 , \tag{23}$$

and involve both the external vector sources $\mathcal{V}, \mathcal{A}$ and the dynamical vector fields V, a. To diagonalize the kinetic terms of the dynamical vector fields require the introduction of the physical tilde vector fields $\tilde{v}, \tilde{a}$ which are

$$\widetilde{v}_\mu^n \equiv v_\mu^n + a_{\mathcal{V}v^n}\mathcal{V}_\mu , \tag{24}$$

$$\widetilde{a}_\mu^n \equiv a_\mu^n + a_{\mathcal{A}a^n}\mathcal{A}_\mu , \tag{25}$$

in terms of which the gauge fields now read

$$A_\mu = \mathcal{V}_\mu\,\psi_v + \mathcal{A}_\mu\psi_a + \sum_{n=1}^{\infty} \widetilde{v}_\mu^n\,\psi_{2n-1} + \sum_{n=1}^{\infty} \widetilde{a}_\mu^n\,\psi_{2n} , \tag{26}$$

with

$$\psi_v \equiv 1 - \sum_{n=1}^{\infty} a_{\mathcal{V}v^n}\psi_{2n-1} , \quad \psi_a \equiv \psi_0 - \sum_{n=1}^{\infty} a_{\mathcal{A}a^n}\psi_{2n} .$$

$$a_{\mathcal{V}v^n} \equiv \alpha^n \equiv \kappa \int dz\, K^{-1/3}\psi_{2n-1} , \quad a_{\mathcal{A}a^n} \equiv \kappa \int dz\, K^{-1/3}\psi_{2n}\psi_0 \tag{27}$$

The DBI action (7) in terms of the physical vector fields to third order is

$$\begin{aligned} &\kappa \int dz \ \mathrm{tr}\left[\frac{1}{2}K^{-1/3}F_{\mu\nu}^2\right] \\ &= \mathrm{tr}\Bigg[\frac{1}{2e^2}\left((F_{\mu\nu}^{A_L})^2 + (F_{\mu\nu}^{A_R})^2\right) + \\ &\quad + \frac{1}{2}(\partial_\mu \widetilde{v}_\nu^n - \partial_\nu \widetilde{v}_\mu^n)^2 + \frac{1}{2}(\partial_\mu \widetilde{a}_\nu^n - \partial_\nu \widetilde{a}_\mu^n)^2 + \\ &\quad + (\partial_\mu \widetilde{v}_\nu^n - \partial_\nu \widetilde{v}_\mu^n)([\widetilde{v}^{p\mu}, \widetilde{v}^{q\nu}]\, g_{v^n v^p v^q} + [\widetilde{a}^{p\mu}, \widetilde{a}^{q\nu}]\, g_{v^n a^p a^q}) + \\ &\quad + (\partial_\mu \widetilde{a}_\nu^n - \partial_\nu \widetilde{a}_\mu^n)([\widetilde{v}^{p\mu}, \widetilde{a}^{q\nu}] - [\widetilde{v}^{q\nu}, \widetilde{a}^{p\mu}])\, g_{v^p a^n a^q} \Bigg] . \end{aligned} \tag{28}$$

Note that all the couplings between the external gauge fields (A_L, A_R) and the vector meson fields $(\widetilde{v}^n, \widetilde{a}^n)$ vanish in the first term of the effective action (7). The second term in the effective action (7)

$$\begin{aligned} &\kappa \int dz \ \mathrm{tr}\left[K F_{z\nu}^2\right] \\ &= \mathrm{tr}\Bigg[m_{v^n}^2(\widetilde{v}_\mu^n - a_{\mathcal{V}v^n}\mathcal{V}_\mu)^2 + m_{a^n}^2(\widetilde{a}_\mu^n - a_{\mathcal{A}a^n}\mathcal{A}_\mu)^2 + (i\partial_\mu \Pi + f_\pi \mathcal{A}_\mu)^2 + \\ &\quad + 2i g_{a^m v^n \pi}\, \widetilde{a}_\mu^m[\Pi, \widetilde{v}^{n\mu}] - 2 g_{v^n \pi\pi}\, \widetilde{v}_\mu^n[\Pi, \partial^\mu \Pi]\Bigg] , \end{aligned} \tag{29}$$

The mesons couple to the external gauge fields only through $\widetilde{v}^n \to \mathcal{V}$ and $\widetilde{a}^n \to \mathcal{A}$ transitions in (29). Finally, inserting (22) and (23) into the Chern-Simons action gives

$$\begin{aligned} S_{\mathrm{D8}}^{\mathrm{CS}} = &-\frac{N_c}{4\pi^2}\frac{i}{f_\pi}\int_{M^4} \mathrm{tr}\Big[\Pi\, dB^n dB^m\, c_{nm} + \\ &+ \Pi\,(dB^m B^n B^p + B^m B^n dB^p)\, c_{mnp} + \Pi\, B^m B^n B^p B^q\, c_{mnpq}\Big] + \\ &+ \frac{N_c}{24\pi^2}\int_{M^4} \mathrm{tr}\left[B^m B^n dB^p\, d_{mn|p} - \frac{3}{2}B^m B^n B^p B^q\, d_{mnp|q}\right] , \end{aligned} \tag{30}$$

where $B^{2n-1} \equiv \widetilde{v}^n$, $B^{2n} \equiv \widetilde{a}^n$ It shows complete vector meson dominace in the WZW term. The terms with two or more pion fields are absent.

3. Induced action at finite density

For simplicity we work in Euclidean space, in which case the DBI action reads

$$S^E_{DBI} = \widetilde{T} \int d^4x^E dz\, U_z^2$$
$$\mathrm{tr}\sqrt{1 + (2\pi\alpha')^2 \frac{R^3}{2U_z^3} F^E_{\mu\nu} F^E_{\mu\nu} + (2\pi\alpha')^2 \frac{9}{4}\frac{U_z}{U_{\mathrm{KK}}} F^E_{\mu z} F^E_{\mu z} + \mathcal{O}(F^3)}\ , \qquad (31)$$

From now on we omit the superscript E. From the arguments to follow, we will see that due to vector dominance no *matter term* is generated through the Chern-Simons part. In a way this is expected; the anomaly in the Chern-Simons form is an ultraviolet effect that after regularization shows up in the infrared. Since matter dwells mostly in the infrared, the scale decoupling insures the insensitivity of the Chern-Simons terms therefore of the anomaly. This result is usually referred to as the non-renormalization theorem.

The density will be introduced through the boundary and via the external vector field $\mathcal{V}$ through

$$\mathcal{V}_\mu = -i\,\mu\,\delta_{\mu 0}\, 1_{N_f\times N_f}, \qquad \mathcal{A}_\mu = 0\ . \qquad (32)$$

The master gauge field now reads from (22)

$$A_\mu = -i\mu\delta_{\mu 0}\,(1 - \sum_{n=1}^{\infty} \alpha^n \psi_{2n-1}) + \sum_{n=1}^{\infty} \widetilde{v}^n_\mu\, \psi_{2n-1} + \sum_{n=1}^{\infty} a^n_\mu\, \psi_{2n}\ , \qquad (33)$$

For calculational convenience below, we split $\widetilde{v}^n_0$ in two parts: a constant $U(1)$ part $(-i\widetilde{v}^n_{0C})$ and the rest $(\overline{v^n_{0C}})$,

$$\begin{aligned}
\widetilde{v}^n_0 &= -iv^n_0 - i\alpha^n\mu = -iv^n_{0C} + \overline{v^n_{0C}} - i\alpha^n\mu \\
&= \underbrace{-iv^n_{0C} - i\alpha^n\mu}_{=:-i\widetilde{v}^n_{0C}} + \overline{v^n_{0C}}, \\
\widetilde{v}^n_i &\equiv v^n_i
\end{aligned}$$

where $-iv^n_{0C}$ and $\overline{v^n_{0C}}$ are the constant $U(1)$ part and the rest part of v^n_0 respectively. Thus

$$\begin{aligned}
A_0 &= -i\mu + \sum_{n=1}^{\infty} i(\mu\alpha^n - \widetilde{v}^n_{0C})\psi_{2n-1} + \sum_{n=1}^{\infty} \overline{v^n_{0C}}\, \psi_{2n-1} + \sum_{n=1}^{\infty} a^n_0\, \psi_{2n}\ , \\
A_i &= \sum_{n=1}^{\infty} v^n_i\, \psi_{2n-1} + \sum_{n=1}^{\infty} a^n_i\, \psi_{2n}\ ,
\end{aligned} \qquad (34)$$

Defining

$$
\begin{aligned}
&X^{2n-1} \equiv \mu\alpha^n - \widetilde{v}_{0C}^n \ , \qquad X^{2n} \equiv 0 \ , \\
&\bar{B}_0^{2n-1} \equiv \overline{v_{0C}^n} \ , \qquad \bar{B}_i^{2n-1} \equiv v_i^n \ , \qquad \bar{B}_\mu^{2n} \equiv a_\mu^n \ ,
\end{aligned} \tag{35}
$$

leads

$$
\begin{aligned}
A_\mu(x^\mu, z) &= -i\mu\delta_{\mu 0} + \sum_{n=1}^{\infty}\{iX^n\delta_{\mu 0} + \bar{B}_\mu^n(x^\mu)\}\psi_n(z) \ , \\
A_z(x^\mu, z) &= -i\Pi(x)\phi_0(z) \ .
\end{aligned} \tag{36}
$$

The corresponding field strengths are

$$
F_{\mu\nu} = \sum_{n=1}^{\infty}(\partial_\mu \bar{B}_\nu^n - \partial_\nu \bar{B}_\mu^n)\psi_n + \sum_{n,m=1}^{\infty}[\bar{B}_\mu^n, \bar{B}_\nu^m]\psi_n\psi_m \ , \tag{37}
$$

$$
F_{z\mu} = i\partial_\mu \Pi\, \phi_0 + \sum_{n=1}^{\infty}\{iX^n\delta_{\mu 0} + \bar{B}_\mu^n(x^\mu)\}\dot{\psi}_n + i\sum_{n=1}^{\infty}[\Pi, \bar{B}_\mu^n]\psi_n\phi_0 \ . \tag{38}
$$

Notice that $iX^n\delta_{\mu 0}$ does not contribute to $F_{\mu\nu}$. The additional terms due to $iX^n\delta_{\mu 0}$ come from $F_{\mu z}$. They contribute to the induced action through

$$
\begin{aligned}
F_{\mu\nu}F_{\mu\nu} =& (\partial_\mu \bar{B}_\nu^n - \partial_\nu \bar{B}_\mu^n)(\partial_\mu \bar{B}_\nu^m - \partial_\nu \bar{B}_\mu^m)\psi_n\psi_m \\
&+ \left\{[\bar{B}_\mu^n, \bar{B}_\nu^m]\ , (\partial_\mu \bar{B}_\nu^l - \partial_\nu \bar{B}_\mu^l)\right\}\psi_n\psi_m\psi_l \ , \qquad (39) \\
F_{\mu z}F_{\mu z} =& -(\partial_\mu\Pi\partial_\mu\Pi)\phi_0^2 + \bar{B}_\mu^n \bar{B}_\mu^m \dot{\psi}_n\dot{\psi}_m - X^n X^m \dot{\psi}_n\dot{\psi}_m \\
&+ \left(i\{\partial_\mu\Pi\ , \bar{B}_\mu^n\} - 2(\partial_0\Pi)X^n\right)\phi_0\dot{\psi}_n + 2iX^n\bar{B}_0^m\dot{\psi}_n\dot{\psi}_m \\
&+ \left(-2X^n[\Pi\ , \bar{B}_0^m] + i\{\bar{B}_\mu^n\ , [\Pi\ , \bar{B}_\mu^m]\}\right)\dot{\psi}_n\psi_m\phi_0 \\
&+ \left\{\partial_\mu\Pi\ , [\bar{B}_\mu^n\ , \Pi]\right\}{\phi_0}^2\psi_n \ , \qquad (40) \\
\mathcal{O}(F^3) =& (2\pi\alpha')^4 \frac{9}{4}\frac{R^6}{U_{KK}U_z^2}\left\{i(\partial_0\Pi)\phi_0 + (iX^n + \bar{B}_0^n)\dot{\psi}_n\right\} \times \\
&\times \frac{1}{2}F_{ij}F_{ij}\left\{i(\partial_0\Pi)\phi_0 + (iX^m + \bar{B}_0^m)\dot{\psi}_m\right\}
\end{aligned}
$$

For notational simplicity we have omitted $\sum_{n=1}^{\infty}$.

The induced DBI action at finite density can be separated into field independent and dependent parts P_0 and P_1 respectively

$$
\begin{aligned}
P_0 \equiv& \ 1 - (2\pi\alpha')^2\frac{9}{4}\frac{U_z}{U_{\mathrm{KK}}}X^n X^m\dot{\psi}_n\dot{\psi}_m \\
P_1 \equiv& \ (2\pi\alpha')^2\frac{R^3}{2U_z^3}F_{\mu\nu}F_{\mu\nu} + (2\pi\alpha')^2\frac{9}{4}\frac{U_z}{U_{\mathrm{KK}}}(F_{\mu z}F_{\mu z} + X^n X^m\dot{\psi}_n\dot{\psi}_m) \\
&+[(F)^4] + [(F)^5]
\end{aligned} \tag{41}
$$

so that

$$
\begin{aligned}
S_{DBI} &= \tilde{T} \int d^4x dz\, U_z^2 \text{ tr } \sqrt{P_0 + P_1} \\
&= \tilde{T} \int d^4x dz\, U_z^2 \text{ tr } \left[\sqrt{P_0} + \frac{1}{2}\frac{P_1}{\sqrt{P_0}} - \frac{1}{8}\frac{P_1^2}{\sqrt{P_0}^3} + \cdots \right] \\
&= S_1(X^n) + S_2[\Pi, v_\mu; X^n] + \cdots ,
\end{aligned}
$$

with

$$
S_1(X^n) \equiv \tilde{T} \int d^4x dz\, U_z^2 \text{ tr } \sqrt{1 - (2\pi\alpha')^2 \frac{9}{4}\frac{U_z}{U_{\rm KK}} X^n X^m \dot{\psi}_n \dot{\psi}_m} \tag{42}
$$

$$
S_2[\Pi, v_\mu; X^n] \equiv \tilde{T} \int d^4x dz\, U_z^2 \text{ tr } \left[\frac{1}{2}\frac{P_1}{\sqrt{P_0}} - \frac{1}{8}\frac{P_1^2}{\sqrt{P_0}^3} + \cdots \right] . \tag{43}
$$

We note that the dependence on the chemical potential μ in S_1 is of the the type $\sqrt{1 - \#\mu^2}$. There is no term linear in μ(except through the Chern-Simons term which is odd under $t \to -t$). The reason is that the QCD partition function is even under $\mu \to -\mu$ since the matter spectrum is symmetric around zero quark virtuality. We believe that this behavior maybe derived geometrically from a change in the underlying metric, but we are unable to show it.

The chemical potential μ as defined in (22) with (32)(i.e. $\mathcal{V}_\mu = -i\,\mu\,\delta_{\mu 0}\, 1_{N_f \times N_f}$, $\mathcal{A}_\mu = 0$) yields $F_{0z} = 0$. Thus there is no contribution to the DBI action. The only contribution stems from the Chern-Simons action as can be checked explicitly. However the definition (22) is at odd with VMD as explained in detail in.[2] In other words the physical vector field are $\tilde{v}_\mu^n$ defined in (24) and not v_μ^n in (22). So (22) has to be substituted by (33). In this case $F_{0z} \neq 0$ and μ contributes to the DBI action but not to the Chern-Simons action.

4. Mesons in Holographic Dense Matter

The DBI effective action in matter fixes completly the meson dispersion laws and interactions. Indeed, in matter $S_2[v_\mu; X^n]$ is

$$
S_2[\Pi, v_\mu; X^n] = \int d^4x dZ \text{ tr } \left[Q\tilde{T} U_{KK} U_z^2 P_1 - \frac{1}{4} Q^3 \tilde{T} U_{KK} U_z^2 P_1^2 + \cdots \right] . \tag{44}
$$

where

$$Q(Z;X^n) \equiv \frac{1}{\sqrt{P_0}} = \frac{1}{\sqrt{1-\Delta K^{\frac{1}{3}}\dot{\Psi}_1^2}}$$
$$\Delta \equiv \frac{m_{v0}^4 N_c^2 (X^1)^2}{C_2 f_{\Pi 0}^6} = \frac{27^3\pi^5 (X^1)^2}{N_c M_{KK}^2 \lambda^3}$$
$$\sim \frac{1}{N_c} \;\text{ fixed }\; \mu_B \quad (\sim 1 \;\text{ fixed }\; \mu) \tag{45}$$

Only the term $n = m = 1$ in the summation over the vector meson species was retained. In this section we treat A_μ as anti-Hermitiaon matrices with generators t^a normalized as $\mathrm{tr}(t^a t^b) = \frac{1}{2}\delta_{ab}$. The effective action is rotated back to Minkowski space. Throughout, only terms up to the third order in the fields are retained, i.e.$\mathcal{O}((\Pi, v_\mu)^4)$. The expansion can be simplified by using the notations

$$\bar{v}_0 := \overline{v_{0C}^1}, \quad X := X^1\ , \quad v_i := v_i^1 \tag{46}$$

so that

$$\begin{aligned} S_2[\Pi, v_\mu; \mu] = \int d^4x \Big[& a_{\Pi^2}^T \,\mathrm{tr}\,(\partial_0\Pi\partial_0\Pi) - a_{\Pi^2}^S \,\mathrm{tr}\,(\partial_i\Pi\partial_i\Pi) \\ & - a_{v^2}^T \,\mathrm{tr}(\partial_0 v_i - \partial_i\bar{v}_0)^2 + \frac{1}{2} a_{v^2}^S \,\mathrm{tr}(\partial_i v_j - \partial_j v_i)^2 \\ & - m_v^{2^T} \,\mathrm{tr}\,\bar{v}_0^2 + m_v^{2^S} \,\mathrm{tr}\, v_i^2 \\ & - a_{v^3}^T \,\mathrm{tr}\,\Big([\bar{v}_0, v_i](\partial_0 v_i - \partial_i\bar{v}_0)\Big) + a_{v^3}^S \,\mathrm{tr}\,\Big([v_i, v_j](\partial_i v_j - \partial_j v_i)\Big) \\ & - a_{v\Pi^2}^T \,\mathrm{tr}\,\Big(\partial_0\Pi[\bar{v}_0, \Pi]\Big) + a_{v\Pi^2}^S \,\mathrm{tr}\,\Big(\partial_i\Pi[v_i, \Pi]\Big) + \cdots\,, \end{aligned} \tag{47}$$

where we keep the terms up to third order. All the parameters appearing in (47) are tabulated below for non-zero density. Their respective values at zero density are recorded on the right-most column for comparison. We note that $Q, \Delta \sim N_c^0$ for fixed μ, so that all coefficients are of order N_c^0 for fixed μ at large N_c.

Coefficients	nonzero X	$X = 0(Q = 1, \Delta = 0)$
$a^T_{\Pi^2}$	$\frac{1}{\pi}\int dZK^{-1}Q\left(1+\Delta Q^2 K^{\frac{1}{3}}\dot{\Psi}_1^2\right)$	1
$a^S_{\Pi^2}$	$\frac{1}{\pi}\int dZK^{-1}Q$	1
$a^T_{v^2}$	$\int dZ\,K^{-\frac{1}{3}}\Psi_1^2 Q$	1
$a^S_{v^2}$	$\int dZ\,K^{-\frac{1}{3}}\Psi_1^2 Q^{-1}$	1
$m_v^{2\,T}$	$\frac{m^2_{v0}}{\lambda_1}\int dZ\,K\dot{\Psi}_1^2 Q\left(1+\Delta Q^2 K^{\frac{1}{3}}\dot{\Psi}_1^2\right)$	m^2_ρ
$m_v^{2\,S}$	$\frac{m^2_{v0}}{\lambda_1}\int dZ\,K\dot{\Psi}_1^2 Q$	m^2_ρ
$a^T_{v^3}$	$\frac{2m_{v0}}{\sqrt{\pi\lambda_1}f_{\Pi 0}}\int dZ\,K^{-\frac{1}{3}}\Psi_1^3 Q$	$\frac{2m_{v0}}{\sqrt{\pi\lambda_1}f_{\Pi 0}}\cdot 0.45$
$a^S_{v^3}$	$\frac{2m_{v0}}{\sqrt{\pi\lambda_1}f_{\Pi 0}}\int dZ\,K^{-\frac{1}{3}}\Psi_1^3 Q^{-1}$	$\frac{2m_{v0}}{\sqrt{\pi\lambda_1}f_{\Pi 0}}\cdot 0.45$
$a^T_{v\Pi^2}$	$\frac{2}{\sqrt{\kappa\pi}}\int dZ\,K^{-1}\Psi_1^2 Q\left(1+\Delta Q^2 K^{\frac{1}{3}}\dot{\Psi}_1^2\right)$	$\frac{2}{\sqrt{\kappa\pi}}\cdot 0.63$
$a^S_{v\Pi^2}$	$\frac{2}{\sqrt{\kappa\pi}}\int dZ\,K^{-1}\Psi_1^2 Q$	$\frac{2}{\sqrt{\kappa\pi}}\cdot 0.63$

4.1. *Meson Velocities*

For $N_c = 3$, $m_{vo} = 776\text{MeV}$, $f_{\pi 0} = 93\text{MeV}$, we show the pion in Fig. 1 and vector meson in Fig. 2 renormalization constants, both for the time and space-components. The resulting velocity of the pion and the rho meson in holographic dense matter is shown in Fig. 3. The pion velocity

$$v_\pi = \sqrt{\frac{a^S_{\pi^2}}{a^T_{\pi^2}}} = \frac{f^S_\pi}{f^T_\pi} \tag{48}$$

with $f^{S,T}_\pi$, the spatial (S) and temporal (T) pion decay constants, approaches zero when hadronic matter freezes. The vector meson velocity

$$v_v = \sqrt{\frac{a^S_{v^2}}{a^T_{v^2}}} \tag{49}$$

is about half.

4.2. *Vector Screening Masses*

The pion is massless throughout as the current quark masses are set to zero in this work. The vector meson masses follow by Higgsing, and change in matter both space-like and time-like. Space-like, the vector meson masses are screening masses. The unrenormalized masses increase with increasing baryon density as shown in Fig. 4. The physical screening masses(M_T) are obtained by normalizing with the pertinent matter dependent *space-like* wavefunction renormalizations. Both the vector and isovector screening masses increase with increasing baryon density as they should. Fig. 5 show the screening masses for longitudinal (left) and transverse (right) vectors. They are defined as

$$M_L^2 = \frac{m_v^{2T}/m_{v0}^2}{a_{v^2}^T}\,, \qquad M_T^2 = \frac{m_v^{2S}/m_{v0}^2}{a_{v^2}^S}\,. \tag{50}$$

4.3. *Vector Meson Masses*

The time-like vector masses are obtained similarly to the screening masses by instead using the *time-like* wavefunction renormalizations. That is

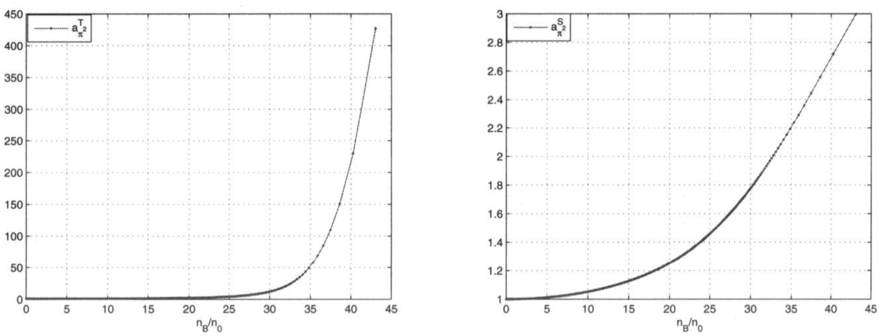

Fig. 1. $a^T_{\pi^2}$ and $a^S_{\pi^2}$.

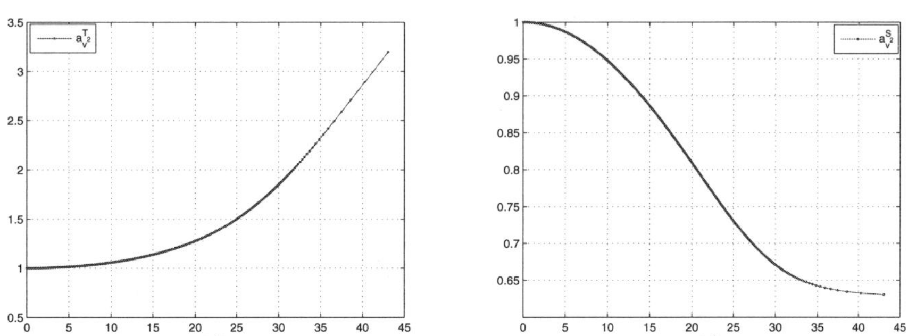

Fig. 2. $a^T_{v^2}$ and $a^S_{v^2}$.

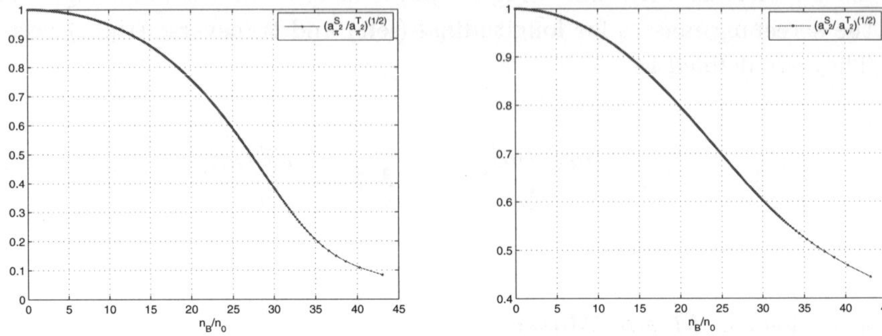

Fig. 3. Velocity: pion(left) and ρ(right).

$$M_V^2 = \frac{m_v^{2S}/m_{v0}^2}{a_{v^2}^T} \tag{51}$$

The dependence of M_V^2 on the baryon density is shown in Fig. 6. The rho meson mass drops by 20% when holographic dense matter freezes.

4.4. *$V\pi\pi$ and VVV Couplings in Matter*

The $V\pi\pi$ coupling is modified in matter. The longitudinal and transverse couplings are shown in Fig. 7 (left). The longitudinal coupling drops by 40% at the freezing point, while the transverse coupling *vanishes* when dense matter freezes. In Fig. 7 (right) VVV coupling also decrease in matter. The transverse couplings drop by 90%, while the longitudinal ones by 30%.

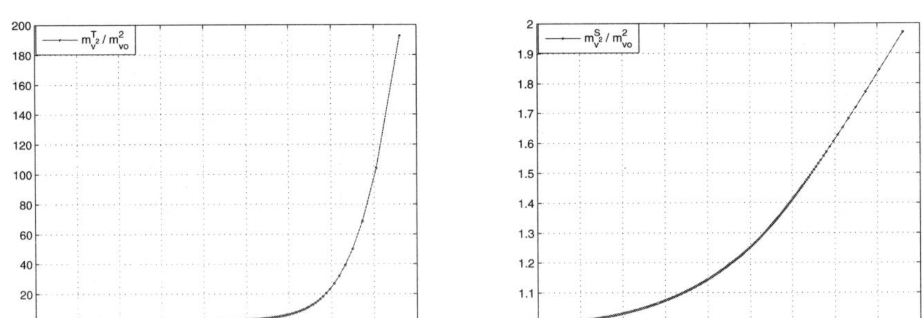

Fig. 4. $\frac{m_v^{2\,T}}{m_{v0}^2}$ and $\frac{m_v^{2\,S}}{m_{v0}^2}$.

Many of the results presented here in matter bears similarities (and differences) with arguments presented at finite temperature using phenomenological models with hidden local symmetry.[8]

5. Conclusions

We have provided a minimal extension of holographic QCD[2] to dense matter. The induced DBI effective action on D8 in the presence of a constant external $U(1)_v$ source $\mathcal{V}_0$ is the effective action we expect from the gravity dual theory from general principles. Indeed, if we were to solve for the effect of the boundary $\mathcal{V}_0$ on the D4-brane background and the resulting DBI action on the D8-probe brane, the answer is the *externally gauged* DBI induced action in the vacuum. The effective action develops an imaginary

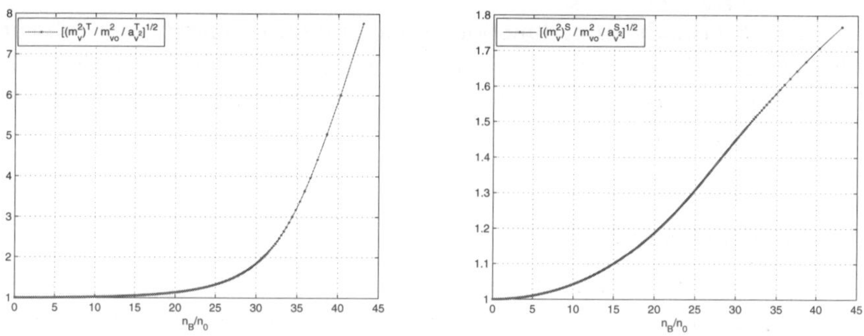

Fig. 5. Screening masses: Longitudinal (left) and Transverse (right).

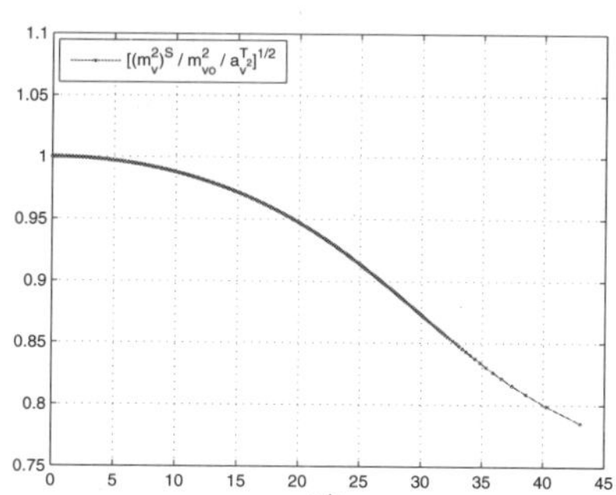

Fig. 6. Time-like Vector mass vs baryon density.

time at large $\mathcal{V}_0$, a signal that the D8-probe-brane craks in the external field.

Holographic matter describes dense QCD at large N_c with baryons as solitons. Its bulk pressure asymptotes a constant at large density, signalling total freezing with zero mean kinetic energy. Before freezing the matter is dominated by two-body repulsion at low density and three-body attraction at intermediate densities. The two-body effects are 100 times stronger than the three-body effects.

In holographic matter the pions stall to almost a stop, while the vector mesons only slow down. The vector masses drop by about 20%, while vector screening becomes increasingly large. The transverse vector mesons completely decouple from pions at large densities. Many of the current results bear similarities with known results from effective models at large N_c. In a way, they are new as they provide first principle calculations to large N_c and strongly coupled QCD.

The thermodynamic functions such as internal energy and pressure in holographic dense matter have been worked out in.[9] In the work to follow, we will present results for holographic matter including finite quark masses and temperature.

Acknowledgments

The work of KYK and IZ was supported in part by US-DOE grants DE-FG02-88ER40388 and DE-FG03-97ER4014. The work of SJS was supported by KOSEF Grant R01-2004-000-10520-0 and by SRC Program of the KOSEF with grant number R11 - 2005- 021.

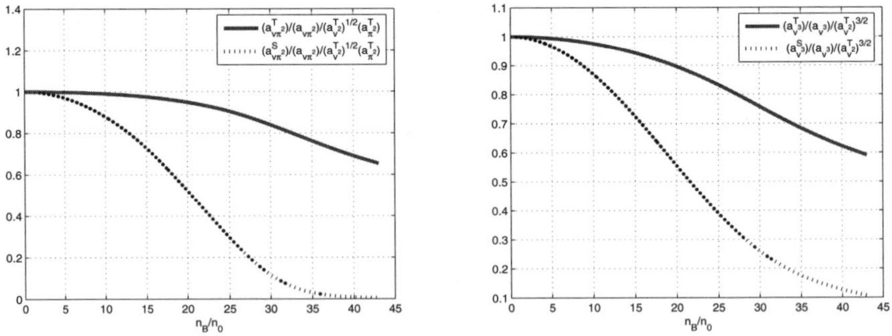

Fig. 7. $V\pi\pi$ and VVV couplings.

References

1. To mention some of them:
   J. Polchinski and M. J. Strassler, Phys. Rev. Lett. **88** (2002) 031601; hep-th/0109174;
   A. Karch and E. Katz, JHEP **0206**, 043 (2002) [arXiv:hep-th/0205236];
   J. Babington, J. Erdmenger, N. J. Evans, Z. Guralnik and I. Kirsch, Phys. Rev. D **69**, 066007 (2004) [arXiv:hep-th/0306018].
   M. Kruczenski, D. Mateos, R. C. Myers and D. J. Winters, JHEP **0405**, 041 (2004) [arXiv:hep-th/0311270].
2. T. Sakai and S. Sugimoto, Prog. Theor. Phys. **113**, 843 (2005) [arXiv:hep-th/0412141];
   T. Sakai and S. Sugimoto, Prog. Theor. Phys. **114**, 1083 (2006) [arXiv:hep-th/0507073].
3. J. M. Maldacena, *Adv. Theor. Math. Phys.* ***2**, 231 (1998)*, *Int. J. Theor. Phys.* ***38**, 1113 (1999)* [arXiv:hep-th/9711200], For review, see:
   O. Aharony, S. S. Gubser, J. M. Maldacena, H. Ooguri and Y. Oz, Phys. Rept. **323**, 183 (2000) [arXiv:hep-th/9905111].
4. G. Policastro, D. T. Son and A. O. Starinets, Phys. Rev. Lett. **87** (2001) 081601;
   S. J. Sin and I. Zahed, Phys. Lett. **B608** (2005) 265;
   E. Shuryak, S. J. Sin and I. Zahed, arXiv:hep-th/0511199;
   C.P. Herzog, A. Karch, P. Kotvun, C. Kozcaz, and L. Yaffe, hep-th/0605158;
   S. Gubser, hep-th/0605182.
5. R. Rapp, T. Schafer, E. Shuryak and M. Velkovsky, Phys. Rev. Lett. **81** (1998) 53; M. Alford, K. Rajagopal and F. Wilczek, Phys. Lett. **B422** (1998) 247.
6. B-Y. Park, M. Rho, A. Wirzba and I. Zahed, Phys. Rev. **D62** (2000) 034015.
7. To list some of them;
   A. Chamblin, R. Emparan, C. V. Johnson and R. C. Myers, Phys. Rev. D **60**, 064018 (1999) [arXiv:hep-th/9902170];
   K. Maeda, M. Natsuume and T. Okamura, Phys. Rev. D **73**, 066013 (2006) [arXiv:hep-th/0602010];
   D. T. Son and A. O. Starinets, JHEP **0603**, 052 (2006) [arXiv:hep-th/0601157];
   S. S. Gubser and J. J. Heckman, JHEP **0411**, 052 (2004) [arXiv:hep-th/0411001].
8. M. Harada and K. Yamawaki, Phys. Rep. **381** (2003) 1; G. Brown and M. Rho, Phys. Rep. **398** (2004) 301.
9. K. Y. Kim, S. J. Sin and I. Zahed, arXiv:hep-th/0608046.

STRONGLY INTERACTING MATTER AT RHIC

Chiho Nonaka

Department of Physics, Nagoya University,
Nagoya 464-8602, Japan
E-mail: nonaka@hken.phys.nagoya-u.ac.jp

We give a short review of hydrodynamic models at heavy ion collisions from the point of view of initial conditions, an equation of states (EoS) and freezeout process. Then we show our latest results of a combined fully three-dimensional macroscopic/microscopic transport approach. In this model for the early, dense, deconfined stage relativistic 3D-hydrodynamics of the reaction and a microscopic non-equilibrium model for the later hadronic stage where the equilibrium assumptions are not valid anymore are employed. Within this approach we study the dynamics of hot, bulk QCD matter, which is being created in ultra-relativistic heavy ion collisions at RHIC.

Keywords: Quark-gluon plasma; Hydrodynamic models.

1. Hydrodynamic Models at RHIC

The first five years of RHIC operations at $\sqrt{s_{NN}} = 130$ GeV and $\sqrt{s_{NN}} = 200$ GeV have yielded a vast amount of interesting and sometimes surprising results. There exists mounting evidence that RHIC has created a hot and dense state of deconfined QCD matter with properties similar to that of an ideal fluid[1] — this state of matter has been termed the *strongly interacting Quark-Gluon-Plasma* (sQGP). One of the evidence is the success of ideal hydrodynamic models in various physical observables. Especially, for the first time, in elliptic flow the hydrodynamic limit shows good agreement with the experimental data at RHIC, though at AGS and SPS hydrodynamic models give larger value as compared to experimental data.

The sophisticated 3D ideal hydrodynamic calculations, however, reveal that many of experimental data have not yet been fully evaluated or understood.[3] For example, elliptic flow at forward and backward rapidity [a] and at

[a]Brazil group shows improved results of v_2 at forward/backward rapidity by using event-by-event fluctuated initial conditions.[2]

peripheral collisions is overestimated by ideal hydrodynamical models. The Hanbury-Brown Twiss (HBT) puzzle is not completely understood from the point of view of hydrodynamics. These distinctions between hydrodynamic calculation and experimental data suggest that the ideal hydrodynamic picture is not applicable to all physical observables even in low transverse momentum region. At present our main interest in hydrodynamic models is the following: *How perfect the sQGP is?*

Here, first, we shall give a short review of hydrodynamic models at heavy ion collisions. In addition to a numerical procedure for solving the relativistic hydrodynamic equation, hydrodynamic models are characterized by initial conditions, EoSs and freezeout process.

The necessity of input of initial conditions for hydrodynamic models is one of the largest limitations of them. Because an initial condition is not be able to determined in the framework of hydrodynamic model itself, usually a parametrization of energy density and baryon number density based on Glauber type is used and parameters in it are determined by comparison with experimental data.[3–5] Recently there are some studies in which more basic approaches, Color Glass Condensate (CGC),[6] pQCD + saturation model[7] are used for construction of initial conditions.

The most important advantage of hydrodynamic models is that it directly incorporates an EoS as input and thus is so far the only dynamical model in which a phase transition can explicitly be incorporated. In the ideal fluid approximation – and once an initial condition has been specified – the EoS is the *only* input to the equations of motion and relates directly to properties of the matter under consideration. In this sense a hydrodynamic model is a bridge between QCD theory and experimental data and indispensable to describe heavy ion physics. However in usual practical hydrodynamic simulations, an EoS with 1st order phase transition (Bag model) is used. In fact, there are few studies on effect of order of QCD phase transition on physical observables.[8]

Conventional hydrodynamic calculations need to assume a *freezeout* temperature at which the hydrodynamic evolution is terminated and a transition from the zero mean-free-path approximation of a hydrodynamic approach to the infinite mean-free-path of free streaming particles takes place. The freezeout temperature usually is a free parameter which can be fitted to measured hadron spectra. There are several approaches for dealing with freezeout process: chemical equilibrium,[5,8] partial chemical equilibrium,[3] continuous emission model[9] and construction of a hybrid model of a hydro + cascade model.[4,10,11]

2. Hydro+UrQMD Model

We calculate hadron distribution at switching temperature from the 3D hydrodynamic model[12] using Cooper-Frye formula[13] and produce initial conditions for UrQMD model by Monte Carlo from it. Such hybrid macro/micro transport calculations (Fig. 1) are to date the most successful approaches for describing the soft physics at RHIC.

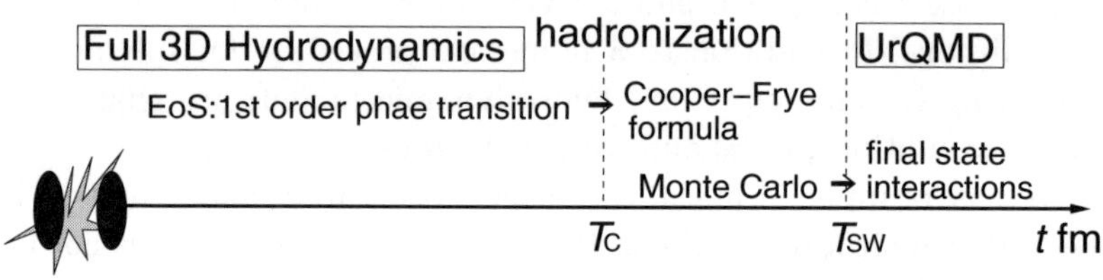

Fig. 1. Schematic sketch of 3D hydro+UrQMD model. $T_c(= 160$ MeV) and $T_{SW}(= 150)$ MeV are critical temperature and switching temperature from hydrodynamics to UrQMD model, respectively.

Figure 2 shows the P_T spectra of π^+, K and p at $\sqrt{s_{NN}} = 200$ GeV central collisions. The most compelling feature is that the hydro+micro approach is capable of accounting for the proper normalization of the spectra for all hadron species without any additional correction as is performed in the pure hydrodynamic model. The introduction of a realistic freezeout process provides therefore a natural solution to the problem of separating chemical and kinetic freeze-out in a pure hydrodynamic approach.

In Fig. 3 centrality dependence of P_T spectra of π^+ is shown. The impact parameter for each centrality is determined simply by the collision geometry. The separation between model results and experiment appears at lower transverse momentum in peripheral collisions compared to central collisions, just as in the pure hydrodynamic calculation. The 3D hydro + micro model does not provide any improvement for this behavior, since the hard physics high P_T contribution to the spectra occurs at early reaction times before the system has reached the QGP phase and is therefore neither included in the pure 3D hydrodynamic calculation nor in the hydro+micro approach.

Figure 4 shows the centrality dependence of the pseudorapidity distribution of charged hadrons compared to PHOBOS data.[16] Solid circles stand for model results and open circles denote data taken by the PHOBOS collaboration.[16] The impact parameters are set to $b = 2.4, 4.5, 6.3, 7.9$ fm for 0-6 %, 6-15 %, 15-25 % and 25-35 % centralities, respectively. Our results

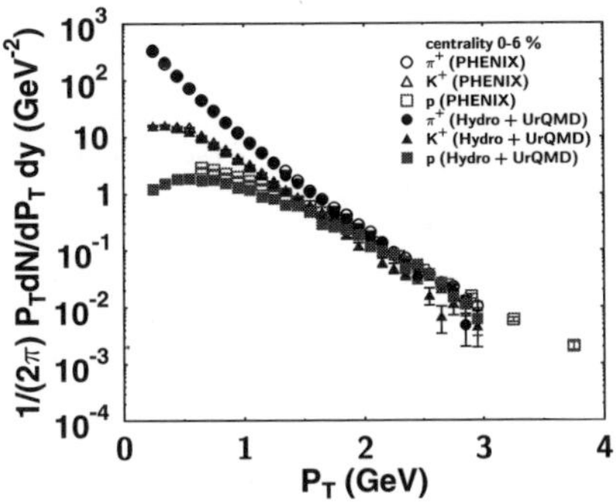

Fig. 2. P_T spectra for π^+, K^+ and p at central collisions with PHENIX data.[15]

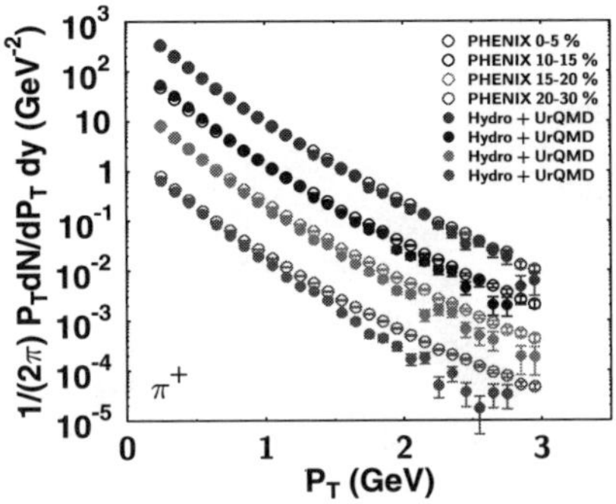

Fig. 3. Centrality dependence of P_T spectra of π^+ with PHENIX data.[15] The P_T spectra at 10–15 %, 15–20 % and 20–30 % are divided by 5, 25 and 200, respectively.

are consistent with experimental data over a wide pseudorapidity region. There is no distinct difference between 3-D ideal hydrodynamic model and the hydro + UrQMD model in the centrality dependence of the psuedorapidity distribution, indicating that the shape of psuedorapidity distribution is insensitive to the detailed microscopic reaction dynamics of the hadronic final state.[4]

In Fig. 5 we analyze the P_T spectra of multistrange particles. Our results show good agreement with experimental data for Λ, Ξ, Ω for centralities 0–5 %. In this calculation the additional procedure for normalization is not needed. Recent experimental results suggest that at thermal freezeout multistrange baryons exhibit less transverse flow and a higher temperature closer to the chemical freezeout temperature compared to non- or single-strange baryons.[17,18] This behavior can be understood in terms of the flavor dependence of the hadronic cross section, which decreases with increasing strangeness content of the hadron. The reduced cross section of multi-strange baryons leads to a decoupling from the hadronic medium at an earlier stage of the reaction, allowing them to provide information on the properties of the hadronizing QGP less distorted by hadronic final state interactions

In Fig. 6 the mean transverse momentum $\langle P_T \rangle$ as a function of hadron mass is shown. Open symbols denote the value at $T_{\mathrm{sw}} = 150$ MeV, corrected for hadronic decays. Not surprisingly, in this case the $\langle P_T \rangle$ follow a straight line, suggesting a hydrodynamic expansion. However if hadronic rescatter-

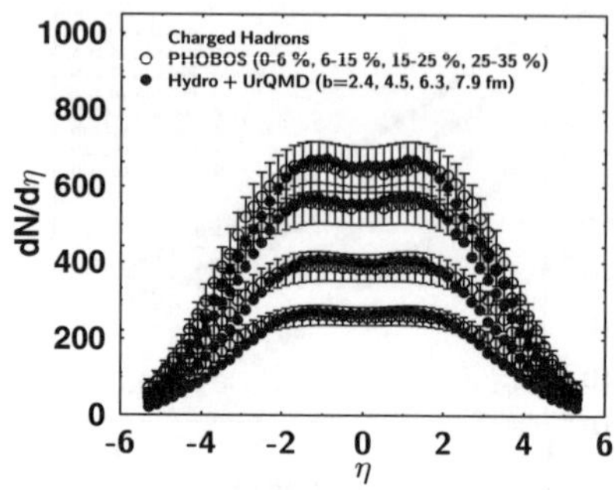

Fig. 4. Centrality dependence of pseudorapidity distribution of charged particles with PHOBOS data.[16]

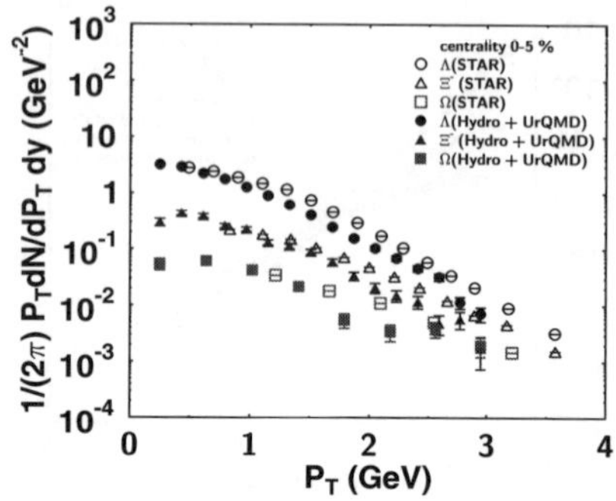

Fig. 5. P_T spectra of multi-strange particles at centralities 0–5 % and 10–20 % with STAR data.[17]

ing is taken into account (solid circles) the $\langle P_T \rangle$ do not follow the straight line any more: the $\langle P_T \rangle$ of pions is actually reduced by hadronic rescattering (they act as a heat-bath in the collective expansion), whereas protons actually pick up additional transverse momentum in the hadronic phase. RHIC data by the STAR collaboration is shown via the solid triangles – overall the proper treatment of hadronic final state interactions significantly improves the agreement of the model calculation with the data.

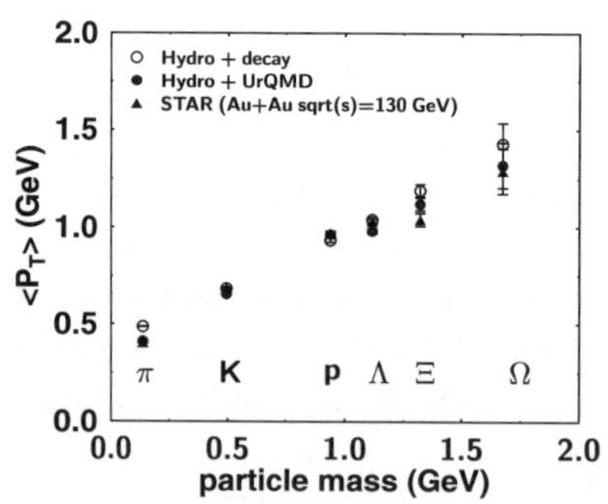

Fig. 6. Mean P_T as a function of mass with STAR data (Au+Au $\sqrt{s_{NN}}$ = 130 GeV).[18]

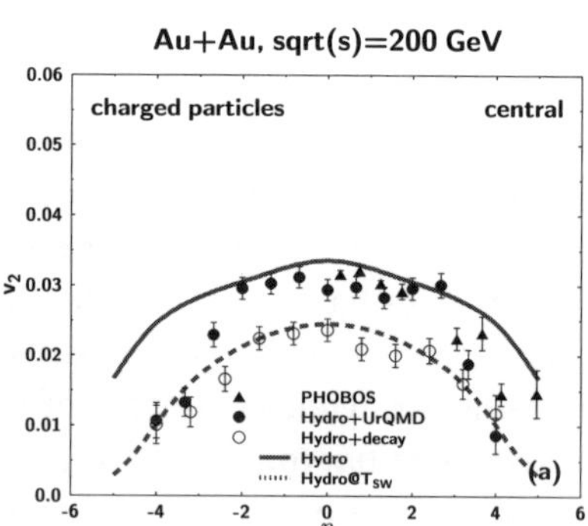

Fig. 7. Elliptic flow as a function of η of charged particles with PHOBOS data.[19]

Figure 7 shows the elliptic flow as a function of η: the pure hydrodynamic calculation is shown by the solid curve, the hydrodynamic contribution at T_{sw} is denoted by the dashed line and the full hydro+micro calculation is given by the solid circles, together with PHOBOS data (solid triangles). The shape of the elliptic flow in the pure hydrodynamic calculation at T_{sw} is quite different from that of the full hydrodynamic one terminated at a freeze-out temperature of 110 MeV. Apparently the slight bump at forward and backward rapidities observed in the full hydrodynamic calculation develops first in the later hadronic phase, since it is not observed in the calculation terminated at T_{sw}. Evolving the hadronic phase in the hydro+micro approach will increase the elliptic flow at central rapidities, but not in the projectile and target rapidity domains. As a result, the elliptic flow calculation in the hydro+micro approach is closer to the experimental data when compared to the pure hydrodynamic calculation.

We developed a novel implementation of the well known hybrid macroscopic/microscopic transport approach, combining a newly developed relativistic 3+1 dimensional hydrodynamic model for the early deconfined stage of the reaction and the hadronization process with a microscopic non-equilibrium model for the later hadronic stage. Within this approach we have dynamically calculated the freezeout of the hadronic system, accounting for the collective flow on the hadronization hypersurface generated by the QGP expansion. We have compared the results of our hybrid model and of a calculation utilizing our hydrodynamic model for the full evolution of the reaction to experimental data. This comparison has allowed us to quantify the strength of dissipative effects prevalent in the later hadronic phase of the reaction, which cannot be properly treated in the framework of ideal hydrodynamics.

3. Summary

The full 3D relativistic hydrodynamics + cascade model is one of successful and realistic models for description of dynamics of hot QCD bulk matter at RHIC, which helps us to understand medium property at RHIC in detail. Using this model, we can explore interesting phenomena which are caused by interactions between medium and jets.[20,21] One of proposed interesting physical observables is mach cone[21] for the wake of them, from which we can know medium property in detail, too.

Furthermore, viscosity effect in medium also starts to be discussed actively.[22] Due to serious difficulty in construction not only of a viscous hydrodynamic code but also of framework of viscous hydrodynamics without the causality problem, the progress of study of viscous hydrodynamics has been slow. However, by virtue of recent rapid development, the practical calculations with viscous hydrodynamics which are comparable to experimental data will be achieved in the near future.[22]

Finally, at LHC heavy ion collisions experiments will start in a year. Upcoming LHC data shall bring a lot of interesting, fruitful and even unexpected results for QGP physics, where hydrodynamic description will be helpful and useful for understanding of it.

References

1. T. Ludlam, Nucl. Phys. A **750**, 9 (2005); M. Gyulassy and L. McLerran, Nucl. Phys. A **750**, 30 (2005).
2. R. Andrade et al., Phys. Rev. Lett. **97**, 202302 (2006).
3. T. Hirano and K. Tsuda, Phys. Rev. **C66**, 054905 (2002).
4. C. Nonaka, S. A. Bass, Phys. Rev. **C75** 014902 (2007).
5. P. F. Kolb, P. Huovinen, U. Heinz, H. Heselberg, Phys. Lett. **B500**, 232 (2001); P. Huovinen, P. F. Kolb, U. Heinz, P. V. Ruusukanen, S. A. Voloshin, Phys. Lett. **B503**, 58 (2001).
6. T. Hirano and Y. Nara, Nucl. Phys. **A743**, 305 (2004).
7. K. J. Eskola et al., Phys. Rev. **C72**, 044904 (2005).
8. P. Huovinen, Nucl. Phys. **A761**, 29 (2005).
9. F. Grassi, Y. Hama and T. Kodama, Phys. Lett. **355**, 9 (1995); Z. Phys. **C73**, 153 (1996).
10. S. A. Bass, A. Dumitru, Phys. Rev. **C61**, 064909 (2000); D. Teaney, J. Lauret,E. V. Shuryak, Phys. Rev. Lett. **86**, 4783 (2001); nucl-th/0110037.
11. T. Hirano, U. Heinz, D. Kharzeev, R. Lacey, Y. Nara, Phys. Lett. **B636**, 299 (2006).
12. C. Nonaka, E. Honda, S. Muroya, Eur. Phys. J. **C17**, 663 (2000).
13. F. Cooper and G. Frye, Phys. Rev. **D10**, 186 (1974).
14. S. A. Bass et al., Progr. Part. Nucl. Physics Vol. **41**, 225 (1998) Progr. Part. Nucl. Physics Vol. **41**, 225 (1998); M. Bleicher et al., J. Phys. **G25**, 1859 (1999).
15. S. S. Adler et al. (PHENIX Collaboration), Phys. Rev. **C69**, 034909 (2004).
16. B. B. Back et al. (PHOBOS Collaboration), Phys. Rev. Lett. **91**, 052303 (2003).
17. M. Estienne (for the STAR Collaboration), J. Phys. **G31**, S873 (2005).
18. J. Adames et al. (STAR Collaboration), Phys. Rev. Lett. **92**, 182301 (2004).
19. B. B. Back et al. (PHOBOS Collaboration), Phys. Rev. **C72**, 051901(R) (2005).
20. T. Renk, J. Ruppert, C. Nonaka, S. A. Bass, nucl-th/0611027, to appear in Phys. Rev. C.
21. Casalderrey-Solana, hep-ph/0701257; T. Renk and J. Ruppert, hep-ph/0702102.
22. D. Teaney, Phys. Rev. **C68**, 034913 (2003). Ph. Mota et al., hep-ph/0701162. A. Muronga, nucl-th/0611090, nucl-th/0611091; U. Heinz, H. Song, A. K. Chaudhuri, Phys. Rev. **C73**,034904,2006.

Chromomagnetic Instability and Gluonic Phase

Michio Hashimoto

Department of Physics, Nagoya University,
Nagoya, 464-8602, Japan
E-mail: michioh@eken.phys.nagoya-u.ac.jp

We briefly report on a recent development in studies of a phase with vector condensates of gluons (gluonic phase) in dense two-flavor quark matter.

1. Introduction

It has been suggested that quark matter might exist inside central regions of compact stars.[1] At sufficiently high baryon density, cold quark matter is expected to be in a color superconducting (CSC) state.[2] This is one of the reasons why the color superconductivity has been intensively studied.[3]

Bulk matter in the compact stars must be in β-equilibrium and be electrically and color neutral. The electric and color neutrality conditions play a crucial role in the dynamics of quark pairing.[4] In addition, the strange quark mass cannot be neglected in moderately dense quark matter. Then a mismatch $\delta\mu$ between the Fermi surfaces of the pairing quarks is induced.

As the mismatch $\delta\mu$ grows, the CSC state tends to be destroyed. However, the dynamics is not yet solved completely. It is one of the central issues in this field to reveal the phase structure.

The problem is that the (gapped/gapless) two-flavor color superconducting phase (2SC/g2SC) suffers from the chromomagnetic instability connected with tachyonic Meissner screening masses of gluons.[5] While the Meissner mass for the 8th gluon is imaginary in the g2SC phase $\delta\mu > \Delta$, where Δ is a diquark gap, the chromomagnetic instability for the 4-7th gluons appears in $\delta\mu > \Delta/\sqrt{2}$. Such a chromomagnetic instability has been found also in the gapless color-flavor locked (gCFL) phase.[6–8]

Since the Meissner masses are defined at zero momentum and thus *not* pole ones, the physical origin of the chromomagnetic instability is not obvious. We then study the spectrum of plasmons with nonzero energy and momenta.[9] We also analyze the dispersion relations of the diquark fields.[10]

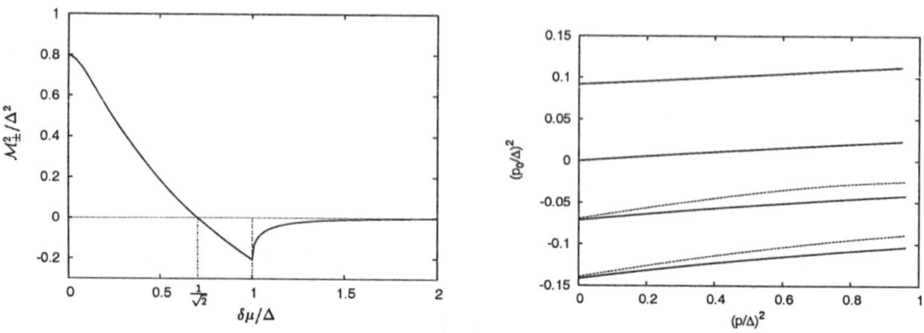

Fig. 1. Mass-gap-squared of the magnetic and electric plasmons for the 4-7th gluons (l.h.s.) and dispersion relations for the 4-7th magnetic plasmons (r.h.s.). In the r.h.s., the bold lines are for $\delta\mu/\Delta = 0.6, 1/\sqrt{2}$, 0.8, and 0.9, (2SC region) from top to bottom, while the dashed lines are for $\delta\mu/\Delta = 1.065$ and 1.009, (g2SC region) from top to bottom.

Then we will find that certain instabilities appear both in $\delta\mu > \Delta/\sqrt{2}$ and $\delta\mu > \Delta$, corresponding to the chromomagnetic instability.

How can we resolve the problem? Actually, numbers of ideas have been proposed by several authors.[11–16]

As a candidate of the genuine vacuum, we introduce a phase with vectorial gluon condensates (gluonic phase).[15] We also show that the single plane wave Larkin-Ovchinnikov-Fulde-Ferrell (LOFF) phase[11,17,18] suffers from a chromomagnetic instability.[19]

2. Spectra of the plasmons and the diquark excitations

Let us study two point functions of gluons and the diquark fields.

First, we analyze the polarization functions of gluons.[9]

For the 4-7th gluons, we find that the mass gaps for magnetic and electric modes coincide. We depict the results in the l.h.s. of Fig. 1. In the 4-7th channel, there exist the light plasmons in the whole region of $\delta\mu/\Delta$. While in $\delta\mu < \Delta/\sqrt{2}$ the light plasmons have the positive mass-gap-squared, $0 < \mathcal{M}^2_\pm \lesssim \Delta^2$, in $\delta\mu > \Delta/\sqrt{2}$ the plasmons become tachyons with $\mathcal{M}^2_\pm < 0$. It is noticeable that the instability occurs *both* for the magnetic and electric modes. This is essentially different from the chromomagnetic instability where the Debye mass for the electric mode remains real.[5]

The dispersion relations for the magnetic and electric modes do not coincide in general. We plot the dispersion relations of the magnetic modes for the 4-7th channels in the r.h.s. of Fig. 1. The dispersion relations qualitatively read $p_0^2 = m^2 + v^2p^2$, $(p \equiv |\vec{p}|)$ with $m^2 > 0$ for $\delta\mu/\Delta < 1/\sqrt{2}$ and $m^2 < 0$ for $\delta\mu/\Delta > 1/\sqrt{2}$. The velocity v is always real and less than 1.

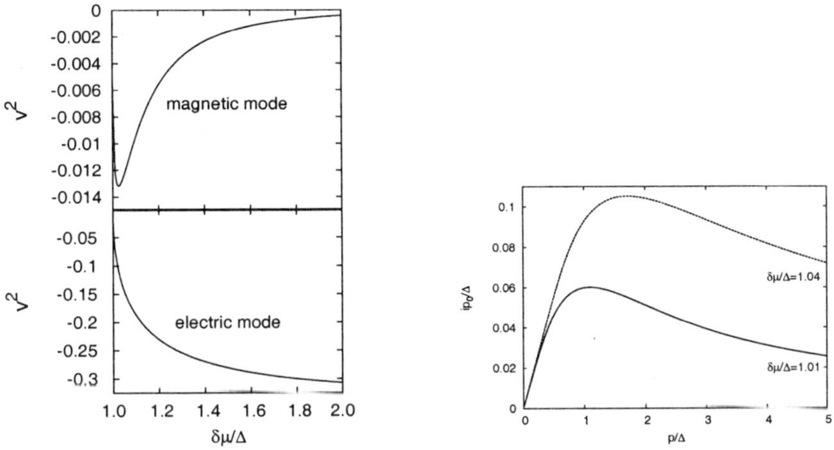

Fig. 2. Velocity squared for gapless tachyons in the magnetic and electric modes of the 8th channel (l.h.s.) and dispersion relations for the gapless magnetic tachyon (r.h.s.).

For the 8th channel, there is no light plasmon in $\delta\mu < \Delta$. On the other hand, in the g2SC region $\delta\mu > \Delta$ there appear gapless tachyons having the dispersion relation $p_0^2 = v^2 p^2$, $(p \simeq 0)$ with $v^2 < 0$, as shown in the l.h.s. of Fig. 2. This instability occurs *both* for the magnetic and electric modes, so that it also differs from the chromomagnetic instability in the 8th channel.[5] For the dispersion relations of the magnetic mode, see the r.h.s. of Fig. 2.

The appearance of the gapless tachyons in the 8th channel seems to be counter-intuitive, because the Meissner mass squared is nonzero and negative in the g2SC phase.[5] The origin is a singular dependence on p_0/p in the polarization function of the 8th gluon.[9]

Let us turn to discuss the two point functions of the diquark fields in the framework of a Nambu-Jona-Lasinio (NJL) model,[a] based on Ref. 10.

While we omit the results for $\Phi^{r,g}$, we do not those for ϕ_{b2} for clarity.

We depict the dispersion relations for ϕ_{b1} (dashed curve) and ϕ_{b2} (bold curves) in Fig. 3. We find that the velocity squared for ϕ_{b2} is $v^2 = 1/3$ in $\delta\mu < \Delta$, while there are two branches in $\delta\mu > \Delta$; one is, say, an ultra-relativistic branch with $1/3 < v^2 < 1$ and the other is a tachyonic one

[a]As usual, we take the anti-blue direction for the vacuum expectation value (VEV) of the diquark fields Φ^α, $(\alpha = r, g, b)$, i.e., $\langle \Phi^b \rangle = \Delta$. If the model is gauged, the anti-red and green diquark fields $\Phi^{r,g}$ are eaten by the 4-7th gluons, while the imaginary part ϕ_{b2} of the anti-blue diquark field Φ^b becomes the longitudinal mode of the 8th gluon. Only the real part ϕ_{b1} is left as a physical mode.

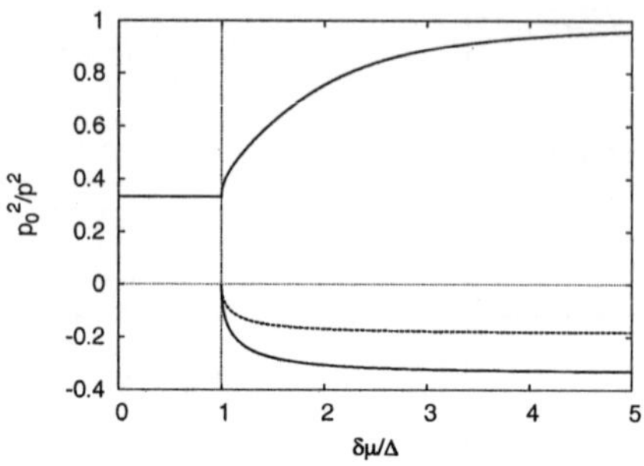

Fig. 3. Velocity squared for the anti-blue diquark field Φ^b. The bold dashed and solid curves are for ϕ_{b1} and ϕ_{b2}, respectively, where ϕ_{b1} and ϕ_{b2} are the real and imaginary parts of Φ^b, respectively. We ignored heavy excitations for ϕ_{b1}.

with $v^2 < 0$. For ϕ_{b1}, we ignore a heavy mode,[20–22] because it should be irrelevant to any instabilities. Actually, there is no light excitation of ϕ_{b1} in $\delta\mu < \Delta$. Surprising is that even for ϕ_{b1} a gapless tachyon with $v^2 < 0$ emerges in $\delta\mu > \Delta$,[b] as shown in the dashed curve in Fig. 3. (Another branch for ϕ_{b1} corresponding to a heavy mode may exist in $\delta\mu > \Delta$.[22])

We here comment that a singular dependence on p_0/p in the two point function of Φ^b causes the peculiar behaviors of the dispersion relations for ϕ_{b1} and ϕ_{b2} in $\delta\mu > \Delta$.[10]

3. Gluonic phase

What do the instabilities imply? Our answer is the existence of the vectorial gluon condensates (gluonic phase).

In order to clarify the problem, let us consider the $SU(2)_c$ decomposition owing to the symmetry breaking $SU(3)_c \to SU(2)_c$ in the presence of the diquark condensate Δ. The adjoint representation of $SU(3)_c$, i.e., the gluon field A_μ^a, $(a = 1, 2, \cdots, 8)$, is decomposed into $\mathbf{3} \oplus \mathbf{2} \oplus \bar{\mathbf{2}} \oplus \mathbf{1}$, namely, $\{A_\mu^a\} = (A_\mu^1, A_\mu^2, A_\mu^3) \oplus \phi_\mu \oplus \phi_\mu^* \oplus A_\mu^8$. Here we introduced the complex $SU(2)_c$ doublets of the vectorial "matter" fields,

$$\phi_\mu \equiv \begin{pmatrix} \phi_\mu^r \\ \phi_\mu^g \end{pmatrix} = \frac{1}{\sqrt{2}} \begin{pmatrix} A_\mu^4 - iA_\mu^5 \\ A_\mu^6 - iA_\mu^7 \end{pmatrix}, \qquad \phi_\mu^* \equiv \begin{pmatrix} \phi_\mu^{*r} \\ \phi_\mu^{*g} \end{pmatrix}. \tag{1}$$

In the gluonic phase, the spatial component of ϕ_μ develops the VEV, so that the role is quite similar to the Higgs doublet in the standard model.

[b]A similar instability is also discussed in Refs. 23,24.

Let us assume nonvanishing VEVs as

$$B \equiv g\langle A_z^6 \rangle, \quad C \equiv g\langle A_z^1 \rangle, \quad D \equiv \mu_3 = g\langle A_0^3 \rangle, \tag{2}$$

where C is required for consistency with the Ginzburg-Landau (GL) approach which we will employ. For a more general ansatz, see Ref.25.

Neglecting irrelevant terms, we obtain a reduced GL potential,[15]

$$\tilde{V}_{\text{eff}} = V_\Delta + \frac{1}{2}M_B^2 B^2 + TDB^2 + \frac{1}{2}\lambda_{BC}B^2C^2 + \frac{1}{2}\lambda_{CD}C^2D^2, \tag{3}$$

where V_Δ is the 2SC part and the parameter M_B^2 is expressed through the Meissner mass. The negative M_B^2 essentially dictates a nonvanishing gluon condensate $B \neq 0$. The coefficients T, λ_{BC} and λ_{CD} are calculated in the fermion one-loop approximation.[15] We then find that λ_{CD} is definitely negative and $\lambda_{BC} > 0$ in the vicinity of the critical point $\delta\mu \approx \Delta/\sqrt{2}$. The free energy at the stationary point is found as

$$\tilde{V}_{\text{eff}} = V_\Delta - \frac{(-M_B^2)^3}{54T^2}\left(-\frac{\lambda_{CD}}{\lambda_{BC}}\right) < V_\Delta. \tag{4}$$

Therefore the gluonic vacuum is stabler than the 2SC one. We can check also that the solution corresponds to a minimum.

It is noticeable that the above gluonic solution describes *non-Abelian* constant chromoelectric fields, $F_{0j}^a \neq 0$. In this sense, the gluonic phase enjoys a non-Abelian nature. We emphasize that while an Abelian constant electric field always leads to an instability, non-Abelian one does not in many cases.[26] The difference seems to be connected with the fact that a constant Abelian electric field is derived only from a vector potential depending on spatial and/or time coordinates, while a constant non-Abelian chromoelectric field can be expressed through constant vector potentials owing to nonzero commutators. Thus energy and momentum can be left as good quantum numbers in the non-Abelian case.

In the gluonic phase, both the rotational $SO(3)$ and the electromagnetic $U(1)$ symmetries are spontaneously broken down. Therefore, this phase describes an anisotropic medium with the color and electromagnetic Meissner effects. There also exist exotic hadrons in the medium.

4. The single plane wave LOFF state and its instability

In order to demonstrate how the neutrality conditions work and dramatically change the situation, we analyze the single plane wave LOFF state.[c,19]

[c]The order parameter of the single plane wave LOFF state has the form $\langle \Phi^b(x) \rangle = \Delta e^{2i\vec{q}\cdot\vec{x}}$ with a constant phase vector $\vec{q}$, instead of $\langle \Phi^b \rangle = \Delta$ in the 2SC/g2SC phase.

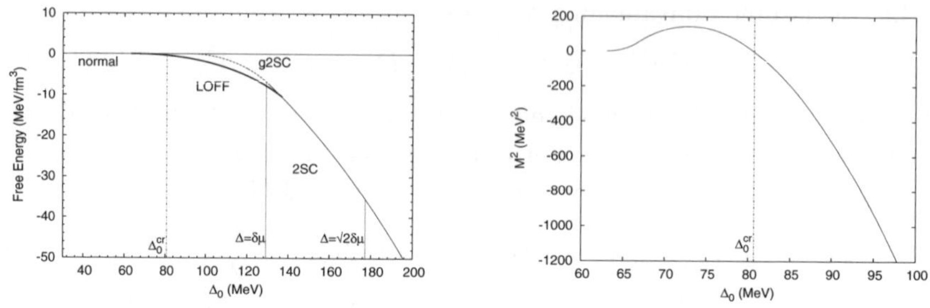

Fig. 4. Free energy of the neutral LOFF state [bold line], and the neutral 2SC/g2SC state [solid/dashed line] (l.h.s.) and Meissner mass squared M^2 for the 4-7th gluons in the LOFF phase (r.h.s.). In the l.h.s. the free energy of the neutral normal state is chosen as a reference point. In the r.h.s. we took the QCD coupling constant $\alpha_s = 1$.

For a numerical analysis, we fix the quark chemical potential μ=400 MeV and the cutoff $\Lambda = 653.3$ MeV, and vary the value of Δ_0, which is the 2SC gap parameter at $\delta\mu = 0$. We show the free energy differences in the l.h.s. of Fig. 4 (the reference point is the free energy of the normal phase with $\Delta = 0$). The results are not sensitive to the choice of μ (= 300–500 MeV).

One can see that the neutral LOFF phase is energetically stabler than the neutral normal phase and the neutral g2SC/2SC one in the whole region of the g2SC plus a narrow region of the 2SC near the edge, i.e., 63 MeV $< \Delta_0 <$ 136 MeV. Since the chromomagnetic instability in the 2SC phase occurs in the region $\Delta < \sqrt{2}\delta\mu$, which corresponds to $\Delta_0 = 177$ MeV in the l.h.s. of Fig. 4, the neutral LOFF solution cannot cure the instability.

Furthermore, by applying the Meissner mass formula in the second paper in Ref.18 to our LOFF solution, we find that the neutral LOFF state itself suffers from a chromomagnetic instability in $\Delta_0 > \Delta_0^{\rm cr} = 81$ MeV and thereby it should not be the genuine ground state. (See the r.h.s. of Fig. 4.)

5. Summary and discussion

We showed that in the 2SC/g2SC phase there appear several instabilities other than the chromomagnetic one.

We introduced the gluonic phase to resolve the instabilities. We also found that the neutral LOFF state is not free from the instabilities. It indicates that the gluonic phase is relevant for curing the problem.[d]

This phase is gauge equivalent to the phase with $\langle\Phi^b\rangle = \Delta$ and $\langle\vec{A}^8\rangle \neq 0$.[15,19]

[d]See also a numerical approach of the gluonic phase.[27]

It is worthwhile to examine the multiple plane-wave 2SC LOFF state.[28] It would be also interesting to figure out whether or not a phase with vectorial gluon condensates exists in three-flavor quark matter.

References

1. D. Ivanenko, *et al.*, Astrofiz. **1** 479 (1965); Lett. Nuovo Cim. **2**, 13 (1969); N. Itoh, Prog. Theor. Phys. **44**, 291 (1970); F. Iachello, *et al.*, Nucl. Phys. A **219**, 612 (1974); J. C. Collins, *et al.*, Phys. Rev. Lett. **34**, 1353 (1975).
2. D. Bailin and A. Love, Phys. Rept. **107**, 325 (1984); M. Iwasaki and T. Iwado, Phys. Lett. B **350**, 163 (1995). M. G. Alford, *et al.*, *ibid* **422**, 247 (1998); R. Rapp, *et al.*, Phys. Rev. Lett. **81**, 53 (1998).
3. K. Rajagopal, *et al.*, hep-ph/0011333; M. G. Alford, hep-ph/0102047; D. K. Hong, hep-ph/0101025; S. Reddy, nucl-th/0211045; T. Schäfer, hep-ph/0304281; D. H. Rischke, nucl-th/0305030; M. Buballa, Phys. Rept. **407**, 205 (2005); M. Huang, hep-ph/0409167; I. A. Shovkovy, nucl-th/0410091.
4. K. Iida and G. Baym, Phys. Rev. D **63**, 074018 (2001) [Erratum-ibid. D **66**, 059903 (2002)]; M. Alford and K. Rajagopal, JHEP **0206**, 031 (2002).
5. M. Huang, *et al.*, Phys. Rev. D **70**, 051501(R) (2004); *ibid* **70**, 094030 (2004).
6. R. Casalbuoni, *et al.*, Phys. Lett. B **605**, 362 (2005); *ibid* **615**, 297 (2005).
7. M. Alford and Q. Wang, J. Phys. G **31**, 719 (2005).
8. K. Fukushima, Phys. Rev. D **72**, 074002 (2005).
9. E. V. Gorbar, M. Hashimoto, V. A. Miransky and I. A. Shovkovy, Phys. Rev. D **73**, 111502(R) (2006).
10. M. Hashimoto, Phys. Lett. B **642**, 93 (2006).
11. M. G. Alford, J. A. Bowers and K. Rajagopal, Phys. Rev. D **63**, 074016 (2001); R. Casalbuoni and G. Nardulli, Rev. Mod. Phys. **76**, 263 (2004).
12. S. Reddy and G. Rupak, Phys. Rev. C **71**, 025201 (2005).
13. M. Huang, Phys. Rev. D **73**, 045007 (2006).
14. D. K. Hong, hep-ph/0506097.
15. E. Gorbar, M. Hashimoto and V. Miransky, Phys. Lett. B **632**, 305 (2006).
16. A. Kryjevski, hep-ph/0508180; T. Schäfer, Phys.Rev.Lett.**96**, 012305 (2006).
17. A. I. Larkin, *et al.*, Zh. Eksp. Teor. Fiz. **47**, 1136 (1964) [Sov. Phys. JETP **20**, 762 (1965)]; P. Fulde, *et al.*, Phys. Rev. **135** A550 (1964).
18. I. Giannakis and H. C. Ren, Phys. Lett. B **611**, 137 (2005); Nucl. Phys. B **723**, 255 (2005); I. Giannakis, *et al.*, Phys. Lett. B **631**, 16 (2005).
19. E. Gorbar, M. Hashimoto, V. Miransky, Phys. Rev. Lett. **96**, 022005 (2006).
20. D. Ebert, *et al.*, Phys. Rev. C **72**, 015201 (2005); *ibid* D **72**, 056007 (2005).
21. L. y. He, M. Jin and P. f. Zhuang, hep-ph/0504148.
22. D. Ebert, *et al.*, Phys. Rev. D **75**, 025024 (2007).
23. K. Iida and K. Fukushima, Phys. Rev. D **74**, 074020 (2006).
24. I. Giannakis, *et al.*, Phys. Rev. D **75**, 011501(R) (2007); *ibid* 014015 (2007).
25. E. V. Gorbar, M. Hashimoto and V. A. Miransky, hep-ph/0701211.
26. L. S. Brown, *et al.*, Nucl. Phys. B **157**, 285 (1979); *ibid* **172**, 544 (1980).
27. O. Kiriyama, *et al.*, Phys. Lett. B **643**, 331 (2006).
28. J. A. Bowers and K. Rajagopal, Phys. Rev. D **66**, 065002 (2002).

PROGRESS ON CHIRAL SYMMETRY BREAKING IN A STRONG MAGNETIC FIELD

SHANG-YUNG WANG

Department of Physics, Tamkang University,
Tamsui, Taipei 25137, Taiwan
E-mail: sywang@taos.phys.tku.edu.tw

The problem of chiral symmetry breaking in QED in a strong magnetic field is briefly reviewed. Recent progress on issues regarding the gauge fixing independence of the dynamically generated fermion mass is discussed.

Gauge theories play an important role in our understanding of a wide variety of phenomena in many areas of physics, ranging from the descriptions of fundamental interactions in elementary particle physics to the study of high temperature superconductivity in condensed matter physics. While the usual perturbative approach based on the loop expansion is sufficient in most circumstances, there are many interesting phenomena that can be understood only through a nonperturbative analysis. Examples include strong coupling gauge theories (such as QCD and hadron physics) as well as gauge fields under the influence of extreme conditions (such as high temperature and/or high density, strongly out-of-equilibrium instabilities, and strong external fields).

Several methods have been developed to study nonperturbative phenomena in gauge theories. Lattice approach based on discretized descriptions is well suited for studying strong coupling gauge theories, especially the properties in vacuum or at finite temperature. New approach based on the Anti-de Sitter/conformal field theory (AdS/CFT) correspondence offers the possibility to study strong coupling gauge theories by performing perturbative calculations in their gravity duals. Yet, field theoretical continuum approaches such as the Schwinger–Dyson (SD) equations and the effective action provide a natural framework for studies of nonperturbative phenomena in gauge theories. In particular, continuum descriptions are capable of describing dynamical real-time quantities in gauge fields under the influence of extreme conditions. A consistent formulation with gauge fields is important to the development of approximation techniques that will complement investigations utilizing lattice gauge theory and the AdS/CFT correspondence.

In this contribution we focus on the SD equations approach and discuss the important issues regarding the consistency and gauge independence of the truncation schemes therein. The SD equations are an infinite set of coupled integral equations among the Green's functions in a field theory, and form equations of motion of the corresponding theory. Nevertheless, practical calculations necessitate the use of approximated, or truncated, versions of the exact SD equations. A truncation of the SD equations corresponds diagrammatically to a selected resummation of an infinite subset of diagrams arising from every order in the loop expansion. Hence, in gauge theories the gauge independence of physical observables cannot be guaranteed unless consistent schemes are employed in truncating the SD equations.

We take as an example the simplest bare vertex approximation (BVA) to the SD equations in the simplest gauge theory, i.e., QED with massless Dirac fermions, in the presence of a background gauge field. Specifically, we consider the problem of chiral symmetry breaking in QED in a strong, external magnetic field. This problem was originally motivated by a proposal[1] to explain the correlated e^+e^- peaks observed in heavy ion collision experiments, and has subsequently received a lot of attention over the past decade largely because of its applications in astrophysics, condensed matter physics and cosmology.

Since the dynamics of fermion pairing in a strong magnetic field is dominated by the lowest Landau level (LLL),[2] it is a common practice to consider the propagation of, as well as radiative corrections originating only from, fermions occupying the LLL. This is referred to in the literature as the lowest Landau level approximation (LLLA).[2–4] In Ref. 3 chiral symmetry breaking in QED in a strong magnetic field was first studied in the so-called (quenched) rainbow approximation, in which the bare vertex and the free photon propagator were used in truncating the SD equations. These studies provided a preliminary affirmation of the phenomenon of chiral symmetry breaking in QED in a strong magnetic field.

More recently, this phenomenon has been studied in the so-called improved rainbow approximation,[4,5] in which the bare vertex and the full photon propagator were used. The authors of Ref. 4 claimed that (i) in covariant gauges there are one-loop vertex corrections arising from the longitudinal components of the full photon propagator that are not suppressed by powers of the gauge coupling constant and hence need to be accounted for; (ii) there exists a noncovariant and nonlocal gauge in which, and only in which, the BVA is a reliable truncation of the SD equations that consistently resums these one-loop vertex corrections; (iii) the phenomenon of chiral symmetry breaking is universal in that it takes place for any number of the fermion flavors. In Ref. 5 the authors included contributions to the vacuum polarization from higher Landau levels in an unspecified gauge, and asserted that (i) in QED with N_f fermion flavors a critical number N_{cr} exists for any value of the gauge coupling constant, such that chiral symmetry remains unbroken for $N_f > N_{cr}$; (ii) the dynamical fermion mass is generated with a double splitting for $N_f < N_{cr}$. The results of Refs. 4,5 are clearly in contradiction, leading to a controversy[6] over the correct calculation of the dynamical fermion mass generated through chiral symmetry breaking in a strong magnetic field.

The controversy was later resolved in Ref. 7 by establishing the gauge independence of the dynamical fermion mass calculated in the SD equations approach. In particular, it was shown that (i) the BVA is a consistent truncation of the SD equations in the LLLA; (ii) within this consistent truncation scheme the physical dynamical fermion mass, obtained as the solution of the truncated fermion SD equations evaluated on the mass shell, is manifestly gauge independent.

We take the constant external magnetic field of strength H in the x_3-direction. The corresponding vector potential is given by $A_\mu = (0, 0, Hx_1, 0)$, where $\mu = 0, 1, 2, 3$. In our convention, the metric has the signature $g_{\mu\nu} = \mathrm{diag}(-1, 1, 1, 1)$. In the LLLA, the SD equations for the full fermion propagator $G(x, y)$ are given by

$$G^{-1}(x,y) = S^{-1}(x,y) + \Sigma(x,y), \tag{1}$$

$$\Sigma(x,y) = ie^2 \int d^4x' d^4y' \, \gamma^\mu \, G(x,x') \, \Gamma^\nu(x',y,y') \, \mathcal{D}_{\mu\nu}(x,y'), \tag{2}$$

where $S(x, y)$ is the bare propagator for the LLL fermion in the external field A_μ, $\Sigma(x, y)$ is the LLL fermion self-energy, and $\Gamma^\nu(x, y, z)$ is the full LLL fermion-photon vertex. The full photon propagator $\mathcal{D}_{\mu\nu}(x, y)$ satisfies the SD equations

$$\mathcal{D}^{-1}_{\mu\nu}(x,y) = D^{-1}_{\mu\nu}(x,y) + \Pi_{\mu\nu}(x,y), \tag{3}$$

$$\Pi_{\mu\nu}(x,y) = -ie^2 \, \mathrm{tr} \int d^4x' d^4y' \, \gamma_\mu \, G(x,x') \, \Gamma_\nu(x',y',y) \, G(y',x), \tag{4}$$

where $D_{\mu\nu}(x, y)$ is the free photon propagator and $\Pi_{\mu\nu}(x, y)$ is the vacuum polarization. In the BVA to the SD equation one replaces the full vertex by the bare one, viz, $\Gamma^\mu(x,y,z) = \gamma^\mu\, \delta^{(4)}(x-z)\, \delta^{(4)}(y-z)$.

A consistent truncation of the SD equations is that which respects the Ward–Takahashi (WT) identity satisfied by the truncated vertex and inverse fermion propagator. The WT identity in the BVA within the LLLA was first studied in Ref. 8. It was shown that in order to satisfy the WT identity in the BVA within the LLLA, the LLL fermion self-energy in momentum space has to be a momentum independent constant. As per the WT identity in the BVA, the LLL fermion self-energy takes the form $\Sigma(p_\parallel) = m_\xi$, where $p_\parallel$ is the momentum of the LLL fermion and m_ξ is a momentum independent *but gauge dependent* constant, with ξ being the gauge fixing parameter. Here and henceforth, the subscript $\parallel$ ($\perp$) refers to the longitudinal: $\mu = 0, 3$ (transverse: $\mu = 1, 2$) components. It is noted that m_ξ depends implicitly on ξ through the full photon propagator $\mathcal{D}_{\mu\nu}$ in (2). We emphasize that because of its ξ-dependence, m_ξ should *not* be taken for granted to be the dynamical fermion mass, which is a gauge independent physical observable. This is one of the subtle points that has been overlooked in the literature.

We now show that the BVA within the LLLA is a consistent truncation of the SD equations (1)–(4), in which m_ξ is ξ-independent and hence can be identified unambiguously as the physical dynamical fermion mass if, and only if, the truncated SD equation for the fermion self-energy is evaluated on the fermion mass shell. First recall that, as proved in Ref. 9, in gauge theories the singularity structures (i.e., the positions of poles and branch singularities) of gauge boson and fermion propagators are gauge independent when all contributions of a given order of a systematic expansion scheme are accounted for. Consequently, the physical dynamical fermion mass has to be determined by the pole of the full fermion propagator obtained in a consistent truncation scheme of the SD equations. Assume for the moment that the BVA is a consistent truncation in the LLLA, such that the position of the pole of the LLL fermion propagator is gauge independent. In accordance with the WT identity in the BVA, we have

$$\Sigma(p_\parallel) = m, \tag{5}$$

where the constant m is the gauge independent, physical dynamical fermion mass, yet to be determined by solving the truncated SD equations self-consistently. What remains to be verified is the following statements: (i) the truncated vacuum polarization is transverse; (ii) the truncated fermion self-energy is gauge independent when evaluated on the mass shell, $p_\parallel^2 = -m^2$. We highlight that the fermion mass shell condition is one of the important points that has gone unnoticed in the literature, where the truncated fermion self-energy used to be evaluated off the mass shell at $p_\parallel^2 = 0$.[3–5]

The vacuum polarization $\Pi_{\mu\nu}(q)$ in the BVA is found to be given by

$$\Pi^{\mu\nu}(q) = -\frac{ie^2}{2\pi} N_f\, |eH| \exp\left(-\frac{q_\perp^2}{2|eH|}\right) \mathrm{tr} \int \frac{d^2 p_\parallel}{(2\pi)^2} \gamma_\parallel^\mu \frac{1}{\gamma_\parallel \cdot p_\parallel + m} \gamma_\parallel^\nu \\ \times \frac{1}{\gamma_\parallel \cdot (p-q)_\parallel + m} \Delta[\mathrm{sgn}(eH)], \qquad (6)$$

where $\Delta[\mathrm{sgn}(eH)] = [1 + i\gamma^1\gamma^2\, \mathrm{sgn}(eH)]/2$ is the projection operator on the fermion states with the spin parallel to the external magnetic field. The presence of $\Delta[\mathrm{sgn}(eH)]$ in (6) is a consequence of the LLLA, which implies an effective dimensional reduction from $(3+1)$ to $(1+1)$ in the fermion sector.[3]

The WT identity in the BVA guarantees that the vacuum polarization $\Pi_{\mu\nu}(q)$ is transverse, viz, $q^\mu \Pi_{\mu\nu}(q) = 0$. An explicit calculation yields $\Pi^{\mu\nu}(q) = \Pi(q_\parallel^2, q_\perp^2)(g_\parallel^{\mu\nu} - q_\parallel^\mu q_\parallel^\nu / q_\parallel^2)$, which in turn implies that the full photon propagator takes the following form in covariant gauges ($\xi = 1$ is the Feynman gauge):

$$\mathcal{D}^{\mu\nu}(q) = \frac{1}{q^2 + \Pi(q_\parallel^2, q_\perp^2)} \left(g_\parallel^{\mu\nu} - \frac{q_\parallel^\mu q_\parallel^\nu}{q_\parallel^2} \right) + \frac{g_\perp^{\mu\nu}}{q^2} + \frac{q_\parallel^\mu q_\parallel^\nu}{q^2 q_\parallel^2} + (\xi - 1) \frac{1}{q^2} \frac{q^\mu q^\nu}{q^2}. \qquad (7)$$

In the above expressions, the polarization function $\Pi(q_\parallel^2, q_\perp^2)$ is given by

$$\Pi(q_\parallel^2, q_\perp^2) = \frac{2\alpha}{\pi} N_f\, |eH| \exp\left(-\frac{q_\perp^2}{2|eH|}\right) F\left(\frac{q_\parallel^2}{4m^2}\right), \qquad (8)$$

where $\alpha = e^2/4\pi$. The function $F(u)$ has the following asymptotic behavior: $F(u) \simeq 0$ for $|u| \ll 1$ and $F(u) \simeq 1$ for $|u| \gg 1$. This implies that photons of momenta $m^2 \ll |q_\parallel^2| \ll |eH|$ and $q_\perp^2 \ll |eH|$ are screened with a characteristic length $L = (2\alpha N_f |eH|/\pi)^{-1/2}$. This screening effect renders the rainbow approximation[3] completely unreliable in this problem.

The fermion self-energy in the BVA, when evaluated on the fermion mass shell, $p_\parallel^2 = -m^2$, is given by

$$m\, \Delta[\mathrm{sgn}(eH)] = ie^2 \int \frac{d^4 q}{(2\pi)^4} \exp\left(-\frac{q_\perp^2}{2|eH|}\right) \gamma_\parallel^\mu \frac{1}{\gamma_\parallel \cdot (p-q)_\parallel + m} \gamma_\parallel^\nu \\ \times \mathcal{D}_{\mu\nu}(q)\, \Delta[\mathrm{sgn}(eH)] \Big|_{p_\parallel^2 = -m^2}, \qquad (9)$$

where $\mathcal{D}_{\mu\nu}(q)$ is given by (7). The WT identity in the BVA guarantees that this *would-be* gauge dependent contribution to the fermion self-energy (denoted symbolically as Σ_ξ) is proportional to $(\gamma_\parallel \cdot p_\parallel + m)$. We find through an explicit calculation that

$$\Sigma_\xi = \alpha\,(\xi - 1)\,(\gamma_\parallel \cdot p_\parallel + m) \int_0^1 dx\,(1-x) \int \frac{d^2 q_\perp}{(2\pi)^2} \exp\Bigl(-\frac{q_\perp^2}{2|eH|}\Bigr) \times \frac{(1-x)q_\perp^2 - xm(\gamma_\parallel \cdot p_\parallel - m)}{[(1-x)q_\perp^2 + x(1-x)p_\parallel^2 + xm^2]^2}\,\Delta[\mathrm{sgn}(eH)]. \tag{10}$$

Hence, Σ_ξ vanishes identically on the fermion mass shell $p_\parallel^2 = -m^2$ or, equivalently, $\gamma_\parallel \cdot p_\parallel + m = 0$. This, together with the transversality of the vacuum polarization, completes our proof that the BVA is a consistent truncation of the SD equations. Consequently, the dynamical fermion mass, obtained as the solution of the truncated fermion SD equations evaluated on the fermion mass shell, is gauge independent.

It can be verified in a similar manner that contributions to the fermion self-energy from the longitudinal components in $\mathcal{D}^{\mu\nu}(q)$ that are proportional to $q_\parallel^\mu q_\parallel^\nu / q_\parallel^2$ also vanish when evaluated on the fermion mass shell. Therefore, only the first term in $\mathcal{D}^{\mu\nu}(q)$ proportional to $g_\parallel^{\mu\nu}$ contributes to the on-shell fermion self-energy. As a result, the matrix structures on both sides of (9) are consistent. Using the mass shell condition $p_\parallel^\mu = (m, 0)$, corresponding to a LLL fermion at rest, we find (9) in Euclidean space to be given by

$$m = \frac{\alpha}{2\pi^2} \int d^2 q_\parallel \frac{m}{q_3^2 + (q_4 - m)^2 + m^2} \int_0^\infty dq_\perp^2 \frac{\exp(-q_\perp^2/2|eH|)}{q_\parallel^2 + q_\perp^2 + \Pi(q_\parallel^2, q_\perp^2)}, \tag{11}$$

where $q_\parallel^2 = q_3^2 + q_4^2$. Numerical analysis shows that the solution of (11) can be fit by the following analytic expression:

$$m = a\,\sqrt{2|eH|}\,\beta(\alpha)\,\exp\left[-\frac{\pi}{\alpha \log(b/N_f\,\alpha)}\right], \tag{12}$$

where a is a constant of order one, $b \simeq 2.3$, and $\beta(\alpha) \simeq N_f\,\alpha$. From (12) it follows that in a strong magnetic field chiral symmetry is broken regardless of the number of the fermion flavors.

The results of Refs. 4,5 can be attributed to gauge dependent artifacts. Had the authors of Ref. 4 calculated properly the on-shell, physical dynamical fermion mass, they would not have found the "large vertex corrections" they obtained, and therefore their claim that the BVA is a good approximation only in the special noncovariant and nonlocal gauge they invoke is not valid. In fact, that special gauge was invoked by hand such that the gauge dependent contribution cancels contributions from terms proportional to

$q_{\|}^{\mu} q_{\|}^{\nu} / q_{\|}^{2}$ in $\mathcal{D}^{\mu\nu}(q)$. Our gauge independent analysis in the BVA reveals clearly that such a gauge fixing not only is ad hoc and unnecessary, but also leaves the issue of gauge independence unaddressed.

The truncation used in Ref. 5 is *not* a consistent truncation of the SD equations because the WT identity in the BVA can be satisfied only within the LLLA.[7] Their result suggests that in the inconsistent truncation as well as in the unspecified gauge, the gauge dependent unphysical contributions from higher Landau levels become dominant over the gauge independent physical contribution from the LLL, thus leading to the authors' incorrect conclusions. Therefore we emphasize that the LLL dominance in a strong magnetic field should be understood in the context of a consistent truncation. Namely, contributions to the dynamical fermion mass from higher Landau levels that are obtained in a (yet to be determined) consistent truncation of the SD equations are subleading when compared to that from the LLL obtained in the consistent BVA truncation. Research along this line will be the subject of further investigations.

I would like to thank the organizers of the conference for their invitation and hospitality. I am grateful to C. N. Leung for a careful reading of the manuscript. This work was done in collaboration with C. N. Leung, and was supported in part by the National Science Council of Taiwan.

References

1. D. G. Caldi and A. Chodos, Phys. Rev. D **36**, 2876 (1987); Y. J. Ng and Y. Kikuchi, *ibid.*, 2880 (1987); and references therein.
2. V. P. Gusynin, V. A. Miransky, and I. A. Shovkovy, Phys. Rev. Lett. **73**, 3499 (1994); Phys. Lett. B **349**, 477 (1995).
3. V. P. Gusynin, V. A. Miransky, and I. A. Shovkovy, Phys. Rev. D **52**, 4747 (1995); Nucl. Phys. **B462**, 249 (1996); C. N. Leung, Y. J. Ng, and A. W. Ackley, Phys. Rev. D **54**, 4181 (1996); D.-S. Lee, C. N. Leung, and Y. J. Ng, *ibid.* **55**, 6504 (1997).
4. V. P. Gusynin, V. A. Miransky, and I. A. Shovkovy, Phys. Rev. Lett. **83**, 1291 (1999); Nucl. Phys. **B563**, 361 (1999); Phys. Rev. D **67**, 107703 (2003).
5. A. V. Kuznetsov and N. V. Mikheev, Phys. Rev. Lett. **89**, 011601 (2002).
6. V. P. Gusynin, V. A. Miransky, and I. A. Shovkovy, Phys. Rev. Lett. **90**, 089101 (2003); A. V. Kuznetsov and N. V. Mikheev, *ibid.*, 089102 (2003).
7. C. N. Leung and S.-Y. Wang, Nucl. Phys. **B747**, 266 (2006); Ann. Phys. (N.Y.) **322**, 701 (2007).
8. E. J. Ferrer and V. de la Incera, Phys. Rev. D **58**, 065008 (1998).
9. R. Kobes, G. Kunstatter, and A. Rebhan, Nucl. Phys. **B355**, 1 (1991).

Electroweak theory on the lattice with exact gauge invariance

Y. Kikukawa
Institute of Physics, University of Tokyo,
Tokyo, 153-8902, Japan
E-mail: kikukawa@hep1.c.u-tokyo.ac.jp

We present a gauge-invariant lattice formulation of the Glashow-Weinberg-Salam model based on the lattice Dirac operator satisfying the Ginsparg-Wilson relation. Our construction covers all topologically non-trivial SU(2) sectors with vanishing U(1) magnetic flux and would be usable for a description of the baryon number non-conservation.

Keywords: Lattice field theory; Chiral Symmetry; the Ginsparg-Wilson relation.

1. Introduction

Chiral gauge theories have several interesting possibilities in their own dynamics: fermion number non-conservation due to chiral anomaly,[1,2] various realizations of the gauge symmetry and global flavor symmetry,[3,4] the existence of massless composite fermions suggested by 't Hooft's anomaly matching condition[5] and so on. Unfortunately, very little is known so far about the actual behavior of chiral gauge theories beyond perturbation theory. It is desirable to develop a formulation to study the non-perturbative dynamics of chiral gauge theories.

Lattice gauge theory can now provide a framework for non-perturbative formulation of chiral gauge theories. The clue to this development is the construction of gauge-covariant and local lattice Dirac operators satisfying the Ginsparg-Wilson relation.[6–11] By this relation, it is possible to realize an exact chiral symmetry on the lattice.[12] It is also possible to introduce Weyl fermions on the lattice and this opens the possibility to formulate anomaly-free chiral lattice gauge theories.[13–26] Although it is believed that a chiral gauge theory is a difficult case for numerical simulations because the effective action induced by Weyl fermions has a non-zero imaginary part, still it would be interesting and even useful to develop a formulation of chiral lattice gauge theories by which one can work out fermionic observables numerically as the functions of link field with exact gauge invariance.

In the case of U(1) chiral gauge theories, Lüscher[14] proved rigorously that it is possible to construct the fermion path-integral measure which depends smoothly on the gauge field and fulfills the fundamental requirements such as locality, gauge-invariance, integrability and lattice symmetries. In this article, we show that it is possible to extend this lattice formulation to the case of SU(2)×U(1) chiral gauge theory in the Glashow-Weinberg-Salam model. We give a simple and closed expression of the fermion measure (term) for the SU(2)×U(1) chiral lattice gauge theory defined on the finite-volume lattice. Our formulation provides the first gauge-invariant (non-perturbative) regularization of the electroweak theory, which would be usable both in perturbative and non-perturbative analyses. In particular, since our construction covers all topologically non-trivial SU(2) sectors with vanishing U(1) magnetic flux, it would be usable for a description of the baryon number non-conservation.

This article is organized as follows. In section 2, we introduce our lattice formulation of the Glashow-Weinberg-Salam model at the classical level. In section 3, we define the path-integral measure of chiral fermion fields and formulate the reconstruction theorem which asserts that if there exist local currents satisfying cetain properties, it is possible to reconstruct the fermion measure which depends smoothly on the gauge field and fulfills the fundamental requirements such as locality,[a] gauge-invariance, integrability and lattice symmetries.[b] In section 4, we give an explicit formula of the local currents (the measure term) which fulfills all the required properties for the reconstruction theorem.

2. The Glashow-Weinberg-Salam model on the lattice

In this section, we describe a construction of the Glashow-Weinberg-Salam model on the lattice within the formulation of chiral lattice gauge theories based on the lattice Dirac operator satisfying the Ginsparg-Wilson relation. We assume a local, gauge-covariant lattice Dirac operator D which satisfies the Ginsparg-Wilson relation. An explicit example of such lattice Dirac operator is given by the overlap Dirac operator.[7,9] In this case, our formulation is equivalent to the overlap formalism for chiral lattice gauge theories.[28–30]

[a] We adopt the generalized notion of locality on the lattice given in[11,13,14] for Dirac operators and composite fields. See also[27] for the case of the finite volume lattice.
[b] The lattice symmetries mean translations, rotations, reflections and charge conjugation.

2.1. *SU(2)× U(1) Gauge fields*

We consider the four-dimensional lattice of the finite size L and choose lattice units,

$$\Gamma = \left\{x = (x_1, x_2, x_3, x_4) \in \mathbb{Z}^4 \mid 0 \leq x_\mu < L \, (\mu = 1,2,3,4)\right\}. \quad (1)$$

Adopting the compact formulation for U(1) lattice gauge theory, the SU(2) and U(1) gauge fields on Γ may be represented through periodic link fields on the infinite lattice:

$$U^{(1)}(x,\mu) \in \mathrm{U}(1), \qquad x \in \mathbb{Z}^4, \quad (2)$$

$$U^{(1)}(x + L\hat{\nu},\mu) = U^{(1)}(x,\mu) \quad \text{for all } \mu,\nu, \quad (3)$$

and

$$U^{(2)}(x,\mu) \in \mathrm{SU}(2), \qquad x \in \mathbb{Z}^4, \quad (4)$$

$$U^{(2)}(x + L\hat{\nu},\mu) = U^{(2)}(x,\mu) \quad \text{for all } \mu,\nu. \quad (5)$$

We require the so-called admissibility condition on the gauge fields,

$$|F_{\mu\nu}(x)| < \epsilon_1, \qquad F_{\mu\nu}(x) \equiv \frac{1}{i}\ln P^{(1)}(x,\mu,\nu) \in (-\pi,\pi], \quad (6)$$

$$|1 - P^{(2)}(x,\mu,\nu)| < \epsilon_2, \quad (7)$$

for all x, μ,ν, where the plaquette variables are defined by

$$P^{(i)}(x,\mu,\nu) = U^{(i)}(x,\mu)U^{(i)}(x+\hat{\mu},\nu)U^{(i)}(x+\hat{\nu},\mu)^{-1}U^{(i)}(x,\nu)^{-1} \quad (8)$$

for $i = 1,2$. This condition ensures that the overlap Dirac operator[7,9] is a smooth and local function of the gauge field if $\epsilon_i < 1/30$ $(i = 1,2)$.[11]

To impose the admissibility condition dynamically, we adopt the following action for the gauge fields:

$$S_G = \frac{1}{g_2^2}\sum_{x\in\Gamma}\sum_{\mu,\nu} \mathrm{tr}\{1 - P^{(2)}(x,\mu,\nu)\}\left[1 - \mathrm{tr}\{1 - P^{(2)}(x,\mu,\nu)\}/\epsilon_2^2\right]^{-1}$$

$$+\frac{1}{4g_1^2}\sum_{x\in\Gamma}\sum_{\mu,\nu} \left[F_{\mu\nu}(x)\right]^2\left\{1 - \left[F_{\mu\nu}(x)\right]^2/\epsilon_1^2\right\}^{-1}. \quad (9)$$

2.2. *Quarks and Leptons*

Right- and left- handed (Weyl) fermions are introduced on the lattice based on the Ginsparg-Wilson relation. Let us first consider a generic gauge group

G and a Dirac field $\psi(x)$ coupled the gauge field $U(x,\mu)$ in a certain representaion R of G. Given a local, gauge-covariant lattice Dirac operator D which acts on $\psi(x)$ and satisfies the Ginsparg-Wilson relation,

$$\gamma_5 D + D\gamma_5 = 2D\gamma_5 D, \tag{10}$$

one can introduce a chiral operator as

$$\hat{\gamma}_5 \equiv \gamma_5(1-2D), \qquad \{\hat{\gamma}_5\}^2 = \mathbb{I}. \tag{11}$$

Then, the right- and left-handed (Weyl) fermions can be defined by the eigenstates of the chiral operator $\hat{\gamma}_5$ (and γ_5 for the anti-fields). Namely,

$$\psi_\pm(x) = \hat{P}_\pm \psi(x), \qquad \bar{\psi}_\pm(x) = \bar{\psi}(x) P_\mp, \tag{12}$$

where $\hat{P}_\pm$ and $P_\pm$ are the chiral projection operators given by

$$\hat{P}_\pm = \left(\frac{1\pm\hat{\gamma}_5}{2}\right), \quad P_\pm = \left(\frac{1\pm\gamma_5}{2}\right). \tag{13}$$

Now we consider quarks and leptons in the Glashow-Weinberg-Salam model. For simplicity, we consider the first family. We adopt the convention for the normalization of the hyper-charges such that the Nishijima-Gell-man relation reads $Q = T_3 + \frac{1}{6}Y$. To describe the left-handed quarks and leptons, which are SU(2) doublets, we introduce a left-handed fermion $\psi_-(x)$ with the flavor index $\alpha(=1,\cdots,4)$, each component of which couples to the SU(2)×U(1) gauge field, $U^{(2)}(x,\mu)\otimes\{U^{(1)}(x,\mu)\}^{Y_\alpha}$, with the hyper-charge Y_α ($Y_{1,2,3}=1$ and $Y_4=-3$). Namely,

$$\psi_-(x) = {}^t\left(q_-^1(x), q_-^2(x), q_-^3(x), l_-(x)\right). \tag{14}$$

Similarly, to describe the right-handed quarks and leptons, which are SU(2) singlets, we introduce a right-handed fermion $\psi_+(x)$ with the flavor index $\beta(=1,\cdots,8)$, each component of which couples to the U(1) gauge field, $\{U^{(1)}(x,\mu)\}^{Y_\beta}$, with the hyper-charge Y_β ($Y_{1,3,5}=4$, $Y_{2,4,6}=-2$, $Y_7=0$ and $Y_8=-6$). Namely,

$$\psi_+(x) = {}^t\left(u_+^1(x), d_+^1(x), u_+^2(x), d_+^2(x), u_+^3(x), d_+^3(x), \nu_+(x), e_+(x)\right). \tag{15}$$

Then the action of quarks and leptons may be given by

$$S_F = \sum_{x\in\Gamma} \bar{\psi}_-(x) D \psi_-(x) + \sum_{x\in\Gamma} \bar{\psi}_+(x) D \psi_+(x). \tag{16}$$

2.3. *Higgs field and Yukawa-couplings*

Higgs field is a SU(2) doublet with the hyper-charge $Y_h = +6$. The action of the Higgs field may be given by

$$S_H = \sum_x \left[\sum_\nu \left(\nabla_\nu \phi(x)\right)^\dagger \nabla_\nu \phi(x) + \frac{\lambda}{2} \left(\phi(x)^\dagger \phi(x) - v^2\right)^2 \right], \qquad (17)$$

where $\phi(x)$ couples to the gauge field $U^{(2)}(x,\mu) \otimes \{U^{(1)}(x,\mu)\}^{Y_h}$ and ∇_ν is the SU(2)× U(1) gauge-covariant difference operator. Yukawa couplings of the Higgs field to the quarks and leptons may also be introduced as follows:

$$\begin{aligned} S_Y = \sum_x \Big[& y_u\, \bar{q}^i_-(x)\tilde{\phi}(x)u^i_+(x) + y_u^*\, \bar{u}^i_+(x)\tilde{\phi}(x)^\dagger q^i_-(x) \\ & + y_d\, \bar{q}^i_-(x)\phi(x)d^i_+(x) + y_d^*\, \bar{d}^i_+(x)\phi(x)^\dagger q^i_-(x) \\ & + y_l\, \bar{l}_-(x)\phi(x)e_+(x) + y_l^*\, \bar{e}_+(x)\phi(x)^\dagger l_-(x) \Big], \end{aligned} \qquad (18)$$

where $\tilde{\phi}(x)$ is the SU(2) conjugate of $\phi(x)$.

Thus the total lattice action,

$$S = S_G + S_F + S_H + S_Y, \qquad (19)$$

defines a classical theory of the Glashow-Weinberg-Salam model on the lattice. In the classical action, locality, gauge-invariance and lattice symmetries such as translations and rotations are manifest. CP symmetry, however, is not manifest in this formulation even with the real Yukawa couplings. But it is possible to show that at the quantum level both the partition function and the on-shell amplitudes respect the CP symmetry.[31] With the three families, then, the breaking of CP symmetry comes from the Kobayashi-Maskawa phase as in the continuum theory.

3. Construction of the path-integral measure

In this section, we consider a construction of the path-integral measure of the chiral fermions in the lattice Glashow-Weinberg-Salam model.[c] We show that, as in the case of the U(1) chiral gauge theories,[14] it is possible to formulate a reconstruction theorem of the fermion measure in the topological sectors of the admissible SU(2)×U(1) gauge fields with vanishing U(1) magnetic fluxes. The reconstruction theorem asserts that if there

[c]The path-integral measure of SU(2)×U(1) gauge fields and Higgs field may be defined as usual.

exist local currents which satisfy cetain properties, it is possible to reconstruct the fermion measure (the bases $\{u_j(x), \bar{u}_k(x)\}$ and $\{v_j(x), \bar{v}_k(x)\}$) which depends smoothly on the gauge fields and fulfills the fundamental requirements such as locality. gauge-invariance, integrability and lattice symmetries.

3.1. *Path-integral measure of chiral fermions*

The path-integral measure of quark fields and lepton fields may be defined by the Grassmann integrations,

$$\mathcal{D}[\psi_+]\mathcal{D}[\bar{\psi}_+]\mathcal{D}[\psi_-]\mathcal{D}[\bar{\psi}_-] = \prod_j db_j \prod_k d\bar{b}_k \prod_j dc_j \prod_k d\bar{c}_k, \tag{20}$$

where $\{b_j, \bar{b}_k\}$ and $\{c_j, \bar{c}_k\}$ are the grassman coefficients in the expansion of the chiral fields,

$$\psi_+(x) = \sum_j u_j(x) b_j, \quad \bar{\psi}_+(x) = \sum_k \bar{b}_k \bar{u}_k(x), \tag{21}$$

$$\psi_-(x) = \sum_j v_j(x) c_j, \quad \bar{\psi}_-(x) = \sum_k \bar{c}_k \bar{v}_k(x), \tag{22}$$

in terms of the chiral (orthonormal) bases defined by

$$\hat{P}_+ u_j(x) = u_j(x), \quad \bar{u}_k(x) P_- = \bar{u}_k(x). \tag{23}$$

$$\hat{P}_- v_j(x) = v_j(x), \quad \bar{v}_k(x) P_+ = \bar{v}_k(x). \tag{24}$$

Since the projection operators $\hat{P}_\pm$ depend on the gauge fields through D, the fermion measure also depends on the gauge fields. In this gauge-field dependence of the fermion measure, there is an ambiguity by a pure phase factor, because any unitary transformations of the bases,

$$\tilde{u}_j(x) = \sum_l u_l(x) \left(\mathcal{Q}_+{}^{-1}\right)_{lj}, \qquad \tilde{b}_j = \sum_l \left(\mathcal{Q}_+\right)_{jl} b_l, \tag{25}$$

$$\tilde{v}_j(x) = \sum_l v_l(x) \left(\mathcal{Q}_-{}^{-1}\right)_{lj}, \qquad \tilde{c}_j = \sum_l \left(\mathcal{Q}_-\right)_{jl} c_l, \tag{26}$$

induces a change of the measure by the pure phase factor $\det \mathcal{Q}_+ \cdot \det \mathcal{Q}_-$. This ambiguity should be fixed so that it fulfills the fundamental requirements such as locality, gauge-invariance, integrability and lattice symmetries.

3.2. *Topology of SU(2)× U(1) gauge fields*

The admissibility condition ensures that the overlap Dirac operator[7,9] is a smooth and local function of the gauge field.[11] Then, using the lattice Dirac operator D, it is possible to define a topological charge. Namely, for the admissible SU(2) and U(1) gauge fields, one has[d]

$$Q^{(i)} = \mathrm{Tr}\gamma_5(1-D)|_{U=U^{(i)}} = \sum_{x\in\Gamma} \mathrm{tr}\left\{\gamma_5(1-D)\right\}(x,x)|_{U=U^{(i)}} \tag{27}$$

for $i = 1, 2$, where $D(x, y)$ is the kernel of the lattice Dirac operator D. For $0 < \epsilon_1 < \pi/3$, the admissible U(1) gauge fields can also be classified by the magnetic fluxes,

$$m_{\mu\nu} = \frac{1}{2\pi}\sum_{s,t=0}^{L-1} F_{\mu\nu}(x + s\hat{\mu} + t\hat{\nu}), \tag{28}$$

which are integers independent of x. $m_{\mu\nu}$ is related to $Q^{(1)}$ by $Q^{(1)} = (1/2)\sum_{\mu\nu} m^2_{\mu\nu}$. Then the admissible SU(2) and U(1) gauge fields may be classified by the topological numbers Q_2 and $m_{\mu\nu}$, respectively. We denote the space of the admissible SU(2) gauge fields with a given topological charge $Q^{(2)}$ by $\mathfrak{U}^{(2)}[Q]$ and the space of the admissible U(1) gauge fields with a given magnetic fluxes $m_{\mu\nu}$ by $\mathfrak{U}^{(1)}[m]$.

3.3. *Reconstruction theorem of the fermion measure*

The properties of the fermion measure can be characterized by the so-called measure term which is given in terms of the chiral basis and its variation with respect to the gauge fields, $\delta_\eta U^{(i)}(x,\mu) = i\eta^{(i)}_\mu(x)U^{(i)}(x,\mu)$ $(i = 1, 2)$ as

$$\mathfrak{L}_\eta = i\sum_j (u_j, \delta_\eta u_j) + i\sum_j (v_j, \delta_\eta v_j). \tag{29}$$

Similar to the case of U(1) chiral lattice gauge theories,[14] one can establish the following reconstruction theorem for the topological sectors with vanishing U(1) magnetic flux, $\mathfrak{U}^{(2)}[Q] \otimes \mathfrak{U}^{(1)}[0]$:

Theorem 3.1. *If there exist local currents $j^a_\mu(x)(a = 1,2,3)$, $j_\mu(x)$ which satisfy the following four properties, it is possible to reconstruct the fermion*

[d]Throughout this paper, Tr stands for the trace over the lattice index $x\,(\in\Gamma)$, the flavor index $\alpha\,(= 1,\cdots,N)$ and the spinor index. tr stands for the trace over the flavor and spinor indices only.

measure (the bases $\{u_j(x)\}$, $\{v_j(x)\}$) which depends smoothly on the gauge fields and fulfills the fundamental requirements such as locality, gauge-invariance, integrability and lattice symmetries:

(1) $j^a_\mu(x)$, $j_\mu(x)$ *are defined for all admissible SU(2)× U(1) gauge fields in the topological sectors* $\mathfrak{U}^{(2)}[Q] \otimes \mathfrak{U}^{(1)}[0]$ *and depends smoothly on the link variables.*

(2) $j^a_\mu(x)$ *and* $j_\mu(x)$ *are gauge-covariant and -invariant, respectively and both transform as axial vector currents under the lattice symmetries.*

(3) The linear functional $\mathfrak{L}_\eta = \sum_{x\in\Gamma}\{\eta^a_\mu(x) j^a_\mu(x) + \eta_\mu(x) j_\mu(x)\}$ *is a solution of the integrability condition*

$$\delta_\eta \mathfrak{L}_\zeta - \delta_\zeta \mathfrak{L}_\eta + \mathfrak{L}_{[\eta,\zeta]} = i\mathrm{Tr}\left\{\hat{P}_-[\delta_\eta \hat{P}_-, \delta_\zeta \hat{P}_-]\right\} \tag{30}$$

for all periodic variations $\eta^a_\mu(x), \eta_\mu(x)$ *and* $\zeta^a_\mu(x), \zeta_\mu(x)$.

(4) The anomalous conservation laws hold:

$$\{\nabla^*_\mu j_\mu\}^a(x) = \mathrm{tr}\{T^a \gamma_5 (1-D)(x,x)\}, \tag{31}$$

$$\partial^*_\mu j_\mu(x) = \mathrm{tr}\{Y_- \gamma_5 (1-D)(x,x)\} - \mathrm{tr}\{Y_+ \gamma_5 (1-D)(x,x)\}, \tag{32}$$

where $Y_- = diag(1,1,1,-6)$ *and* $Y_+ = diag(4,-2,\cdots,0,-6)$.

3.4. *Proof*

3.4.1. *Global integrability condition*

Given the local currents $j^a_\mu(x)(a = 1,2,3)$, $j_\mu(x)$ which satisfy all four properties required for the reconstruction theorem, the basis vectors of the fermion fields may be constructed along a one-parameter family of the link fields $U^{(2)}_t(x,\mu) \otimes U^{(1)}_t(x,\mu)$ $t \in [0,1]$ as follows:[15]

$$v_j(x) = \begin{cases} Q_{-1} v^0_1 W^{-1} & \text{if } j = 1, \\ Q_{-1} v^0_j & \text{otherwise,} \end{cases} \tag{33}$$

$$u_j(x) = Q_{+1} u^0_j, \tag{34}$$

where W is defined by

$$W \equiv \exp\left\{ i \int_0^1 dt\, \mathfrak{L}_\eta \right\}, \qquad \eta_\mu(x) = i\partial_t U_t(x,\mu)\, U_t(x,\mu)^{-1}, \tag{35}$$

$Q_{\pm t}$ is defined by the evolution operator of the projector $P_{\pm t} = \hat{P}_\pm\Big|_{U=U_t}$ satisfying

$$\partial_t Q_{t\pm} = [\partial_t P_{t\pm}, P_{t\pm}]\, Q_{t\pm}, \quad Q_{0\pm} = 1, \tag{36}$$

and u_j^0, v_j^0 are the basis vectors for a certain reference field at $t = 0$, $U_0^{(2)}(x,\mu) \otimes U_0^{(1)}(x,\mu)$. The fermion measure defined with these basis vectors is smooth if (and only if) the following condition holds true for any closed curve in the space of the admissible SU(2)×U(1) gauge fields, $U_t^{(2)}(x,\mu) \otimes U_t^{(1)}(x,\mu)$ $(t \in [0,1])$:

$$W = \det(1 - P_{0+} + P_{0+}Q_{1+})\det(1 - P_{0-} + P_{0-}Q_{1-}). \tag{37}$$

We refer this condition as global integrability condition.

3.4.2. *Topology of the space of SU(2)× U(1) gauge fields*

Any admissible U(1) gauge field in a given topological sector $\mathfrak{U}^{(1)}[m]$ can be expressed as

$$U(x,\mu) = \tilde{U}(x,\mu)\, V_{[m]}(x,\mu), \tag{38}$$

where

$$V_{[m]}(x,\mu) = \mathrm{e}^{-\frac{2\pi i}{L^2}\left[L\delta_{\tilde{x}_\mu, L-1}\sum_{\nu>\mu} m_{\mu\nu}\tilde{x}_\nu + \sum_{\nu<\mu} m_{\mu\nu}\tilde{x}_\nu\right]}. \tag{39}$$

Here the abbreviation $\tilde{x}_\mu = x_\mu \bmod L$ has been used. $V_{[m]}(x,\mu)$ has the constant field tensor equal to $2\pi m_{\mu\nu}/L^2 (< \epsilon)$ and may be regarded as a reference field of $\mathfrak{U}^{(1)}[m]$. Then $\tilde{U}(x,\mu)$ stands for the dynamical degrees of freedom in the given topological sector. It can be parametrized with the three degrees of freedom:

$$\tilde{U}(x,\mu) = \Lambda(x)\, \mathrm{e}^{iA_\mu^T(x)}\, U_{[w]}(x,\mu)\, \Lambda(x+\hat{\mu})^{-1}, \tag{40}$$

where $A_\mu^T(x)$ is the transverse vector potential satisfying

$$\partial_\mu^* A_\mu^T(x) = 0, \qquad \sum_{x\in\Gamma} A_\mu^T(x) = 0, \tag{41}$$

$$\partial_\mu A_\nu^T(x) - \partial_\nu A_\mu^T(x) + 2\pi m_{\mu\nu}/L^2 = F_{\mu\nu}(x). \tag{42}$$

$U_{[w]}(x,\mu)$ represents the degrees of freedom of the Wilson lines defined by

$$U_{[w]}(x,\mu) = \begin{cases} w_\mu = \prod_{s=0}^{L-1} U(0+s\hat{\mu},\mu) & \text{if } x_\mu = L-1, \\ 1 & \text{otherwise,} \end{cases} \tag{43}$$

and $\Lambda(x)$ is the gauge function statisfying $\Lambda(0) = 1$.

By this parametrization, one can see that the topological structure of $\mathfrak{U}^{(1)}[m]$ is a $(4 + L^4 - 1)$-dimensional torus times a contractible space.

$$\mathfrak{U}^{(1)}[m] \simeq U(1)^4 \times \mathfrak{G}_0 \times \mathfrak{A}. \tag{44}$$

There exist two kinds of non-contractible loops ($0 \leq t \leq 1$) in $\mathfrak{U}[m]$. The first one is the gauge loops given by

$$U_t(x,\mu) = \Lambda_t(x) V_{[m]}(x,\mu) \Lambda_t(x+\hat{\mu})^{-1}, \quad \Lambda_t(x) = \exp\{i2\pi t \delta_{\tilde{x}\tilde{y}}\}. \tag{45}$$

The second one is the non-gauge loops given by

$$U_t(x,\mu) = V_{[m]}(x,\mu) \exp\{i2\pi t \delta_{\mu\nu} \delta_{\tilde{x}0}\}. \tag{46}$$

On the other hand, the topological structure of the space of the admissible SU(2) gauge fields, $\mathfrak{U}^{(2)}[Q]$, is not known.

3.4.3. *Use of the pseudo reality of SU(2)*

In the topological sectors with vanishing U(1) magnetic flux, $\mathfrak{U}^{(2)}[Q] \otimes \mathfrak{U}^{(1)}[0]$, any admissible U(1) link field can be continuously deformed to the trivial configuration, $U^{(1)}(x,\mu) = 1$. There, only the SU(2) gauge field couples to the left-handed fermion $\psi_-(x)$. In this limit, one may choose the basis vectors for the left-handed fermion $\psi_-(x) = {}^t\left(q_-^1(x), q_-^2(x), q_-^3(x), l_-(x)\right)$ as follows:

$$q_-^1(x) = \sum_j v_j(x) c_j^1, \tag{47}$$

$$q_-^2(x) = \sum_j \left(\gamma_5 C^{-1} \otimes i\sigma_2\right) \left[v_j(x)\right]^* c_j^2, \tag{48}$$

$$q_-^3(x) = \sum_j v_j(x) c_j^3, \tag{49}$$

$$l_-(x) = \sum_j v_j(x) \left(\gamma_5 C^{-1} \otimes i\sigma_2\right) \left[v_j(x)\right]^* c_j^4, \tag{50}$$

where $\{v_j(x)\}$ is the basis for a single SU(2) doublet. By virture of the pseudo reality of SU(2), the measure term of this basis vanishes identically. Therefore the measure of the chiral fermion $\psi_-(x)$ in this limit can be defined globally and the global integrability condition holds true for any closed curves in $\mathfrak{U}^{(2)}[Q]$. Then we should consider the global integrability condition only for the U(1) gauge fields.

3.4.4. *Proof of the global integrability condition for U(1) gauge fields*

The global integrability condition can be proved for the non-contractible loops ($0 \leq t \leq 1$) in the topological sectors $\mathfrak{U}[Q] \otimes \mathfrak{U}[0]$ in the following manner.

For gauge loops, we have $i\eta_\mu(x) = U_t(x,\mu)^{-1}\partial_t U_t(x,\mu) = -i\partial\delta_{\tilde{x}\tilde{y}}$ and $\mathfrak{L}_\eta = \mathrm{tr}\{\gamma_5(1-aD)(y,y)\}$. Then the l.h.s. is evaluated as

$$W = \exp\{i2\pi[\mathrm{tr}\{Y_-\gamma_5(1-D)(y,y)\} - \mathrm{tr}\{Y_+\gamma_5(1-D)(y,y)\}]|_{t=0}\}. \quad (51)$$

On the other hand, the r.h.s. is evaluated as

$$\begin{aligned} \det\{1-P_{0\pm}+P_{0\pm}Q_{1\pm}\} &= \lim_{n\to\infty}\det\left\{1-P_{0\pm}+(P_{0\pm}\Lambda_{\Delta t}^{-1}P_{0\pm})^n\right\} \\ &= \exp\left\{-i2\pi\mathrm{Tr}[\omega Y_\pm P_{0\pm}]\right\}, \end{aligned} \quad (52)$$

where $\Delta t = 2\pi/n$ and $\omega(x) = \delta_{\tilde{x}\tilde{y}}$ and therefore

$$\begin{aligned} &\det\{1-P_{0+}+P_{0+}Q_{1+}\}\det\{1-P_{0-}+P_{0-}Q_{1-}\} \\ &= \exp\left\{-i2\pi\mathrm{Tr}[\omega Y_\pm P_{0+}]\right\}\exp\left\{-i2\pi\mathrm{Tr}[\omega Y_- P_{0-}]\right\} = W. \end{aligned} \quad (53)$$

Thus the global integrability condition holds ture for the gauge loops.

For non-gauge loops, we have $i\eta_\mu(x)_{(\nu)} = U_{[w]}(x,\mu)^{-1}\partial_{t_\nu}U_{[w]}(x,\mu) = i2\pi\delta_{\mu\nu}\delta_{\tilde{x}0}$ and $\mathfrak{L}_\eta = 2\pi j_\nu(0)$. We also note the charge conjugation property of the SU(2)×U(1) gauge field and the local current: $U_{[w]} \to U^*_{[w]}$, $U^{(2)} \to U^{(2)*} \simeq U^{(2)}$ and $j_\mu(x)|_{U^*} = +j_\mu(x)|_U$. Using this property, the l.h.s. is evaluated as

$$W = \exp\left\{i\int_0^{2\pi}dt j_\nu(0)\right\} = \exp\left\{i\int_0^{\pi}dt j_\nu(0) - i\int_0^{-\pi}dt j_\nu(0)\right\} = 1 \quad (54)$$

On the other hand, the r.h.s. is evaluated as $(n=2r)$

$$\begin{aligned} &\det\{1-P_{0\pm}+P_{0\pm}Q_{1\pm}\} \\ &= \lim_{n\to\infty}\det\left\{1-P_{0\pm}+P_{0\pm}(\mathcal{C})^{-1}P_{t_1\pm}\mathcal{C}\cdots(\mathcal{C})^{-1}P_{t_r\pm}\mathcal{C}\right. \\ &\qquad\qquad \left.\times P_{t_{r-1}\pm}P_{t_{r-2}\pm}\cdots P_{t_1\pm}P_{t_0\pm}\right\} \\ &= \det\left\{1-P_{0\pm}+P_{0\pm}(\mathcal{C})^{-1}P_{0\pm}\right\}\det\left\{1-P_{\pi\pm}+P_{\pi\pm}(\mathcal{C})^{-1}P_{\pi\pm}\right\} \end{aligned}$$

where $\mathcal{C} = \left(\gamma_5 C^{-1}\otimes i\sigma_2\right)$. The extra factors are unity for even numbers of component fields. Thus the global integrability condition holds ture also for the non-gauge loops.

This completes the proof of the reconstruction theorem for the topological sectors, $\mathfrak{U}^{(2)}[Q]\otimes\mathfrak{U}^{(1)}[0]$.

4. An explicit construction of the measure term

It is possible to show by an explicit construction that there exist *local* currents $j^a_\mu(x), j_\mu(x)$ which satisfy the properties required in the reconstruction theorem. In this section, we describe our construction of the measure term on the finite volume lattice. We first state a few results which hold true in the finite volume lattice without proof. Then, using these results, we write down a closed formula of the measure term.

4.1. *Gauge anomaly cancellation in the finite volume*

In the finite volume, it is possible to show that the U(1) gauge anomaly,

$$q^{(1)}(x) = \mathrm{tr}\{Y_- \gamma_5 (1-D)(x,x)\} - \mathrm{tr}\{Y_+ \gamma_5 (1-D)(x,x)\}, \qquad (55)$$

has the form

$$\begin{aligned} q^{(1)}(x) = \; & \mathrm{tr}\{Y_- \gamma_5 (1-D)(x,x)\}|_{U=U^{(2)}} \\ & + \left(\mathrm{tr}\{Y_-^3\} - \mathrm{tr}\{Y_+^3\}\right) \gamma \, \epsilon_{\mu\nu\lambda\rho} F_{\mu\nu}(x) F_{\lambda\rho}(x+\hat{\mu}+\hat{\nu}) \\ & + \partial_\mu^* \bar{k}_\mu(x), \end{aligned} \qquad (56)$$

where γ is a constant which takes the value $\gamma = \frac{1}{32\pi^2}$ for the overlap Dirac operator[32] and $\bar{k}_\mu(x)$ is a local, gauge-invariant current which can be constructed so that it transforms as the axial vector current under the lattice symmetries. We emphasize that this is the result of *the gauge anomaly in the finite volume*, which is obtained by combining the result in the infinite lattice,[13,33,34] and the result of the analysis of the finite volume correction.[35] See also.[27] Then, for the anomaly-free multiplet, the cohomologically non-trivial part of the gauge anomaly cancels exactly at a finite lattice spacing and the total U(1) gauge anomaly is cohomologically trivial:

$$q^{(1)}(x) = \partial^* \bar{k}_\mu(x), \qquad (L < \infty). \qquad (57)$$

4.2. *Vector potential in the finite volume and the Wilson line degrees of freedom*

In the finite volume, the admissible U(1) link field in a given topological sector $\mathfrak{U}^{(1)}[m]$ can be expressed by the vector potential of the following properties:

Lemma 4.2 There exists a periodic vector potential $\tilde{A}_\mu(x)$ such that

$$U^{(1)}(x,\mu) = \mathrm{e}^{i\tilde{A}_\mu(x)} \, U_{[w]}(x,\mu) \, V_{[m]}(x,\mu), \qquad (58)$$

$$F_{\mu\nu}(x) = \partial_\mu \tilde{A}_\nu(x) - \partial_\nu \tilde{A}_\mu(x) + \frac{2\pi m_{\mu\nu}}{L^2}, \qquad (59)$$

$$\begin{cases} |\tilde{A}_\mu(x)| \le \pi(1 + 4\|x - x_0\|) & \text{for } |(x-x_0)_\nu| \neq L/2 - 1 \ (\nu = 1,2,3,4), \\ |\tilde{A}_\mu(x)| \le \pi(1 + 6L^2) & \text{otherwise.} \end{cases} \qquad (60)$$

(L is assumed to be an even number for simplicity and x_0 is a reference point which may be chosen arbitrarily. The abbreviation $\tilde{x}_\mu = x_\mu \bmod L$ has been used.) Moreover, if $\tilde{A}'_\mu(x)$ is any other field with these properties we have

$$\tilde{A}'_\mu(x) = \tilde{A}_\mu(x) + \partial_\mu \omega(x), \tag{61}$$

where the gauge function $\omega(x)$ is *periodic* and takes values that are integer multiples of 2π.

This property makes sure that any gauge invariant function of $\tilde{A}_\mu(x)$ in the finite-volume is smooth with respect to the link field. The locality property of this mapping can be argued as in the case of the lemma 5.1 in Ref. 13: the locality properties of the gauge invariant fields are the same independently of whether they are considered to be functions of the link variables or the vector potential.

4.3. *A closed formula of the measure term*

We now construct the measure term for the generic admissible SU(2)⊗U(1) gauge fields in $\mathfrak{U}^{(2)}_{[Q]} \otimes \mathfrak{U}^{(1)}_{[0]}$. For this purpose, we choose a one-parameter family of the gauge fields as

$$U_s(x,\mu) = U^{(2)}(x,\mu) \otimes \left[\mathrm{e}^{is\tilde{A}_\mu(x)}\, U_{[w]}(x,\mu) \right], \qquad 0 \le s \le 1. \tag{62}$$

Then we consider the linear functional of the variational parameters $\eta^{(2)}_\mu(x)$, $\eta^{(1)}_\mu(x)$ which is given in terms of quantities defined on the finite-volume lattice:

$$\begin{aligned} \mathfrak{L}^\diamond_\eta = {} & i \int_0^1 ds\, \mathrm{Tr}\left\{ \hat{P}_- [\partial_s \hat{P}_-, \delta_\eta \hat{P}_-] \right\} \\ & + \delta_\eta \int_0^1 ds \sum_{x\in\Gamma_4} \left\{ \tilde{A}_\mu(x)\, \bar{k}_\mu(x) \right\} + \bar{\mathfrak{L}}_\eta |_{U=U^{(2)}\otimes U_{[w]}}, \end{aligned} \tag{63}$$

where $\bar{k}_\mu(x)$ is the gauge-invariant local current which satisfies $\partial^*_\mu \bar{k}_\mu(x) = q^{(1)}(x)$ and transforms as an axial vector field under the lattice symmetries. $\bar{\mathfrak{L}}_\eta|_{U=U^{(2)}\otimes U_{[w]}}$ is the additional measure term at the gauge field $U = U^{(2)} \otimes U_{[w]}$ with the variational parameters in the directions of the Wilson lines for the U(1) gauge field: $\eta^{(2)}_\mu(x), \eta^{(1)}_\mu = \eta_{\mu(\nu)}$.

The currents defined by eq. (63),

$$\mathfrak{L}^\diamond_\eta = \sum_{x\in\Gamma} \{ \eta^a_\mu(x) j^{a\diamond}_\mu(x) + \eta_\mu(x) j^\diamond_\mu(x) \}, \tag{64}$$

are local functionals of the link variables and satisfy all the properties required for the reconstruction theorem. The proof of the properties of the currents and the detail of the construction of $\bar{\mathfrak{L}}_\eta|_{U=U^{(2)}\otimes U_{[w]}}$ will be given in the force coming papers.[37,38]

Acknowledgments

The authors would like to thank M. Lüscher for valuable comments. Y.K. is grateful to Ting-Wai Chiu for his kind hospitality at 2005 Taipei Summer Institute on Strings, Particles and Fields and D. Adams, K. Fujikawa, H. Suzuki for discussions. Y.K. is supported in part by Grant-in-Aid for Scientific Research No. 17540249.

References

1. G. 't Hooft, Phys. Rev. Lett. **37**, 8 (1976).
2. G. 't Hooft, Phys. Rev. D **14**, 3432 (1976) [Erratum-ibid. D **18**, 2199 (1978)].
3. S. Raby, S. Dimopoulos and L. Susskind, Nucl. Phys. B **169**, 373 (1980).
4. S. Dimopoulos, S. Raby and L. Susskind, Nucl. Phys. B **173**, 208 (1980).
5. G. 't Hooft, PRINT-80-0083 (UTRECHT) *Lecture given at Cargese Summer Inst., Cargese, France, Aug 26 - Sep 8, 1979.*
6. P. H. Ginsparg and K. G. Wilson, Phys. Rev. D **25**, 2649 (1982).
7. H. Neuberger, Phys. Lett. B **417**, 141 (1998) [arXiv:hep-lat/9707022].
8. P. Hasenfratz, V. Laliena and F. Niedermayer, Phys. Lett. B **427**, 125 (1998) [arXiv:hep-lat/9801021].
9. H. Neuberger, Phys. Lett. B **427**, 353 (1998) [arXiv:hep-lat/9801031].
10. P. Hasenfratz, Nucl. Phys. B **525**, 401 (1998) [arXiv:hep-lat/9802007].
11. P. Hernandez, K. Jansen and M. Lüscher, Nucl. Phys. B **552**, 363 (1999) [arXiv:hep-lat/9808010].
12. M. Lüscher, Phys. Lett. B **428**, 342 (1998) [arXiv:hep-lat/9802011].
13. M. Lüscher, Nucl. Phys. B **538**, 515 (1999) [arXiv:hep-lat/9808021].
14. M. Lüscher, Nucl. Phys. B **549**, 295 (1999) [arXiv:hep-lat/9811032].
15. M. Lüscher, Nucl. Phys. B **568**, 162 (2000) [arXiv:hep-lat/9904009].
16. M. Lüscher, Nucl. Phys. Proc. Suppl. **83**, 34 (2000) [arXiv:hep-lat/9909150].
17. M. Lüscher, arXiv:hep-th/0102028.
18. H. Suzuki, Prog. Theor. Phys. **101**, 1147 (1999) [arXiv:hep-lat/9901012].
19. H. Neuberger, Phys. Rev. D **63**, 014503 (2001) [arXiv:hep-lat/0002032].
20. D. H. Adams, Nucl. Phys. B **589**, 633 (2000) [arXiv:hep-lat/0004015].
21. H. Suzuki, Nucl. Phys. B **585**, 471 (2000) [arXiv:hep-lat/0002009].
22. H. Igarashi, K. Okuyama and H. Suzuki, arXiv:hep-lat/0012018.
23. M. Lüscher, JHEP **0006**, 028 (2000) [arXiv:hep-lat/0006014].
24. Y. Kikukawa and Y. Nakayama, Nucl. Phys. B **597**, 519 (2001) [arXiv:hep-lat/0005015].
25. Y. Kikukawa, Phys. Rev. D **65**, 074504 (2002) [arXiv:hep-lat/0105032].
26. T. Aoyama and Y. Kikukawa, arXiv:hep-lat/9905003.

27. D. Kadoh, Y. Kikukawa and Y. Nakayama, JHEP **0412**, 006 (2004) [arXiv:hep-lat/0309022].
28. R. Narayanan and H. Neuberger, Nucl. Phys. B **412**, 574 (1994) [arXiv:hep-lat/9307006].
29. R. Narayanan and H. Neuberger, Phys. Rev. Lett. **71**, 3251 (1993) [arXiv:hep-lat/9308011].
30. R. Narayanan and H. Neuberger, Nucl. Phys. B **443**, 305 (1995) [arXiv:hep-th/9411108].
31. K. Fujikawa, M. Ishibashi and H. Suzuki, JHEP **0204**, 046 (2002) [arXiv:hep-lat/0203016].
32. Y. Kikukawa and A. Yamada, Phys. Lett. B **448**, 265 (1999) [arXiv:hep-lat/9806013].
33. T. Fujiwara, H. Suzuki and K. Wu, Nucl. Phys. B **569**, 643 (2000) [arXiv:hep-lat/9906015].
34. T. Fujiwara, H. Suzuki and K. Wu, Phys. Lett. B **463**, 63 (1999) [arXiv:hep-lat/9906016].
35. H. Igarashi, K. Okuyama and H. Suzuki, Nucl. Phys. B **644**, 383 (2002) [arXiv:hep-lat/0206003].
36. D. Kadoh and Y. Kikukawa, JHEP **0501**, 024 (2005) [arXiv:hep-lat/0401025].
37. D. Kadoh and Y. Kikukawa, "A simple construction of fermion measure term in U(1) chiral lattice gauge theories with exact gauge invariance", in preparation.
38. D. Kadoh and Y. Kikukawa, "A gauge-invariant construction of the fermion measure in the Glashow-Weinberg-Salam model on the lattice", in preparation.

EQUILIBRIUM THERMODYNAMICS OF LATTICE QCD

D. K. Sinclair

HEP Division, Argonne National Laboratory, 9700 South Cass Avenue, Argonne, IL, 60439, USA

Lattice QCD allows us to simulate QCD at non-zero temperature and/or densities. Such equilibrium thermodynamics calculations are relevant to the physics of relativistic heavy-ion collisions. I give a brief review of the field with emphasis on our work.

1. Introduction

High temperature hadronic matter was certainly present in the early universe. Relativistic heavy-ion colliders such as RHIC and in future the LHC with heavy-ions, produce hot hadronic matter. Lower energy machines can produce hot hadronic matter with an appreciable baryon/quark-number density – hot nuclear matter. At high enough temperatures this hadronic matter is expected to become quark/gluon matter – the quark-gluon plasma. Preliminary evidence for the quark-gluon plasma has been reported from RHIC and CERN.

At high baryon/quark-number density and low temperature such as might exist in the cores of neutron stars, exotic states of matter, such as colour superconducting phases have been proposed.

In Quantum Chromodynamics (QCD), the accepted theory of hadrons and their strong interactions, the most interesting finite-temperature properties are non-perturbative. Lattice QCD simulations are the only systematic approach to the study of such phenomena.

Fig. 1 shows a simplified version of the proposed phase diagram. (For an introduction to the phase structure of QCD and predictions from lattice QCD, see recent reviews[1,2]).

In Lattice QCD, we only really know how to calculate equilibrium thermodynamics. Relativistic heavy-ion collisions are clearly not equilibrium processes, but knowledge of equilibrium thermodynamics and the consequent extraction of thermodynamic quantities and the prediction of the

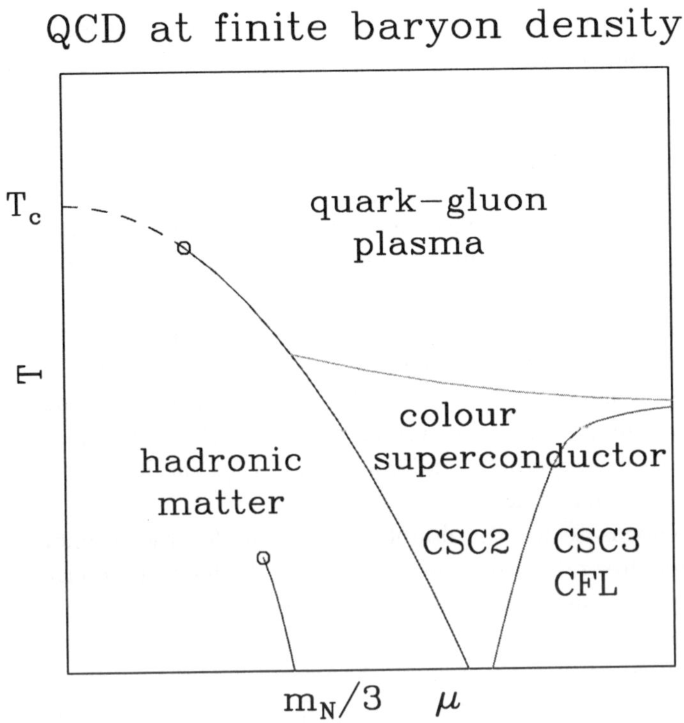

Fig. 1. Simplified phase diagram for QCD at finite temperature and quark-number chemical potential μ.

equation-of-state give us some idea of the type of behaviour we might expect. In addition, it can provide input for hydrodynamic models of such collisions.

QCD at finite temperature and/or densities yields information on the dynamics of confinement and chiral symmetry breaking and thus enhances our understanding of QCD.

2. Lattice QCD at finite temperature

QCD is quantized by functional integral methods. To make these integrals well-defined, we rotate to imaginary time – Euclidean space-time. The quark fields are then defined on the sites of a 4-dimensional hypercubic lattice, the gauge fields $U_\mu = \exp(igA_\mu)$ on its links.

$$Z = \int \mathcal{D}\psi \mathcal{D}\bar{\psi} \mathcal{D}U e^{-S} \tag{1}$$

where the action $S = S_f + S_g$. The simplest lattice implementation (which preserves gauge invariance) has

$$S_g = \beta \sum_{\Box} \left(1 - \frac{1}{3}\mathrm{ReTr}_{\Box} UUUU\right), \tag{2}$$

where $\beta = 6/g^2$, and $S_f = \sum_{sites} \bar{\psi}(\not{D} + m)\psi$.

Simulations are performed by replacing the fermion fields with boson fields (pseudo-fermions) which results in replacing the Dirac operator by its inverse. Adding a 'kinetic' term for the gauge fields, allowing them to evolve in a fictitious 'time', turns Z into the partition function for a system of classical particles. One then uses molecular-dynamics techniques to effectively 'evaluate' the integrals, by numerically integrating the equations of motion for this system. Bringing the system in contact with a heat-bath at regular intervals assures ergodicity and compensates for the lack of dynamics for the pseudo-fermion fields.

To compensate for the doubling problems for the fermion fields frequently requires taking fractional powers of the fermion determinant. This is performed by using rational approximations to fractional powers of the Dirac operator. Use of a global Metropolis accept/reject step removes the discretization errors in the numerical integration of the equations of motion. (For a recent description of simulation methods with dynamical fermions see[3]).

If we use a lattice of temporal extent $1/T$ and a spatial extent much larger than this, and demand periodic boundary conditions for the gauge fields and antiperiodic boundary conditions for the fermion fields in the time direction, $Z = Z(T)$ the partition function for lattice QCD at temperature T.

3. The finite temperature transition

For massless quarks, chiral symmetry is restored at the finite temperature transition. $\langle\bar{\psi}\psi\rangle \neq 0$ below the transition and $\langle\bar{\psi}\psi\rangle = 0$ above the transition. Hence this transition is a phase transition. Arguments based on dimensional reduction predict that this phase transition is a second order phase transition (critical point) for $N_f = 2$, in the universality class of the 3-dimensional $O(4)$ spin model.[4] For non-zero quark mass it is expected to weaken to a crossover with no real phase transition. For $N_f > 2$ this transition is predicted to be a first-order phase transition which is expected to remain first-order for small quark masses, becoming a crossover for larger quark masses.

In the real world where $N_f = 2+1$ (u, d, s), the important question is whether the strange quark mass is light enough for the transition to be first-order or whether it is a crossover. Recent lattice simulations indicate that the strange quark mass is too large, and the transition is a crossover[5] (see Fig 2). This extends earlier work of[8] (NB the $N_f = 3$ point agrees with

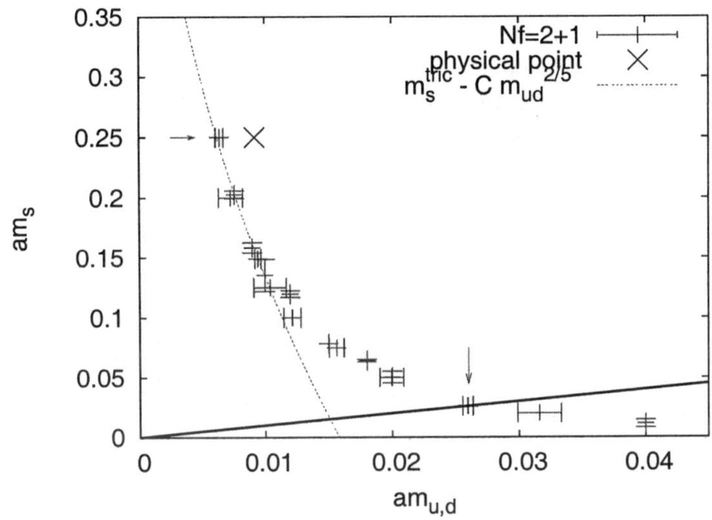

Fig. 2. The chiral critical line at zero baryon-number density (from Ref. 5).

our own simulations.)

Simulations with larger lattices, finer lattice spacings and improved actions have recently been reported by Ref. 6. Here the u, d and s quark masses are chosen by fixing m_K/m_π (and f_K/f_π) at their physical values. Making several choices of masses restricted in this manner, they were able to extrapolate to the physical quark masses and below. They confirm the observation that the physical transition is a crossover and not a phase transition, in simulations which probe much closer to the continuum limit than in Ref. 5. The same simulations give new estimates for the temperature(s) of this crossover.[7]

Since the strange quark appears too massive to control the nature of the transition, it is useful to study the 2-flavour case, to see if the nature of the $m = 0$ phase transition agrees with the above prediction.

Determining the nature of the 2-flavour phase transition has proved difficult, since is expected only at $m = 0$, and finite volume effects appear to be large.[9–12] Standard (including highly improved) lattice actions forbid simulations at $m = 0$, and small mass simulations are very expensive. No such simulations have been performed at masses small enough to uncover the $m = 0$ results.

We have performed $N_f = 2$ simulations using the χQCD action which allows simulations at zero quark mass.[13] Since it is a staggered quark action, the reduced flavour symmetry leads us to predict that the phase transition should be second-order and in the universality class of the 3-dimensional $O(2)$ spin model. To accommodate the finite volume effects, we compare our chiral condensate measurements to the magnetization of the $O(2)$ spin model also on finite volumes, as shown in Fig. 3. The agreement is excellent. Good agreement is also obtained with correlation lengths and susceptibilities.

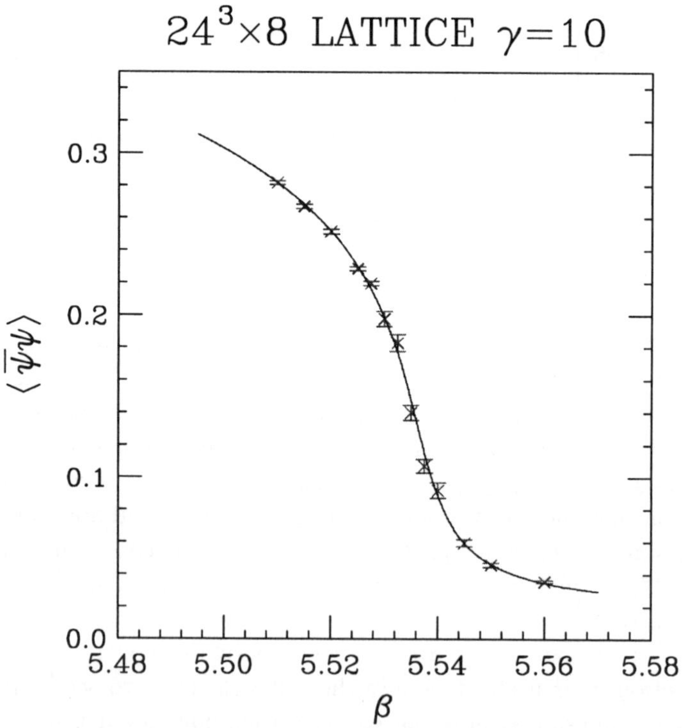

Fig. 3. Chiral condensate (points) near the chiral phase transition for lattice QCD, fitted to the magnetization (solid line) for the $O(2)$ spin model (from Ref.[13]).

Fig. 4 shows recent estimates of the transition temperature T_c from lattice QCD simulations, compared with an experimental estimate from RHIC. All lattice simulations are consistent with the chiral and deconfinement transitions being coincident.

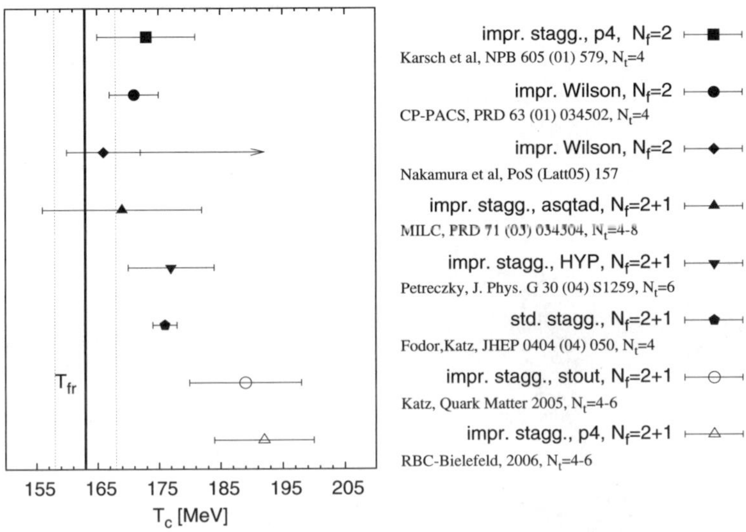

Fig. 4. Recent lattice estimates of T_c compared with experiment (from Ref. 14).

4. The equation-of-state for hot QCD.

The equation-of-state (EOS) expresses the pressure p, the entropy density s and the energy density ϵ as functions of the temperature T. The free energy density, pressure and entropy are given by

$$f = -\frac{T}{V}\ln Z(T), \quad p = -f \quad , s = \frac{dp}{dT}, \tag{3}$$

respectively. The energy density is not an independent quantity but is given by

$$\epsilon = Ts - p. \tag{4}$$

$Z(T)$ is calculated on the lattice by numerically integrating

$$\frac{d\ln Z}{d\beta} = \langle S_g \rangle. \tag{5}$$

To obtain T as a function of β requires knowing the lattice spacing as a function of β. This can be obtained by measuring physical quantities as functions of β at zero temperature.

Knowledge of the EOS is needed as input to models for the evolution of the hot hadronic matter in relativistic heavy-ion collisions. Fig. 5 is a graph of these quantities from the MILC collaboration.[15]

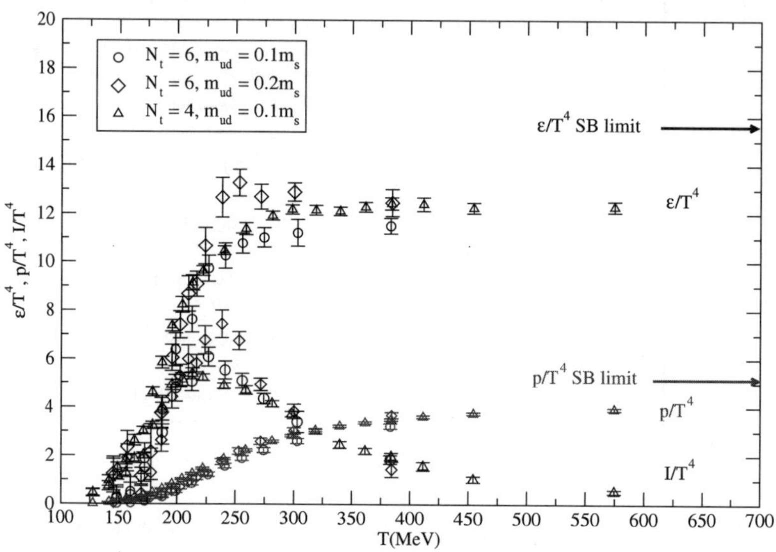

Fig. 5. The EOS from Lattice QCD simulations (MILC preliminary: status Lattice2006[16]). $I = \epsilon - 3p$.

For earlier work on the finite temperature EOS see for example.[17] Karsch *et al.* have shown that the low temperature behaviour is well modeled as a non-interacting gas of hadron resonances.[18] See also Ref. 19 for other work on the QCD EOS.

5. Meson spectral functions at finite T

Meson spectral functions yield information about the propagation of mesons, or excitations with mesonic quantum numbers in hot hadronic matter[20] (This review gives references to earlier works). Even just above T_c, the quark-gluon plasma is a strongly interacting fluid and mesonic states survive. Not only do the spectral functions have information about hadronic stability at high temperatures, but they also have information about transport coefficients and dilepton production. At zero temperature, the spectral function is just the momentum space propagator. The Euclidean-time meson propagator $G(\tau, \boldsymbol{p}, T)$ is related to the spectral function $\sigma(\omega, \boldsymbol{p}, T)$ by

$$G(\tau, \boldsymbol{p}, T) = \int_0^\infty d\omega \sigma(\omega, \boldsymbol{p}, T) K(\tau, \omega, T) \tag{6}$$

where $K(\tau, \omega, T) = \frac{\cosh[\omega(\tau - 1/2T)]}{\sinh(\omega/2T)}$.

The main difficulty has been that, since the spatial extent of the lattice must be much greater than its temporal extent, the temporal extent of the lattice is typically ~ 10 or less in lattice units. This gives a rather poor estimate of σ. Recently simulations have been performed on anisotropic lattices on which the spatial lattice spacing is much greater than the temporal lattice spacing.[21,22] This allows for many more points in the time direction, while still keeping the spatial extent of the lattice much greater than its temporal extent. This leads to better estimates for the spectral functions.

The charmonium spectral functions have been of particular interest. It had been suggested that one signal for the quark-gluon plasma phase could be the 'melting' of charmonium. This would reveal itself as a drop in the charmonium production relative to the production of $D\bar{D}$ pairs and similar states.

Current simulations by the TrinLat collaboration using anisotropic lattices with dynamical light quarks, indicate that the J/ψ and η_c survive above T_c and melt somewhere between $1.3T_c$ and $2T_c$. The χ_c states appear to melt at or below $1.3T_c$ (see Fig. 6).

At the LHC, even higher temperatures are expected, and it has been suggested that these might be high enough to melt bottomonia.[23] We plan to measure NRQCD bottomonium propagators (and hence spectral functions) on the TrinLat configurations to determine these melting temperatures.

6. Transport coefficients

To use hydrodynamic models for hadronic matter in relativistic heavy-ion collisions requires knowledge of the transport coefficients, shear viscosity η, volume viscosity ζ and thermal conductivity κ. These viscosities are expressed in terms of Green's functions of the stress-energy tensor.[24,25]

$$\eta = -\int d^3x' \int_{-\infty}^{t} dt_1 e^{\epsilon(t_1 - t)} \int_{-\infty}^{t_1} dt' \langle T_{12}(\boldsymbol{x}, t) T_{12}(\boldsymbol{x}', t') \rangle_{ret}$$

$$\frac{4}{3}\eta + \zeta = -\int d^3x' \int_{-\infty}^{t} dt_1 e^{\epsilon(t_1 - t)} \int_{-\infty}^{t_1} dt' \langle T_{11}(\boldsymbol{x}, t) T_{11}(\boldsymbol{x}', t') \rangle_{ret} \tag{7}$$

These real-time Green's functions are obtained from their lattice (Euclidean time) counterparts through the spectral function $\sigma(\omega)$. Quenched

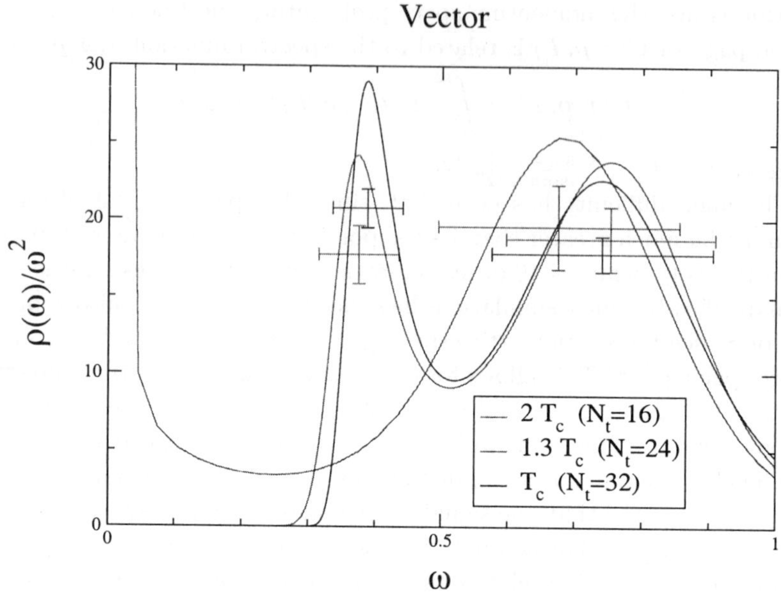

Fig. 6. Spectral function $\rho \equiv \sigma$ for the J/ψ at T_c, $1.3T_c$ and $2T_c$ (from Ref. 22).

lattice results have been obtained for these quantities.[26] In terms of the spectral function, $\eta = \pi \lim_{\omega \to 0} \frac{\sigma(\omega)}{\omega}$.

This means that one needs the spectral function near $\omega = 0$, where it is least well known. Thus the determination is difficult, and systematic as well as statistical errors are large as shown in Fig. 7. The same paper finds $\zeta \approx 0$. Clearly there is a long way to go.

More recently, improved methods have been used to calculate the electrical conductivity of the quark-gluon plasma.[27]

7. Lattice QCD at finite baryon number density

Finite baryon/quark-number density is implemented by introduction of a quark-number chemical potential μ. On the lattice this is achieved by multiplying each of the links in the $+t$ direction by e^{μ} and each of those in the $-t$ direction by $e^{-\mu}$ in the quark action S_f.

Integrating out the fermion fields gives the determinant of the Dirac operator which is complex, with a real part of indefinite sign. Standard simulation methods, which rely on importance sampling, fail for such systems.

Some progress has been made in circumventing these problems for small

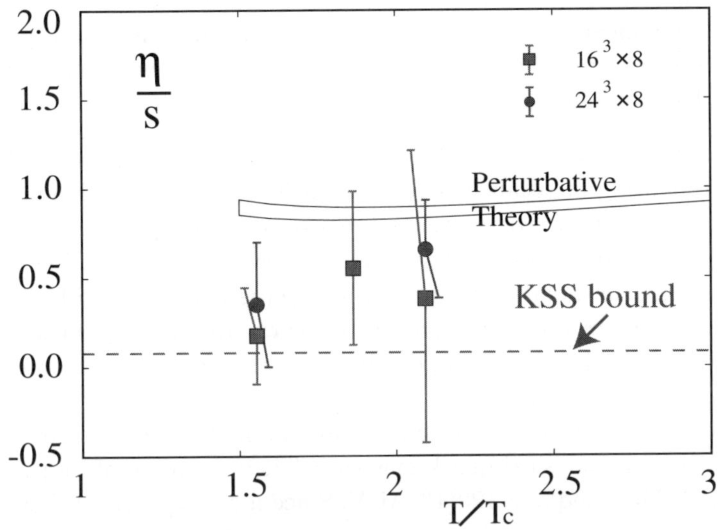

Fig. 7. η/s as a function of temperature. Dashed line is at $1/4\pi$ (from Ref.[26]).

μs close to the finite temperature transition. Methods for doing this fall into several classes.

- Analytic continuation: In the simplest case people simulate at imaginary μs, where the fermion determinant is real and positive. The results are fitted to a power series in μ^2, which allows continuation to real μ.[28,29] Fancier analytic continuation methods have also been used.[30]
- Power series expansions: These are similar in spirit to the analytic continuation methods. The exponential of the action is expanded in powers of μ^2. The coefficients of this expansion are then observables whose expectation values are measured in $\mu = 0$ simulations.[31,32]
- Reweighting methods: These start from a quark action with a positive fermion determinant, and reweight measurements by the ratio of the original fermion determinant to this positive fermion determinant. One then divides by the expectation value of this ratio of determinants.[33]
- Phase quenched methods: One simulates using the magnitude of the fermion determinant ignoring the phase. For small enough μ on a finite lattice, the phase is small enough that this should yield

the same phase structure as the full simulation.[34]

- Canonical ensemble methods: The fermion determinant is projected on to states of fixed quark number. One deals with any sign problems by reweighting.[35]

For 3-flavour QCD, it has been argued that the critical point observed at zero chemical potentials, where the transition weakens from a first-order phase transition to a crossover, would move to larger quark masses as μ is increased from zero, becoming the sort-after critical endpoint. Recent work using analytic continuation from imaginary μ (de Forcrand and Philipsen[28]) and simulations in the phase-quenched theory (Kogut and Sinclair),[34] indicate that this does not happen. Instead, the critical mass appears to decrease with increasing μ.

Fodor and Katz, using the reweighting method, claim to find the critical endpoint, for physical quark masses.[33] Their estimate of its position is $T = 162(2)$ MeV and $\mu = 120(13)$ MeV. Since $\mu > m_\pi/2$, this is beyond the reach of analytic continuation, phase quenched and series expansion methods. It remains to develop other methods which could check this. Fig. 8

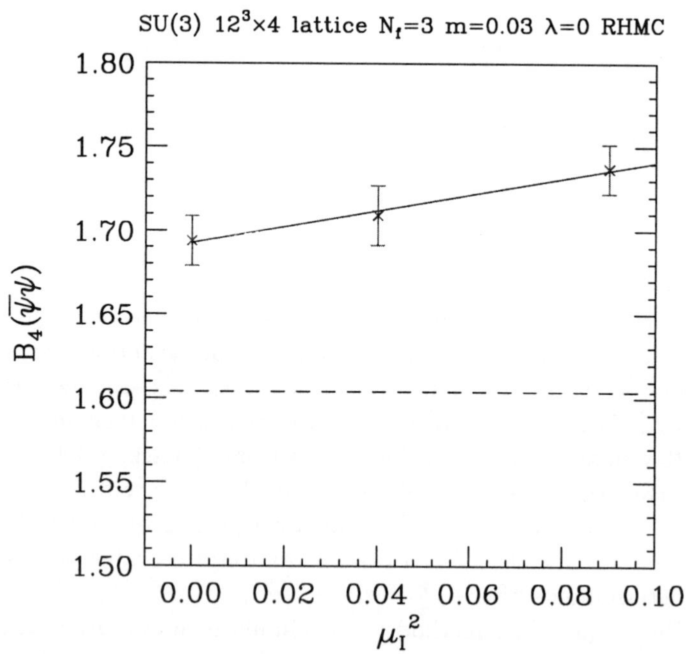

Fig. 8. Binder cumulant for the chiral condensate as a function of μ_I (from Ref.[34]).

shows the behaviour of the Binder cumulant in the phase-quenched theory. If there were a critical endpoint at (small) finite μ_I, this Binder cumulant would decrease through the Ising value 1.604(1) (dashed line). Instead it increases indicating that there is no critical endpoint in this range of μ_I.

8. The EOS at non-zero T and μ

At finite temperature and small μ, the series expansions in terms of μ also enable one to calculate such quantities as p, s and ϵ and hence to study the equation-of-state in terms of T *and* μ.[36] Similar results have been obtained using multi-parameter reweighting techniques.[37]

The pressure p can be obtained from measurement of the quark-number density ρ, since

$$\rho = \frac{\partial p}{\partial \mu}. \tag{8}$$

The pressure at zero μ is calculated as above. That at finite μ can be obtained by integrating the previous equation. ϵ can also be calculated, but requires knowledge of the running of the coupling constant and quark mass. Results from the Bielefeld group are shown in Fig. 9.

9. Summary and Conclusion

Lattice QCD at finite temperature can probe the nature of the phase transition from hadronic matter to a quark-gluon plasma. For physical quark masses this appears to be a crossover without a true phase transition, influenced by the second-order transition at $m_u = m_d = 0$.

The QCD equation-of-state and other equilibrium thermodynamic quantities measured in lattice QCD simulations provide input for an understanding of the non-equilibrium thermodynamics at RHIC and the LHC. Finite temperature (lattice) QCD probes QCD dynamics such as confinement and chiral-symmetry breaking.

Charmonium spectral functions show that the J/ψ remains intact at the finite temperature transition T_c, but dissociates below $2T_c$. This should produce a reduction in J/ψ's from the most energetic processes at RHIC. Bottomonium spectral functions should show similar behaviour at the even higher temperatures of the LHC heavy-ion program.

Just above T_c, the 'deconfined' phase is a strongly-interacting fluid with low viscosity. Transport coefficients (including viscosity) need to be calculated as input for hydrodynamic models. Early attempts have been made in lattice QCD simulations, but such measurements are difficult.

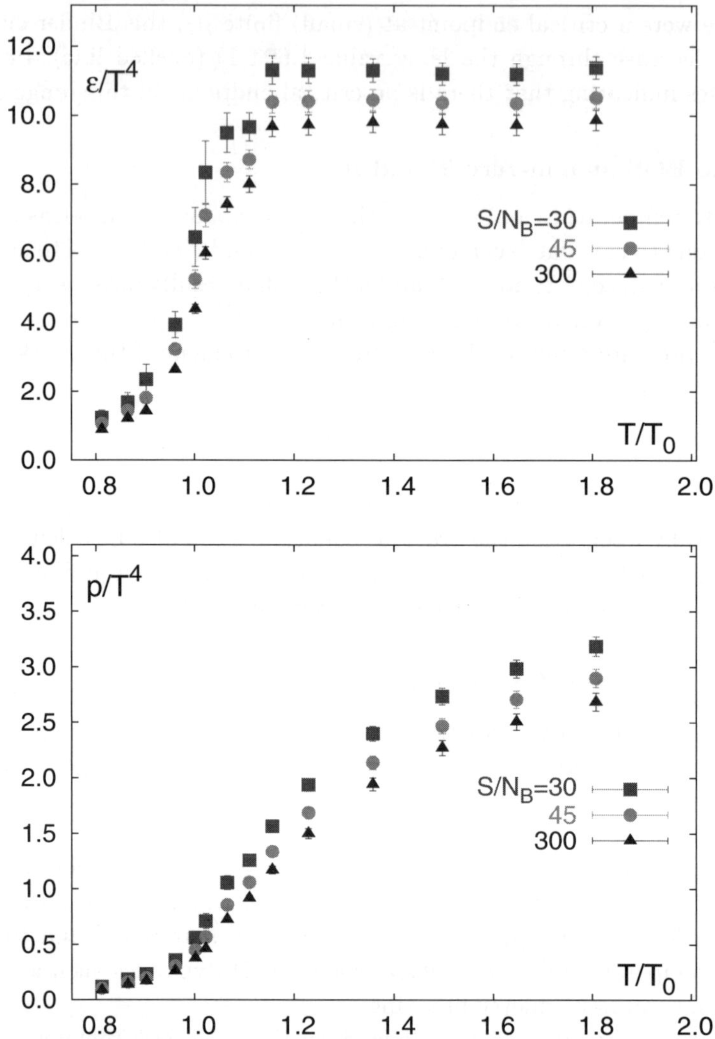

Fig. 9. ϵ and p as functions of T (from Ref. 36).

Sign problems hamper simulations at finite baryon/quark-number density. Some progress has been made for small μs close to the finite temperature transition. The equation-of-state has been determined in this high-temperature low baryon-number-density regime. More work is needed to observe the critical endpoint, which is the most striking feature expected in this region of the QCD phase diagram. This is a region accessible experimentally to lower-energy relativistic heavy-ion collisions.

New methods will be needed if one is ever to reach the high baryon-number densities needed to understand the physics of neutron stars. At such densities one expects such exotic states of matter as colour superconductors.

Acknowledgments

Research was supported by the U.S. Department of Energy contract DE-AC-02-06CH11357. I thank the authors of the manuscripts, from which some of the figures were reproduced, for their permission to incorporate these figures in this manuscript. My own research reported in this talk was performed in collaboration with J. B. Kogut.

References

1. U. M. Heller, PoS **LAT2006**, 011 (2006) [arXiv:hep-lat/0610114].
2. M. G. Alford, Ann. Rev. Nucl. Part. Sci. **51**, 131 (2001) [arXiv:hep-ph/0102047].
3. A. D. Kennedy, arXiv:hep-lat/0607038.
4. R. D. Pisarski and F. Wilczek, Phys. Rev. D **29**, 338 (1984).
5. P. de Forcrand and O. Philipsen, arXiv:hep-lat/0607017.
6. Y. Aoki, G. Endrodi, Z. Fodor, S. D. Katz and K. K. Szabo, Nature **443**, 675 (2006) [arXiv:hep-lat/0611014].
7. Y. Aoki, Z. Fodor, S. D. Katz and K. K. Szabo, Phys. Lett. B **643**, 46 (2006) [arXiv:hep-lat/0609068].
8. F. Karsch, E. Laermann and C. Schmidt, Phys. Lett. B **520**, 41 (2001) [arXiv:hep-lat/0107020].
9. F. Karsch, Phys. Rev. D **49**, 3791 (1994) [arXiv:hep-lat/9309022].
10. F. Karsch and E. Laermann, Phys. Rev. D **50**, 6954 (1994) [arXiv:hep-lat/9406008].
11. S. Aoki *et al.* [JLQCD Collaboration], Phys. Rev. D **57**, 3910 (1998) [arXiv:hep-lat/9710048].
12. C. W. Bernard *et al.* [MILC Collaboration], Phys. Rev. D **55**, 6861 (1997) [arXiv:hep-lat/9612025].
13. J. B. Kogut and D. K. Sinclair, Phys. Rev. D **73**, 074512 (2006) [arXiv:hep-lat/0603021].
14. P. Petreczky, arXiv:hep-lat/0609040.
15. C. Bernard *et al.*, arXiv:hep-lat/0611031.
16. C. Bernard *et al.*, arXiv:hep-lat/0610017.
17. F. Karsch, E. Laermann and A. Peikert, Phys. Lett. B **478**, 447 (2000) [arXiv:hep-lat/0002003].
18. F. Karsch, K. Redlich and A. Tawfik, Eur. Phys. J. C **29**, 549 (2003) [arXiv:hep-ph/0303108].

19. Y. Aoki, Z. Fodor, S. D. Katz and K. K. Szabo, JHEP **0601**, 089 (2006) [arXiv:hep-lat/0510084].
20. P. Petreczky, Nucl. Phys. Proc. Suppl. **140**, 78 (2005) [arXiv:hep-lat/0409139].
21. A. Jakovac, P. Petreczky, K. Petrov and A. Velytsky, arXiv:hep-lat/0611017.
22. G. Aarts, C. R. Allton, R. Morrin, A. P. O. Cais, M. B. Oktay, M. J. Peardon and J. I. Skullerud, arXiv:hep-lat/0610065.
23. J. F. Gunion and R. Vogt, Nucl. Phys. B **492**, 301 (1997) [arXiv:hep-ph/9610420].
24. R. Horsley and W. Schoenmaker, Nucl. Phys. B **280**, 716 (1987).
25. R. Horsley and W. Schoenmaker, Nucl. Phys. B **280**, 735 (1987).
26. A. Nakamura and S. Sakai, Phys. Rev. Lett. **94**, 072305 (2005) [arXiv:hep-lat/0406009].
27. S. Gupta, Phys. Lett. B **597**, 57 (2004) [arXiv:hep-lat/0301006].
28. P. de Forcrand and O. Philipsen, arXiv:hep-lat/0611027.
29. M. D'Elia and M. P. Lombardo, Phys. Rev. D **70**, 074509 (2004) [arXiv:hep-lat/0406012].
30. V. Azcoiti, G. Di Carlo, A. Galante and V. Laliena, Nucl. Phys. B **723**, 77 (2005) [arXiv:hep-lat/0503010].
31. S. Ejiri, C. R. Allton, S. J. Hands, O. Kaczmarek, F. Karsch, E. Laermann and C. Schmidt, Prog. Theor. Phys. Suppl. **153**, 118 (2004) [arXiv:hep-lat/0312006].
32. R. V. Gavai and S. Gupta, Phys. Rev. D **71**, 114014 (2005) [arXiv:hep-lat/0412035].
33. Z. Fodor and S. D. Katz, JHEP **0404**, 050 (2004) [arXiv:hep-lat/0402006].
34. D. K. Sinclair and J. B. Kogut, arXiv:hep-lat/0609041.
35. S. Kratochvila and P. de Forcrand, PoS **LAT2005**, 167 (2006) [arXiv:hep-lat/0509143].
36. S. Ejiri, F. Karsch, E. Laermann and C. Schmidt, Phys. Rev. D **73**, 054506 (2006) [arXiv:hep-lat/0512040].
37. F. Csikor, G. I. Egri, Z. Fodor, S. D. Katz, K. K. Szabo and A. I. Toth, JHEP **0405**, 046 (2004) [arXiv:hep-lat/0401016].

THE CHIRAL LIMIT IN LATTICE QCD

Hidenori Fukaya (for JLQCD collaboration)

Theoretical Physics Laboratory, RIKEN
Wako, Saitama 351-0198, Japan
E-mail: hfukaya@riken.jp

It has been a big challenge for lattice QCD to simulate dynamical quarks near the chiral limit. Theoretically, it is well-known that the naive chiral symmetry cannot be realized on the lattice (the Nielsen-Ninomiya theorem). Also practically, the computational cost rapidly grows as the quark mass is reduced. The JLQCD collaboration started a project to perform simulations with exact but modified chiral symmetry using the the overlap-Dirac operator and the topology conserving action. The latter is helpful to reduce the numerical cost of the dynamical quarks. Our simulation of two-flavor QCD has been successful to reduce the sea quark mass down to a few MeV.

Keywords: Lattice QCD; chiral symmetry.

1. Introduction

Lattice QCD, has been successful to analyze the low-energy dynamics of mesons and baryons. Non-perturbative quantities, such as the hadron spectrum, matrix elements, the chiral phase transition *etc.* have been investigated by large-scale calculations often using supercomputers.

The lattice regularization, however, violates some important symmetries that the continuum theory has. For instance, the translational invariance is violated except for its discrete subgroup. The chiral symmetry, one of the most important symmetries of QCD, is also difficult to realize on the lattice. This is known as the fermion-doubling problem that any chiral Dirac operator must have unphysical poles.[1,2] In order to eliminate the doubler modes, one has to give up the chiral symmetry, or to employ complicated Dirac operators satisfying the Ginsparg-Wilson relation,[3–8] for which the locality is less obvious and more importantly much larger numerical cost is required compared to the explicitly local (the so-called ultra-local) Dirac operators. Most of the previous QCD simulations with the dynamical quarks were, therefore, limited to those with the Dirac operator which explicitly breaks the chiral symmetry. Moreover, the sea quark mass in such simulations were much larger than the physical values. Their results may contain systematic effects due to additive quark mass renormalizaiton, unwanted operator mixings with opposite chirality, additional symmetry breaking terms in the chiral perturbation theory, chiral extrapolation from rather heavy quark masses, and so on.

The JLQCD collaboration started a new project to simulate QCD near the chiral limit. It uses a new supercomputer system installed at KEK in early 2006. We employ the overlap-Dirac operator,[6,7] which satisfies the Ginsparg-Wilson relation[8] and thus realizes the exact chiral symmetry.[9] In order to avoid gauge configurations that are too rough and the topology is not well-defined, we use the Iwasaki action[10,11] combined with the topology conserving action[12–14] for the gauge part of the action. It turned out that keeping topology is also helpful to reduce numerical costs of the dynamical overlap fermions. On a $16^3 \times 32$ lattice with the lattice spacing $a \sim 0.11$–0.12 fm, we have simulated two-flavor QCD with the quark mass down to $\sim$ 3 MeV, which is even smaller than the physical up and down quark masses. We found that the locality of the overlap-Dirac operator is good enough compared to the QCD scale (See Ref. [15]). The preliminary results from the project have been reported in Refs. [16–18].

The outline of this article is as follows. In Sec. 2, We give a brief review about how the exact chiral symmetry and its topological properties can be realized on the lattice. The results of numerical simulations are reported in Sec. 3. Since our simulations are limited in a fixed topological sector on a fixed volume lattice $\sim 2^3 \times 4$ fm^4, the effects from finite volume and topology could be significant after reaching the chiral limit. We discuss this in Sec. 4. Summary and discussion are given in Sec. 5.

2. Chiral symmetry and topology

The difficulty of chiral symmetry on the lattice is summarized by the Nielsen-Ninomiya theorem:[1,2] any local Dirac operator which satisfies $D\gamma_5 + \gamma_5 D = 0$ must have unphysical poles. One can easily see this in free fermions. By the discretization, the continuum Dirac operator in the momentum space, $D = i\gamma_\mu p_\mu$, is replaced by the lattice counterpart

$$D^{\text{naive}} = \frac{\gamma_\mu}{2}(\partial_\mu + \partial_\mu^*) \sim i\gamma_\mu \sin(ap_\mu). \tag{1}$$

It has unphysical poles at $p_\mu = \pi/a$. Here, ∂_μ and ∂_μ^* denote the forward and backward subtraction, respectively. Wilson's prescription[19] to avoid the unphysical poles is to add a higher derivative term

$$D_W = \frac{\gamma_\mu}{2}(\partial_\mu + \partial^*_\mu) - \frac{a}{2}\partial_\mu\partial^*_\mu \sim \gamma_\mu \sin(ap_\mu) + \frac{2}{a}\sum_\mu \sin^2(p_\mu a/2). \quad (2)$$

This additional term, known as the Wilson term, gives the doublers a mass of the cut-off scale and let them decoupled from the theory. but it explicitly breaks the chiral symmetry. Since the Wilson term contains the covariant derivative, the eigenvalue distribution of the Dirac operator is largely deformed from the continuum limit. As a result, one loses the identification of the zero-modes, and thus the clear definition of the quark mass.

Neuberger's overlap-Dirac operator[6,7] is defined by

$$D = \frac{1}{\bar{a}}\left(1 + \gamma_5 \frac{aH_W}{\sqrt{a^2 H_W^2}}\right), \bar{a} = \frac{a}{1+s}, aH_W = \gamma_5(aD_W - 1 - s), \quad (3)$$

where a constant s is taken in a range $0 < s < 1$. It satisfies the Ginsparg-Wilson relation,[8] $\gamma_5 D + D\gamma_5 = \bar{a}D\gamma_5 D$, and the fermion action $S_F = \sum_x \bar{\psi}(x)D\psi(x)$, is exactly invariant under the chiral rotation,[9]

$$\psi \to e^{i\alpha\hat{\gamma}_5}\psi, \quad \bar{\psi} \to e^{i\alpha\gamma_5}, \quad \hat{\gamma}_5 = \gamma_5(1 - \bar{a}D), \quad (4)$$

at finite lattice spacings. Since $|\gamma_5 \frac{aH_W}{\sqrt{a^2H_W^2}}|^2 = 1$, the eigenvalues of the overlap Dirac operator are distributed on a circle with a radius $(1+s)/a$. Note that this circle exactly passes the origin where we can count the number of the chiral zero-modes (all the zero-modes commute with γ_5), and the quark mass can be simply defined by just adding m to D. Because of the relation[20]

$$n_+ - n_- = \mathrm{Tr}\hat{\gamma}_5/2, \quad (5)$$

where $n_\pm$ denotes the number of zero-modes with $\pm$ chirality, and the perturbative expansion of $\hat{\gamma}_5$,[21,22]

$$\hat{\gamma}_5(x,x)/2 = \frac{1}{32\pi^2}\mathrm{tr}\epsilon_{\mu\nu\rho\sigma}F_{\mu\nu}(x)F_{\rho\sigma}(x) + O(a^2), \quad (6)$$

one can identify the index $n_+ - n_-$ as the topological charge.

One should note, however, that Eq.(3) is ill-defined when H_W has zero-modes. From Eq.(5),

$$Q \equiv n_+ - n_- = \mathrm{Tr}\hat{\gamma}_5/2 = \mathrm{Tr}\gamma_5(1-\bar{a}D)/2 = \mathrm{Tr}\gamma_5\bar{a}D/2 = \frac{1}{2}\mathrm{Tr}\frac{H_W}{\sqrt{H_W^2}}, \quad (7)$$

one can see that the topological charge Q changes when the eigenvalue of H_W crosses zero. In other words, the equation $H_W = 0$ forms the topology boundaries in the configuration space. On the topology boundaries,

$H_W = 0$, the overlap-Dirac operator is not smooth and its locality is not obvious.[23] Near zero-modes of H_W also cause practical problems that the numerical cost of approximating $1/\sqrt{H_W^2}$ increases as the inverse of the lowest eigenvalue and the discontinuity of D requires a huge extra numerical cost for the dynamical quarks.

In order to achieve $H_W \neq 0$ at a finite lattice spacing (note that $H_W \neq 0$ is automatically satisfied in $a \to 0$ limit), which is known as the admissibility condition,[23,24] a transparent way is to add extra fields which produces a determinant,[12–14]

$$\det H_W^2/(H_W^2 + m_t^2) \tag{8}$$

to the theory. It is nothing but the determinant of two-flavor Wilson fermions and twisted-mass ghosts with a twisted mass m_t. Both of the additional fields have a cut-off scale mass and do not affect the low-energy physics. With this determinant, the overlap-Dirac operator D is smooth, and its locality is guaranteed. Furthermore, the numerical costs are largely reduced. Since this determinant does not allow the topology change during the hybrid Monte Carlo updates, we call it the topology conserving action.

3. Lattice simulation

We generate the gauge configurations of two-flavor QCD according to the Boltzmann weight

$$\exp(-\beta S_G)\det((1-\bar{a}m/2)D+m)^2 \det H_W^2/(H_W^2+m_t^2), \tag{9}$$

where βS_G denotes the Iwasaki gauge action[10,11] with the coupling $\beta = 2.3$ and 2.35, and $(1-\bar{a}m/2)D+m$ is an expression of the massive overlap fermion. We take $\bar{a} = a/1.6$ and $am_t = 0.2$ for all the simulations. On a $16^3 \times 32$ lattice, we have performed $O(1000)$ trajectories of the hybrid Monte Carlo updates in the $Q = 0$ topological sector. The lattice spacing is estimated to be $a \sim 0.11$-0.12 fm from the Sommer scale r_0 ($= 0.49$ fm). Details are seen in Refs. [16–18]

The new supercomputer system at KEK and some algorithmic improvements[28–30] enable us to reduce the quark mass down to $ma = 0.002(\sim 3$ MeV), which is even less than the physical value. In fact, we observe that the numerical cost, or the number of multiplication of D_W, does not depend too strongly on the sea quark mass m. However, we note that this insensitivity to m indicates that the lowest eigenvalue of D is larger than the quark mass, due to the finite volume effects, as discussed in the next section.

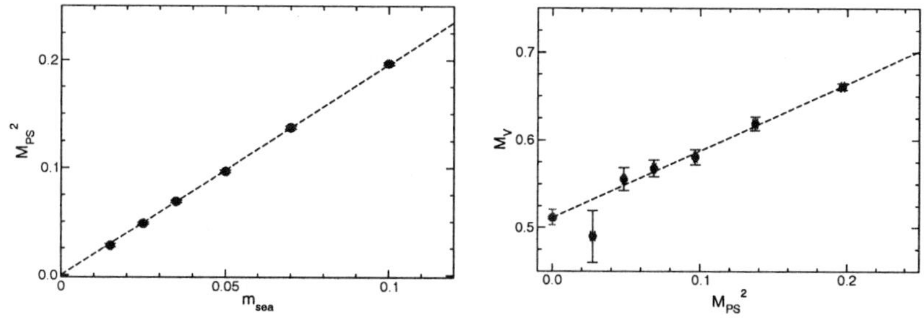

Fig. 1. Chiral extrapolation of pseudo-scalar (left) and vector meson masses (right).

Meson masses are evaluated from the correlation functions measured with smeared source and local sink operators. Figure 1 shows the chiral extrapolation of the pion and the vector meson masses. The pion mass agrees well with the leading ChPT formula, $m_\pi^2 \propto m\Sigma/F_\pi^2$, where the bare quark mass m needs no additive renormalization, owing to the exact chiral symmetry. With our present statistics, we find no significant deviation from a simple linear fit

$$m_\pi^2 = b_{\rm PS}\, m, \qquad m_{\rm V} = a_{\rm V} + b_{\rm V}\, m_\pi^2, \tag{10}$$

which gives $\chi^2/\text{dof} \lesssim 1.0$. With this linear fit, we obtain $a = 0.1312(23)$ fm using the ρ meson mass as an input. This is consistent with the estimate from r_0 within 10% accuracy.

4. Finite V and fixed Q effects

Since our simulations are done on a fixed volume lattice ($\sim$ 1.8-2 fm)^4 in a fixed topological sector, any observable may systematic errors due to the finite volume and fixed topology. Although the simplest solution to eliminate these errors is to go to larger volumes, we can take an alternative way that we treat finite V and Q effects using an effective theory.

The chiral perturbation theory (ChPT)[25,26] and the chiral Random Matrix theory (ChRMT)[27] are valid in estimating the finite V effects on pions, especially when the quark mass is so small that the pion Compton wave length is larger than the lattice size. This set-up is known as the ϵ-regime. In the ϵ-regime, ChRMT describes the eigenvalue spectrum of the Dirac operator with the chiral condensate Σ as an unique parameter. The pion correlator in the ϵ-regime is no more exponential but quadratic function of t, of which curvature depends on the pion decay constant F_π.

We compare our numerical result at the lightest quark mass with the

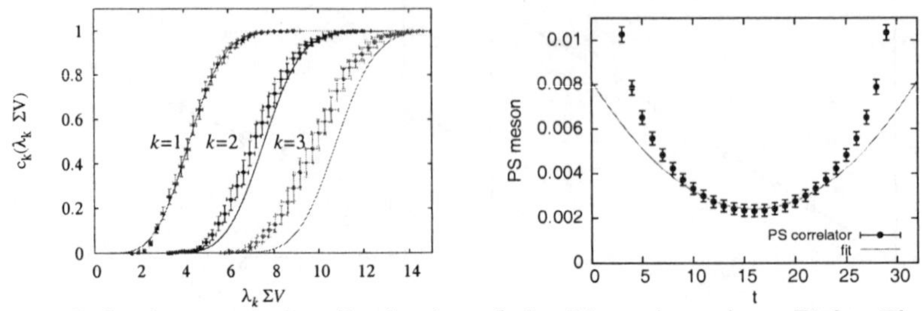

Fig. 2. Left: the cumulative distribution of the Dirac eigenvalues. Right: The pion correlator.

prediction of ChRMT and ChPT. The left panel of Fig. 2 shows the cumulative eigenvalue distribution of the overlap Dirac operator, from which we extract the chiral condensate as $\Sigma = (251(7)(11)\ \text{MeV})^3$, where the errors are statistical and an estimate of the higher order effects in the ϵ-expansion. This value is consistent with earlier works[31,32] which are done with heavier quark masses. The right panel shows the pion correlator and the quadratic fit curve, which gives $F_\pi = 86(7)$ MeV. Note that this value is obtained near the chiral limit without doing any chiral extrapolations.

5. Summary and discussion

The QCD simulation in the chiral regime is feasible with the exact chiral symmetry respected. The topology conserving action is very helpful to reduce the numerical cost of the dynamical overlap quarks. We can reduce the sea quark mass to the physical up and down quark masses or even lower. Near the chiral limit, the finite V and fixed Q effects are important since the pion is sensitive to these effects. Through the chiral perturbation theory or the chiral Random Matrix Theory, these effects can even be used to extract the low energy constants, such as the chiral condensate and the pion decay constant. To extend our lattice size is, however, important to confirm them in the future works.

Numerical simulations are performed on Hitachi SR11000 and IBM System Blue Gene Solution at High Energy Accelerator Research Organization (KEK) under a support of its Large Scale Simulation Program (No. 06-13). This work is supported in part by the Grant-in-Aid of the Ministry of Education (No. 18840045).

References

1. H. B. Nielsen and M. Ninomiya, Nucl. Phys. B **185**, 20 (1981) [Erratum-ibid. B **195**, 541 (1982)].

2. H. B. Nielsen and M. Ninomiya, Nucl. Phys. B **193**, 173 (1981).
3. D. B. Kaplan, Phys. Lett. B **288**, 342 (1992) [arXiv:hep-lat/9206013].
4. Y. Shamir, Nucl. Phys. B **406**, 90 (1993) [arXiv:hep-lat/9303005].
5. V. Furman and Y. Shamir, Nucl. Phys. B **439**, 54 (1995) [arXiv:hep-lat/9405004].
6. H. Neuberger, Phys. Lett. B **417**, 141 (1998) [arXiv:hep-lat/9707022].
7. H. Neuberger, Phys. Lett. B **427**, 353 (1998) [arXiv:hep-lat/9801031].
8. P. H. Ginsparg and K. G. Wilson, Phys. Rev. D **25**, 2649 (1982).
9. M. Luscher, Phys. Lett. B **428**, 342 (1998) [arXiv:hep-lat/9802011].
10. Y. Iwasaki, Nucl. Phys. B **258**, 141 (1985).
11. Y. Iwasaki and T. Yoshie, Phys. Lett. B **143**, 449 (1984).
12. P. M. Vranas, arXiv:hep-lat/0001006.
13. T. Izubuchi and C. Dawson [RBC Collaboration], Nucl. Phys. Proc. Suppl. **106**, 748 (2002).
14. H. Fukaya, S. Hashimoto, K. I. Ishikawa, T. Kaneko, H. Matsufuru, T. Onogi and N. Yamada [JLQCD Collaboration], Phys. Rev. D **74**, 094505 (2006) [arXiv:hep-lat/0607020].
15. N. Yamada *et al.* [JLQCD Collaboration], arXiv:hep-lat/0609073.
16. H. Fukaya *et al.* [JLQCD Collaboration], arXiv:hep-lat/0702003.
17. T. Kaneko *et al.* [JLQCD Collaboration], arXiv:hep-lat/0610036.
18. H. Matsufuru *et al.* [JLQCD Collaboration], PoS **LAT2006**, 031 (2006) [arXiv:hep-lat/0610026].
19. K. G. Wilson, "Quarks And Strings On A Lattice," *New Phenomena In Subnuclear Physics. Part A. Plenum Press, New York, 1977.*
20. P. Hasenfratz, V. Laliena and F. Niedermayer, Phys. Lett. B **427**, 125 (1998) [arXiv:hep-lat/9801021].
21. Y. Kikukawa and A. Yamada, Phys. Lett. B **448**, 265 (1999) [arXiv:hep-lat/9806013].
22. D. H. Adams, Annals Phys. **296**, 131 (2002) [arXiv:hep-lat/9812003].
23. P. Hernandez, K. Jansen and M. Luscher, Nucl. Phys. B **552**, 363 (1999) [arXiv:hep-lat/9808010].
24. M. Luscher, Nucl. Phys. B **568**, 162 (2000) [arXiv:hep-lat/9904009].
25. J. Gasser and H. Leutwyler, Phys. Lett. B **188**, 477 (1987).
26. P. H. Damgaard, M. C. Diamantini, P. Hernandez and K. Jansen, Nucl. Phys. B **629**, 445 (2002) [arXiv:hep-lat/0112016].
27. P. H. Damgaard and S. M. Nishigaki, Phys. Rev. D **63**, 045012 (2001) [arXiv:hep-th/0006111].
28. Z. Fodor, S. D. Katz and K. K. Szabo, JHEP **0408**, 003 (2004) [arXiv:hep-lat/0311010].
29. N. Cundy, J. van den Eshof, A. Frommer, S. Krieg, T. Lippert and K. Schafer, Comput. Phys. Commun. **165**, 221 (2005) [arXiv:hep-lat/0405003].
30. M. Hasenbusch, Phys. Lett. B **519**, 177 (2001) [arXiv:hep-lat/0107019].
31. T. DeGrand, Z. Liu and S. Schaefer, Phys. Rev. D **74**, 094504 (2006) [Erratum-ibid. D **74**, 099904 (2006)] [arXiv:hep-lat/0608019].
32. C. B. Lang, P. Majumdar and W. Ortner, arXiv:hep-lat/0611010.

QED CORRECTIONS TO HADRON AND QUARK MASSES

Y. Namekawa *
Department of Physics, Nagoya University, Nagoya 464-8602, Japan

Y. Kikukawa
Institute of Physics, University of Tokyo, Komaba, Tokyo 153-8902, Japan

We calculate electromagnetic mass differences by lattice simulations. Electromagnetic and isospin breaking contribution to light hadrons are estimated. Violation of Dashen's theorem as well as proton-neutron mass difference are obtained. We also extract quark masses from the spectrum and examined QED corrections to them.

1. Introduction

Lattice QCD simulations allow us to calculate physical quantities non-perturbatively. We can investigate dynamics of QCD from the first principle. For example, light hadron spectrum are obtained in high accuracy.[1] However, electromagnetic and isospin-violating effects are ignored in usual lattice QCD simulations, and mass differences such as $m_{\pi^+} - m_{\pi^0}$ and $m_{proton} - m_{neutron}$ are not obtained. For up and down quark masses, only their averaged value is calculated. It is important to take into account electromagnetic and isospin-violating effects for more realistic predictions.

There is an attempt to include electromagnetic effects in lattice QCD. Duncan *et al.* computed hadron masses on the background of gluons and photons in the quenched approximation, ignoring dynamical quarks.[2] In the quenched case, one can apply Monte Carlo method to photons and gluons independently. $\pi^+ - \pi^0$ mass splitting is found to be 4.9(3) MeV at $\beta = 5.7$ with the Wilson action. Though their results show good agreement with experiments, there are unsatisfactory points. One is the discretization effect. Their simulations were performed at a single lattice spacing with

*Present address : Center for Computational Sciences, University of Tsukuba, Tsukuba, Ibaraki 305-8577, Japan

unimproved actions. Scaling violations of their results may not be small. Another is the finite size effect. Finite size effects were estimated only by a model. It is desirable to evaluate finite size effects from a first principle. RBC collaboration also studies electromagnetic effects with $N_f = 2$ domain wall fermions but at single lattice spacing and single spatial volume.[3]

We study electromagnetic mass splittings of hadrons in the quenched approximation.[4] We use improved actions and perform chiral and continuum extrapolations. We also evaluate finite size effects using lattices with spatial extents of 2.4 and 3.2 fm.

2. Method

For $SU(3)$ gauge part, we employ the RG improved gauge action.

$$S_g = \frac{\beta}{6}\left\{c_0 \sum_{x,\mu\nu} W_{\mu\nu}^{1\times 1}(x) + c_1 \sum_{x,\mu\nu} W_{\mu\nu}^{1\times 2}(x)\right\}, \tag{1}$$

where $\beta = 6/g^2$. $c_0 = 3.648$ and $c_1 = -0.331$ are determined by a renormalization group analysis. For the $U(1)$ gauge part we employ a non-compact Abelian gauge action.

$$S_{EM} = \frac{1}{4e^2}\sum_{n,\mu\nu}(\nabla_\mu A_\nu(n) - \nabla_\nu A_\mu(n))^2, \tag{2}$$

where e is a bare electric coupling.

We generate $SU(3)$ and $U(1)$ fields independently and compute hadron masses on the combined $SU(3) \times U(1)$ configurations. $SU(3)$ link fields are generated by Monte Carlo simulations. We use the pseudoheat bath algorithm. $U(1)$ gauge configurations are constructed from the Fourier transformed photon propagators in the momentum space.[5] The photon field $A_\mu(n)$ is made up of a noise vector multiplied by the propagator weight. For the quark with its charge qe, the electromagnetic link variable is $U_\mu^{EM}(n) = \exp(iqA_\mu(n))$. We assign $q = +2/3$ for the up-type quark and $q = -1/3$ for the down-type quark as in the case of our real world.

Quark propagators are computed on the $SU(3) \times U(1)$ configurations. We use the clover quark action defined by

$$S_q = \sum_{x,y} \overline{q}_x D_{x,y} q_y, \tag{3}$$

$$D_{x,y} = \delta_{xy} - \kappa \sum_{\mu}\left\{(1-\gamma_\mu)U_{x,\mu}\delta_{x+\hat{\mu},y} + (1+\gamma_\mu)U_{x,\mu}^\dagger \delta_{x,y+\hat{\mu}}\right\}, \tag{4}$$

$$-\delta_{xy} c_{\rm SW} \kappa \sum_{\mu<\nu} \sigma_{\mu\nu} F_{\mu\nu}, \tag{5}$$

Table 1. Simulation parameters.

	β	κ	κ_u	κ_d	N_{conf}
$12^3 \times 24$	2.187	0.1365-0.1390	0.1369-0.1394	0.1366-0.1391	900
$16^3 \times 24$		0.1390	0.1394	0.1391	400
$16^3 \times 32$	2.334	0.1349-0.1374	0.1353-0.1378	0.1350-0.1375	600
$24^3 \times 48$	2.575	0.1337-0.1353	0.1341-0.1357	0.1338-0.1354	300

where κ is the hopping parameter, $F_{\mu\nu}$ is the clover-shaped lattice discretization of the field strength and $\sigma_{\mu\nu} = (i/2)[\gamma_\mu, \gamma_\nu]$. For the clover coefficient we adopt a meanfield value $c_{SW} = u_0^{-3}$ where $u_0 = \left(W^{1\times1}\right)^{1/4} = (1 - 0.8412/\beta)^{1/4}$, using the plaquette $W^{1\times1}$ in one-loop perturbation theory. We fix the gauge to the Coulomb one and use a gaussian smearing function to enhance signals.

Simulations are performed at $a = 0.2$ fm($\beta = 2.187$), $a = 0.16$ fm($\beta = 2.334$) and $a = 0.1$ fm($\beta = 2.575$). At $\beta = 2.187$, we also employ $16^3 \times 24$ lattices with $L = 3.2$ fm to examine the finite size effects. The hopping parameters are chosen so that pi-rho ratios in the neutral channel are $m_{PS}/m_V = 0.76$–0.5. Configurations for measurements are stored at each 100 sweeps. Our simulation parameters are summarized in Table 1.

3. Simulation results

3.1. *Masses*

We compute hadron masses from quark propagators on the $SU(3) \times U(1)$ configurations. Masses are obtained by χ^2 fits to the correlators, taking account of correlations among different time slices. The fitting range is $t = N_t/4 - N_t/2$, which is determined by inspecting stability of the resulting mass. Statistical errors of masses are estimated with the jack-knife method.

3.2. *Finite-size effects*

Finite size effects may be enhanced in the presence of electromagnetic fields because electromagnetic fields have a long interaction range. It is necessary to check magnitude of finite size corrections. In Fig. 1, we plot charged pseudoscalar and vector meson masses as a function of the spatial volume. The results obtained on $12^3 \times 24$ and $16^3 \times 24$ lattices are mutually consistent within errors. Finite size effects are not observed in baryon masses, either. In the quenched approximation, $L = 2.4$ fm seems to be enough for spectrum calculations, even if there is an electromagnetic interaction.

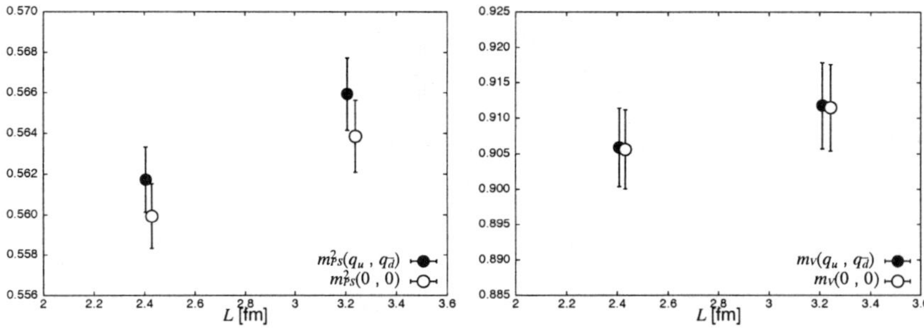

Fig. 1. Volume dependence of charged pseudoscalar (left panel) and vector meson masses (right panel) at $m_{PS}/m_V = 0.60$ on $12^3 \times 24$ lattice. For comparison, masses in the pure QCD (open symbols) are also plotted.

3.3. *Chiral extrapolations*

Since lattice QCD simulations are performed at relatively heavy quark mass region, lattice data must be extrapolated to the physical point. For chiral extrapolations, we fit hadron masses as functions of quark masses and charges. We employ vector Ward identity quark masses,

$$m_q = \frac{1}{2}\left(\frac{1}{\kappa_q} - \frac{1}{\kappa_c^q}\right), m_{\overline{q}} = \frac{1}{2}\left(\frac{1}{\kappa_{\overline{q}}} - \frac{1}{\kappa_c^{\overline{q}}}\right), \tag{6}$$

where κ_q is a hopping parameter for a quark and $\kappa_{\overline{q}}$ is for an antiquark. $\kappa_c^{q,\overline{q}}$ are the corresponding critical hopping parameters.

We use the following function to fit pseudoscalar meson masses.

$$m_{PS}^2 = A_{PS}(q_q + q_{\overline{q}})^2 + (B_0^{PS} + B_2^{PS}(q_q + q_{\overline{q}})^2)(m_q + m_{\overline{q}}), \tag{7}$$

where A_{PS}, B_0^{PS} and B_2^{PS} and $\kappa_c^{q,\overline{q}}$ in $m_q, m_{\overline{q}}$ are fitting parameters. q_q is an electric charge for a particle in units of e and $q_{\overline{q}}$ is for an antiparticle. This function only depends on the total charge and quark masses, inspired by the chiral perturbation theory. Strictly speaking, the quenched chiral logarithm term must be added. However, the logarithmic curvature is not seen in our data within $m_{PS}/m_V = 0.76$–0.5. Smaller quark mass data are needed for a more precise extrapolation. A fit result is presented in Fig. 2. This function reproduces our data well.

We fit vector meson masses using the following function.

$$m_V = (A_0^V + A_2^V(q_q + q_{\overline{q}})^2) + (B_0^V + B_2^V(q_q + q_{\overline{q}})^2)(m_q + m_{\overline{q}}), \tag{8}$$

where A_0^V, A_2^V, B_0^V and B_2^V are fitting parameters. $\rho^+ - \rho^0$ mass difference is found to be very small as shown in Fig. 2, as predicted by the hidden local symmetry formulation, $\Delta m_\rho = -1$ MeV.[6]

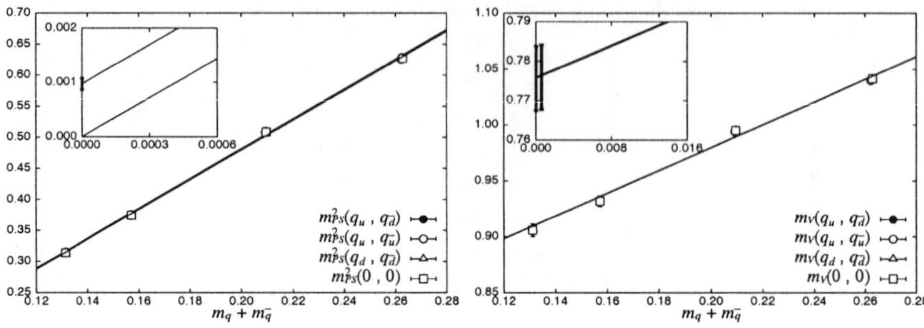

Fig. 2. Chiral extrapolations of pseudoscalar (left panel) and vector meson masses (right panel) on $12^3 \times 24$ lattice.

Similar to meson masses, proton and neutron masses are extrapolated to the chiral limit by use of the linear functions in quark masses.

$$\begin{aligned} m_{proton} &= (A_B^{(0)} + A_B^{(2)}(2q_u + q_{\bar{d}})^2) \\ &\quad + (B_B^{(0)} + B_B^{(2)}(2q_u + q_{\bar{d}})^2)(2m_u + m_d), \qquad (9) \\ m_{neutron} &= A_B^{(0)} + B_B^{(0)}(m_u + 2m_d), \qquad (10) \end{aligned}$$

where $A_B^{(0)}$, $A_B^{(2)}$, $B_B^{(0)}$ and $B_B^{(2)}$ are fitting parameters. In the chiral limit, proton becomes heavier than neutron, as expected. On the other hand, neutron becomes heavier than proton at the physical point due to mass difference of up and down quarks.

3.4. *Continuum extrapolations*

After performing chiral extrapolations, we take continuum limits. We use rho meson mass to determine lattice spacing at each β. Since we do not find any significant lattice spacing dependence in hadron mass splittings, continuum extrapolations are performed with a constant fit. For electromagnetic mass differences of mesons, we obtained

$$m_{\pi^+} - m_{\pi^0} = 4.03(14) \text{ MeV}, \quad m_{\rho^+} - m_{\rho^0} = 0.24(18) \text{ MeV}. \qquad (11)$$

For proton–neutron mass difference, we have

$$m_{proton} - m_{neutron} = -2.93(67) \text{ MeV}. \qquad (12)$$

They are comparable to their experimental values.

In addition to electromagnetic mass differences, we can evaluate electromagnetic and isospin contributions of each mass difference. Then, we calculate violation of Dashen's theorem, Δ_{EM} :

$$\Delta_{EM} = \frac{(m^2_{K^+} - m^2_{K^0})_{EM} - (m^2_{\pi^+} - m^2_{\pi^0})_{EM}}{(m^2_{\pi^+} - m^2_{\pi^0})_{EM}}. \tag{13}$$

As in the case of mass splittings, we find mild lattice spacing dependence in Δ_{EM} and perform continuum extrapolations with a constant fit. In the continuum limit, we have

$$\Delta_{EM} = 0.518(37). \tag{14}$$

This value is close to the results by NJL model combined with perturbative QCD, $\Delta_{EM} = 0.84(25)$.[7]

We also evaluate electromagnetic contribution to proton–neutron mass difference. Electromagnetic contributes positively and isospin breaking contributes negatively to the mass difference.

$$(m_{proton} - m_{neutron})_{EM} = 0.45(66), \tag{15}$$

$$(m_{proton} - m_{neutron})_{isospin} = -3.36(11). \tag{16}$$

A sum of both contributions gives a negative value of proton–neutron mass difference. Within 2σ level, our value is consistent with an estimate from heavy baryon chiral perturbation theory combined with partially quenched calculations, $(m_{proton} - m_{neutron})_{isospin} = -2.26(72)$ MeV.[8]

Quark masses can be extracted from the hadron spectrum. Up, down and strange quark masses are determined by requiring that hadron masses reproduce the experimental values. We employ m_{π^0}, m_{K^+} and m_{K^0} as inputs. We obtain quark masses using 1–loop renormalization factors, and extrapolate them linearly to the continuum limit. Fig. 3 illustrates the continuum extrapolations of quark masses. Up and down quark masses can be determined separately by including QED effects. We found the QED correction to the strange quark mass is tiny. Our values are

$$m_u^{\overline{MS}}(\mu = 2 \text{ GeV}) = 2.80(9) \text{ MeV}, \tag{17}$$

$$m_d^{\overline{MS}}(\mu = 2 \text{ GeV}) = 4.76(13) \text{ MeV}, \tag{18}$$

$$m_s^{\overline{MS}}(\mu = 2 \text{ GeV}) = 100.3(23) \text{ MeV}. \tag{19}$$

4. Conclusions

We calculated electromagnetic mass splittings using the RG-improved gauge action and the clover-improved Wilson quark action. In the quenched approximation, our results for electromagnetic mass splittings at the continuum limit are comparable with experimental values. We confirmed that

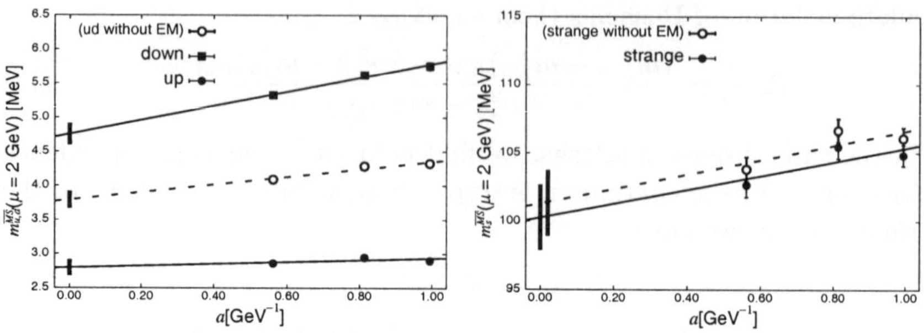

Fig. 3. Continuum extrapolations of up, down (left panel) and strange quark masses(right panel).

these values obtained with $L = 2.4$ fm are not shifted by finite size effects. We also evaluated electromagnetic and isospin breaking contributions to meson mass splittings. Our value for the violation of Dashen's theorem is $\Delta_{EM} = 0.518(37)$. Electromagnetic and isospin breaking contributions to proton–neutron mass difference are estimated as $(m_{proton} - m_{neutron})_{EM} = 0.45(66)$ MeV, $(m_{proton} - m_{neutron})_{isospin} = -3.36(11)$ MeV. Finally, we extracted quark masses from the spectrum using experimental values of m_{π^0}, m_{K^+} and m_{K^0} as inputs. Our values are $m_u^{\overline{MS}}(\mu = 2 \text{ GeV}) = 2.80(9)$ MeV, $m_d^{\overline{MS}}(\mu = 2 \text{ GeV}) = 4.76(13)$ MeV, $m_s^{\overline{MS}}(\mu = 2 \text{ GeV}) = 100.3(23)$ MeV.

Clearly, the next step is to perform unquenched simulations. Calculations on the $N_f = 2$ configurations are ongoing.

Acknowledgments

We thank M. Harada, M. Hayakawa, M. Tanabashi, U.-J. Wiese, N. Yamada and K. Yamawaki for valuable discussions. This work is supported in part by the 21st Century COE Program, Nagoya University and by Grants-in-Aid of the Ministry of Education (Nos. 18740139).

References

1. For a recent review, see L. Giusti, PoS (LAT2006) 009 (2007).
2. A. Duncan *et al.*, Phys. Rev. Lett. **76** 3894 (1996).
3. N. Yamada *et al.* (RBC Collaboration), PoS (LAT2005) 092 (2006).
4. Y. Namekawa and Y. Kikukawa, PoS (LAT2005) 090 (2006).
5. E. Dagotto *et al.*, Nucl .Phys. B **331** 500 (1990).
6. M. Harada and K. Yamawaki, Phys. Rept. **381** 1 (2003).
7. J. Bijnens *et al.*, Nucl. Phys. B **490** 239 (1997).
8. S. R. Beane *et al.* (NPLQCD Collaboration), Nucl. Phys. B **768** 38 (2007).

PHASE STRUCTURE OF NJL MODEL WITH FINITE QUARK MASS AND QED CORRECTION

Takahiro Fujihara*#, Tomohiro Inagaki†‡, Daiji Kimura#

#*Department of physics, Hiroshima University, HigashiHiroshima, 739-8526, JAPAN*
‡ *Information Media Center, Hiroshima University HigashiHiroshima, 739-8521, JAPAN*
* *E-mail: fujihara@hiroshima-u.ac.jp*
† *E-mail: inagaki@hiroshima-u.ac.jp*

We study QED corrections to the chiral symmetry breaking in Nambu-Jona-Lasinio (NJL) model with two flavors of quarks. In the model the isospin symmetry is broken by differences of the current quark masses and the electromagnetic charges between up and down quarks. In the leading order of the 1/N expansion we calculate the effective potential of the model with one-loop QED corrections. Evaluating the effective potential, we study an influence of the isospin symmetry breaking on the orientation of chiral symmetry breaking. The current quark mass has an important contribution for the orientation of chiral symmetry breaking.

1. Introduction

The broken chiral symmetry is restored in certain extreme environments, e.g. high temperature, high density and so on. The phase transition of chiral symmetry is understood as a non-perturbative phenomenon in QCD. Thus the phase structure of QCD has been able to study in some low energy effective theories or through numerical calculations in lattice QCD. These results show that the chiral symmetry should be restored above a critical temperature. It is conjectured in a lot of models that the critical temperature is observed blow 200MeV, $T_{\rm cr} < 200$MeV. Recently, an experimental study at RHIC founds an evidence of the symmetry restoration from a high temperature hadronic matter to deconfined partonic matter.

The effect of an electromagnetic field on chiral symmetry breaking is not so simple. Low energy effective theories provide possible approaches for studying non-perturbative QCD phenomena in an external electromagnetic field. The NJL model is one of the simplest models in which the chiral symmetry is broken dynamically. The dynamical origin of symmetry breaking in the NJL model has been studied under an external electromagnetic field. Then it was found that the external electromagnetic fields can induce a variety of phases.[1–3]

Radiative QED corrections play an essential role for phenomena with the isospin symmetry breaking of quarks and hadrons, even if we consider a system with no external electromagnetic field. The difference between the electromagnetic charges of up and down quarks breaks the SU(2) isospin symmetry. The mass difference between up and down quarks also breaks the isospin symmetry. Because of these isospin breaking, it is found that an interesting phase is realized inside quark matter through QED corrections. If the sum of the up and down quark masses, $m_u + m_d$, is small enough, it is possible that a pion field develops a non-vanishing vacuum expectation value.

To study the effect of these isospin breaking we introduce terms for the current quark mass and QED interactions into the NJL model, and evaluate the effective potential under some assumptions. We observe the minimum of the effective potential and determine the phase. Then we evaluate the pion mass difference in our model.

2. Gauged NJL Model

We introduce the current quark mass and QED interactions into the NJL model. This model is called the gauged NJL model. The Lagrangian for the gauged NJL model[4,5] is

$$\mathcal{L} = \mathcal{L}_m + \mathcal{L}_{photon} + \mathcal{L}_{QED}, \tag{1}$$

$$\mathcal{L}_m = \bar{\psi}\left(i\gamma^\mu\partial_\mu - M\right)\psi + \frac{G}{2N}\left[(\bar{\psi}\psi)^2 + (\bar{\psi}i\gamma_5\tau^a\psi)^2\right], \tag{2}$$

$$\mathcal{L}_{photon} = -\frac{1}{4}F_{\mu\nu}F^{\mu\nu} - \frac{1}{2\xi}(\partial_\mu A^\mu)^2, \tag{3}$$

$$\mathcal{L}_{QED} = -\bar{\psi}eQ\gamma^\mu A_\mu\psi, \quad Q = \mathrm{diag}(2/3, -1/3). \tag{4}$$

Because of electric charges for up and down quarks are different, the isospin symmetry is explicitly broken. We also introduce the isospin breaking mass term, $M = \mathrm{diag}(m_u, m_d)$, $m_u \neq m_d$.

We evaluate the path integral for the quark fields by using the auxiliary field method, $\sigma \sim \bar{\psi}\psi$ and $\pi^a \sim \bar{\psi}i\gamma_5\tau^a\psi$. The generating functional is given by

$$Z = \int \mathcal{D}\sigma \mathcal{D}\pi \mathcal{D}A \exp \left[i \int dx^4 \mathcal{L}_{photon} + iN \left\{ -\frac{1}{2G} \int d^4x \left(\sigma^2 + (\pi^a)^2\right) \right.\right.$$
$$\left.\left. - i \ln \det \left[i\gamma^\mu(\partial_\mu + ieQA_\mu) - M - \sigma - i\gamma_5 \tau^a \pi^a \right] \right\} \right] . \tag{5}$$

We expand the log determinant in terms of the small current quark mass M and the small electric charge e,

$$i \ln \det \left[i\gamma^\mu \left(\partial_\mu + ieQA_\mu\right) - M - \sigma - i\gamma_5 \tau^a \pi^a \right]$$
$$= i \mathrm{tr} \ln \left(i\gamma^\mu \partial_\mu \sigma - i\gamma_5 \tau^a \pi^a \right) + \mathrm{tr} \sum_{n=1}^{\infty} J_n, \tag{6}$$
$$J_n = \frac{1}{in} \left(\frac{M + eQ\gamma^\mu A_\mu}{i\gamma^\mu \partial_\mu - \sigma - i\gamma_5 \tau^a \pi^a + i\epsilon} \right)^n . \tag{7}$$

We evaluate the effective potential up to the second order of M and e, i.e. up to $n = 2$. As is shown in Fig. 1, in this order the photon self-energy contributes to the effective potential. Evaluating these diagrams in the $\overline{\mathrm{MS}}$ scheme, we obtain the effective potential,

$$V(\sigma, \pi^a) = V_m(\sigma, \pi^a) + V_{gauge}(\sigma, \pi^a), \tag{8}$$
$$V_m(\sigma, \pi^a) = \frac{1}{4G}\sigma'^2 - \frac{1}{8\pi^2} f(\sigma'^2; \Lambda_f^2)$$
$$- \frac{1}{4\pi^2}(m_u + m_d)\sigma g(\sigma'^2; \Lambda_f^2) - \frac{1}{8\pi^2}(m_u^2 + m_d^2) g(\sigma'^2; \Lambda_f^2)$$
$$- \frac{1}{4\pi^2} \left[(m_u^2 + m_d^2)\sigma^2 + (m_u - m_d)^2 \pi^+ \pi^- \right]$$
$$\times \left(\frac{\Lambda_f}{\sqrt{\Lambda_f^2 + \sigma'^2}} - \ln \frac{\Lambda_f + \sqrt{\Lambda_f^2 + \sigma'^2}}{\sqrt{\sigma'^2}} \right), \tag{9}$$

$$V_{gauge}(\sigma, \pi^a) = \frac{1}{192\pi^2} \left[3f\left(4\sigma'^2(1 + 4\alpha N/27\pi) - (4\alpha N/3\pi)\pi^+\pi^-; \Lambda_p^2\right) \right.$$
$$\left. + f\left(4\sigma'^2(1 + 4\alpha N/27\pi); \Lambda_p^2\right) - 4f\left(4\sigma'^2; \Lambda_p^2\right) \right], \tag{10}$$
$$f(s^2; t^2) \equiv (2t^2 + s^2)\sqrt{t^2(t^2 + s^2)} - s^4 \ln \frac{\sqrt{t^2} + \sqrt{t^2 + s^2}}{\sqrt{s^2}}, \tag{11}$$
$$g(s^2; t^2) \equiv \sqrt{t^2(t^2 + s^2)} - s^2 \ln \frac{\sqrt{t^2} + \sqrt{t^2 + s^2}}{\sqrt{s^2}}, \tag{12}$$

where $\sigma'^2 \equiv \sigma^2 + (\pi^a)^2$, Λ_f is 3-dimensional momentum UV cutoff, Λ_p is photon 3-dimensional momentum UV cutoff and $\alpha \equiv \frac{e^2}{4\pi^2}$. In this paper we set $\Lambda_p = \Lambda_f$. The ground state of the model is found by observing the minimum of the effective potential (8).

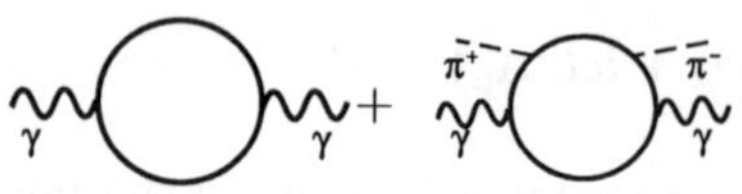

Fig. 1. Diagrams of photon self-energy.

Parameters in the model are determined to reproduce the realistic pion decay constant $f_\pi = 91.9\text{MeV}$ and pion mass $m_\pi = 135\text{MeV}$. For $e = 0$ and $m_u = m_d$, the pion wave function renormalization Z_π is given by

$$Z_\pi^{-1} = \frac{N}{2\pi^2}\left[\ln\frac{\Lambda_f + \sqrt{\Lambda_f^2 + (\sigma + m_q)^2}}{\sigma + m_q} - \frac{\Lambda_f}{\sqrt{\Lambda_f^2 + (\sigma + m_q)^2}}\right]. \quad (13)$$

Using Eq.(13) and Gell-Mann-Oakes-Renner relation,[6] we fix parameters $\Lambda = 0.697\text{GeV}$, $G = 24.8\text{GeV}^{-2}$ with $m_u = m_d = 4.5\text{MeV} \equiv m_q$. Below we use these parameters even in cases of $m_u \neq m_d$ and $e \neq 0$.

First we evaluate the effective potential without QED corrections. Evaluating the isospin breaking contribution from the current quark mass, we put $m_u = m_q - \delta$ for the up quark mass and $m_d = m_q + \delta$ for the down quark mass, as the sum of quark masses, $m_u + m_d$, is fixed ($m_u + m_d = 9\text{MeV}$). We show the result for, $\delta = 2\text{MeV}$ in Fig. 2.

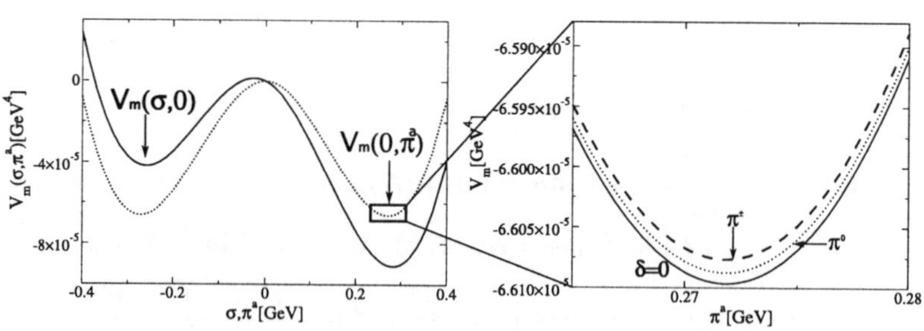

Fig. 2. Behavior of the effective potential (9) without QED corrections. In the left figure, the solid line shows the effective potential along the sigma axis and the dotted line shows the one along the pion axis. In the right figure, we zoom the rectangle part in the left one. The solid line shows the $\delta = 0$ case, the dotted line shows the effective potential along the π^0 axis and the dashed line shows the one along the $\pi^\pm$ axis.

We look for the minimum of the effective potential in sigma σ, neutral pion π^0 and charged pion $\pi^\pm$ hyper plane. The contribution of the isospin breaking effect is extremely small in Fig. 2. We can not distinguish the neutral pion direction and the charged pion direction in the left figure of Fig. 2. The right figure of Fig. 2 shows the local minimum of the effective potential along the pion axis. The effect of δ splits curves of the effective potential for π^0 and $\pi^\pm$ a little. We see only a small effect of the isospin breaking for the difference of the dotted and the dashed lines in Fig. 2.

In this model pion mass difference is given by

$$\begin{aligned}\Delta m_\pi &\equiv m_{\pi^\pm} - m_{\pi^0} \\ &= \sqrt{2NZ_\pi}\left[\left(\frac{\partial^2 V}{\partial\pi^\pm\partial\pi^\pm}\right)^{1/2} - \left(\frac{\partial^2 V}{\partial\pi^0\partial\pi^0}\right)^{1/2}\right]. \end{aligned} \tag{14}$$

For $\delta = 2\text{MeV}$, we get the pion mass difference $\Delta m_\pi = 0.06\text{MeV}$. It is experimentally observed about 5 MeV. In Fig. 3 we plot Δm_π as a function of δ. To reproduce a realistic Δm_π another contribution have to be introduced.

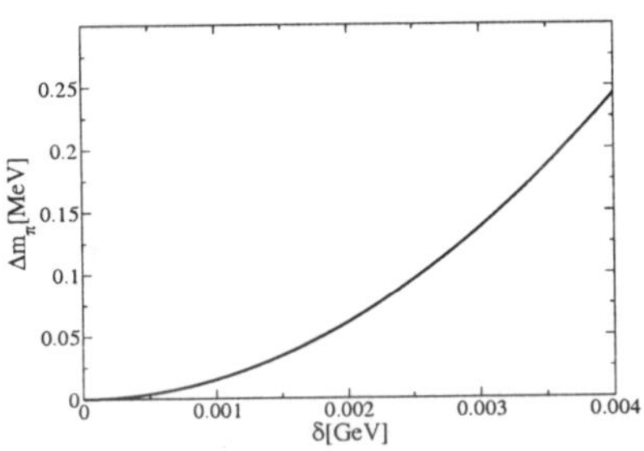

Fig. 3. Pion mass difference as a function of δ without QED corrections.

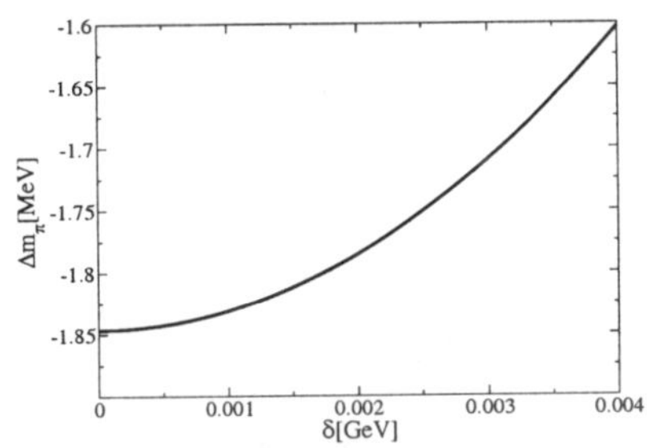

Fig. 4. Pion mass difference as a function of δ with QED corrections.

Next we consider the QED corrections. We numerically evaluate the effective potential (8) and illustrate the behaviors near the local minimum of the effective potential along the pion axis in Fig. 5. A larger isospin breaking effect is observed in this figure. The effective potential for the charged pion is smaller than that for neutral pion. Therefore the ground state is found at $\sigma \neq 0$ and $\pi^a = 0$. In Fig. 6 we plot the expectation value of σ as a function of δ. QED corrections decrease this expectation value, i.e. the chiral symmetry breaking is suppressed.

In Fig. 7 we show the local minimum of the effective potential on the

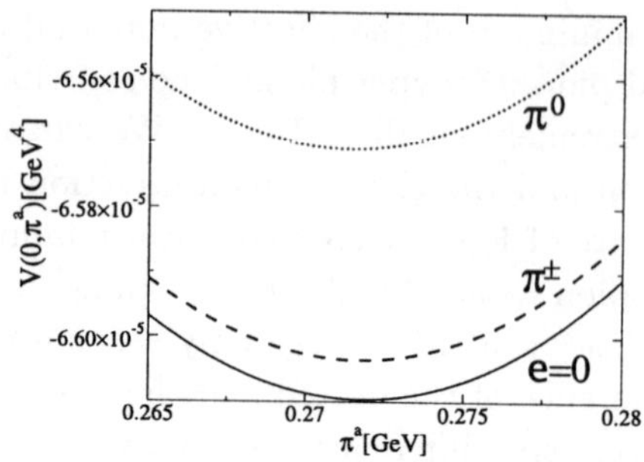

Fig. 5. Local minimum of the effective potential along the pion axis with $\delta = 0$. The solid line shows the effective potential without QED corrections. The dotted line and the dashed line show the effective potential with QED corrections along the neutral pion and the charged pion axes, respectively.

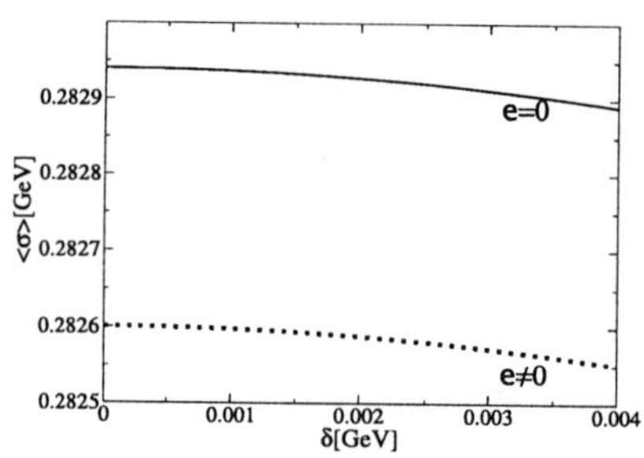

Fig. 6. The global minimum of the effective potential as a function of δ.

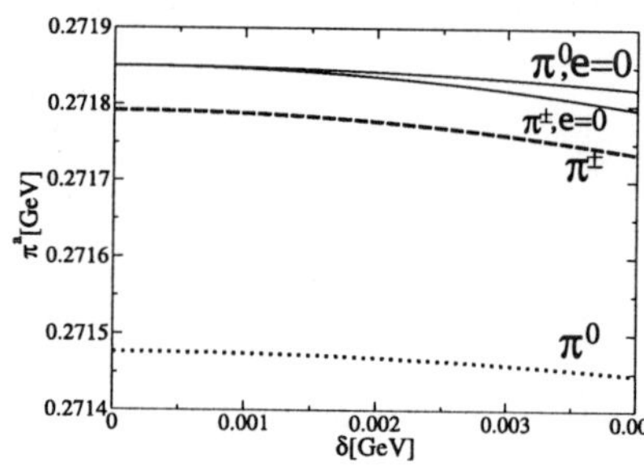

Fig. 7. The local minimum of the effective potential along the pion axis as a function of δ.

pion axis. The solid lines show the one without QED corrections. The dotted and dashed lines mean the one with QED corrections along the neutral pion and the charged pion axes, respectively. QED corrections and mass difference decrease the local minimum. These isospin breaking split the neutral and the charged pion fields dependencies of the effective potential. A contribution from the isospin breaking of the current quark mass slightly suppresses the chiral symmetry breaking. QED corrections have much larger effect and also suppress the chiral symmetry breaking.

As is shown in Fig. 4, the pion mass difference has a negative value. In next section we consider the phenomenological model to obtain the experimental pion mass difference.

3. Phenomenological model for Δm_π

As is well-known, a pion-photon interaction plays an important role in determining the pion mass difference.[7] We introduce kinetic terms for mesons,

$$\mathcal{L} = \mathcal{L}_{\rm m} + \mathcal{L}_{\rm photon} + \mathcal{L}_{\rm QED} + \mathcal{L}_{\rm s}, \tag{15}$$

$$\mathcal{L}_{\rm s} = \frac{1}{2N}\left[\frac{1}{2}(\partial_\mu\sigma)^2 + \frac{1}{2}(\partial_\mu\pi^0)^2 + (\partial_\mu + ieA_\mu)\pi^+(\partial_\mu - ieA_\mu)\pi^-\right]. \tag{16}$$

We evaluate the effective potential again and calculate the pion mass difference. Then we obtain more realistic value, $\Delta m_\pi \sim 8$MeV.

4. Summary

We have evaluated the effect of the isospin symmetry breaking due to current quark masses and electric charges. The effective potential has been evaluated up to $O(e^2)$ and $O(m^2)$. The isospin breaking effects split neutral and charged pion masses. The current quark mass difference slightly suppresses the chiral symmetry breaking. QED corrections also suppress the chiral symmetry breaking. The effect of the current quark mass differences is extremely small for the pion mass difference. The contribution of the QED corrections gives the opposite sign for the pion mass difference in our model. To obtain a realistic pion mass difference we have introduced the kinetic term for meson fields. Then we have obtained a larger pion mass difference $\Delta m_\pi \sim 8$MeV. The long distance interaction between photons and the charged pions makes the kinetic term for mesons necessary.

The temperature dependence of the pion mass difference is discussed in Ref. 8.

References

1. S. P. Klevansky and R. H. Lemmer, *Phys. Rev.* **D39**, 3478 (1989).
2. H. Suganuma and T. Tatsumi, *Prog. Theor. Phys.* **90**, 379 (1993).
3. M. Ishi-i, T. Kashiwa and N. Tanimura, *Prog. Theor. Phys.* **100**, 353 (1998).
4. V. A. Miransky and K. Yamawaki, *Mod. Phys. Lett.* **A4**, 129 (1989).
5. K.-i. Kondo, M. Tanabashi and K. Yamawaki, *Prog. Theor. Phys.* **89**, 1249 (1993).
6. M. Gell-Mann, R. J. Oakes and B. Renner, *Phys. Rev.* **175**, 2195 (1968).
7. V. Dmitrasinovic, R. H. Lemmer and R. Tegen, *Phys. Lett.* **B284**, 201 (1992).
8. T. Fujihara, T. Inagaki and D. Kimura, *Prog. Theor. Phys.* **177**, 139 (2007).

PHASE STRUCTURE OF THERMAL QED BASED ON THE HARD THERMAL LOOP IMPROVED LADDER DYSON-SCHWINGER EQUATION –A "GAUGE INVARIANT" SOLUTION–

HISAO NAKKAGAWA, HIROSHI YOKOTA and KOJI YOSHIDA

Institute for Natural Sciences, Nara University,
Nara 631-8502, Japan
E-mail: {nakk, yokotah, yoshidak}@daibutsu.nara-u.ac.jp

Based on the hard-thermal-loop resummed improved ladder Dyson-Schwinger equation for the fermion mass function, we study how we can get the gauge invariant solution in the sense it satisfies the Ward identity. Properties of the "gauge-invariant" solutions are discussed.

Keywords: Thermal QED; Hard Thermal Loop; Dyson-Schwinger equation; Gauge invariance.

1. Introduction and summary

The Dyson-Schwinger equation (DSE) is a powerful tool to investigate with the analytic procedure the nonperturbative structure of field theories, such as the phase structure of gauge theories. However, the full DSEs are coupled integral equtions for several unknown functions, thus are hard to be solved without introducing appropriate approximations. We usually adopt the step-by-step approach to this problem, firstly approximate the integration kernel by the tree, or, ladder kernel, next use the improved ladder one, etc. Advantage of the DSE analysis lies in the possibility of such an systematic improvement through the analytic investigation.

Analyses of the DSE have proven to be successful in studying the phase structure of vacuum gauge theories [1-3]. In the Landau gauge DSE with the ladder kernel for the fermion mass function in the vacuum QED, the fermion wave function renormalization constant is guaranteed to be unity [1], satisfying the Ward identity. Thus irrespective of the problem of the ladder approximation, the results obtained would be gauge invariant.

Same analyses have been carried out in the finite temperature/density case with the ladder kernel [4-6], and with the hard-thermal-loop (HTL) resummed improved ladder kernel [7]. Results of Ref. [7] show that at finite temperature/density it is important to correctly analyse the physical mass function Σ_R, the mass function of the "unstable" quasiparticle in thermal field theories, and also to correctly take the dominant thermal effect into the interaction kernel.

All the preceding analyses [4-7], however, suffer from the serious problem coming from the ladder approximation of the interaction kernel. Although in the vacuum case, despite the use of ladder kernel, in the analysis in the Landau gauge the Ward identity is guaranteed to be satisfied, at finite temperature /density there is no such guarantee. In fact, even in the Landau gauge the fermion wave function renormalization constant largely deviates from unity [7,8], being not even real. At finite temperature/density the results obtained from the ladder DSE explicitly violate the Ward identity, thus depend on the gauge, their physical meaning being obscure.

In this paper, we worked out, in the analysis of the HTL resummed improved ladder DS equation for the fermion mass function in thermal QED, the procedure to get the gauge invariant solution in the sense it satisfies the Ward identity, and investigated the properties of the "gauge invariant" solution.

Results of the present analysis are summarized as follows:

(1) We can determine the solution that satisfies the Ward identity, namely the fermion wave function renormalization constant being almost equal to unity. To get such a solution it is essential that the gauge parameter ξ depends on the momentum of the gauge boson.

(2) The chiral phase transition in the massless thermal QED is confirmed to occur through the second order transition.

(3) Two critical exponents ν and η are consistent with constant within the range of values of temperatures and couplingconstants under analysis: $\nu = 0.395, \eta = 0.518$.

(4) The effect of thermal fluctuation on the chiral symmetry breaking and/or restoration is smaller than that expected in the previous analysis in the Landau gauge [7].

2. The Dyson-Schwinger equation for the fermion self-energy function Σ_R

The fermion self-energy function Σ_R appearing in the fermion propagator

$$S_R(P) = [\not{P} + i\epsilon\gamma^0 - \Sigma_R(P)]^{-1} \tag{1}$$

can be decomposed at finite temperature and/or density as

$$\Sigma_R(P) = (1 - A(P))p_i\gamma^i - B(P)\gamma^0 + C(P) \tag{2}$$

with $A(P)$, $B(P)$ and $C(P)$ being the three scalar invariants to be determined. In the present analysis, we use the HTL resummed form ${}^*G^{\mu\nu}$ for the gauge boson propagator $G^{\mu\nu}$

$${}^*G^{\mu\nu}(K) = \frac{1}{{}^*\Pi_T - K^2 - i\epsilon k_0}A^{\mu\nu} + \frac{1}{{}^*\Pi_L - K^2 - i\epsilon k_0}B^{\mu\nu} - \frac{\xi}{K^2 + i\epsilon k_0}D^{\mu\nu} \tag{3}$$

where ${}^*\Pi_{L/T}$ is the HTL resummed longitudinal/transverse photon self-energy function [9], and $A^{\mu\nu}$, $B^{\mu\nu}$ and $D^{\mu\nu}$ are the projection tensors [10],

$$A^{\mu\nu} = g^{\mu\nu} - B^{\mu\nu} - D^{\mu\nu}, B^{\mu\nu} = -\tilde{K}^\mu\tilde{K}^\nu/K^2, D^{\mu\nu} = K^\mu K^\nu/K^2, \tag{4}$$

where $\tilde{K} = (k, k_0\hat{\mathbf{k}})$, $k = \sqrt{\mathbf{k}^2}$ and $\hat{\mathbf{k}} = \mathbf{k}/k$ denotes the unit three vector along $\mathbf{k}$. The parameter ξ appearing in the term proportional to the projection tensor $D_{\mu\nu}$ represents the gauge-fixing parameter ($\xi = 0$ in the Landau gauge). This gauge term plays an important role in the present analysis.

The vertex function is approximated by the tree (point) vertex. With the instantaneous exchange approximation for the longitudinal photon mode, we get the DSEs for the three invariant functions $A(P)$, $B(P)$ and $C(P)$

$$\begin{aligned}
-p^2[1 - A(P)] = -e^2 \int \frac{d^4K}{(2\pi)^4} \Big[&\{1 + 2n_B(p_0 - k_0)\}\mathrm{Im}[\,{}^*G_R^{\rho\sigma}(P - K)] \times \\
\Big[\{K_\sigma P_\rho + K_\rho P_\sigma &- p_0(K_\sigma g_{\rho 0} + K_\rho g_{\sigma 0}) - k_0(P_\sigma g_{\rho 0} + P_\rho g_{\sigma 0}) + pkz g_{\sigma\rho} \\
+2p_0k_0g_{\sigma 0}g_{\rho 0}\}&\frac{A(K)}{[k_0 + B(K) + i\epsilon]^2 - A(K)^2k^2 - C(K)^2} + \{P_\sigma g_{\rho 0} \\
+P_\rho g_{\sigma 0} - 2p_0g_{\sigma 0}g_{\rho 0}\}&\frac{k_0 + B(K)}{[k_0 + B(K) + i\epsilon]^2 - A(K)^2k^2 - C(K)^2}\Big] \\
+\{1 - 2n_F(k_0)\}\,{}^*G_R^{\rho\sigma}&(P - K)\mathrm{Im}\Big[\{K_\sigma P_\rho + K_\rho P_\sigma - p_0(K_\sigma g_{\rho 0} + K_\rho g_{\sigma 0}) \\
-k_0(P_\sigma g_{\rho 0} + P_\rho g_{\sigma 0}) &+ pkz g_{\sigma\rho} + 2p_0k_0g_{\sigma 0}g_{\rho 0}\} \times \\
\frac{A(K)}{[k_0 + B(K) + i\epsilon]^2 - A(K)^2k^2 - C(K)^2} &+ \{P_\sigma g_{\rho 0} + P_\rho g_{\sigma 0} \\
-2p_0g_{\sigma 0}g_{\rho 0}\}&\frac{k_0 + B(K)}{[k_0 + B(K) + i\epsilon]^2 - A(K)^2k^2 - C(K)^2}\Big]\Big] \,,
\end{aligned} \tag{5}$$

$$-B(P) = -e^2 \int \frac{d^4K}{(2\pi)^4} \Bigg[\{1 + 2_B(p_0 - k_0)\} \mathrm{Im}[\ {}^*G_R^{\rho\sigma}(P-K)] \times$$
$$\Big[\{K_\sigma g_{\rho 0} + K_\rho g_{\sigma 0} - 2k_0 g_{\sigma 0} g_{\rho 0}\} \frac{A(K)}{[k_0 + B(K) + i\epsilon]^2 - A(K)^2 k^2 - C(K)^2}$$
$$+ \{2g_{\rho 0} 2g_{\sigma 0} - g_{\sigma\rho}\} \frac{k_0 + B(K)}{[k_0 + B(K) + i\epsilon]^2 - A(K)^2 k^2 - C(K)^2} \Big]$$
$$+ \{1 - 2n_F(k_0)\}\ {}^*G_R^{\rho\sigma}(P-K) \mathrm{Im} \Big[\frac{A(K)}{[k_0 + B(K) + i\epsilon]^2 - A(K)^2 k^2 - C(K)^2}$$
$$\times \{K_\sigma g_{\rho 0} + K_\rho g_{\sigma 0} - 2k_0 g_{\sigma 0} g_{\rho 0}\}$$
$$+ \frac{k_0 + B(K)}{[k_0 + B(K) + i\epsilon]^2 - A(K)^2 k^2 - C(K)^2} \{2g_{\rho 0} 2g_{\sigma 0} - g_{\sigma\rho}\} \Big] \Bigg] , \tag{6}$$

$$C(P) = -e^2 \int \frac{d^4K}{(2\pi)^4} g_{\sigma\rho} \{1 + 2_B(p_0 - k_0)\} \mathrm{Im}[\ {}^*G_R^{\rho\sigma}(P-K)] \times$$
$$\Big[\frac{C(K)}{[k_0 + B(K) + i\epsilon]^2 - A(K)^2 k^2 - C(K)^2} + \{1 - 2n_F(k_0)\} \times$$
$${}^*G_R^{\rho\sigma}(P-K) \mathrm{Im} \Big[\frac{C(K)}{[k_0 + B(K) + i\epsilon]^2 - A(K)^2 k^2 - C(K)^2} \Big] \Big] . \tag{7}$$

The function $A(P)$ is nothing but the inverse of the fermion wave function renormalization constant Z_2, thus must be unity in order to satisfy the Ward identity in the ladder DSE analysis, where the vertex function receives no renormalization effect, $Z_1 = 1$.

We must solve the above DSEs and get the solution satisfying the Ward identity $Z_2 = Z_1 (= 1)$, where $Z_2 = A(P)^{-1}$. The procedure to get the "gauge invariant" solution is as follows;

(1) Assume the nonlinear gauge such that the gauge parameter ξ to be a function of the photon momentum $K = (k_0, \mathbf{k})$, and parametrize ξ as

$$\xi(k_0, k) = \sum \xi_{mn} H_m(k_0) L_n(k), \tag{8}$$

where ξ_{mn} are unknown parameters to be determined, H_m the Hermite functions and L_n the Laguerre functions.

(2) In solving DSEs iteratively, impose the condition $A(P) = 1$ by constraint for the input-functions at each step of the iteration.

(3) Determine ξ_{mn} so as to minimize $|A(P) - 1|^2$ for the out-put functions and find the solutions for $B(P)$ and $C(P)$.

3. Results of the analysis "gauge invariant" solution

Here we present the results obtained by allowing the gauge parameter ξ to be a complex value. Number of parameters ξ_{mn} to minimize $|A(P) - 1|^2$

is $2 \times 5 \times 2 = 20$ (i.e., $m = 1 \sim 4$ and $n = 1, 2$). All the quantities with the mass dimension are evaluated in the unit of Λ, the cut-off parameter introduced as usual to regularize the DSEs.

Now we present the solution consistent with the Ward identity, i.e., the "auge invariant" solution. Firstly in Fig.1 we show Re$A(P)$. For comparison, results in the constant ξ analyses are also shown in the same figure.

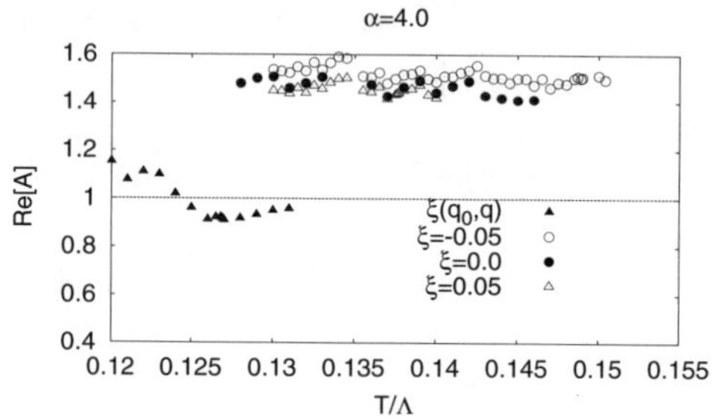

Fig. 1. Comparison of the renormalization constant Re$[A]$ at the coupling constant $\alpha = 4.0$ evaluated at $p_0 = 0$, $p = 0.1$, see, text.

Next let us study the property of the phase transition. Fig.2(a) shows the real part of the fermion mass Re$M(P)$, $M(P) \equiv C(P)/A(P)$, obtained from the "gauge invariant" solution, as a function of the temperature T. The mass is evaluated at $p_0 = 0$, $p = 0.1$, to be consistent with the standard prescription to define the mass in the static limit, $p_0 = 0$, $p \to 0$.

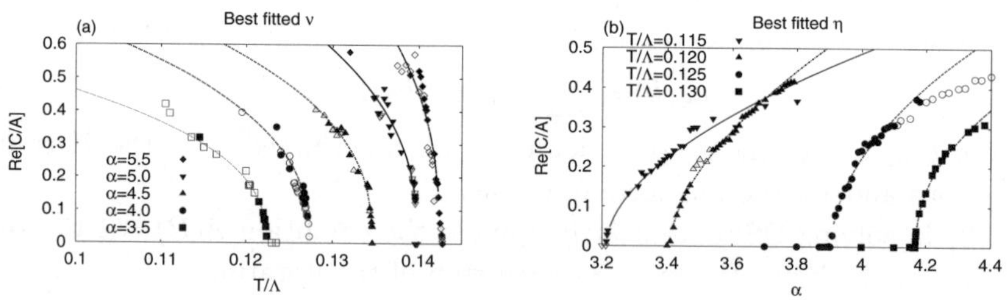

Fig. 2. (a)Temperature-dependence of the fermion mass Re$[M(P)]$ for various values of the coupling α. (b) Coupling constant-dependence of Re$[M(P)]$ for various values of the temperature T. Both are evaluated at $p_0 = 0$ and $p = 0.1$. As for the various curves, see text.

The curves in the figure show the best-fit curves in determining the critical temperature T_c and the critical exponent ν, by fitting, at each coupling constant α and near the critical temperature T_c , the temperature-dependent data of Re$M(P)$ to the functional form

$$\mathrm{Re}M(P) = C_T(T_c - T)^{\nu}. \tag{9}$$

Also shown in Fig.2(b) is the Re$M(P)$ obtained from the "gauge invariant" solution, as a function of the coupling constant α. The mass is evaluated at $p_0 = 0$, $p = 0.1$ as above. The curves in the figure show the best-fit curves in determining the critical coupling α_c and the critical exponent η, by fitting, at each temperature T and near the critical coupling α_c, the coupling-dependent data of Re$M(P)$ to the functional form

$$\mathrm{Re}M(P) = C_{\alpha}(\alpha - \alpha_c)^{\eta}. \tag{10}$$

The determined critical exponents are given in Table 1.

Table 1. Critical exponent ν for various values of the coupling constant α and critical exponent η for various values of the temperature T.

α	ν	T	η
3.5	0.42800	0.115	0.54718
4.0	0.38126	0.120	0.57872
4.5	0.36420	0.125	0.51430
5.0	0.40579	0.130	0.46153

The averaged value of ν over the various coupling is $<\mu>= 0.395$, which fits to all the data Re$M(P)$ in Fig.2(a) irrespective of the value of the coupling constant. The averaged value of η over various temperatures is $<\eta>= 0.518$, which fits to all the data Re$M(P)$ in Fig.2(b) irrespective of the value of the temperature.

The phase boundary curve in the (T, α)-plane thus determined shows that the region of the symmetry broken phase shrinks to the low-temperature-strong- coupling side compared with that of the Landau gauge. This fact means that the effect of thermal fluctuation on the chiral symmetry breaking/restoration is smaller than that expected in the previous analysis in the Landau gauge [7].

4. Discussion and comments

Results presented in the present paper are preliminary, because of the rough analysis of the data processing. We are now refining the data analysis and soon get the results of the thorough reanalysis. Though the main conclusion will not be altered, several remarks should be added.

(1) Present analysis was performed by allowing the gauge parameter ξ to be a complex value. Such a choice of gauge may correspond to studying the non-hermite dynamics, thus may cause some troubles. What happens if we restrict the gauge parameter to the real value? We are studying this case, finding a remarkable result: In both cases results completely agree, thus getting a solution totally independent of the choice of gauges.

(2) In the present analysis, the consistency of the solution with the Ward identity is respected only by imposing the condition $A(P) \approx 1$. Needless to say, in solving the (improved) ladder Dyson-Schwinger equation, there are no solutions totally consistent with the Ward identity. Despite that fact, following point should be closely examined: At least around or in the static limit, $p_0 = 0$, $p \to 0$, where we calculated (defined) the mass, each invariant function $A(P)$, $B(P)$ or $C(P)$ should not have big momentum dependence. This condition may be important in connection with the consistency of the obtained solution with the gauge invariance. Result of the present analysis shows that at least $B(P)$ and $C(P)$ satisfy this condition.

References

1. T. Maskawa and H. Nakajima, *Prog. Theor. Phys.* **52**, 1326 (1974); **54**, 860 (1975); R. Fukuda and T. Kugo, *Nucl. Phys.* **B117**, 250 (1976).
2. K. Yamawaki, M. Bando and K. Matumoto, *Phys. Rev. Lett.* **56**, 1335 (1986); K.-I. Kondo, H. Mino and K. Yamawaki, *Phys. Rev.* **D39**, 2430 (1989).
3. W. A. Bardeen, C. N. Leung and S. T. Love, *Phys. Rev. Lett.* **56**, 1230 (1986); C. N. Leung, S. T. Love and W. A. Bardeen, *Nucl. Phys. B* **273**, 649 (1986); W. A. Bardeen, C. N. Leung and S. T. Love, *Nucl. Phys. B* **323**, 493 (1989).
4. K.-I. Kondo and K. Yoshida, *Int. J. Mod. Phys. A* **10**, 199 (1995).
5. M. Harada and A. Shibata, *Phys. Rev. D* **59**, 014010 (1998).
6. K. Fukazawa, T. Inagaki, S. Mukaigawa and T. Muta, *Prog. Theor. Phys.* **105**, 979 (2001).
7. Y. Fueki, H. Nakkagawa, H. Yokota and K. Yoshida, *Prog. Theor. Phys.* **110**, 777 (2003); **107**, 759 (2002); H. Nakkagawa, H. Yokota, K. Yoshida and Y. Fueki, *Pramana – J.of Phys.* **60**, 1029 (2003).
8. A. Rebhan, *Phys. Rev. D* **46**, 4779 (1992).
9. V. V. Klimov, *Sov. J. Nucl.Phys.* **33**, 934 (1981); *Sov. Phys. JETP* **55**, 199 (1982); H. A. Weldon, *Phys. Rev. D* **26**, 1394 (1982); *Phys. Rev. D* **26**, 2789 (1982).
10. H. A. Weldon, *Ann. Phys. (N.Y.)* **271**, 141 (1999).

SOME RECENT RESULTS ON MODELS OF DYNAMICAL ELECTROWEAK SYMMETRY BREAKING

R. Shrock*

C. N. Yang Institute for Theoretical Physics
State University of New York
Stony Brook, New York 11794, USA
** email: robert.shrock@sunysb.edu*

We review some recent results on models of dynamical electroweak symmetry breaking involving extended technicolor.

1. Introduction

The origin of electroweak symmetry breaking (EWSB) is an outstanding unsolved question in particle physics. In the Standard Model with gauge group $G_{SM} = \mathrm{SU}(3)_c \times \mathrm{SU}(2)_L \times \mathrm{U}(1)_Y$, this symmetry breaking is produced by postulating a Lorentz scalar Higgs field $\phi = \binom{\phi^+}{\phi^0}$ with weak isospin $I = 1/2$ and weak hypercharge $Y = 1$ and assuming that its potential, $V = \mu^2\phi^\dagger\phi + \lambda(\phi^\dagger\phi)^2$, has $\mu^2 < 0$, thereby leading to a nonzero vacuum expection value for ϕ. However, this mechanism is unsatisfying for several reasons: (i) the EWSB is put in by hand and no explanation is provided as to why μ^2 is negative when, *a priori*, it could be positive; (ii) μ^2 and hence $m_H^2 = -2\mu^2 = 2\lambda v^2$ are unstable to large radiative corrections from much higher energy scales (the gauge hierarchy problem), so that extreme fine-tuning is needed to keep the Higgs mass of order the electroweak scale; (iii) the SM accomodates, but does not explain, fermion masses and mixing, via Yukawa couplings of the generic form (suppressing the matrix structure) $m_f \simeq y_f v/\sqrt{2}$; the y_f values and generational hierarchy are put in by hand with some y_f's ranging down to 10^{-5} with no explanation.

These facts have motivated an alternative approach based on dynamical electroweak symmetry breaking driven by a strongly coupled vectorial gauge interaction, associated with an exact gauge symmetry, called technicolor (TC).[1] The EWSB is produced by the formation of bilinear condensates of technifermions. To communicate this symmetry breaking to the SM fermions (which are technisinglet), one embeds technicolor in a larger, extended technicolor (ETC) theory.[2,3] In this talk we will review some of our recent results in this area[4–14] obtained in collaboration with T. Appelquist, M. Piai, N. Christensen, and M. Kurachi. (The work with Kurachi is also discussed in his talk.[15])

As further motivation, one may recall that in both of two major previous cases where fundamental scalar fields were used to model spontaneous symmetry breaking, the actual underlying physics did not involve fundamental scalar fields but instead a bilinear fermion condensate. The first of these was superconductivity, where the phenomenological Landau-Ginzburg model made use of a complex scalar field $\phi = \rho e^{i\theta}$ with a free energy functional of the form $V = c_2|\phi|^2 + c_4|\phi|^4$, where $c_2 \propto (T - T_c)$ as $T \to T_c$ (with T_c being the critical temperature). Hence, for $T < T_c$, $c_2 < 0$ and $\langle \phi \rangle \neq 0$. However, the true origin of superconductivity is the dynamical formation of a condensate of Cooper pairs, $\langle ee \rangle$. A second example is spontaneous chiral symmetry breaking in hadronic physics. In the Gell-Mann Lévy σ model for this phenomenon, the chiral symmetry breaking is a result of the coefficient of the quadratic term in the potential being arbitrarily chosen to be negative, producing a nonzero vev of the σ field, quite analogous to the Higgs mechanism. However, the real origin of spontaneous chiral symmetry breaking in quantum chromodynamics (QCD) is the dynamical formation of a bilinear quark condensate $\langle \bar{q}q \rangle$. Perhaps these previous examples might serve as a guide for thinking about the physics underlying EWSB.

Actually, one already knows of a source of dynamical electroweak symmetry breaking, namely QCD. Consider, for simplicity, QCD with $N_f = 2$ quarks, u, d, taken to be massless. The quark condensate $\langle \bar{q}q \rangle = \langle \bar{q}_L q_R \rangle + \langle \bar{q}_R q_L \rangle$, transforms as $I_w = 1/2$, $|Y| = 1$. By itself this theory would produce nonzero W and Z masses $m_W^2 \simeq (g^2/4) f_\pi^2$, $m_Z^2 \simeq (1/4)(g^2 + g'^2) f_\pi^2$. With $f_\pi = 92$ MeV, this yields $m_W \simeq 30$ MeV, $m_Z \simeq 34$ MeV. These W and Z masses satisfy the tree-level relation $\rho = 1$, where $\rho = m_W^2/(m_Z^2 \cos^2 \theta_W)$. While the scale here is too small by $\sim 10^3$ to explain the observed W and Z masses, it suggests how to construct a model with dynamical EWSB.

We shall consider a technicolor theory with gauge group $SU(N_{TC})$ and a set of technifermions, generically denoted F, with zero Lagrangian masses, transforming according to the fundamental representation of the group. The scale of confinement and chiral symmetry breaking in this theory is denoted Λ_{TC} and is of order the electroweak scale. One assigns the $SU(2)_L$ representations of the technifermions so that their left and right components form $SU(2)_L$ doublets and singlets, respectively. A minimal choice is the "one-doublet" model, with $\binom{F_u^\tau}{F_d^\tau}_L$ and F_{uR}^τ, F_{dR}^τ, where τ is a TC gauge index and $Y = 0$ ($Y = \pm 1$) for the $\mathrm{SU}(2)_L$ doublet (singlets). Since the technicolor theory is asymptotically free, it follows that, as the energy scale decreases, α_{TC} increases, eventually producing condensates $\langle \bar{F}_u F_u \rangle$ and $\langle \bar{F}_d F_d \rangle$ transforming as $I_w = 1/2$, $|Y| = 1$, breaking EW sym. at Λ_{TC}. It follows that, to leading order, $m_W^2 = g^2 f_{TC}^2 N_D/4$ and $m_Z^2 = (g^2 + g'^2) f_{TC}^2 N_D/4$, where $f_{TC} \lesssim \Lambda_{TC}$ is the TC analogue of $f_\pi \lesssim \Lambda_{QCD}$ and N_D is the number of $\mathrm{SU}(2)_L$ technidoublets ($N_D = 1$ in this case). These masses satisfy the tree-level relation $\rho = 1$.[16] For this minimal example, $f_{TC} \simeq 250$ GeV. Another class of TC models uses one SM family of technifermions,[17]

$$\begin{pmatrix} U^{a\tau} \\ D^{a\tau} \end{pmatrix}_L , \qquad U_R^{a\tau}, \;\; D_R^{a\tau} \tag{1}$$

$$\begin{pmatrix} N^{\tau} \\ E^{\tau} \end{pmatrix}_L , \qquad N_R^{\tau}, \;\; E_R^{\tau} \tag{2}$$

(where a and τ are color and technicolor indices, respectively) with the usual Y assignments. Again, there is technifermion condensate formation, with approximately equal condensates $\langle \bar{F} F \rangle$ for $F = U^a$, D^a, N, E, generating dynamical technifermion masses $\Sigma_{TC} \sim \Lambda_{TC}$. In this class of models $N_D = N_c + 1 = 4$, so $f_{TC} \simeq 125$ GeV.

Technicolor has the potential to solve/explain various problematic and/or mysterious features of the Standard Model: (i) given the asymptotic freedom of the TC theory, the condensate formation and hence EWSB are automatic and do not require any *ad hoc* parameter choice like $\mu^2 < 0$ in the SM; (ii) because TC has no fundamental scalar field, there is no hierarchy problem; (iii) because $\langle \bar{F} F \rangle = \langle \bar{F}_L F_R \rangle + \langle \bar{F}_R F_L \rangle$, technicolor explains why the chiral part of G_{SM} is broken and the residual exact gauge symmetry, $\mathrm{SU}(3)_c \times \mathrm{U}(1)_{em}$, is vectorial. The fact that this latter symmetry is vectorial is, of course, crucial in allowing nonzero mass terms for fermions that are nonsinglets under color or charge.

In order to give masses to quarks and leptons, one must communicate the EWSB in the technicolor sector to these SM fermions, and hence, as mentioned above, one must embed technicolor in the larger ETC theory, with ETC gauge bosons V_τ^i transforming (technisinglet) SM fermions into technifermions and vice versa. To satisfy constraints on flavor-changing neutral current (FCNC) processes, the ETC gauge bosons must have large masses. These masses can arise from self-breaking (tumbling) of the ETC chiral gauge symmetry. The ETC theory is arranged to be asymptotically free, so as the energy decreases from a high scale, the ETC coupling α_{ETC} grows and eventually becomes large enough to form condensates which sequentially break the ETC symmetry group down at scales Λ_j, $j = 1,2,3$ to a residual exact TC subgroup, where

$$\Lambda_1 \simeq 10^3 \text{ TeV}, \qquad \Lambda_2 \simeq 50 - 100 \text{ TeV}, \qquad \Lambda_3 \simeq \text{few TeV}, \tag{3}$$

We will mainly focus on SU(N_{ETC}) ETC models with one-family TC. These gauge the Standard Model fermion generation index and combine it with the TC index τ, so

$$N_{ETC} = N_{gen.} + N_{TC} \;=\; 3 + N_{TC} \;. \tag{4}$$

To be viable, modern TC models are designed to have a coupling g_{TC} that gets large, but runs slowly ("walks") over an extended interval of energy (WTC).[18–24] For sufficiently many technifermions, the TC beta function has a second zero (approximate infrared fixed point of the renormalization group) at a certain $\alpha_{TC} = \alpha_{IR} \neq 0$. As the number of technifermions, N_f, increases, α_{IR} decreases. In WTC, one arranges so that α_{IR} is slightly greater than the critical value, α_{cr}, for $\langle \bar{F}F \rangle$ formation. As $N_f \nearrow N_{f,cr}$, $\alpha_{IR} \searrow \alpha_{cr}$. Combining the calculation of α_{IR} from the two-loop beta function and an estimate of α_{cr} from the Schwinger-Dyson equation, one finds $N_{f,cr} = 2N_{TC}(50N_{TC}^2 - 33)/[5(5N_{TC}^2 - 3)]$.[22] Hence, for $N_{TC} = 2$, one has $N_{f,cr} \simeq 8$. Thus, if $N_{TC} = 2$, a one-family TC theory, with its $N_f = N_w(N_c + 1) = 8$ technifermions plausibly exhibits walking behavior.

In a walking TC theory, as energy scale decreases, α_{TC} grows, but its rate of increase, $|\beta|$, decreases toward zero as $\alpha_{TC} \nearrow \alpha_{IR}$, where $\beta = 0$. Eventually, α_{TC} exceeds α_{cr}, $\langle \bar{F}F \rangle$ forms, and the technifermions gain dynamical masses $\Sigma_{TC} \simeq \Lambda_{TC}$. WTC has several advantages: (i) SM fermion masses are enhanced by the factor $\eta = \exp\left[\int_{\Lambda_{TC}}^{\Lambda_w} \mu^{-1} d\mu \; \gamma(\alpha_{TC}(\mu))\right]$; (ii) hence, one can increase ETC scales Λ_i for a fixed m_{f_i}, reducing FCNC effects; (iii) η also enhances masses of pseudo-Nambu Goldstone bosons (PNGB's); and (iv) the walking can reduce the value of the S parameter,[25] given by

$$\frac{\alpha_{em} S}{\sin^2(2\theta_W)} = \frac{\Pi_{ZZ}(m_Z^2) - \Pi_{ZZ}(0)}{m_Z^2} \;. \tag{5}$$

The value $N_{TC} = 2$ is also motivated by the fact that it makes possible a mechanism to explain light neutrinos in (E)TC.[4,5] Substituting this value $N_{TC} = 2$ into eq. (4) yields $N_{ETC} = 5$.

2. ETC Models

TC/ETC models are very ambitious, since a successful model would explain not only EWSB but also the spectrum of Standard Model fermion masses. Thus it is perhaps not surprising that no fully realistic model of this type has been constructed yet. These models are subject to several strong constraints from neutral flavor-changing current processes and precision electroweak data. We have studied the properties of several types of ETC models in our work. One of these has a gauge group $G = \mathrm{SU}(5)_{ETC} \times \mathrm{SU}(2)_{HC} \times G_{SM}$. In addition to ETC, this has another gauge interaction, hypercolor (HC), which helps to produce the desired ETC symmetry-breaking pattern. The SM fermions and corresponding technifermions transform according to the representations

$$Q_L : \ (5,1,3,2)_{1/3,L} \ , \quad u_R : \ (5,1,3,1)_{4/3,R} \ , \quad d_R : \ (5,1,3,1)_{-2/3,R} \ , \tag{6}$$

$$L_L : \ (5,1,1,2)_{-1,L}, \qquad e_R : \ (5,1,1,1)_{-2,R} \ , \tag{7}$$

where the subscripts denote Y. For example, writing out the components of e_R, one has $(e^1, e^2, e^3, e^4, e^5)_R \equiv (e, \mu, \tau, E^4, E^5)_R$, where the last two entries are the charged technileptons. In addition, the model includes the SM-singlet ETC-nonsinglet fermions

$$\psi_{ij,R} : \ (\overline{10},1,1,1)_{0,R} \ , \qquad \zeta_R^{ij,\alpha} : \ (10,2,1,1)_{0,R} \ , \qquad \omega_{p,R}^{\alpha} : \ 2(1,2,1,1)_{0,R} \tag{8}$$

where here $1 \le i,j \le 5$ are $\mathrm{SU}(5)_{ETC}$ indices, $\alpha = 1,2$ is an $\mathrm{SU}(2)_{HC}$ index, and $p = 1,2$ is a copy number. In this model the SM fermions and corresponding technifermions have vectorial couplings to ETC gauge bosons, but the SM-singlet sector makes the full ETC theory a chiral gauge theory. Hence, there are no fermion mass terms in the lagrangian. This theory has no gauge or global anomalies.

Since the ETC theory is asymptotically free, its gauge coupling grows as the energy scale decreases. The ETC breaking occurs because of the formation of ETC-noninvariant bilinear condensates of ETC-nonsinglet, SM-singlet fermions. To analyze the stages of symmetry breaking, we identify plausible preferred condensation channels using a generalized most-attractive-channel (MAC) approach. We envision that as the energy decreases from high values down to $E \sim \Lambda_1 \sim 10^3$ TeV, the coupling $\alpha_{_{ETC}}$ becomes sufficiently large to produce condensation in the attractive channel $(\overline{10},1,1,1)_{0,R} \times (\overline{10},1,1,1)_{0,R} \rightarrow (5,1,1,1)_0$, breaking $\mathrm{SU}(5)_{ETC} \rightarrow \mathrm{SU}(4)_{ETC}$. With respect to the unbroken $\mathrm{SU}(4)_{ETC}$, we have

$(\overline{10}, 1, 1, 1)_{0,R} = (\bar{4}, 1, 1, 1)_{0,R} + (\bar{6}, 1, 1, 1)_{0,R}$; we denote the $(\bar{4}, 1, 1, 1)_{0,R}$ as $\alpha_{1i,R} \equiv \psi_{1i,R}$ for $2 \le i \le 5$ and the $(\bar{6}, 1, 1, 1)_{0,R}$ as $\xi_{ij,R} \equiv \psi_{ij,R}$ for $2 \le i, j \le 5$. The associated condensate is then

$$\langle \epsilon^{1ijk\ell} \xi^T_{ij,R} C \xi_{k\ell,R} \rangle = 8 \langle \xi^T_{23,R} C \xi_{45,R} - \xi^T_{24,R} C \xi_{35,R} + \xi^T_{25,R} C \xi_{34,R} \rangle \ . \quad (9)$$

The six fields $\xi_{ij,R}$, $2 \le i, j \le 5$, involved in this condensate gain dynamical masses $\simeq \Lambda_1$.

At energy scales below Λ_1, depending on relative strengths of gauge couplings, different symmetry-breaking sequences can occur. Again, these arise via the dynamical formation of condensates involving SM-singlet ETC-nonsinglet fermions. For example, one sequence, S1, leads to the breaking $\mathrm{SU}(4)_{ETC} \to \mathrm{SU}(3)_{ETC}$ at $\Lambda_2 \simeq 50 - 100$ TeV and finally $\mathrm{SU}(3)_{ETC} \to \mathrm{SU}(2)_{TC}$ at $\Lambda_3 \simeq$ few TeV, with $\mathrm{SU}(2)_{HC}$ unbroken. Another sequence, S2, involves a breaking $\mathrm{SU}(4)_{ETC} \to \mathrm{SU}(2)_{ETC}$ at $\Lambda_{23} \simeq 50$ TeV and $\mathrm{SU}(2)_{HC} \to \mathrm{U}(1)_{HC}$ at a scale $\Lambda_{BHC} \lesssim \Lambda_1$. In all cases, $\mathrm{SU}(2)_{TC}$ is an exact gauge symmetry. At the lowest scale, Λ_{TC}, the technifermion condensates form, breaking electroweak symmetry and giving masses to the W and Z.

Certain fermion condensates contribute to nondiagonal ETC gauge boson propagator corrections and hence mixing. For example, in sequence S_1, the mixing $V_1^\tau \to V_3^\tau$, $\tau = 4, 5$ is induced by the graph in Fig. 1, with the

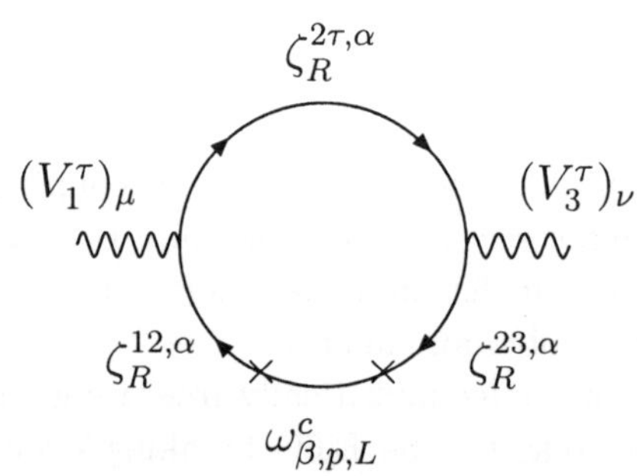

Fig. 1. Graph for $V_1^\tau \to V_3^\tau$ with $\tau = 4, 5$, for sequence S1.

result

$$ {}^\tau_3\Pi^\tau_1(0) \simeq \frac{\alpha_{ETC}}{4\pi} \int (k^2 dk^2) \frac{k^4\, \Sigma_3(k)^2}{[k^2 + \Sigma_3(k)^2]^4} \ , \quad (10)$$

where $\Sigma_3 \simeq \Lambda_3$. This yields ${}^\tau_3\Pi^\tau_1(0) \simeq \mathrm{const.} \times \Lambda_3^2$. In sequence S_2, the mixing $V_2^\tau \to V_3^\tau$, $\tau = 4, 5$ is induced by the graph in Fig. 2, giving ${}^\tau_3\Pi^\tau_2(0) \simeq$

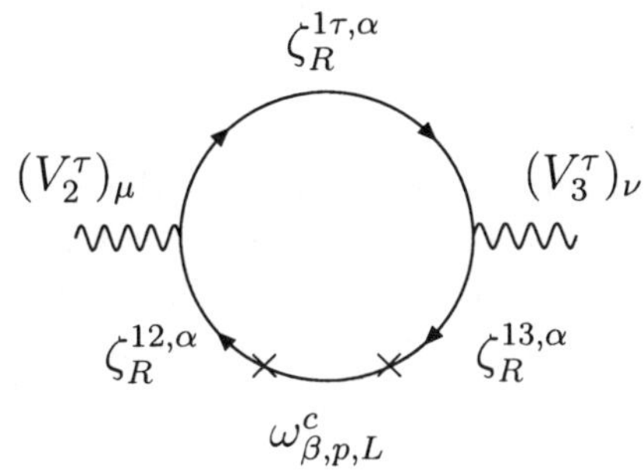

Fig. 2. Graph for $V_2^\tau \to V_3^\tau$ with $\tau = 4, 5$, for sequence S2.

const. $\times\ \Lambda_{23}^2$. We find that the feature that nondiagonal ETC gauge boson propagator corrections ${}^\tau_i\Pi^\tau_j(0)$ are proportional to the square of the lowest ETC scale (or smaller) is generic in this type of ETC model, reflecting a type of approximate generational symmetry. Other ETC gauge boson mixings are similarly suppressed.

3. Fermion Masses and Mixing

Figure 3 shows a one-loop graph contributing to the mass matrix element $M_{ij}^{(f)}$ for a SM fermion f (up- or down-type quark or charged lepton) appearing in the operator $\bar{f}_{i,L} M_{ij}^{(f)} f_{j,R} + h.c.$. In this figure we distinguish the first three ETC indices, which refer to SM fermion generations, and the indices 4,5 which are TC, by denoting the latter as $\tau = 4, 5$. An estimate

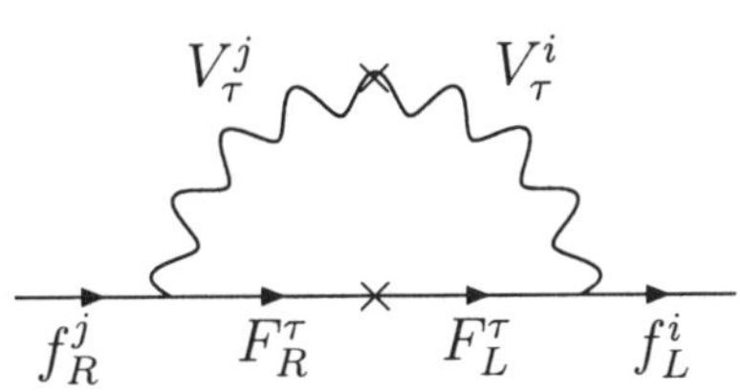

Fig. 3. Graph contributing to the fermion mass matrix element $M_{ij}^{(f)}$.

for diagonal entries is (with no sum on i)

$$M_{ii}^{(f)} \simeq \frac{\kappa\,(N_{TC}/2)\,\eta\,\Lambda_{TC}^3}{\Lambda_i^2}\,, \tag{11}$$

where $\kappa \simeq O(10)$ is a numerical factor from the integral and in WTC, $\eta \simeq \Lambda_3/\Lambda_{TC}$. This is only a rough estimate, since ETC coupling is strong, so higher-order diagrams are also important. The sequential breaking of the ETC symmetry at the highest scale Λ_1, the intermediate scale Λ_2, and the lowest scale Λ_3, thus produces the generational hierarchy in the fermion masses.

Insertions of the nondiagonal ETC propagator corrections (indicated by the cross on the gauge boson line in Fig. 3) give rise to off-diagonal elements of the $M^{(f)}$. These have the form

$$M_{ij}^{(f)} \simeq \frac{\kappa\,\eta\,\Lambda_{TC}^3\,{}_{\tau}^{j}\Pi_{\tau}^{i}}{\Lambda_i^2\Lambda_j^2}\,. \tag{12}$$

Although the resultant mixing is not fully realistic, this model shows how not just diagonal but also off-diagonal elements of SM fermion mass matrices $M^{(f)}$ and hence CKM mixing, could arise dynamically.[6] Further corrections to $M^{(f)}$ arise from SM gauge interactions, in particular, $SU(3)_c$ and $U(1)_Y$, and from direct diagonal ETC gauge boson exchanges, in particular, V_{d3} (having mass $\sim \Lambda_3$ and corresponding to the $SU(5)_{ETC}$ generator $T_{d3} = (2\sqrt{3})^{-1}\mathrm{diag}(0,0,-2,1,1)$).

4. Mechanism for Light Neutrinos

An old puzzle in TC/ETC theories was how to explain light neutrino masses. A solution to this was given in[4] and analyzed further in.[5,6] The $\alpha_{1j,R}$ with $j = 2,3$ are right-handed electroweak-singlet neutrinos and get induced Dirac neutrino mass terms connecting with $(n^1, n^2, n^3)_L = (\nu_e, \nu_\mu, \nu_\tau)_L$. These Dirac masses $\bar{n}_{i,L} M_D \alpha_{1j,R} + h.c.$ cannot be generated by the usual one-loop ETC graphs that produce diagonal quark and charged lepton masses and are thus suppressed. The $\alpha_{1j,R}$ also have induced Majorana mass terms $\alpha_{1i,R}^T C r_{ij} \alpha_{1j,R} + h.c.$ For example, with sequence S2, denoting the relevant Dirac mass terms as b_{ij}, one finds $b_{ij} \simeq \kappa\Lambda_{TC}^4/\Lambda_{23}^3 \simeq O(0.1)$ MeV for $2 \le i,j \le 3$; further, $r_{23} \simeq \kappa\Lambda_{BHC}^3\Lambda_{23}^3/\Lambda_1^5$. Numerically, $|r_{23}| \simeq O(10^2)$ GeV. The resultant electroweak-nonsinglet neutrinos are, to very good approximation, linear combinations of three mass eigenstates, with ν_3 mass

$$m(\nu_3) \simeq \frac{(b_{23}+b_{22})^2}{r_{23}} \simeq \frac{\kappa\Lambda_{TC}^8\Lambda_1^5}{\Lambda_{23}^9\Lambda_{BHC}^3}\,. \tag{13}$$

With the above-mentioned numerical values and $\Lambda_{BHC} \lesssim \Lambda_1$, we find $m(\nu_3) \simeq 0.05$ eV, consistent with experimental results on neutrino oscillations. Similarly,

$$\frac{m(\nu_2)}{m(\nu_3)} = \left(\frac{b_{23} - b_{22}}{b_{23} + b_{22}}\right)^2 , \tag{14}$$

which is again consistent with experimental data. Since $|r_{23}| >> |b_{ij}|$, this is a seesaw, but quite different from the SUSY GUT seesaw; the Majorana masses r_{ij} that underly the seesaw are not GUT-scale and are actually much smaller than the ETC scales Λ_i.

5. Constraints from Neutral Flavor-Changing Current Processes

An early concern was that ETC interactions would lead to excessively large flavor-changing neutral current processes. A particularly severe constraint arises from $K^0 - \bar{K}^0$ mixing. Early treatments wrote the effective Lagrangian for this as

$$\mathcal{L}_{eff} \simeq \frac{1}{\Lambda_{ETC}^2}[\bar{s}\gamma_\mu d]^2 , \tag{15}$$

where Λ_{ETC} was a generic ETC scale. To suppress FCNC effects adequately, it was thought that Λ_{ETC} had to be so high that there would be excessive suppression of SM fermion masses. However, we have shown, using our reasonably UV-complete theories, that in the present type of ETC theory this old view was too pessimistic; ETC contributions to FCNC processes are smaller than had been inferred with the above naive $\mathcal{L}_{eff}$, because of residual approximate generational symmetries.[6–8]

The coupling of the ETC gauge bosons to the SM fermion mass eigenstates is given by

$$\mathcal{L}_{int} = g_{ETC} \sum_{f,j,k} \bar{f}_j \gamma_\lambda (\mathcal{V}^\lambda)^j_k f^k \equiv g_{ETC} \sum_{f,j,k} \bar{f}_{m,j} \gamma_\lambda (A^\lambda)^j_k f^k_m , \tag{16}$$

where $\mathcal{V}$ is a matrix containing the ETC gauge fields, $A^\lambda \equiv U^{(f)} \mathcal{V}^\lambda U^{(f)\ -1}$, and the $U^{(f)}$ are the unitary transformations that diagonalize $M^{(f)}$ via $U^{(f)} M^{(f)} U^{(f)\ -1} = M^{(f)}_{diag.}$. We parametrize each $U^{(f)}$ with a PDG-type parametrization in terms of $\theta^{(f)}_{12}$, $\theta^{(f)}_{13}$, $\theta^{(f)}_{23}$, and $\delta^{(f)}$. The approximate generational symmetry in the ETC sector can naturally yield relatively small CKM mixing with $\theta^{(f)}_{jk}$ depending on ratios of smaller to larger ETC scales.

We have analyzed ETC contributions to a number of processes, including the neutral meson mixings $K^0 - \bar{K}^0$, $D^0 - \bar{D}^0$, $B^0_d - \bar{B}_d$, $B^0_s - \bar{B}_s$, and decays such as $\mu^+ \to e^+e^+e^-$, involving four-fermion operators.[6,8] As an example, consider $K^0 - \bar{K}^0$ mixing and the resultant $K_L - K_S$ mass difference $\Delta m_{K_L K_S}$. The SM contribution is consistent with the experimental value, $\Delta m_{K_L K_S}/m_K = 0.70 \times 10^{-14}$. A key to the suppression is the fact that, in terms of ETC eigenstates, an $s\bar{d}$ in a $\bar{K}^0$ produces a V_1^2 ETC gauge boson, but this cannot directly yield a $d\bar{s}$ in the final-state K^0; the latter is produced by a V_2^1. Hence, this requires either the ETC gauge boson mixing $V_1^2 \to V_2^1$ or the mixing of ETC quark eigenstates to produce mass eigenstates. The contribution from $V_1^2 \to V_2^1$ yields a coefficient

$$c \sim \frac{1}{\Lambda_1^2}\, {}_2^1\Pi_1^2\, \frac{1}{\Lambda_1^2} \sim \frac{\Lambda_3^2}{\Lambda_1^2}\frac{1}{\Lambda_1^2} \ll \frac{1}{\Lambda_1^2}\,. \tag{17}$$

With above values, $\Lambda_1 \sim 10^3$ TeV, $\Lambda_3 \sim 3$ TeV, the suppression factor is $(\Lambda_3/\Lambda_1)^2 \simeq 10^{-5}$. So, rather than the naive result $\Delta m_{K_L K_S}/m_K \sim \Lambda_{QCD}^2/\Lambda_1^2$, this yields the much smaller result $\Delta m_{K_L K_S}/m_K \sim \Lambda_3^2\Lambda_{QCD}^2/\Lambda_1^4 \sim 10^{-18}$. Hence, the dominant ETC contributions arise from the mixing of ETC eigenstates of quarks to form mass eigenstates. First, $s\bar{d}$ can couple to V_{d2} (having mass $\simeq \Lambda_2$ and corresponding to the $SU(5)_{ETC}$ generator $T_{d2} = (2\sqrt{6})^{-1}\mathrm{diag}(0,-3,1,1,1)$), with coefficient $\theta_{12}^{(d)}$. The resultant diagram with exchange of this gauge boson V_{d2} contributes $\sim (\theta_{12}^{(d)})^2/\Lambda_2^2$ to the amplitude. Requiring this to be small relative to SM contribution yields the constraint $|\theta_{12}^{(d)}| \lesssim 10^{-2}$. Second, $s\bar{d}$ can couple to V_{d3}, with coefficient $\theta_{13}^{(d)}\theta_{23}^{(d)}$. The resultant diagram contributes $\sim (\theta_{13}^{(d)}\theta_{23}^{(d)})^2/\Lambda_3^2$ to the amplitude. This yields the constraint $|\theta_{13}^{(d)}\theta_{23}^{(d)}| \lesssim 0.4 \times 10^{-3}$. A comprehensive analysis of constraints from processes involving dimension-six four-fermion operators was given in.[6,8] A related analysis is.[26]

We have also studied ETC contributions to (dimension-5) diagonal and transition lepton and quark dipole moments, and constraints from limits on $\mu^+ \to e^+\gamma$, $\tau^+ \to \ell^+\gamma$, the measured muon $(g-2)$ and $b \to s\gamma$ decay, and limits on lepton, neutron, atomic electric dipole moments.[7] Again, we found that this type of ETC model can be consistent with existing data.

6. Precision Electroweak Constraints

One may ask what the momentum dependence of a SM fermion mass is in this type of theory. This question was answered in Ref. 9; the running mass $m_{f_j}(p)$ exhibits the power-law decay

$$m_{f_j}(p) \propto \frac{\Lambda_j^2}{p^2} \tag{18}$$

for Euclidean momenta $p \gg \Lambda_j$, where f_j is a fermion of generation j. (Here we neglect logarithmic factors, which are subdominant relative to this power-law falloff.) Thus, $m_t(p)$ and $m_b(p)$ decay like Λ_3^2/p^2 for $p \gg \Lambda_3$, while $m_u(p)$ and $m_d(p)$ are hard up to the much higher scale Λ_1. We have investigated whether precision electroweak data are consistent with these power-law decays of SM fermion masses and have found that they are. The largest effects occur for m_t. Consider, e.g., the t-quark contribution to ρ. The conventional (hard, one-loop) result for this is $(\Delta\rho)_{t,hard} \simeq 3G_f m_t^2/(8\pi^2\sqrt{2})$. The power-law decay of m_t above Λ_3 changes this to $(\Delta\rho)_t = (\Delta\rho)_{t,hard}[1 - a_\rho(m_t^2/\Lambda_3^2)]$, where a_ρ is positive, $\sim O(1)$. The softness of m_t thus slightly reduces the violation of custodial symmetry. We find a similarly small change in the (t, b) contribution to S relative to the conventional hard-mass result.

As noted, the S parameter places a stringent constraint on TC/ETC theories. A naive perturbative calculation gives, for the TC contribution to S, the result $S_{pert.} \simeq N_{TC}\, N_D/(6\pi)$, where N_D denotes the number of technifermion EW doublets (given that the dynamical masses of the technifermions in each EW doublet are nearly degenerate, as should be true to satisfy the ρ-parameter constraint). However, this calculation assumes that the technifermions are weakly interacting, whereas actually, they are strongly interacting, at the relevant scale, $\sim m_Z$; hence, this perturbative formula cannot be expected to be reliable. This is important, since otherwise, even with the minimal value, $N_{TC} = 2$, it would yield an excessively large value of $S_{pert.} = 4/(3\pi) \simeq 0.4$ for a one-family TC model (and the marginally acceptable value of $S_{pert.} = 1/(3\pi) \simeq 0.1$ for a TC model with one EW technifermion doublet. Experimentally, $S \lesssim 0.1$. Several studies have found that walking can reduce S.[13,27–29] In particular, Refs. 28 and 13 have used solutions of Schwinger-Dyson and Bethe-Salpeter equations to evaluate S.[15]

One way to reduce S would be to reconsider TC models with only one technifermion electroweak doublet. But with this technifermion content alone, these models would not have the walking behavior that is needed not just to reduce S but also to generate adequate fermion masses. One can maintain the necessary walking behavior in a TC model with one EW doublet of technifermions by adding a requisite set of SM-singlet technifermions.[11] The $SU(2)_{TC}$ theory should have $N_f \simeq 8$ Dirac technifermions for walking. The electroweak-nonsinglet technifermions comprise two, so we need six more. One adds six SM-singlet, Dirac fermions (i.e., 12 chiral components) that transform as 2's under $SU(2)_{TC}$. These do not contribute to S

or T. Note that in this theory, $[G_{ETC}, G_{SM}] \neq 0$, so the ETC gauge bosons carry color and charge. Embedding TC in ETC is thus more complicated than for one-family TC models. Another approach is to use technifermions in higher-dimensional representations,[30,3111] In general, the S constraint remains a concern for TC/ETC theories.

7. Splitting of t and b Masses

Another challenge for TC/ETC models is to account for the splitting of the t and b masses without excessive contributions to $\rho - 1$ (violation of custodial SU(2) symmetry). One cannot do this by having $\Sigma_{TC,U}$ significantly larger than $\Sigma_{TC,D}$ since this would violate the custodial symmetry too much. One could consider trying to achieve this splitting using a class of ETC models in which left and right components of up-type quarks and techniquarks transform the same way under $SU(5)_{ETC}$, but the left and right components of down-type quarks and techniquarks transform according to relatively conjugate representations. However, we showed that such models have serious problems with excessively large FCNC's.[8] For example, consider a model in which Q_L and u_R are 5's of $SU(5)_{ETC}$ and d_R is a $\bar{5}$. So, e.g., the $\bar{K}^0 - K^0$ transition can proceed directly via $s_L \bar{d}_L \to V_1^2 \to d_R \bar{s}_R$ without ETC gauge boson mixing. This gives too large a value for $\Delta m_{K_L K_S}$.

A different idea would be to use two ETC gauge groups, say $SU(5)_{ETC} \times SU(5)'_{ETC}$ such that the left- and right-handed components of charge $Q = 2/3$ quarks transform under the same ETC group, while left- and right-handed components of charge $-1/3$ quarks and charged leptons transform under different ETC groups.[12,32] These models thereby suppress the masses m_b and m_τ relative to m_t, etc. because generating m_b requires mixing between the two ETC groups, which is suppressed, while m_t does not. However, they tend to produce too much suppression of the masses of first- and second-generation down-type quarks and charged leptons.[12] Thus, a satisfactory explanation of $t - b$ splitting appears to remain a challenge for ETC models.

8. Dynamical Breaking of Higher Gauge Symmetries

To what extent can one embed a TC/ETC in a theory having higher gauge symmetry, using dynamical symmetry breaking to break this higher symmetry? Such higher unification would be desirable in order to explain features not explained by the standard model, including charge quantization, prediction of relative sizes of gauge couplings, and unification of quarks and leptons. In Ref. 5 we constructed asymptotically free gauge theories exhibiting dynamical breaking of the left-right, strong-electroweak gauge group

$$G_{LR} = \mathrm{SU}(3)_c \times \mathrm{SU}(2)_L \times \mathrm{SU}(2)_R \times \mathrm{U}(1)_{B-L} \tag{19}$$

(where B and L denote baryon and lepton number) and its extension to

$$G_{422} = \mathrm{SU}(4)_{PS} \times \mathrm{SU}(2)_L \times \mathrm{SU}(2)_R \ , \tag{20}$$

where $\mathrm{SU}(4)_{PS} \supset \mathrm{SU}(3)_c \times \mathrm{U}(1)_{B-L}$ is the Pati-Salam group. These models technicolor for electroweak breaking, and extended technicolor for the breaking of G_{LR} and G_{422} and the generation of fermion masses, including a seesaw mechanism for neutrino masses. These models explain why G_{LR} and G_{422} break to $\mathrm{SU}(3)_c \times \mathrm{SU}(2)_L \times \mathrm{U}(1)_Y$, and why this takes place at a scale ($\sim 10^3$ TeV) which is large compared to the electroweak scale. In particular, the model with the gauge group G_{422} achieves charge quantization in the context of dynamical breaking of all symmetries.

We have also investigated various more ambitious unification schemes involving TC/ETC.[10] In particular, we have studied the possibility of unifying a one-family technicolor group with the SM gauge group in a simple group, but find that it appears difficult to obtain the requisite symmetry breaking dynamically.

9. Other Phenomenology

There are several other relevant topics for discussion. One has to do with the spectrum of these theories. As a confining gauge theory, technicolor produces a spectrum of TC-singlet techni-hadrons composed of technifermions and technigluons. This spectrum exhibits differences in a walking theory, as compared with a QCD-like theory. For example, while $m_{a_1}/m_\rho = 1.6$ in QCD, the analogous ratio $m_{(a_1)_{TC}}/m_{\rho_{TC}}$ is close to unity in the walking limit.[33] This is natural, since a theory with walking behavior has N_f near to, although slightly less than, the critical value $N_{f,cr}$ at which the theory would go over from the confined phase with spontaneous chiral symmetry breaking to a chirally symmetric phase. It is of interest to investigate how these and other hadron masses change as one decreases N_f (hence increases α_{IR}) to move from the walking regime near $N_{f,cr}$ toward the QCD-like regime at smaller N_f; this was done in.[14] Searches for the production and decays of these technihadrons at the CERN Large Hadron Collider will be of great importance in testing technicolor theories. For example, by analogy with $\rho \to \pi\pi$ in QCD, since the technipions are absorbed to become the longitudinal components of the W and Z, one would have $\rho^0_{TC} \to W^+_L W^-_L$,

$\rho_{TC}^{+} \to W_L^{+} Z_L^0$, etc. A related issue concerns pseudo-Nambu Goldstone bosons. While walking raises the masses of many of these PNGB's, they can present a phenomenological challenge for the model. It will be natural to search further for these at the LHC.

10. Summary

The question of the origin of electroweak symmetry is still not answered, and dynamical EWSB via technicolor and extended technicolor remains an interesting possibility. This approach is strongly constrained by flavor-changing neutral current data and precision electroweak measurements. We have reviewed here some recent progress on TC/ETC models. Clearly, there are a number of challenges for such TC/ETC models. Soon, experiments at the LHC will show whether these ideas are realized in nature.

Acknowledgments

I would like to thank my collaborators on works discussed here, T. Appelquist and M. Piai and, in subsequent papers, N. Christensen and M. Kurachi. I also thank M. Harada, M. Tanabashi, and K. Yamawaki, for organizing this SCGT06 conference. Our research was partially supported by the grant NSF-PHY-03-54776.

References

1. S. Weinberg, Phys. Rev. D **19**, 1277 (1979); L. Susskind, Phys. Rev. D **20**, 2619 (1979); see also S. Weinberg, Phys. Rev. D **13**, 974 (1976).
2. S. Dimopoulos and L. Susskind, Nucl. Phys. B **155**, 237 (1979); E. Eichten and K. Lane, Phys. Lett. B **90**, 125 (1980).
3. Some recent reviews are K. Lane, hep-ph/0202255; C. Hill and E. Simmons, Phys. Rep. **381**, 235 (2003); R. S. Chivukula, M. Narain, J. Womersley, in http://pdg.lbl.gov
4. T. Appelquist and R. Shrock, Phys. Lett. **B548**, 204 (2002).
5. T. Appelquist and R. Shrock, Phys. Rev. Lett. **90**, 201801 (2003).
6. T. Appelquist, M. Piai, and R. Shrock, Phys. Rev. D **69**, 015002 (2004).
7. T. Appelquist, M. Piai, and R. Shrock, Phys. Lett. B **593**, 175 (2004); *ibid.*, **595**, 442 (2004).
8. T. Appelquist, N. Christensen, M. Piai, and R. Shrock, Phys. Rev. D **70**, 093010 (2004).
9. N. Christensen and R. Shrock, Phys. Rev. Lett. **94**, 241801 (2005).
10. N. Christensen and R. Shrock, Phys. Rev. D **72**, 035013 (2005).
11. N. Christensen and R. Shrock, Phys. Lett. B **632**, 92 (2006).
12. N. Christensen and R. Shrock, Phys. Rev. D **74**, 015004 (2006).

13. M. Kurachi and R. Shrock, Phys. Rev. D **74**, 056003 (2006).
14. M. Kurachi and R. Shrock, JHEP **12**, 034 (2006).
15. M. Kurachi, in these SCGT06 proceedings.
16. P. Sikivie, L. Susskind, M. Voloshin, and V. Zakharov, Nucl. Phys. B **173**, 189 (1980).
17. See, e.g., E. Farhi and L. Susskind, Phys. Rept. **74**, 277 (1981).
18. B. Holdom, Phys. Lett. B **150**, 301 (1985).
19. K. Yamawaki, M. Bando, and K. Matumoto, Phys. Rev. Lett. **56**, 1335 (1986).
20. T. Appelquist, D. Karabali, and L. C. R. Wijewardhana, Phys. Rev. Lett. **57**, 957 (1986); T. Appelquist and L. C. R. Wijewardhana, Phys. Rev. D **35**, 774 (1987); Phys. Rev. D **36**, 568 (1987).
21. T. Appelquist and J. Terning, Phys. Rev. D **50**, 2116 (1994).
22. T. Appelquist, J. Terning, and L. C. R. Wijewardhana, Phys. Rev. Lett. **77**, 1214 (1996); T. Appelquist, A. Ratnaweera, J. Terning, and L. C. R. Wijewardhana, Phys. Rev. D **58**, 105017 (1998).
23. V. Miransky and K. Yamawaki, Phys. Rev. D **55**, 5051 (1997).
24. R. S. Chivukula, Phys. Rev. D **55**, 5238 (1997).
25. M. Peskin and T. Takeuchi, Phys. Rev. Lett. **65**, 964 (1990); Phys. Rev. D **46**, 381 (1992).
26. K. Lane and A. Martin, Phys. Rev. D **71**, 015011 (2005).
27. R. Cahn and M. Suzuki, Phys. Rev. D **44**, 3641 (1991); T. Appelquist and G. Triantaphyllou, Phys. Lett. B **278**, 345 (1992); R. Sundrum and S. Hsu, Nucl. Phys. B **391**, 127 (1993); T. Appelquist and F. Sannino, Phys. Rev. D **59**, 067702 (1999); S. Ignjatovic, L. C. R. Wijewardhana, and T. Takeuchi, Phys. Rev. D **61**, 056006 (2000).
28. M. Harada, M. Kurachi, and K. Yamawaki, Prog. Theor. Phys. **115**, 765 (2006).
29. D. K. Hong and H.-U. Yee, Phys. Rev. D **74**, 015011 (2006); J. Hirn and V. Sanz, Phys. Rev. Lett. **97**, 121803 (2006); M. Piai, hep-ph/0608241.
30. E. Eichten and K. Lane, Phys. Lett. B **222**, 274 (1988).
31. D. Hong, S. Hsu, and F. Sannino, Phys. Lett. B **597**, 89 (2004); F. Sannino and K. Tuominen, Phys. Rev. D **71**, 051901 (2005).
32. T. Appelquist, N. Evans, and S. Selipsky, Phys. Lett. B **374**, 145 (1996).
33. M. Harada, M. Kurachi, and K. Yamawaki, Phys. Rev. D **68**, 076001 (2003).

HIGGSLESS MODELS and DECONSTRUCTION

Masaharu TANABASHI*

Department of Physics, Tohoku University
Sendai 980-8578, JAPAN

We give a brief review on the deconstructed Higgsless models (linear moose models). The Fermi-constant G_F and the low energy ρ parameter are evaluated, and the physics behind the delay of unitarity violation is clarified. In order to construct a viable model consistent with the precision electroweak data, left-handed quarks and leptons need to be delocalized over the deconstructed extra dimension. We discuss "ideal delocalization" of quarks and leptons with which deviations in precision electroweak parameters are minimized. As contributions to the S parameter vanish to leading order with the ideal delocalization, current constraints on these models arise from limits on deviations in triple gauge boson vertices. By using the block-spin transformation, we argue that low energy effective theory of any consistent Higgsless model can be described by highly deconstructed linear moose model with only a fewer number of deconstructed sites. We propose the three-site model as the lowest order effective theory of Higgsless models.

1. Introduction

The Higgs boson, the agent of electroweak symmetry breaking in the standard model, has not been discovered yet. The origin of electroweak symmetry breaking still remains one of the most significant mysteries waiting unraveled at future experiments. The one-doublet Higgs in the standard model leaves many unanswered theoretical questions and is thus unsatisfactory. It is an interesting theoretical research avenue to investigate models of the electroweak symmetry breaking without introducing fundamental Higgs scalar particle ("Higgsless models").

The technicolor model,[1–3] which is arranged to trace the dynamical chiral symmetry breaking in QCD, is an elegant example in this class of models. Unfortunately, however, it turned out very difficult to explain quark/lepton masses without inducing excessive flavor-changing-neutral-currents in the

*Present address: Department of Physics, Nagoya University, Nagoya 464-8602, Japan.

QCD-like technicolor models. Moreover, the simplest QCD-like technicolor model gives large contribution to the electroweak oblique corrections,[4–7] which is clearly inconsistent with the present precision measurements of these parameters. Walking technicolor model[8–14] was introduced so as to remedy these drawbacks. In the walking technicolor, however, we are not able to make firm predictions because of the lack of our knowledge on the non-perturbative walking dynamics.

Recently, a new class of Higgsless models has been proposed.[21] Based on five-dimensional gauge theories compactified on an interval, these models achieve unitarity of electroweak boson self-interactions[15–20] through the exchange of a tower of massive vector bosons,[22–24] rather than the exchange of a scalar Higgs boson.[25] Various models were constructed in this direction.[26–31] Motivated by gauge/gravity duality,[32–35] models of this kind may be viewed as "dual" to more conventional models of dynamical symmetry breaking such as the walking technicolor. Unlike the walking technicolor, however, the electroweak boson self-interaction ($W_L W_L$ scattering) remains perturbative at $\sqrt{s} \simeq 1$TeV. We may be thus able to make firm predictions in these five dimensional Higgsless models.

This talk is based on a series of our papers.[36–47] The 5D Higgsless models are deconstructed[48,49] so as to obtain their effective 4D descriptions (linear moose[50] theories).[36,37,39,51–53] The collective nature of electroweak symmetry breaking is shown to play an essential role in these linear moose theories. The low energy Fermi-constant G_F and the ρ parameter are evaluated,[37,39] and the physics behind the delay of unitarity violation is clarified. In order to construct a viable model consistent with the precision electroweak data, the left-handed quarks and leptons need to be delocalized over the deconstruction lattice.[40,41,43] We discuss "ideal delocalization" of quarks and leptons with which deviations in precision electroweak parameters are minimized.[41,43] As contributions to the S parameter vanish to leading order, current constraints on these models arise from limits on deviations in triple gauge boson vertices.[42] By using the block-spin transformation, we argue that low energy effective theory of any consistent Higgsless model can be described at low energy by highly deconstructed linear moose model with only a few number of deconstructed sites.[45] A three-site Higgsless model is then proposed as the lowest order effective theory.[44]

2. Deconstruction of boundary conditions

Let us first start with a brief review on the deconstruction (latticization) of gauge theories with an extra dimension.[48,49] We consider a five dimensional

$SU(2)$ gauge theory with a flat extra dimension compactified on an interval $[0, \pi R]$,

$$S = -\frac{1}{2g_5^2}\int_0^{\pi R} dy \int d^4x \mathrm{tr}\left[F_{MN}F^{MN}\right], \tag{1}$$

with

$$F_{MN} = \partial_M A_N - \partial_N A_M + i[A_M, A_N]. \tag{2}$$

The endpoints $y = 0, \pi R$ correspond to the branes introduced in this setup. We impose the Dirichlet boundary condition (BC) for A_μ $(\mu = 0, 1, 2, 3)$,

$$A_\mu|_{y=0} = A_\mu|_{y=\pi R} = 0, \tag{3}$$

and the Neumann BC for A_y,

$$\partial_y A_y|_{y=0} = \partial_y A_y|_{y=\pi R} = 0. \tag{4}$$

Latticizing the extra dimension coordinate y in Eq.(1), we obtain a deconstructed action

$$\mathcal{L} = \sum_{j=1}^{N+1} \frac{f_j^2}{4}\mathrm{tr}\left[(D_\mu U_j)(D^\mu U_j)^\dagger\right] - \sum_{j=1}^{N} \frac{1}{2g_j^2}\mathrm{tr}\left[F_{\mu\nu}^j F^{j\mu\nu}\right], \tag{5}$$

where $F_{\mu\nu}^j$ is the gauge field strength at the j-th site in the latticized extra dimension,

$$F_{\mu\nu}^j = \partial_\mu A_\nu^j - \partial_\nu A_\mu^j. \tag{6}$$

In Eq.(5) we also introduced non-linear sigma model link field (Nambu-Goldstone boson field of $[SU(2) \times SU(2)]/SU(2)$),

$$U_j = \exp\left[2i\frac{\pi_j^a T^a}{f}\right], \qquad T^a = \frac{\tau^a}{2}, \tag{7}$$

which corresponds to the A_y degree of freedom in the 5D action. Covariant derivatives of U_j are given by

$$D_\mu U_j = \partial_\mu U_j + iA_\mu^{(j-1)} U_j - U_j A_\mu^j. \tag{8}$$

Since we are considering the flat extra dimension, we assume position independent (flat) g_j and f_j, $g_j = g, f_j = f$. Taking the continuum limit $N \to \infty$

$$g_j = \sqrt{N}\tilde{g}, \qquad f_j = \sqrt{N+1}v, \tag{9}$$

with fixed $\tilde{g}$ and v, and identifying

$$\frac{\pi R}{g_5^2} = \frac{1}{\tilde{g}^2}, \qquad \pi R = \frac{2}{\tilde{g}v}, \tag{10}$$

we see that the deconstructed Lagrangian Eq.(5) actually reproduces the continuum 5D action Eq.(1). Here Eq.(10) plays the role of the dictionary between deconstruction and continuum languages.

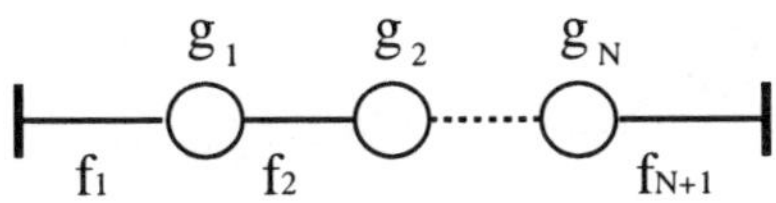

Fig. 1. The moose diagram corresponding to the deconstructed Lagrangian Eq.(5). $SU(2)$ gauge fields are shown as open circles. Lines connecting open circles denote the non-linear sigma model link fields. Flat extra dimension corresponds to $f = f_1 = \cdots = f_{N+1}$ and $g = g_1 = \cdots = g_N$.

The deconstructed Lagrangian Eq.(5) can be expressed intuitively by a moose diagram shown in Fig. 1. In the moose diagram, $SU(2)$ gauge fields are shown as open circles. Lines connecting open circles denote the non-linear sigma model link fields. Note that the A_μ degree of freedom is frozen by the the Dirichlet condition Eq.(3) at the endpoints (branes) of the interval ($y = 0, \pi R$). This corresponds to the absence of the gauge degree of freedom at the endpoints in the linear moose diagram. Detailed analyses of the deconstruction of boundary conditions are given in Ref. 55.

"Eating" Nambu-Goldstone bosons, all of N gauge fields A_μ^j acquire their masses through the Higgs mechanism. A tower of these massive gauge bosons corresponds to the Kaluza-Klein (KK) tower of spin-1 resonances in the original continuum 5D gauge theory. There also exists one massless uneaten Nambu-Goldstone field in the deconstructed model Eq.(5). The deconstructed model Eq.(5) and thus the original 5D theory with BCs (3) and Eq.(4) can be regarded as a model of spontaneous global symmetry breaking.

Taking $N = 1$ in the deconstructed Lagrangian Eq.(5), we notice that this model coincides with the Georgi's vector limit[56,57] in the hidden local symmetry (HLS) model,[58–62] which successfully describes properties of π and ρ mesons in the low energy hadron dynamics. We also notice that the $N = 2$ model with $f_1 = f_3 = \sqrt{2} f_2$ reproduces the generalized HLS model of π, ρ and a_1 mesons.[60] Motivated by AdS/QCD duality, further generalization ($N > 2$) has been discussed in Ref. 68 in the context of low energy hadron dynamics.

We are now ready to turn to the Higgsless models, in which electroweak gauge symmetry is broken by its boundary conditions. The most popular setup of 5D Higgsless model introduces $SU(2) \times SU(2) \times U(1)$ gauge group

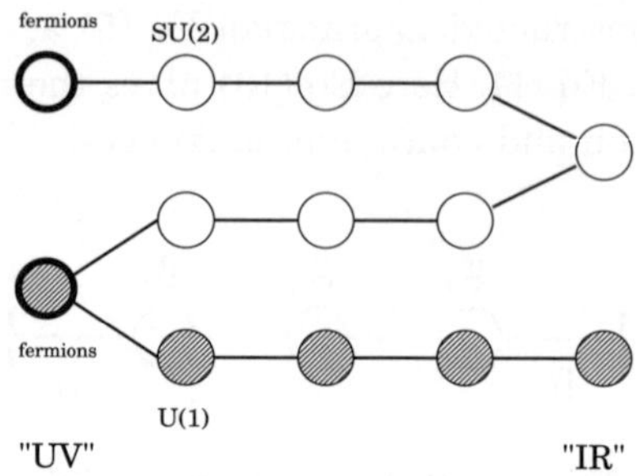

Fig. 2. Moose diagrams for Higgsless models with bulk gauge group $SU(2) \times SU(2) \times U(1)$.

in the extra dimension compactified on an interval. Motivated by models in a warped background, we interpret the two ends of the interval as ultraviolet (UV) and infrared (IR) "branes". The $SU(2) \times SU(2)$ bulk gauge group is broken by boundary conditions to a diagonal custodial $SU(2)$ at the IR brane, and the $SU(2) \times U(1)$ bulk gauge group to $U(1)$ at the UV brane. The light quarks and leptons are assumed to be localized on the UV brane. Figure 2 shows the moose diagram of the deconstructed 4D Lagrangian corresponding to this setup. $SU(2)$ gauge groups are shown in open circles, while $U(1)$ gauge groups in shaded circles. By unfolding this moose, we see that the deconstructed version of this model corresponds to a special case of that shown in Fig.3.

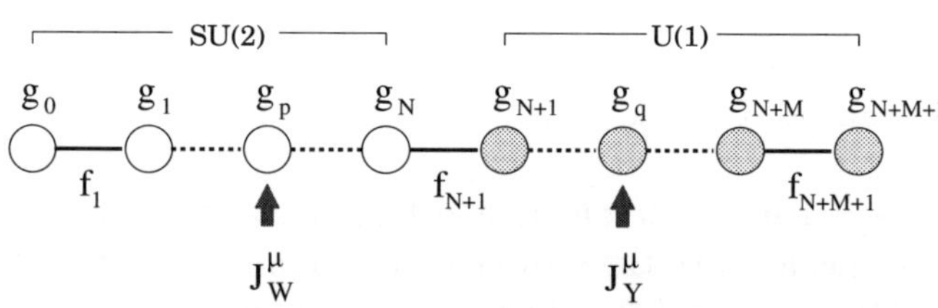

Fig. 3. The general linear moose model.

The linear moose model of Fig.3 incorporates an $SU(2)^{N+1} \times U(1)^{M+1}$ gauge group, and $N+1$ nonlinear $[SU(2) \times SU(2)]/SU(2)$ sigma models adjacent to M $[U(1) \times U(1)]/U(1)$ sigma models in which the global symmetry groups in adjacent sigma models are identified with the corresponding factors of the gauge group. The Lagrangian for this model at the lowest order is given by

$$\mathcal{L} = \frac{1}{4} \sum_{j=1}^{N+M+1} f_j^2 \mathrm{tr}\left[(D_\mu U_j)^\dagger (D^\mu U_j)\right] - \sum_{j=0}^{N+M+1} \frac{1}{2g_j^2} \mathrm{tr}\left(F^j_{\mu\nu} F^{j\mu\nu}\right), \quad (11)$$

with

$$D_\mu U_j = \partial_\mu + i A_\mu^{j-1} U_j - i U_j A_\mu^j, \quad (12)$$

where all gauge fields A_μ^j ($j = 0, 1, 2, \cdots, N+M+1$) are dynamical. The first $N+1$ gauge fields ($j = 0, 1, \cdots, N$) correspond to $SU(2)$ gauge groups; the other $M+1$ gauge fields ($j = N+1, N+2, \cdots, N+M+1$) correspond to $U(1)$ gauge groups. The symmetry breaking between the A_μ^N and A_μ^{N+1} follows an $[SU(2)_L \times SU(2)_R]/SU(2)_V$ symmetry breaking pattern with the $U(1)$ embedded as the T_3 generator of $SU(2)_R$. In what follows, we will denote $N+M$ by K for brevity.

The quarks and leptons in this model take their weak interactions from the $SU(2)$ group at $j = p$ and their hypercharge interactions from the $U(1)$ group with $j = q$. The neutral-current couplings to the fermions are thus written as

$$J_3^\mu A_\mu^{p,3} + J_Y^\mu A_\mu^q, \quad (13)$$

while the charged-current couplings arise from

$$\frac{1}{\sqrt{2}} J_\pm^\mu A_\mu^{p,\mp}. \quad (14)$$

We find a fairly compact expression for the charged-current Fermi-constant G_F

$$\sqrt{2} G_F = \frac{1}{v_{\mathrm{cc}}^2} \equiv \sum_{j=p+1}^{N+1} \frac{1}{f_j^2}. \quad (15)$$

Note that Eq.(15) is a consequence of the collective electroweak symmetry breaking

$$v_{\mathrm{cc}}^2 \propto f_{p+1}^2 f_{p+2}^2 \cdots f_{N+1}^2. \quad (16)$$

The property of the collective symmetry breaking Eq.(16) can be understood easily by considering a hypothetical limit in which one of f_j vanishes. In this limit the charged-current mass-squared matrix M_W^2 is separated into two parts. Note here that this separated matrix contains a zero eigenvalue. For $p+1 \le j \le N+1$ this zero-mass charged boson couples with the weak current $J_\pm^\mu$, implying the divergent G_F and thus $v_{\mathrm{cc}}^2 = 0$.

Equation (15) is a key formula to understand the physics behind the delay of unitarity violation. Equivalence theorem[15–19] says that violation of the unitarity in the longitudinal gauge boson scattering takes place at

$$\sqrt{8\pi}\,\min(f_j), \tag{17}$$

Taking large $N-p$ we see Eq.(17) can be pushed up above the usual scale of unitarity violation,

$$\sqrt{8\pi}\,\min(f_j) \gg \sqrt{8\pi} v_{\rm cc} \simeq 1.2 \text{ TeV}. \tag{18}$$

It should be emphasized, however, we cannot arrange the unitarity violation scale arbitrarily high. This is because the QED coupling is restricted by

$$\frac{1}{e^2} = \sum_{j=0}^{K+1} \frac{1}{g_j^2}. \tag{19}$$

In order to arrange

$$\min(f_j) \sim \sqrt{N-p} v_{\rm cc} \gg v_{\rm cc} \tag{20}$$

we need to consider a linear moose model with large N. Eq.(19) shows, however, that the gauge coupling strength of such a linear moose with large N becomes non-perturbative

$$\max(g_j) \sim \sqrt{K} e, \qquad K = N + M \geq N. \tag{21}$$

Clearly we need to optimize the number N so as to get the maximum delay of the unitarity violation.

We next turn to the ρ parameter. It would be convenient to define the ρ parameter

$$\rho = \frac{1/v_{\rm nc}^2}{1/v_{\rm cc}^2} \tag{22}$$

at the zero-momentum. Here $1/v_{\rm nc}^2$ is proportional to the neutral current Fermi-constant. From the analysis of Ref. 39 we know

$$\frac{1}{v_{\rm nc}^2} = \sum_{j=p+1}^{q} \frac{1}{f_j^2}. \tag{23}$$

Again, the collective symmetry breaking property determines the structure of Eq.(23). Combining Eq.(15), Eq.(22) and Eq.(23) we find a simple expression for $\Delta\rho$,

$$\Delta\rho = \rho - 1 = \frac{1}{\sqrt{2} G_F} \sum_{j=N+2}^{q} \frac{1}{f_j^2} \geq 0. \tag{24}$$

We see that the low-energy ρ parameter is identically 1 when $q = N+1$ and is greater than 1 otherwise. As shown in Ref. 63, the current limit of $\Delta\rho$ is quite severe, $\Delta\rho < 4 \times 10^{-3}$. In this talk we therefore restrict ourselves within the "case-I" model ($q = N+1$) of Fig. 4, in which $\Delta\rho = 0$ is automatically satisfied.

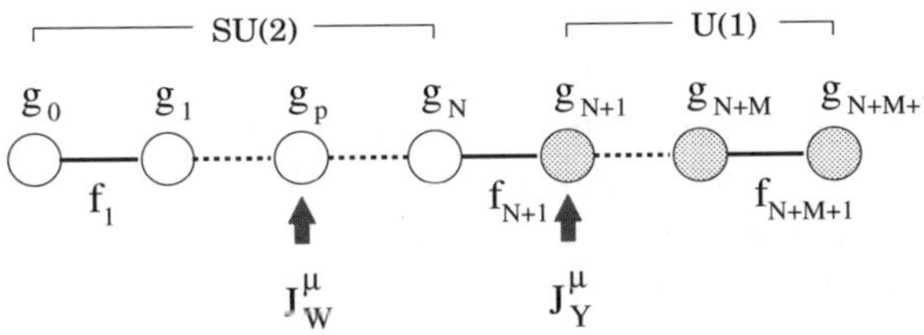

Fig. 4. The general case-I ($q = N+1$) linear moose model.

3. Ideal delocalization

As we emphasized in the previous section, the case-I model satisfies the condition $\Delta\rho = 0$, which is, however, not enough to make the model fully consistent with the present precision electroweak measurements. Barbieri *et al.* introduced a four-parameter formalism to express the deviation of electroweak interactions from their standard model forms in "universal" theories.[63] The parameters $(\hat{S}, \hat{T}, W, Y)$ introduced by these authors are defined by the properties of the correlation functions at zero momentum, and differ from the quantities directly measured through the properties of the on-shell electroweak gauge bosons. These parameters $(\hat{S}, \hat{T}, W, Y)$ can be related with the on-shell parameters $(S, T, \delta, \Delta\rho)$ as[38]

$$\alpha S = 4s^2(\hat{S} - Y - W), \quad \alpha T = \hat{T} - \frac{s^2}{c^2}Y, \quad \alpha\delta = 4s^2c^2W, \quad \Delta\rho = \hat{T}, \tag{25}$$

with s^2 (c^2) being the Weinberg angle. Here S and T are Peskin-Takeuchi parameters defined at the on-shell Z boson. $\Delta\rho$ is already defined in the previous section. In order to construct a viable model consistent with the precision electroweak measurements, all of these four parameters need to be sufficiently small. Unfortunately, however, in any unitary Higgsless theory with localized fermions, it has been shown that the electroweak parameter $\hat{S}$ is bounded by

$$\hat{S} \gtrsim 4 \times 10^{-3}, \tag{26}$$

which is disfavored by precision electroweak data. (See Ref. 39 and references therein.) It should be emphasized that this bound is true for arbitrary background 5D geometry, spatially dependent gauge-couplings, and brane kinetic energy terms. We thus need to delocalize quarks and leptons over the (deconstructed) extra dimension.

In Refs. 40 and 41 we examined the properties of deconstructed Higgsless models for the case of a fermion whose $SU(2)$ properties arise from delocalization over many sites of deconstruction lattice. (See also Refs. 64–67.) We found that if the probability distribution of the delocalized fermions is appropriately related to the W wavefunction ("ideal fermion delocalization"), then deviations in the precision electroweak parameters are minimized.[41] In particular, we have shown that the three electroweak parameters $\hat{S}$, $\hat{T}$ and W are exactly zero at tree level with the ideal fermion delocalization. Also the size of Y parameter is shown to be sufficiently small.

It is interesting to note Higgsless models with ideally-delocalized fermions cannot be constrained by direct collider searches for new bosons that rely on the bosons' couplings to fermions. This is because that the higher KK resonances are fermiophobic to leading order in these models. Existing searches for W' bosons assume that the W' quarks and leptons generally with SM strength. Likewise, existing direct searches for Z' bosons rely on $Z'f\bar{f}$ couplings.

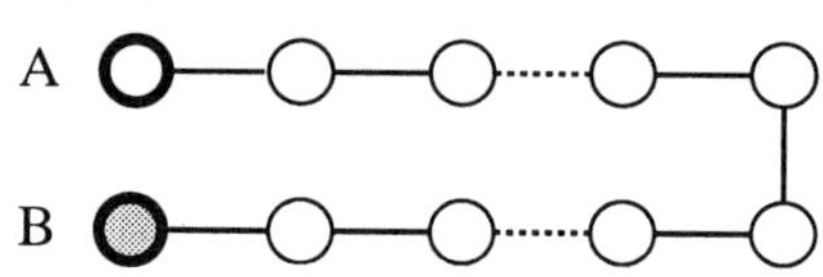

Fig. 5. Moose diagram for the deconstruction corresponding to the $SU(2)_A \times SU(2)_B$ models analyzed in Ref.[42] $SU(2)$ gauge groups are shown as open circles; $U(1)$ gauge group is shown as a shaded circle. The brane kinetic energy terms are indicated by the thick circles. As drawn, this model does not have an interpretation as a local higher-dimensional gauge theory; however, the results of this model agree (up to corrections suppressed by $M_W^4/M_{W'}^4$) with a consistent higher dimensional theory based on an $SU(2) \times SU(2) \times U(1)$ gauge group in the bulk.[43]

As contributions to the precision electroweak parameters sufficiently small in the ideal delocalization, current constraints arise from limits on deviations in multi-gauge-boson vertices. To leading order, in the absence of CP-violation, triple electroweak gauge boson vertices may be written in the Hagiwara-Peccei-Zeppenfeld-Hikasa triple-gauge-vertex notation.[69] In Ref. 42 we calculated the Hagiwara-Peccei-Zeppenfeld-Hikasa parameters

$\Delta\kappa_\gamma, \Delta\kappa_Z, \Delta g_1^Z$ in various Higgsless models. We found

$$\Delta\kappa_\gamma = 0, \qquad \Delta\kappa_Z = \Delta g_1^Z. \tag{27}$$

For an $SU(2)_A \times SU(2)_B$ flat-space model shown in the moose diagram Fig. 5, the form of Δg_1^Z is given by

$$\Delta g_1^Z = \frac{\pi^2}{12c^2}\frac{M_W^2}{M_{W'}^2}\left[\frac{1}{4}\cdot\frac{7+\kappa}{1+\kappa}\right] \tag{28}$$

with κ being the ratio of bulk $SU(2)_{A,B}$ gauge couplings, $\kappa \equiv g_{5B}^2/g_{5A}^2$. Here $M_{W'}$ is the mass of the lightest KK resonance. The ideal delocalization of fermion wave-function is assumed. In an $SU(2)_A \times SU(2)_B$ model in warped space, we found

$$\Delta g_1^Z = \frac{3x_1^2}{16c^2}\frac{M_W^2}{M_{W'}^2}\left[\frac{2}{9}\cdot\frac{7+2\kappa}{1+\kappa}\right] \tag{29}$$

where $x_1 \simeq 2.4$ is the first zero of the Bessel function J_0.

Currently, the strongest bounds on Δg_1^Z and $\Delta\kappa_Z$ come from LEP II. The 95%CL. upper limit is $\Delta g_1^Z \leq 0.028$, which implies a lower bound of 500 (570) GeV on the first KK resonance in flat (warped) space models for $\kappa = 1$ and lower bounds of 250–650 (380–700) GeV as κ varies from ∞ to 0.

4. Block-spin and *f*-flat deconstruction of a warped space

Motivated by AdS/CFT duality, a warped extra dimension is widely used in various 5D Higgsless models. It is also known that the warped bulk background ensures a mass hierarchy between W' and W, $M_{W'} \gg M_W$, as desired from a phenomenological viewpoint. What is the physics behind this hierarchy, then? How can we interpret it in the deconstruction language? The key to understand this issue lies in the effective field theory viewpoint of the deconstructed gauge theories. As we argue next, we can construct effective field theories of any Higgsless models by using the block-spin transformation in the decontructed lattice.[45]

In order to illustrate the block-spin transformation, in this section, we consider a toy five dimensional gauge theory in a warped space. The extra dimension is assumed to be compactified on a warped interval $[R, R']$. The endpoint $y = R'$ ($y = R$) corresponds to the IR (UV) brane. The boundary conditions are taken as

$$A_\mu\big|_{y=R'} = \partial_y A_y\big|_{y=R'} = 0, \qquad \partial_y A_\mu\big|_{y=R} = A_y\big|_{y=R} = 0. \tag{30}$$

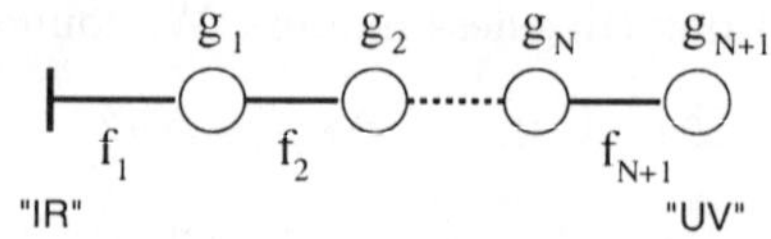

Fig. 6. Moose diagram for the warped gauge theory with the boundary conditions Eq.(30).

The deconstruction of this model is given by a moose diagram shown in Fig. 6. The authors of Refs. 49, 70 and 71 argued that it is possible to realize the warped geometry in the continuum limit if we take

$$\frac{1}{f_{n+1}^2} = \frac{r}{f_n^2}, \qquad \frac{1}{g_{n+1}^2} = \frac{1}{g_n^2}, \qquad (0 < r < 1) \tag{31}$$

for $n = 1, 2, \cdots, N$. Note here that the fundamental scale at link-n is given by the decay constant f_n. Eq.(31) thus implies a small "warp"-factor $f_{n+1}/f_n = 1/\sqrt{r}$ between links n and $n+1$. The total warp-factor is given by

$$e^{b/2} = \frac{f_{N+1}}{f_1} = \frac{1}{r^{N/2}} \gg 1. \tag{32}$$

Note that the gauge coupling strength g_n is independent of the lattice site n in Eq.(31), which is therefore refereed to as a "g-flat" deconstruction.

The large mass scale splitting $f_{N+1} \gg f_1$ makes it impossible to interpret the g-flat deconstruction as an effective field theory below certain low energy scale μ, however. In order to formulate a sensible effective theory, we need to construct a deconstruction model without such a large mass scale splitting. In Ref. 45 we introduced a block-spin transformation to make an "f-flat" deconstruction in which all of the f-constant f_j of the linear moose model are identical. Figure 7 shows the block-spin transformation in a schematic figure.

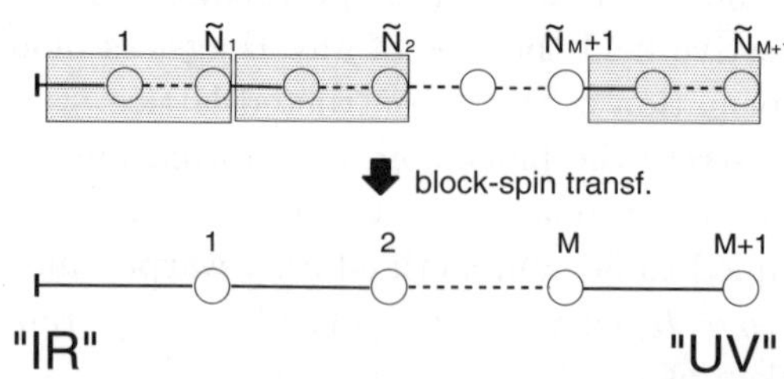

Fig. 7. Schematic explanation of the block-spin transformation.

After the block-spin transformation,[a] we obtain a deconstruction lattice with $M+1$ sites. The decay constants and the gauge couplings after the block-spin transformation are given by

$$\frac{1}{\tilde{f}_1^2} = \sum_{i=1}^{\tilde{N}_1} \frac{1}{f_i^2}, \quad \cdots, \qquad \frac{1}{\tilde{g}_1^2} = \sum_{i=1}^{\tilde{N}_1} \frac{1}{g_i^2}, \quad \cdots. \tag{33}$$

An f-flat deconstruction can be realized by arranging $\tilde{N}_1$, $\tilde{N}_2$, $\cdots$ $\tilde{N}_{M+1}$ to satisfy $\tilde{f}^2 = \tilde{f}_1^2 = \tilde{f}_2^2 = \cdots = \tilde{f}_{M+1}^2$. Without having a large mass scale splitting in the f-constants, the f-flat deconstruction can be regarded as a sensible effective field theory valid below $4\pi\tilde{f}$. We find

$$\frac{1}{\tilde{g}_n^2} = \frac{1}{bg^2} \ln\left(\frac{M+2-n}{M+1-n}\right), \qquad (n = 1, 2, \cdots M), \tag{34}$$

and

$$\frac{1}{\tilde{g}_{M+1}^2} = \frac{1}{g^2}\left[1 - \frac{1}{b}\ln(M+1)\right], \tag{35}$$

with b being the warp-factor defined in Eq.(32). For $b \gg 1$ we find

$$\tilde{g}_1^2 \simeq \tilde{g}_2^2 \simeq \cdots \simeq \tilde{g}_M^2 \gg \tilde{g}_{M+1}^2. \tag{36}$$

As clarified by the spring analogy,[54] the mass of the lightest vector particle is proportional to $\tilde{g}_{M+1}$, while other heavier KK particles are not. Eq.(36) is thus responsible for the desired mass hierarchy $M_{W'} \gg M_W$ in the warped Higgsless models.

5. Three-site model

As we found in the previous section, the block-spin transformation technique enables us to construct an f-flat deconstruction effective theory starting from any continuum Higgsless model. In Ref. 44 we introduced a highly-deconstructed model with only three sites (three-site Higgsless model) as the lowest order effective field theory. The corresponding moose diagram is shown in Fig. 8. The three-site model contains sufficient complexity to incorporate interesting physics issues, such as ideal fermion delocalization and the generation of fermion masses. The gauge sector of this model is equivalent to that of the BESS model;[72,73] the new physics of interested here lies in the fermion sector.

[a]The block-spin transformation corresponds to the freedom to perform coordinate transformation in the continuum.

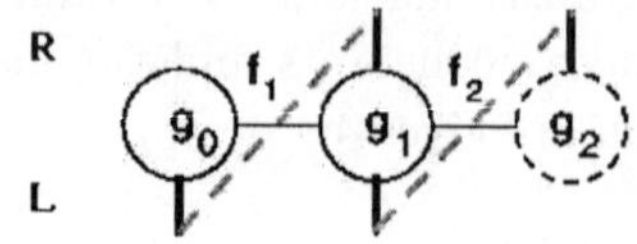

Fig. 8. The moose diagram for the three site model. The solid circles represent $SU(2)$ gauge groups, with coupling strengths g_0 and g_1, and the dashed line is a $U(1)$ gauge group with coupling g_2. The left-handed fermions, denoted by the lower vertical lines, are located at sites 0 and 1, and the right-handed fermions, denoted by the upper vertical lines, at sites 1 and 2. The dashed green lines correspond to Yukawa couplings. We take $f_1 = f_2 = \sqrt{2}\,v$, denote $g_0 = g$, $g_1 = \tilde{g}$, $g_2 = g'$ and take $\tilde{g} \gg g, g'$.

We may write the Yukawa couplings and fermion mass term

$$\mathcal{L}_f = \lambda f_1\, \bar{\psi}_{L0} U_1 \psi_{R1} + M \bar{\psi}_{R1} \psi_{L1} + f_2\, \bar{\psi}_{L1} U_2 \begin{pmatrix} \lambda'_u & \\ & \lambda'_d \end{pmatrix} \begin{pmatrix} u_{R2} \\ d_{R2} \end{pmatrix} + h.c. \quad (37)$$

Here the matrices $U_{1,2}$ are the nonlinear sigma model fields associated with the $f_{1,2}$ links of the moose. The Yukawa couplings introduced here are of precisely the correct form required to implement a deconstruction of a five-dimensional fermion with chiral boundary conditions. We take f-flat deconstruction $f_1 = f_2 = \sqrt{2}v$. The fermion mass matrix is given by

$$\begin{pmatrix} m & 0 \\ M & m'_f \end{pmatrix} \equiv \sqrt{2}\tilde{\lambda} v \begin{pmatrix} \varepsilon_L & 0 \\ 1 & \varepsilon_{fR} \end{pmatrix}, \qquad \varepsilon_L \equiv \frac{\lambda}{\tilde{\lambda}}, \quad \varepsilon_{fR} \equiv \frac{\lambda'_f}{\tilde{\lambda}}, \quad (38)$$

with

$$m = \sqrt{2}\lambda v, \quad m'_f = \sqrt{2}\lambda'_f v, \quad M = \sqrt{2}\tilde{\lambda} v. \quad (39)$$

It is straightforward to calculate the light fermion mass m_f

$$m_f \simeq \frac{m m'_f}{\sqrt{M^2 + m'^2_f}} = \frac{\sqrt{2}\tilde{\lambda} v \varepsilon_L \varepsilon_{fR}}{\sqrt{1 + \varepsilon^2_{fR}}}, \quad (40)$$

and its wave-function in the deconstruction lattice

$$\psi_L^{f,\text{light}} \simeq -\left(1 - \frac{\varepsilon_L^2}{2}\right) \psi^f_{L0} + \varepsilon_L \psi^f_{L1}, \quad (41)$$

where we assumed $\varepsilon_{fR} \ll 1$. There also exist heavy KK fermions in this model. The KK fermion mass is given by

$$M_{f,KK} \simeq \sqrt{M^2 + m'^2_f} = M\sqrt{1 + \varepsilon^2_{fR}} \simeq M. \quad (42)$$

The S parameter is evaluated as

$$S = \frac{4\pi}{g_1^2}\left(1 - \frac{2g_1^2}{g_0^2}\varepsilon_L^2\right) \quad (43)$$

at the tree level. It is easy to see that Eq.(43) vanishes in the ideal delocalization limit

$$\varepsilon_L^2 = \frac{g_0^2}{2g_1^2}. \tag{44}$$

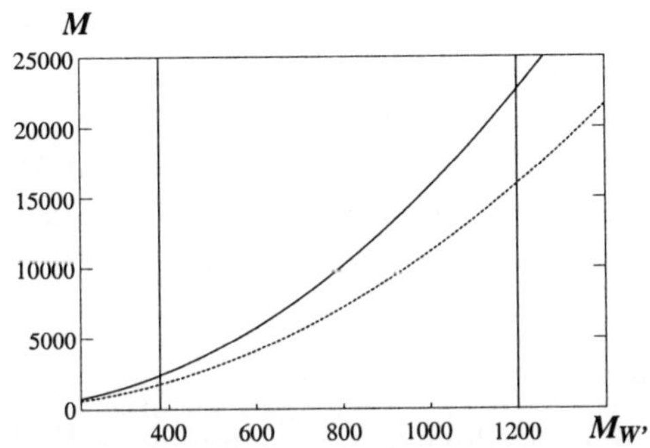

Fig. 9. Phenomenologically acceptable values of M and $M_{W'}$ in GeV or $\Delta\rho = 2.5\times10^{-3}$ (solid curve) and 5×10^{-3} (dashed curve). The region bounded by the lines bounded by 380 GeV $< M_{W'} <$ 1200 GeV and above the curve are allowed.

Phenomenological bounds on this model are summarized in Fig. 9. Here we assumed the ideal fermion delocalization Eq.(44). The bounds on the tripe gauge vertices give

$$M_{W'} \geq 380 \text{ GeV}, \tag{45}$$

while the unitarity of $W_L W_L$ scattering leads to

$$M_{W'} \leq 1200 \text{ GeV}. \tag{46}$$

Figure 9 also shows the bound from the KK fermion one-loop contributions to the ρ parameter. From this figure, we see that the heavy KK fermion mass $M_{f,KK} \simeq M$ should be greater than 1.8 TeV. Bosonic one-loop contributions to S and T parameters are calculated in Refs. 46, 47 and 74.

6. Summary

In this talk, using deconstruction, we have calculated the form of the corrections to the electroweak interactions in a large class of Higgsless models, allowing for arbitrary 5D geometry, position-dependent gauge coupling, and brane kinetic-energy terms.[37–39] We also clarified the physics behind the delay of unitarity violation in the $W_L W_L$ scattering. We discussed the ideal fermion delocalization with which deviations in precision electroweak parameters are minimized. As contributions to the S parameter vanish to leading order with the ideal delocalization, current constraints on these models arise from limits on deviations in triple gauge boson vertices. We have applied the block-spin transformation to a deconstructed gauge theory and showed how this enables us to construct a sensible effective field theory with an f-flat deconstruction. We also considered the three-site Higgsless model as the lowest order effective field theory and explicitly showed the value of the electroweak parameters αS can be minimized with the ideal fermion delocalization. Combining the triple-gauge-vertex bound $M_{W'} \geq 380$ and the KK fermion one-loop contributions to $\Delta\rho$, we found that the KK fermions should be heavier than 1.8 TeV in the three-site model.

Acknowledgments

I would like to thank R. S. Chivukula, E. H. Simmons, H.-J. He, M. Kurachi, B. Coleppa, S. Di Chiara, and S. Matsuzaki for fruitful collaborations on which this talk is based. My work is supported in part by the JSPS Grant-in-Aid for Scientific Research No.16540226.

References

1. S. Weinberg, Phys. Rev. D **19**, 1277 (1979).
2. L. Susskind, Phys. Rev. D **20**, 2619 (1979).
3. For a review, C. T. Hill and E. H. Simmons, Phys. Rept. **381**, 235 (2003) [Erratum-ibid. **390**, 553 (2004)].
4. M. E. Peskin and T. Takeuchi, Phys. Rev. Lett. **65**, 964 (1990).
5. B. Holdom and J. Terning, Phys. Lett. B **247**, 88 (1990).
6. M. Golden and L. Randall, Nucl. Phys. B **361**, 3 (1991).
7. M. E. Peskin and T. Takeuchi, Phys. Rev. D **46**, 381 (1992).
8. B. Holdom, Phys. Rev. D **24**, 1441 (1981).
9. B. Holdom, Phys. Lett. B **150**, 301 (1985).
10. K. Yamawaki, M. Bando and K. i. Matumoto, Phys. Rev. Lett. **56**, 1335 (1986).
11. T. W. Appelquist, D. Karabali and L. C. R. Wijewardhana, Phys. Rev. Lett. **57**, 957 (1986).
12. T. Appelquist and L. C. R. Wijewardhana, Phys. Rev. D **35**, 774 (1987).
13. T. Appelquist and L. C. R. Wijewardhana, Phys. Rev. D **36**, 568 (1987).
14. T. Akiba and T. Yanagida, Phys. Lett. B **169**, 432 (1986).
15. J. M. Cornwall, D. N. Levin and G. Tiktopoulos, Phys. Rev. Lett. **30**, 1268 (1973) [Erratum-ibid. **31**, 572 (1973)].
16. J. M. Cornwall, D. N. Levin and G. Tiktopoulos, Phys. Rev. D **10**, 1145 (1974) [Erratum-ibid. D **11**, 972 (1975)].

17. B. W. Lee, C. Quigg and H. B. Thacker, Phys. Rev. Lett. **38**, 883 (1977).
18. B. W. Lee, C. Quigg and H. B. Thacker, Phys. Rev. D **16**, 1519 (1977).
19. M. S. Chanowitz and M. K. Gaillard, Nucl. Phys. B **261**, 379 (1985).
20. D. A. Dicus and V. S. Mathur, Phys. Rev. D **7**, 3111 (1973).
21. C. Csaki, C. Grojean, H. Murayama, L. Pilo and J. Terning, Phys. Rev. D **69**, 055006 (2004).
22. R. S. Chivukula, D. A. Dicus and H. J. He, Phys. Lett. B **525**, 175 (2002).
23. R. S. Chivukula and H. J. He, Phys. Lett. B **532**, 121 (2002).
24. R. S. Chivukula, D. A. Dicus, H. J. He and S. Nandi, Phys. Lett. B **562**, 109 (2003).
25. P. W. Higgs, Phys. Lett. **12**, 132 (1964).
26. C. Csaki, C. Grojean, L. Pilo and J. Terning, Phys. Rev. Lett. **92**, 101802 (2004).
27. Y. Nomura, JHEP **0311**, 050 (2003).
28. R. Barbieri, A. Pomarol and R. Rattazzi, Phys. Lett. B **591**, 141 (2004).
29. C. Csaki, C. Grojean, J. Hubisz, Y. Shirman and J. Terning, Phys. Rev. D **70**, 015012 (2004).
30. H. Davoudiasl, J. L. Hewett, B. Lillie and T. G. Rizzo, Phys. Rev. D **70**, 015006 (2004).
31. G. Burdman and Y. Nomura, Phys. Rev. D **69**, 115013 (2004).
32. J. M. Maldacena, Adv. Theor. Math. Phys. **2**, 231 (1998) [Int. J. Theor. Phys. **38**, 1113 (1999)].
33. S. S. Gubser, I. R. Klebanov and A. M. Polyakov, Phys. Lett. B **428**, 105 (1998).
34. E. Witten, Adv. Theor. Math. Phys. **2**, 253 (1998).
35. O. Aharony, S. S. Gubser, J. M. Maldacena, H. Ooguri and Y. Oz, Phys. Rept. **323**, 183 (2000).
36. R. S. Chivukula, M. Kurachi and M. Tanabashi, JHEP **0406**, 004 (2004).
37. R. S. Chivukula, E. H. Simmons, H. J. He, M. Kurachi and M. Tanabashi, Phys. Rev. D **70**, 075008 (2004).
38. R. S. Chivukula, E. H. Simmons, H. J. He, M. Kurachi and M. Tanabashi, Phys. Lett. B **603**, 210 (2004).
39. R. S. Chivukula, E. H. Simmons, H. J. He, M. Kurachi and M. Tanabashi, Phys. Rev. D **71**, 035007 (2005).
40. R. S. Chivukula, E. H. Simmons, H. J. He, M. Kurachi and M. Tanabashi, Phys. Rev. D **71**, 115001 (2005).
41. R. S. Chivukula, E. H. Simmons, H. J. He, M. Kurachi and M. Tanabashi, Phys. Rev. D **72** (2005) 015008.
42. R. S. Chivukula, E. H. Simmons, H. J. He, M. Kurachi and M. Tanabashi, Phys. Rev. D **72** (2005) 075012.
43. R. S. Chivukula, E. H. Simmons, H. J. He, M. Kurachi and M. Tanabashi, Phys. Rev. D **72** (2005) 095013.
44. R. S. Chivukula, B. Coleppa, S. Di Chiara, E. H. Simmons, H. J. He, M. Kurachi and M. Tanabashi, Phys. Rev. D **74** (2006) 075011.
45. R. S. Chivukula, E. H. Simmons, H. J. He, M. Kurachi and M. Tanabashi, Phys. Rev. D **75** (2007) 035005.

46. S. Matsuzaki, R. S. Chivukula, E. H. Simmons and M. Tanabashi, Phys. Rev. D **75** (2007) 073002.
47. R. S. Chivukula, E. H. Simmons, S. Matsuzaki and M. Tanabashi, Phys. Rev. D **75** (2007) 075012.
48. N. Arkani-Hamed, A. G. Cohen and H. Georgi, Phys. Rev. Lett. **86**, 4757 (2001).
49. C. T. Hill, S. Pokorski and J. Wang, Phys. Rev. D **64**, 105005 (2001).
50. H. Georgi, Nucl. Phys. B **266**, 274 (1986).
51. R. Foadi, S. Gopalakrishna and C. Schmidt, JHEP **0403**, 042 (2004).
52. J. Hirn and J. Stern, Eur. Phys. J. C **34**, 447 (2004).
53. R. Casalbuoni, S. De Curtis and D. Dominici, Phys. Rev. D **70**, 055010 (2004).
54. H. Georgi, Phys. Rev. D **71**, 015016 (2005).
55. H. J. He, arXiv:hep-ph/0412113.
56. H. Georgi, Phys. Rev. Lett. **63**, 1917 (1989).
57. H. Georgi, Nucl. Phys. B **331**, 311 (1990).
58. M. Bando, T. Kugo, S. Uehara, K. Yamawaki and T. Yanagida, Phys. Rev. Lett. **54**, 1215 (1985).
59. M. Bando, T. Kugo and K. Yamawaki, Nucl. Phys. B **259**, 493 (1985).
60. M. Bando, T. Fujiwara and K. Yamawaki, Prog. Theor. Phys. **79**, 1140 (1988).
61. M. Bando, T. Kugo and K. Yamawaki, Phys. Rept. **164**, 217 (1988).
62. M. Harada and K. Yamawaki, Phys. Rept. **381**, 1 (2003).
63. R. Barbieri, A. Pomarol, R. Rattazzi and A. Strumia, Nucl. Phys. B **703**, 127 (2004).
64. G. Cacciapaglia, C. Csaki, C. Grojean and J. Terning, Phys. Rev. D **71**, 035015 (2005).
65. R. Foadi, S. Gopalakrishna and C. Schmidt, Phys. Lett. B **606**, 157 (2005).
66. R. Casalbuoni, S. De Curtis, D. Dolce and D. Dominici, Phys. Rev. D **71**, 075015 (2005).
67. H. Georgi, arXiv:hep-ph/0508014.
68. D. T. Son and M. A. Stephanov, Phys. Rev. D **69**, 065020 (2004).
69. K. Hagiwara, R. D. Peccei, D. Zeppenfeld and K. Hikasa, Nucl. Phys. B **282** (1987) 253.
70. H. Abe, T. Kobayashi, N. Maru and K. Yoshioka, Phys. Rev. D **67** (2003) 045019.
71. L. Randall, Y. Shadmi and N. Weiner, JHEP **0301** (2003) 055.
72. R. Casalbuoni, S. De Curtis, D. Dominici and R. Gatto, Phys. Lett. B **155**, 95 (1985).
73. R. Casalbuoni, A. Deandrea, S. De Curtis, D. Dominici, R. Gatto and M. Grazzini, Phys. Rev. D **53**, 5201 (1996).
74. S. Dawson and C. B. Jackson, Phys. Rev. D **76** (2007) 015014.

LITTLE HIGGS M-THEORY

Hsin-Chia Cheng*

Department of Physics, University of California, Davis,
Davis, California 95616, USA
**E-mail: cheng@physics.ucdavis.edu*

In recent years there have been many new theories alternative to supersymmetry proposed to address the question of electroweak symmetry breaking, including theories with extra dimensions and little Higgs theories. Most of these theories at low energies (~TeV) can be well represented by a moose model after integrating out degrees of freedom inaccessible to the LHC. We present a universal moose theory which interpolates among various little Higgs and extra-dimensional models. Different models can be obtained by taking different limits of the parameters of a single theory after integrating out states which become heavy in these limits. It provides a good framework for studying LHC phenomenology of a wide class of electroweak theories.

Keywords: Little Higgs; Moose diagram.

1. Introduction

The origin of electroweak symmetry breaking will be directly probed at the Large Hadron Collider (LHC). In Standard Model (SM) it is achieved by a nonzero vacuum expectation value of a scalar Higgs field. However, the Higgs mass-squared receives large radiative corrections from the SM interactions and is unnatural unless the radiative corrections are cut off at the TeV scale. Naturalness therefore predicts new physics at or beneath the TeV scale. Traditionally, possible new physics may roughly be divided into two categories: weakly coupled (*e.g.,* supersymmetry) and strongly coupled (*e.g.,* technicolor). Typically in weakly coupled theories the Higgs is light and the corrections to the electroweak observables from the SM predictions are small. On the other hand, Higgs is often heavy or even absent in strongly coupled theories and one expects large corrections to the electroweak observables.

The current electroweak precision data agree well with the SM predictions with a light Higgs, so the weakly coupled scenario seems to be preferred. Another way to express the constraints from the electroweak precision measurements is to realize that at low energies, new physics can be integrated out and encoded in higher dimensional operators made of SM fields, and the experimental data put constraints on the sizes of the coefficients of these higher dimensional operators. Some operators which violate symmetries of the Standard Model are very strongly constrained, such as the operators which induce proton decays, CP violations, and flavor-changing effects. However, operators that are generically induced by any new physics without violating SM symmetries are also constrained. Electroweak data tell us that the leading higher dimensional operators need to be suppressed by a scale about 10 TeV.[1] This creates a challenge for any theory that attempts to address the naturalness problem of the electroweak symmetry breaking.

For a long time supersymmetry has been the leading candidate of the new physics near the TeV scale. It has many attractive features such as gauge coupling unification and a natural dark matter candidate. However, The fact that none of the Higgs boson and supersymmetric partners has been discovered has put supersymmetry into some fine-tuned corner.[2–5] In recent years, there are many alternative theories proposed as the possible new physics beyond the Standard Model, including models with extra dimensions and/or new gauge interactions. Given the electroweak constraints, a desired feature of the new theory is to be able to push the strong dynamics (if any) up to $\sim$ 10 TeV and to have a weakly coupled description near the 1 TeV scale. This can be achieved by the little Higgs mechanism: Higgs field is a pseudo-Nambu-Goldstone boson (PNGB) plus collective symmetry breaking. In the next section we will give a brief review of little Higgs theories and present a few representative models. We found that many different models in fact can be described as different limits of a single model. In Section 3 we discuss a unified approach to many different theories at the TeV scale including little Higgs theories and theories with extra dimensions. We argue that this can be a useful effective description at energies accessible to the LHC and a suitable framework for phenomenological studies. Conclusions are drawn in Section 4. This talk was based on the work in collaboration with Jesse Thaler and Lian-Tao Wang.[6] I would like to thank the organizers of the SCGT06 workshop for inviting me to participate in this interesting meeting.

2. Little Higgs Theories

Little Higgs theories are a new type of theories which can stabilize the electroweak scale naturally.[7–9] In little Higgs theories, Higgs fiels is a PNGB of some spontaneously broken global symmetry ($G \to H$). In addition, it has the special property of collective symmetry breaking. The global symmetry is explicitly broken by 2 sets of interactions, with each set preserving a subset of the symmetry.

$$\mathcal{L} = \mathcal{L}_0 + \lambda_1 \mathcal{L}_1 + \lambda_2 \mathcal{L}_2. \tag{1}$$

Higgs field is an exact Nambu-Goldstone boson when either set of couplings vanish, so a Higgs mass can only be generated in the presence of both sets of couplings. In this way, Higgs mass-squared are suppressed by two loops relative to the cutoff, which allows us to push the cutoff to ~ 10 TeV while keeping the Higgs light,

$$\delta m_H^2 \sim \left(\frac{\lambda_1^2}{16\pi^2}\right)\left(\frac{\lambda_2^2}{16\pi^2}\right)\Lambda^2. \tag{2}$$

The one-loop quadratic divergences to the Higgs maas-squared from the SM particles are cancelled by new particles which are partners of the SM top quark, gauge bosons and Higgs (Fig. 1). Unlike SUSY, these new particles have the same spins as the SM particles, and the relationship of the couplings are dictated by the non-linear realized global symmetry.

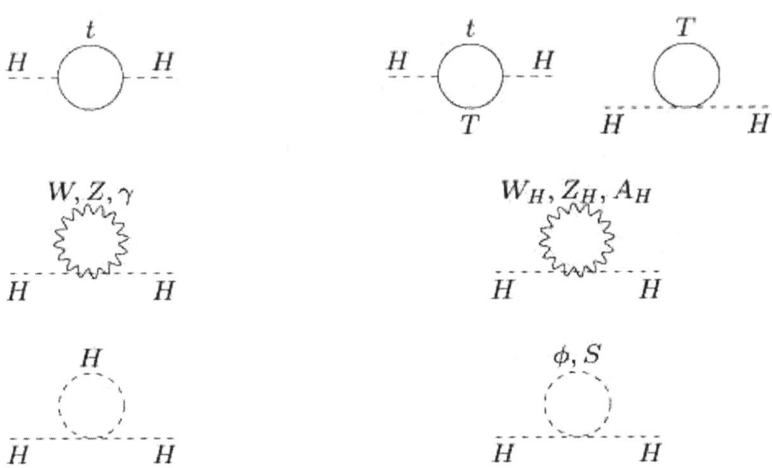

Fig. 1. The cancellation of one-loop quadratic divergences. the SM interactions are shown on the left side and the new particle interactions are on the right side.

There are many different little Higgs models based on various global symmetries, the gauged subgroups, and their breaking patterns. The breaking pattern can be summarized by Fig. 2.

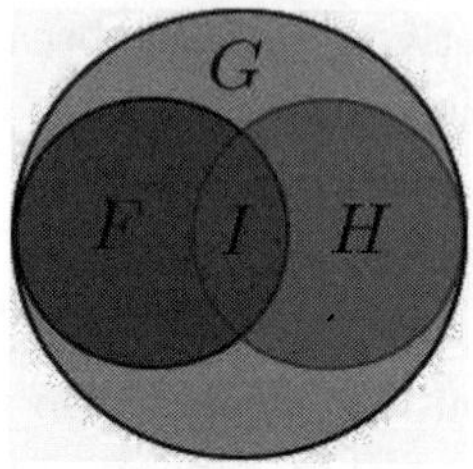

Fig. 2. A global symmetry G is spontaneously broken down to a subgroup H. A subgroup F of G is gauged and it is broken down to the intersection of F and H, $I = F \cap H$, which is identified as the SM electroweak gauge symmetry. The number of uneaten PNGBs is given by the number of generators of $(N(G) - N(H)) - (N(F) - N(I))$. They are identified as the Higgs.

Different types of models can be categorized based on if G and F are simple or product groups. Some representative models are

- Minimal moose:[8] $G/H = SU(3)^2/SU(3)$, $F = [SU(2) \times U(1)]^2$.
- Littlest Higgs:[9] $G/H = SU(5)/SO(5)$, $F = [SU(2) \times U(1)]^2$.
- Simple group little Higgs:[10] $G/H = [SU(3)/SU(2)]^2$, $F = SU(3)[\times U(1)]$.

They all have somewhat different particle contents and spectra, but the generic spectrum for a little Higgs theory looks as follows. At 100 – 200 GeV we have the Standard Model with one or two Higgs doublets. At ~ 1 TeV which corresponds to the global symmetry breaking scale, f, There are new fermions (partners of the top quark), gauge bosons (partners of the SM gauge bosons), and scalars (partners of the Higgs). The cutoff of the effective theory is at $\Lambda \sim 4\pi f \sim 10$ TeV where additional new physics will come in.

As the cutoff is pushed up to the 10 TeV scale, the contributions from the cutoff physics to the electroweak observables are mostly safe. However, the new particles at the 1 TeV scale may still induce large corrections to the electroweak observables. Indeed, electroweak precision data also give strong constraints to the little Higgs models. The constraints are model-dependent but generally require the symmetry breaking scale f to be larger than a few TeV, which would re-introduce the fine-tuning problem.[11–18]

One way to avoid the electroweak precision constraints is to introduce a new symmetry (T-parity) into the little Higgs theories.[19–21] Under the T-parity, all SM fields are even and most of the new particles at the 1 TeV scale (those would induce large corrections to electroweak observables) are odd. The T-odd particles have to appear at least in pair in any interactions. As a result, there cannot be any contribution to the electroweak observables by exchanging the new particles at the tree level. The leading contributions will be loop suppressed and hence safe from the precision constraints. The T-parity can be imposed in many little Higgs models in a similar way that parity is conserved in the QCD chiral Lagrangian.

The introduction of T-parity has dramatic phenomenological consequences in addition to curing the fine-tuning problem of the little Higgs theories. First, the lightest T-odd particle is stable, and can be a good dark matter candidate if it is neutral. Second, the T-odd particles need to be paired produced at the colliders and therefore the search strategies are complete different from the case without T-parity. After produced, the T-odd particles cascade decay down to the lightest T-odd state which escape the detector if it is neutral. Typical collider signals will be jets/leptons + missing energy, similar to SUSY with a conserved R-parity.

3. Little M-theory

As we saw in the previous section, there are many different little Higgs models based on different group structures. This also happens to other alternative candidates for TeV scale physics. For example, for theories with extra dimensions, we can have different fields localized at different places in extra dimesions or have them spread out, and the geometry of the extra dimensions can be flat or curved. The vast amount of possibilities introduces a practical problems: which models deserve more detailed studies and can they well represent the class of models to which they belong?

In fact, many of these new models have similar new particles (new quarks, new spin-1 bosons). It is desirable if there is a unified approach for the LHC studies, in particular, if there is a model that can interpolates among various models and cover most of the features of the non-SUSY theories. We will find that this is possible by observing that most of the non-SUSY models can be well represented or approximated by some moose diagrams at low energies. For example, a gauge field propagating in extra dimensions can be deconstructed into a series of four-dimensional gauge groups broken down to the diagonal group.[22,23] This works for both flat and curved extra dimensions as the warp factor can easily be represented by different Goldstone decay constants of the link fields.[24–26]

For little Higgs theories, some models are also based on moose diagrams, *e.g.*, minimal moose model. On the other hand, there are also models containing simple global or gauged groups such as the littlest Higgs and simple group little Higgs models. At the first sight, they are not described by moose diagrams. However, by using the construction of Callan, Coleman, Wess, and Zumino (CCWZ)[27,28] and related ideas such as hidden local symmetry,[29–32] and AdS/CFT correspondence[33–37] plus deconstruction, they can all be converted into moose models. The procedure is to extend the global group and its symmetry breaking pattern to $(G \times G)/G$, with a subgroup F gauged in the first G group and a subgroup H gauged in the second G group. The gauged subgroups are broken down to the SM electroweak group I. This is described in Fig. 3. One can easily show that the number of uneaten PNGB's are the same as before. The extra degrees of freedom added in this extension decouple if one takes the gauge coupling of the H subgroup to be large. In this limit, one recovers exactly the same low energy effective theory as the original model, but now the new model is described by a moose diagram.

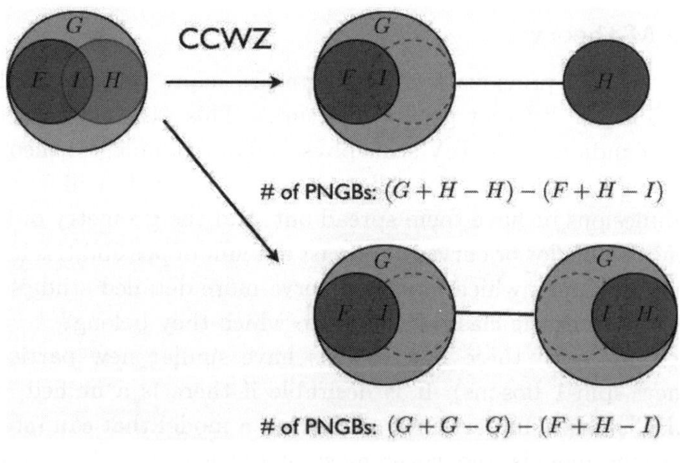

Fig. 3. The graphical representation of converting a non-moose theory into a moose theory.

One can easily apply this procedure to the littlest Higgs model and the simple group little Higgs model. For the littlest Higgs model, we extend the global symmetry to $SU(5) \times SU(5)$, with the $SU(2)^2 \times U(1)$ subgroup gauged in one $SU(5)$ and the $SO(5)$ subgroup gauged in the other $SU(5)$ (Fig. 4). The moosified model also allows a technicolor type UV completion of the littlest Higgs model by considering that the breaking the $SU(5) \times SU(5)$ down to the diagonal $SU(5)$ is due to the condensate of some techni-fermion pairs.[38] Similarly, the simple group little Higgs model

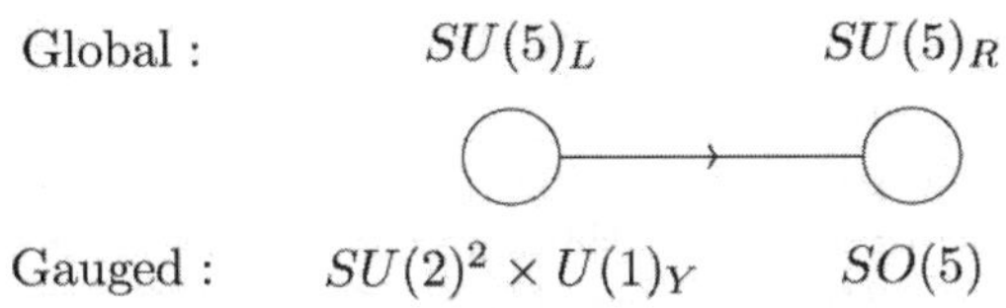

Fig. 4. The moosified littlest Higgs model.

can be moosified in the same way. We obtain a model based on a linear three-site moose diagram. There is an $SU(3)$ global symmetry on each site. The whole $SU(3)$ is gauged at the middle site while at the two boundary sites only $SU(2)[\times U(1)]$ subgroups are gauged (Fig. 5).

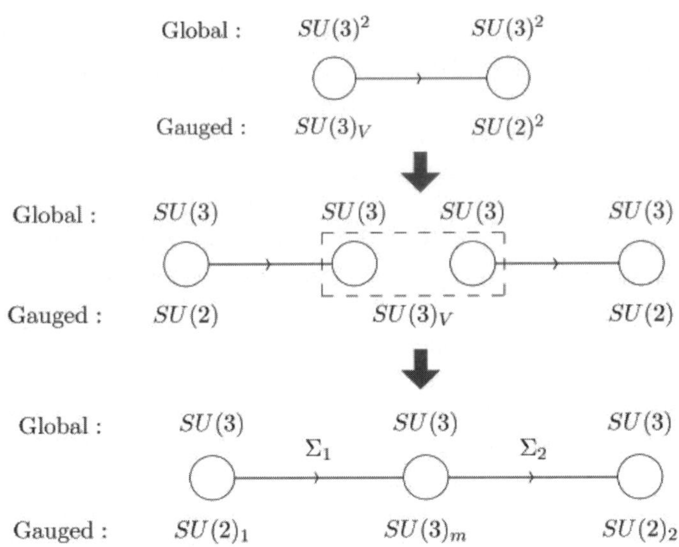

Fig. 5. The moosified simple group little Higgs model.

Actually, the three-site moose model is quite versatile and can describe several different-looking models by taking various limits. The simple group little Higgs limit is obtained by taking the gauge couplings of the two $SU(2)$'s on the boundaries to infinity. On the other hand, if we take the gauge couplings of the two $SU(2)$'s finite but take the gauge coupling of the middle $SU(3)$ to infinity, we can integrate out the middle site and reduce it to the minimal moose little Higgs model. The T-parity can easily be implemented by taking the gauge couplings of the two $SU(2)$'s and the two decay constants associated with the links equal. Furthermore, it can also be a good low energy description of the holographic PNGB Higgs model.[39,40] The holographic PNGB Higgs model is based on the AdS/CFT correspondence. In the AdS description, there is an $SU(3)$ gauge symmetry in the bulk while only an $SU(2)$ is gauged on each of the UV and IR branes. By a simple deconstruction, we can see that the three-site moose gives a good approximation of the first few Kaluza-Klein (KK) modes of this model.

Of course, different models have very different UV behaviors. The three-site model or any other constructions based on the idea discussed in this section can only be the low energy approximations of these various models. However, this construction can be a useful tool for the phenomenological studies in the LHC era. The reason is that the precision electroweak constraints indicate that the scale of the strong dynamics may be out of the reach of the LHC, and in the case of extra dimensions only a few KK modes may be discovered at the LHC at most, so we will only be able to test the low energy part of a fundamental theory. At low energies (accessible to the LHC), most non-SUSY models can be well represented by some simple moose models, as moose diagrams are just a convenient way to categorize spin-1 and spin-0 degrees of freedom. As we found, many models can even be described as various limits of the same moose diagram. Furthermore, this construction also reveals a much larger model space for new physics as it interpolates among various models proposed before. There is a continuous distribution of viable models simply by changing the parameters of the "unified" model.

Obviously, more models can be obtained by more complicated moose diagrams. In practice, in order to have a simple and manageable starting point, we need to make some compromises between versatility and complexity. The desired features of such a unified moose model we consider are as follows. (1) We want a theoretically consistent and calculable theory. That is, the theory should stay perturbative at energies below the cutoff. (2) There should be no large quadratic divergent contributions to the Higgs mass-squared so that the theory is natural. (3) The model should have a custodial $SU(2)$ symmetry to protect the ρ parameter and the Yukawa structure allows for minimal isospin breaking. (4) Minimal flavor violation can be implemented to avoid large flavor-violating effects. (5) To mimic models with extra dimensions we include "KK-partners" of the SM fermions. (6) The model should have a T-symmetric limit because models with or without the T-parity have very different phenomenologies. (7) The models should have a lot of dials which allow enough flexibility to cover a variety of phenomenology.

As we discussed earlier, the simple three-site moose model is already quite versatile and is able to describe many different models in various limits. One problem of that model based on $SU(3)$ groups is that it does not have a custodial $SU(2)$ symmetry to protect the ρ parameter. A simple way to solve this problem is to replace the $SU(3)$ by a group which contains a custodial $SU(2)$. The minimal choice is $SO(5) \simeq Sp(4)$. We therefore construct a three-site model based on $Sp(4)$ groups. The whole $Sp(4)$ is gauged on the middle site while the $SU(2)$ subgroups are gauged at the two boundary sites. An additional $U(1)$ is gauged to account for the hypercharge of the Standard Model. The model is summarized in Fig. 6. Similar to the $SU(3)$ moose, this model can describe several different models by taking various limits, except that these various limits are also based on the $Sp(4)$ symmetry now. This is desirable because they contain a custodial $SU(2)$ symmetry. The detailed discussion of the model can be found in Ref. 6.

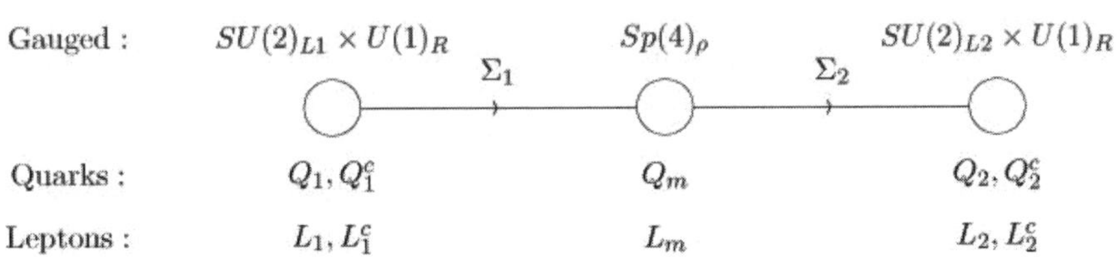

$$Q_i = \begin{pmatrix} q_i \\ 0 \\ 0 \end{pmatrix}, \quad Q_i^c = \begin{pmatrix} q_i^c \\ t_i^c \\ b_i^c \end{pmatrix}, \quad Q_m = \begin{pmatrix} q_m \\ t_m \\ b_m \end{pmatrix}, \quad Q' = \begin{pmatrix} 0 \\ t' \\ b' \end{pmatrix}, \quad Q'^c = \begin{pmatrix} 0 \\ t'^c \\ b'^c \end{pmatrix},$$

$$L_i = \begin{pmatrix} \ell_i \\ 0 \\ 0 \end{pmatrix}, \quad L_i^c = \begin{pmatrix} \ell_i^c \\ \nu_i^c \\ \tau_i^c \end{pmatrix}, \quad L_m = \begin{pmatrix} \ell_m \\ \nu_m \\ \tau_m \end{pmatrix}, \quad L' = \begin{pmatrix} 0 \\ \nu' \\ \tau' \end{pmatrix}, \quad L'^c = \begin{pmatrix} 0 \\ \nu'^c \\ \tau'^c \end{pmatrix}.$$

Fig. 6. A little Higgs M-theory.

The model has a lot of dials and very rich phenomenology. In the vector sector, it contains W', Z' as well as W_R [gauge bosons of right-handed $SU(2)$] and off-diagonal X gauge bosons. By dialing the parameters we can change the spectrum and obtain different phenomenologies. There are also "KK" (excited) quarks and leptons which can involve in the production and decays. There are also extra PNGB scalars. T-parity can easily be implemented to give dark matter and missing energy signals at the colliders. This covers a wide range of possibilities which may appear at the LHC and provides a good framework for collider phenomenology studies. With this experience, if some states like the above ones are discovered at the LHC, it should not be difficult to construct a moose model to describe what we observe, with appropriate couplings and parameters to fit the experimental measurements

4. Conclusions

Little Higgs theories provide a new mechanism to address the naturalness of the weak scale. It allows the strong coupling regime to be pushed up to ~ 10 TeV scale while keeping a weakly coupled description at ~ 1 TeV and below. Thus, it can reconcile the problems with the precision electroweak data. There are many different little Higgs models based on different group structures, which makes the phenomenological studies of little Higgs theories a complicated issue in practice. However, by using the tools we learned from CCWZ, hidden local symmetry, and AdS/CFT correspondence, we can find a universal moose structure which can describe many different little Higgs models together with many other non-SUSY theories (*e.g.*, theories with extra dimensions). Such a construction serves a good framework for collider studies of non-SUSY theories, as it includes a wide range of possible new physics which may appear at the TeV scale. It also reveals a much larger model space than the individual corners proposed before and therefore covers a wider range of possible experimental results which may be discovered.

Acknowledgments

This work is partially supported by the Department of Energy Outstanding Junior Investigator Award.

References

1. R. Barbieri and A. Strumia, Phys. Lett. B **462**, 144 (1999) [arXiv:hep-ph/9905281].
2. L. Giusti, A. Romanino and A. Strumia, Nucl. Phys. B **550**, 3 (1999) [arXiv:hep-ph/9811386].
3. P. Batra, A. Delgado, D. E. Kaplan and T. M. P. Tait, JHEP **0402**, 043 (2004) [arXiv:hep-ph/0309149].
4. R. Harnik, G. D. Kribs, D. T. Larson and H. Murayama, Phys. Rev. D **70**, 015002 (2004) [arXiv:hep-ph/0311349].
5. R. Barbieri, arXiv:hep-ph/0312253.
6. H. C. Cheng, J. Thaler and L. T. Wang, JHEP **0609**, 003 (2006) [arXiv:hep-ph/0607205].
7. N. Arkani-Hamed, A. G. Cohen and H. Georgi, Phys. Lett. B **513**, 232 (2001) [arXiv:hep-ph/0105239].
8. N. Arkani-Hamed, A. G. Cohen, E. Katz, A. E. Nelson, T. Gregoire and J. G. Wacker, JHEP **0208**, 021 (2002) [arXiv:hep-ph/0206020].
9. N. Arkani-Hamed, A. G. Cohen, E. Katz and A. E. Nelson, JHEP **0207**, 034 (2002) [arXiv:hep-ph/0206021].
10. D. E. Kaplan and M. Schmaltz, JHEP **0310**, 039 (2003) [arXiv:hep-ph/0302049].
11. J. L. Hewett, F. J. Petriello and T. G. Rizzo, JHEP **0310**, 062 (2003) [arXiv:hep-ph/0211218].
12. C. Csaki, J. Hubisz, G. D. Kribs, P. Meade and J. Terning, Phys. Rev. D **67**, 115002 (2003) [arXiv:hep-ph/0211124].
13. C. Csaki, J. Hubisz, G. D. Kribs, P. Meade and J. Terning, Phys. Rev. D **68**, 035009 (2003) [arXiv:hep-ph/0303236].
14. T. Han, H. E. Logan, B. McElrath and L. T. Wang, Phys. Rev. D **67**, 095004 (2003) [arXiv:hep-ph/0301040].
15. T. Gregoire, D. R. Smith and J. G. Wacker, Phys. Rev. D **69**, 115008 (2004) [arXiv:hep-ph/0305275].
16. C. Kilic and R. Mahbubani, JHEP **0407**, 013 (2004) [arXiv:hep-ph/0312053].
17. G. Marandella, C. Schappacher and A. Strumia, Phys. Rev. D **72**, 035014 (2005) [arXiv:hep-ph/0502096].
18. Z. Han and W. Skiba, Phys. Rev. D **72**, 035005 (2005) [arXiv:hep-ph/0506206].
19. H. C. Cheng and I. Low, JHEP **0309**, 051 (2003) [arXiv:hep-ph/0308199].
20. H. C. Cheng and I. Low, JHEP **0408**, 061 (2004) [arXiv:hep-ph/0405243].
21. I. Low, JHEP **0410**, 067 (2004) [arXiv:hep-ph/0409025].
22. N. Arkani-Hamed, A. G. Cohen and H. Georgi, Phys. Rev. Lett. **86**, 4757 (2001) [arXiv:hep-th/0104005].
23. C. T. Hill, S. Pokorski and J. Wang, Phys. Rev. D **64**, 105005 (2001) [arXiv:hep-th/0104035].
24. H. C. Cheng, C. T. Hill and J. Wang, Phys. Rev. D **64**, 095003 (2001) [arXiv:hep-ph/0105323].
25. K. Sfetsos, Nucl. Phys. B **612**, 191 (2001) [arXiv:hep-th/0106126].

26. L. Randall, Y. Shadmi and N. Weiner, JHEP **0301**, 055 (2003) [arXiv:hep-th/0208120].
27. S. R. Coleman, J. Wess and B. Zumino, Phys. Rev. **177**, 2239 (1969).
28. C. G. . Callan, S. R. Coleman, J. Wess and B. Zumino, Phys. Rev. **177**, 2247 (1969).
29. M. Bando, T. Kugo, S. Uehara, K. Yamawaki and T. Yanagida, Phys. Rev. Lett. **54**, 1215 (1985).
30. M. Bando, T. Kugo and K. Yamawaki, Nucl. Phys. B **259**, 493 (1985).
31. M. Bando, T. Fujiwara and K. Yamawaki, Prog. Theor. Phys. **79**, 1140 (1988).
32. M. Bando, T. Kugo and K. Yamawaki, Phys. Rept. **164**, 217 (1988).
33. J. M. Maldacena, Adv. Theor. Math. Phys. **2**, 231 (1998) [Int. J. Theor. Phys. **38**, 1113 (1999)] [arXiv:hep-th/9711200].
34. S. S. Gubser, I. R. Klebanov and A. M. Polyakov, Phys. Lett. B **428**, 105 (1998) [arXiv:hep-th/9802109].
35. E. Witten, Adv. Theor. Math. Phys. **2**, 253 (1998) [arXiv:hep-th/9802150].
36. N. Arkani-Hamed, M. Porrati and L. Randall, JHEP **0108**, 017 (2001) [arXiv:hep-th/0012148].
37. R. Rattazzi and A. Zaffaroni, JHEP **0104**, 021 (2001) [arXiv:hep-th/0012248].
38. J. Thaler, JHEP **0507**, 024 (2005) [arXiv:hep-ph/0502175].
39. R. Contino, Y. Nomura and A. Pomarol, Nucl. Phys. B **671**, 148 (2003) [arXiv:hep-ph/0306259].
40. K. Agashe, R. Contino and A. Pomarol, Nucl. Phys. B **719**, 165 (2005) [arXiv:hep-ph/0412089].

Weak Mixing and Rare Decays in the Littlest Higgs Model

William A. Bardeen

Fermilab
Batavia, IL 60510

Little Higgs models have been introduced to resolve the fine-tuning problems associated with the stability of the electroweak scale and the constraints imposed by the precision electroweak analysis of experiments testing the Standard Model of particle physics. Flavor physics provides a sensitive probe of the new physics contained in these models at next-to-leading order.

1. Introduction

The Standard Model has been extraordinarily successful in describing all known phenomena of particle physics with the possible exception of the astrophysical evidence for dark matter. Precision experiments including the results from LEP, the Tevatron and the B-factories have confirmed detailed predictions of the Standard Model and have placed strong constraints on new physics associated with the electroweak scale. Flavor changing processes such as weak mixing and neutral current processes in rare decays are strongly suppressed in the Standard Model and provide a unique window on new physics at scales much above the electroweak scale.

Despite the success of the Standard Model, there remain puzzles concerning the mechanisms that determine parameters of the SM which must be finely adjusted to agree with experiment. These include the theta angle of strong CP violation, the Higgs mass term which determines the electroweak scale and the net vacuum energy density which may be responsible for the dark energy inferred from astrophysical data. In particular, the Higgs mass term gets large radiative corrections from Standard Model processes which would normally be expected to destabilize the electroweak scale. If new physics is introduced to stabilize these radiative corrections, then there can be a tension between the SM fits to precision electroweak data and the new physics contributions.

2. The Littlest Higgs Model

Little Higgs models [1,2] are based on the dynamics of pseudo-Nambu-Goldstone bosons where the Higgs mass term is protected from radiative corrections by the existence of new symmetries that are dynamically broken. In the absence of additional explicit breaking of these new symmetries, the physical Higgs boson remains massless [3]. The Higgs mass will be protected from the quadratically divergent radiative corrections if the explicit symmetry breaking occurs collectively, ie. two or more symmetry breaking terms must be present in order to generate a mass for the physical Higgs boson.

The main ingredients of Little Higgs models are an extended electroweak gauge group, $G_1 \otimes G_2 \rightarrow$SM, which is embedded in a larger global symmetry with dynamical symmetry breaking, and an extended top quark sector needed to accommodate the observed top quark mass. The global symmetries are broken collectively and the physical Higgs boson is a pseudo-Nambu-Goldstone boson. There are typically three important scales in Little Higgs models, Λ ($10 \rightarrow 30$ TeV) – the scale of new dynamics and the effective cutoff, f ($1 \rightarrow 3$ TeV), $f \sim \Lambda/4\pi$ – the scale of dynamical breaking of the global symmetries and the mass scale of the new particles used to cancel the quadratic divergences, and v (175 GeV), $v \sim f/4\pi$ – the electroweak scale and the mass of the SM Higgs boson, the SM gauge bosons and the top quark.

The Littlest Higgs model [4] is based on the Nambu-Goldstone bosons of the coset space of $SU(5)/SO(5)$. The gauge boson dynamics is based on the $[SU(2)\otimes U(1)]_1 \otimes [SU(2)\otimes U(1)]_2$ subgroup of $SU(5)$ which is broken to the electroweak gauge group, $(SU(2)\otimes U(1))_{\rm SM}$, at the scale f. Four of the fourteen Nambu-Goldstone bosons are absorbed by the heavy gauge bosons, $(W_H^{\pm}, Z_H^0, A_H)$. Of the ten remaining Nambu-Goldstone bosons, six form a complex triplet $(\Phi^{++}, \Phi^{+}, \Phi^{0})$ and acquire a mass of order f, and four form a complex doublet (H^{+}, H^{0}) corresponding to the usual Higgs field of the Standard Model. The top quark sector is also modified. A heavy vector-like top quark is needed in addition to the Standard model top quark in order to cancel the quadratic divergences associated with top quark loops.

In the Littlest Higgs model described above, the heavy gauge bosons mix with the light SM gauge bosons at tree level. These mixings can result in significant modifications to the predictions of the Standard Model that are highly constrained by precision electroweak measurements. These constraints will force the symmetry breaking scale, f, to be large [5]. However, a large scale for f can reintroduce issues associated with the fine-tuning of the Higgs mass parameters.

This tension can be relaxed through the introduction of a new symmetry, T-parity [6], which prohibits the mixing between the heavy and light gauge bosons. The heavy gauge bosons and the complex triplet scalars are odd under T-parity while the complex doublet Higgs field and the Standard Model gauge bosons are even. The top quark sector must also be modified to preserve T-parity. Since all interactions are to be invariant under T-parity, the lightest odd parity particle will be a dark matter candidate. The precision electroweak constraints for the Little Higgs model with T-parity have been carefully analyzed [6] and lower scales for the new physics can be tolerated with less tension with the fine-tuning of the Higgs mass parameters.

3. FCNC Processes in the Littlest Higgs Model

Flavor-changing processes are highly suppressed in the Standard Model and, therefore, provide a unique opportunity for the exploration of new physics contributions. Flavor physics has long been focus of Andrzej Buras and his collaborators. I will report on some of their conclusions concerning weak mixing and flavor-changing neutral current processes in the Littlest Higgs model with and without T-parity. The reader is referred to the original papers for detailed predictions and specific results.

Weak mixing amplitudes for $\Delta S = 2$ and $\Delta B = 2$ processes are described by effective weak Hamiltonians involving appropriate four-fermion operators with coefficient functions that are sensitive to the short distance physics. At leading order, the coefficient functions are determined by box diagrams involving both Standard model particles and the new heavy states of the Littlest Higgs model. The calculations can be made in unitary gauge where only physical particles are taken into account. Single box diagrams in unitary gauge are divergent but the divergences all cancel at one loop due to the GIM mechanism.

For the Littlest Higgs model without T-parity, Buras et al. [7,8] find that the new physics contributions to ΔM_s, ΔM_d and ε_K are all positive. This implies a suppression of $|V_{td}|$ and of the angle γ in the unitarity triangle as well as an enhancement of ΔM_s relative to the Standard Model expectations. If the scale f is as small as 1 TeV, the effects amount, at most, to 15-20% corrections and decrease below 5% for $f > 3$–4 TeV as required by the precision electroweak constraints. Contributions of the charged scalars, $\Phi^{\pm}$, turn out to be negligible.

Rare decays are also described by an effective weak Hamiltonian which includes both box diagrams and flavor-changing neutral current processes. In the Littlest Higgs model without T-parity there is considerable mixing between the new heavy states and those of the Standard Model. In unitary gauge, the analytic results can be decomposed into six classes of diagrams. Custodial symmetries are broken at $O(v^2/f^2)$ in contributions at next-to-leading order.

The corrections to the coefficient functions for rare decay processes are at most 15% for a scale $f \sim O(2\text{–}3\ \text{TeV})$. The amplitudes for $K^+ \to \pi^+ \nu\bar{\nu}$, $K_L \to \pi^0 \nu\bar{\nu}$, $B_{s,d} \to \mu^+\mu^-$ and $B \to X_{s,d}\nu\bar{\nu}$ are all calculated [8] but are hard to distinguish from the Standard Model. There are also very small corrections ($\sim 4\%$) to the branching ratio for $B \to X_s\gamma$. The new physics contributions to rare decays are generally suppressed by the large scale required for f due to the lack of custodial symmetry in the Littlest Higgs model without T-parity.

A novel feature of this calculation is the presence of residual ultraviolet divergences that are found in the amplitudes for rare decay processes at order (v^2/f^2). These residual divergences are not an artifact of unitary gauge but are also present in calculations using renormalizable gauges such as Feynman gauge. The divergences represent a true sensitivity to the ultraviolet completion of the Littlest Higgs model.

These divergences are very analogous to the renormalization of G_A in a chiral quark model where chiral loops generate a log divergent contribution to G_A in the nonlinear version of the theory. The linear quark-sigma model is renormalizable and the renormalization of G_A is finite. In this case, the cutoff scale in the log divergent term is replaced by the mass of the scalar partner to the pion and no other divergences survive.

In the calculations of the FCNC processes described above, Buras et al. [8] estimate the divergent terms by using the scale, $\Lambda = 4\pi f$, the ultraviolet cutoff scale of the Littlest Higgs model. Finite corrections could be added reflecting additional "Leutwyler terms" in the effective field theory description of the dynamics.

4. Flavor Physics in the Littlest Higgs Model with T-parity

To avoid the strong precision electroweak constraints, Cheng et al. [6] introduce a T-parity symmetry to the Littlest Higgs model. The presence of the custodial symmetry in this model allows for a lower scale f for the new physics and a more realistic solution to the tension between the Higgs fine-tuning problem and the precision electroweak constraints. Hubisz et al. [9] were the first to study flavor physics and the potential for new physics in weak mixing processes in the Littlest Higgs model with T-parity. These processes were also studied by Buras et al. [10] and extended to FCNC processes and rare decays [11]. The lower scales for the new physics in these models could imply much larger effects for flavor changing processes than were found in the previous analysis of the models without T-parity.

In the Littlest Higgs model with T-parity, the Standard Model particles have even T-parity while the heavy gauge bosons and Higgs triplet scalars are odd under T-parity. To incorporate the T-parity symmetry, the fermion sector becomes more complex [6] with additional fermions having both even and odd T-parity.

Buras et al. [11] consider two scenarios, one with a degenerate mirror formion spectrum corresponding to a minimal flavor violation (MFV) scenario and one with a mirror fermion hierarchy different from CKM corresponding to a non-MFV scenario. A wide range of parameter space is explored for the impact of the new physics contributions on predictions for weak mixing, CP asymmetries and rare decays.

Some general conclusions are possible. There are now regions where the B_s oscillation parameter, ΔM_s, is smaller than the Standard Model value in closer agreement with the recent Tevatron measurements [12]. The $\sin 2\beta$ "problem" can be solved, ie. the difference between the value 0.786 ± 0.052 from tree-level decays and the value 0.675 ± 0.026 from the CP asymmetries [13]. The two processes actually measure somewhat different angles in the Littlest Higgs model. The semileptonic CP asymmetry, A^s_{SL}, can be enhanced by a factor of 10–20 and A^d_{SL} by a factor of 3 above the SM predictions. The CP asymmetry, $S_{\psi\phi}$, can be as high as 0.3 compared to the Standard Model value of 0.04. The rare decays of neutral and charged K-mesons to neutrinos can be enhanced by an order of magnitude over the SM predictions. In fact there are two branches of parameter space, one where the branching ratio for $K^+ \to \pi^+\nu\bar{\nu}$ can be greatly enhanced but the decay $K_L \to \pi^0\nu\bar{\nu}$ is SM-like and one where $K_L \to \pi^0\nu\bar{\nu}$ can be greatly enhanced with only modest enhancements of $K^+ \to \pi^+\nu\bar{\nu}$. I refer to the original papers [10,11] for detailed plots.

5. Conclusions

Little Higgs models offer a novel approach to resolving the tension between the fine-tuning of the Higgs mass parameter and the constraints of precision electroweak experiments. The Littlest Higgs model is a specific realization of these models which can address all aspects of physics from the electroweak scale to scales of order 10–30 TeV. Flavor physics is highly constrained in the Standard Model by the CKM structure of flavor-changing processes.

Weak mixing and rare decay processes provide sensitive probes of the new physics contained in the Littlest Higgs model. Because of the lower new physics scales possible in the Littlest Higgs model with T-parity, large flavor-changing effects are possible in some regions of parameter space providing specific opportunities for probing physics beyond the Standard Model.

A novel aspect of these calculations is the presence of log divergent contributions to the flavor-changing Z-boson vertex. These contributions are gauge invariant and signal an additional sensitivity of flavor physics to the ultraviolet completion of the Littlest Higgs model.

Acknowledgments

The transparencies for this talk were prepared by A. Poschenrieder (Technical University, Munich) based on his lecture at the Ringberg Workshop on Heavy Flavor Physics, October, 2006. I would like to thank Koichi Yamawaki and other members of the Organizing Committee for their hospitality and inviting me to participate in the 2006 Symposium on Strong Coupled Gauge Theories. Fermilab is operated by Fermi Research Alliance, LLC under Contract No. DE-AC02-07CH11359 with the United States Department of Energy.

References

1. N. Arkani-Hamed, A. Cohen, H. Georgi, Phys.Lett. B513:232(2001); N. Arkani-Hamed, A. Cohen, T. Gregoire and J. Wacker, JHEP 0208:020(2002); N. Arkani-Hamed, A. Cohen, E. Katz, A. Nelson, T. Gregoire and J. Wacker, JHEP 0208:021(2002).
2. N. Arkani-Hamid, A. Cohen, E. Katz and A. Nelson, JHEP 0207:034(2002); I. Low, W. Skiba and D. Smith, Phys.Rev. D66:072001(2002).
3. D. Kaplan and H. Georgi, Phys.Lett. B136:183(1984); D. Kaplan, H. Georgi and S. Dimopoulos, Phys.Lett. B136:187(1984).
4. N. Arkani-Hamed, A. Cohen, E. Katz and A. Nelson, JHEP 0207:034(2007).
5. C. Csaki, J. Hubisz, G. Kribs, P. Meade and J. Terning, Phys.Rev. D67:115002(2003); C. Csaki, J. Hubisz, G. Kribs, P. Meade and J. Terning, Phys.Rev. D68:035009(2003); M.C. Chen and S. Dawson, Phys.Rev. D70:015003(2004); Z. Han and W. Skiba, Phys.Rev. D72:035005(2005).
6. H.C. Cheng and I. Low, JHEP 0309:051(2003); H.C. Cheng and I. Low, JHEP 0408:061(2004); I. Low, JHEP 0410:067(2004); J. Hubisz and P. Meade Phys.Rev. D71:035016(2005); J. Hubisz, P. Meade, A. Noble and M. Perelstein, JHEP 0601:135(2006); M. Asano, S. Matsumoto, N. Okada, and Y. Okada, arXiv:hep-ph/0602157.

7. A. Buras, A. Poschenrieder and S. Uhlig, Nucl.Phys. B716:173(2005); S. Choudhury, N. Gaur, A. Goyal and N. Mahajan, Phys.Lett. B601:164(2004); M. Blanke, A. Buras, A. Poschenrieder, C. Tarantino, S. Uhlig and A. Weiler, JHEP 0612:003(2006).
8. A. Buras, A. Poschenrieder, S. Uhlig and W. Bardeen, JHEP 0611:062(2006).
9. J. Hubisz, S. Lee and G. Paz, JHEP 0606:041(2006).
10. M. Blanke A. Buras, A. Poschenrieder, C. Tarantino, S. Uhlig and A. Weiler, JHEP 0612:003(2006).
11. M. Blanke, A. Buras, A. Poschenrieder, S. Recksiegel, S. Tarantino, S. Uhlig and A. Weiler, JHEP 0701:066(2007); M. Blanke, A. Buras, B. Duling, A. Poschenrieder and C. Tarantino, TUM-HEP-657-07 (2007).
12. A. Abulencia et al. [CDF Collaboration], Phys.Rev.Lett. 97:242003(2006); V. Abazov et al. [DØ Collaboration], Phys.Rev.Lett. 97:021802(2006).
13. M. Bona et al. [UTfit Collaboration], JHEP 0603:080(2006); M. Bona et al., Phys.Rev.Lett. 97:151803(2006); J. Charles et al. [CKMfitter Group], Eur.Phys.J. C41:1(2005).

THE EFFECT OF TOPCOLOR ASSISTED TECHNICOLOR, AND OTHER MODELS, ON NEUTRINO OSCILLATION

Minako Honda[1], Yee Kao[2], Naotoshi Okamura[3], Alexey Pronin[2], and Tatsu Takeuchi[2*]

[1] *Physics Department, Ochanomizu Women's University, Tokyo 112-8610, Japan*
[2] *Physics Department, Virginia Tech, Blacksburg VA 24061, USA*
[3] *Yukawa Institute for Theoretical Physics, Kyoto University, Kyoto 606-8502, Japan*

New physics beyond the Standard Model can lead to extra matter effects on neutrino oscillation if the new interactions distinguish among the three flavors of neutrino. In Ref. 1, we argued that a long-baseline neutrino oscillation experiment in which the Fermilab-NUMI beam in its high-energy mode[2] is aimed at the planned Hyper-Kamiokande detector[3] would be capable of constraining the size of those extra matter effects, provided the vacuum value of $\sin^2 2\theta_{23}$ is not too close to one. In this talk, we discuss how such a constraint would translate into limits on the coupling constants and masses of new particles in models such as topcolor assisted technicolor.[4]

1. Introduction

When considering matter effects on neutrino oscillation, it is customary to consider only the W-exchange interaction of the ν_e with the electrons in matter. However, if new interactions beyond the Standard Model (SM) that distinguish among the three generations of neutrinos exist, they can lead to extra matter effects via radiative corrections to the $Z\nu\nu$ vertex which effectively violate neutral current universality, or via the direct exchange of new particles between the neutrinos and matter particles.

For instance, topcolor assisted technicolor[4] treats the third generation differently from the first two and the Z' in this class of models couples more strongly to the ν_τ than to the ν_e or ν_μ. In Extended Technicolor (ETC) Models, such as that of Appelquist, Piai, and Shrock,[5] the neutral technimesons, which mix with the Z, couple to different generation fermions differently, distinguishing among ν_e, ν_μ, and ν_τ. The diagonal ETC gauge bosons also couple to the different generations differently, as well as the

---

*Presenting Author

large variety of leptoquark states in the model. Flavor distinguishing matter effects from diagonal ETC and leptoquarks are induced by ETC gauge boson mixing.

The effective Hamiltonian that governs neutrino oscillations in the presence of neutral-current lepton universality violation, or new physics that couples to the different generations differently, is given by[1]

$$H = \tilde{U}\begin{bmatrix}\lambda_1 & 0 & 0\\ 0 & \lambda_2 & 0\\ 0 & 0 & \lambda_3\end{bmatrix}\tilde{U}^\dagger = U\begin{bmatrix}0 & 0 & 0\\ 0 & \delta m^2_{21} & 0\\ 0 & 0 & \delta m^2_{31}\end{bmatrix}U^\dagger + \begin{bmatrix}a & 0 & 0\\ 0 & 0 & 0\\ 0 & 0 & 0\end{bmatrix} + \begin{bmatrix}b_e & 0 & 0\\ 0 & b_\mu & 0\\ 0 & 0 & b_\tau\end{bmatrix}, \tag{1}$$

where U is the MNS matrix,

$$a = 2EV_{CC}\,, \qquad V_{CC} = \sqrt{2}G_F N_e = N_e\frac{g^2}{4M_W^2}\,, \tag{2}$$

is the usual matter effect due to W-exchange between ν_e and the electrons, and b_e, b_μ, b_τ are the extra matter effects which we assume to be non-equal. We define the parameter ξ as

$$\frac{b_\tau - b_\mu}{a} = \xi\,. \tag{3}$$

Then, the effective Hamiltonian can be rewritten as

$$H = \tilde{U}\begin{bmatrix}\lambda_1 & 0 & 0\\ 0 & \lambda_2 & 0\\ 0 & 0 & \lambda_3\end{bmatrix}\tilde{U}^\dagger = U\begin{bmatrix}0 & 0 & 0\\ 0 & \delta m^2_{21} & 0\\ 0 & 0 & \delta m^2_{31}\end{bmatrix}U^\dagger + a\begin{bmatrix}1 & 0 & 0\\ 0 & -\xi/2 & 0\\ 0 & 0 & +\xi/2\end{bmatrix}, \tag{4}$$

where we have absorbed the extra b-terms in the $(1,1)$ element into a.

The extra ξ-dependent contribution in Eq. (4) can manifest itself when $a > |\delta m^2_{31}|$ (*i.e.* $E \gtrsim 10\,\text{GeV}$ for typical matter densities in the Earth) in the ν_μ and $\bar{\nu}_\mu$ survival probabilities as[1]

$$\begin{aligned} P(\nu_\mu \to \nu_\mu) &\approx 1 - \sin^2\left(2\theta_{23} - \frac{a\xi}{\delta m^2_{31}}\right)\sin^2\frac{\Delta}{2}\,,\\ P(\bar{\nu}_\mu \to \bar{\nu}_\mu) &\approx 1 - \sin^2\left(2\theta_{23} + \frac{a\xi}{\delta m^2_{31}}\right)\sin^2\frac{\Delta}{2}\,, \end{aligned} \tag{5}$$

where

$$\Delta \approx \Delta_{31}c^2_{13} - \Delta_{21}c^2_{12}\,, \qquad \Delta_{ij} = \frac{\delta m^2_{ij}}{2E}L\,, \qquad c_{ij} = \cos\theta_{ij}\,, \tag{6}$$

and the CP violating phase δ has been set to zero. As is evident from these expressions, the small shift due to ξ will be invisible if the value of $\sin^2 2\theta_{23}$ is too close to one. However, if the value of $\sin^2 2\theta_{23}$ is as low as $\sin^2 2\theta_{23} = 0.92$ (the current 90% lower bound), and if ξ is as large as $\xi = 0.025$ (the central value of from CHARM/CHARM II[6]), then the shift in the survival probability at the first oscillation dip can be as large as $\sim 40\%$. If the Fermilab-NUMI beam in its high-energy mode[2] were aimed at a declination angle of 46° toward the planned Hyper-Kamiokande detector[3] in Kamioka, Japan (baseline 9120 km), such a shift would be visible after just one year of data taking, assuming a Mega-ton fiducial volume and 100% efficiency. The absence of any shift after 5 years of data taking would constrain ξ to[1]

$$|\xi| \leq \xi_0 \equiv 0.005 \,, \tag{7}$$

at the 99% confidence level.

In the following, we look at how this potential limit on ξ would translate into constraints on the Z' in topcolor assisted technicolor, and various types of leptoquarks. A more comprehensive analysis will be presented in Ref. 7.

2. Topcolor Assisted Technicolor

Though there are several different versions of topcolor assisted technicolor, we consider here the simplest in which the quarks and leptons transform under the gauge group

$$SU(3)_s \times SU(3)_w \times U(1)_s \times U(1)_w \times SU(2)_L \tag{8}$$

with coupling constants g_{3s}, g_{3w}, g_{1s}, g_{1w}, and g, respectively. It is assumed that $g_{3s} \gg g_{3w}$ and $g_{1s} \gg g_{1w}$. $SU(2)_L$ is the usual weak-isospin gauge group of the SM. The first and second generation fermions are assumed to be charged only under $SU(3)_w \times SU(2)_L \times U(1)_w$, while the third generation fermions are assumed to be charged only under $SU(3)_s \times SU(2)_L \times U(1)_s$. The $U(1)$ charges for both cases are set equal to the SM hypercharge. At scale $\Lambda \sim 1$ TeV, technicolor, which is included in the model to generate the W and Z masses, is assumed to become strong and generate a condensate (of something which is left unspecified) which breaks the two $SU(3)$'s and the two $U(1)$'s to their diagonal subgroups:

$$SU(3)_s \times SU(3)_w \rightarrow SU(3)_c \,, \qquad U(1)_s \times U(1)_w \rightarrow U(1)_Y \,, \tag{9}$$

which we identify with the usual SM color and hypercharge groups. The massless unbroken U(1) gauge boson B_μ and the massive broken U(1) gauge boson Z'_μ are related to the original $U(1)_s \times U(1)_w$ gauge fields $Y_{s\mu}$ and $Y_{w\mu}$ by

$$Z'_\mu = Y_{s\mu} \cos\theta_1 - Y_{w\mu} \sin\theta_1$$

$$B_\mu = Y_{s\mu} \sin\theta_1 + Y_{w\mu} \cos\theta_1 \tag{10}$$

where

$$\tan\theta_1 = \frac{g_{1w}}{g_{1s}} \,. \tag{11}$$

The currents to which the B_μ and Z'_μ couple to are:

$$g_{1s} J^\mu_{1s} Y_{s\mu} + g_{1w} J^\mu_{1w} Y_{w\mu} = g' \left(\cot\theta_1 J^\mu_{1s} - \tan\theta_1 J^\mu_{1w}\right) Z'_\mu + g' \left(J^\mu_{1s} + J^\mu_{1w}\right) B_\mu \,, \tag{12}$$

where

$$\frac{1}{g'^2} = \frac{1}{g^2_{1s}} + \frac{1}{g^2_{1w}} \,. \tag{13}$$

The current $J^\mu_Y = J^\mu_{1s} + J^\mu_{1w}$ is the SM hypercharge current, and g' is the SM hypercharge coupling constant.

The exchange of the Z' leads to the current-current interaction

$$\frac{g'^2}{2M^2_{Z'}} \left(\cot\theta_1 J_{1s} - \tan\theta_1 J_{1w}\right)\left(\cot\theta_1 J_{1s} - \tan\theta_1 J_{1w}\right) \,, \tag{14}$$

the $J_{1s}J_{1s}$ part of which does not contribute to neutrino oscillations on the Earth, while the $J_{1w}J_{1w}$ part is suppressed relative to the $J_{1w}J_{1s}$ part by a factor of $\tan^2\theta_1 \ll 1$. Therefore, we only need to consider the $J_{1s}J_{1w}$ interaction which only affects the propagation of ν_τ. The effective potential felt by ν_τ due to this interaction is[7]

$$V_{\nu_\tau} = \frac{N_n}{8} \frac{g'^2}{M^2_{Z'}} \approx \frac{N_e}{8} \frac{g'^2}{M^2_{Z'}} \,, \tag{15}$$

and the effective ξ is

$$\xi_{TT} = \frac{V_{\nu_\tau} - V_{\nu_\mu}}{V_{CC}} = \frac{1}{2} \frac{(g'/M_{Z'})^2}{(g/M_W)^2} = \frac{1}{2} \tan^2\theta_W \frac{M^2_W}{M^2_{Z'}} = \frac{1}{2} \sin^2\theta_W \frac{M^2_Z}{M^2_{Z'}} \,. \tag{16}$$

The limit $|\xi_{TT}| \leq \xi_0 = 0.005$ then translates into:

$$M_{Z'} \geq M_Z \sqrt{\frac{\sin^2\theta_W}{2\xi_0}} \approx 440\,\mathrm{GeV} \,. \tag{17}$$

Unfortunately, this potential limit from the measurement of ξ is weaker than what is already available from precision electroweak data,[8] and from direct searches for $p\bar{p} \to Z' \to \tau^+\tau^-$ at CDF.[9,10]

3. Generation Non-diagonal Leptoquarks

The interactions of leptoquarks with ordinary matter can be described in a model-independent fashion by an effective low-energy Lagrangian as discussed in Ref. 11. Assuming the fermionic content of the SM, the most general dimensionless $SU(3)_C \times SU(2)_L \times U(1)_Y$ invariant couplings of scalar and vector leptoquarks satisfying baryon and lepton number conservation are given by:

$$\mathcal{L} = \mathcal{L}_{F=2} + \mathcal{L}_{F=0} \, , \tag{18}$$

where

$$\begin{aligned}
\mathcal{L}_{F=2} = & \left[g_{1L}^{ij}(\overline{u_{iL}^c} e_{jL} - \overline{d_{iL}^c} \nu_{jL}) + g_{1R}^{ij}(\overline{u_{iR}^c} e_{jR}) \right] S_1^0 \\
& + \left[g_{2L}^{ij}(\overline{d_{iR}^c} \gamma^\mu e_{jL}) + g_{2R}^{ij}(\overline{d_{iL}^c} \gamma^\mu e_{jR}) \right] V_{2\mu}^+ \\
& + \left[g_{2L}^{ij}(\overline{d_{iR}^c} \gamma^\mu \nu_{jL}) + g_{2R}^{ij}(\overline{u_{iL}^c} \gamma^\mu e_{jR}) \right] V_{2\mu}^- \\
& + \tilde{g}_{2L}^{ij}\left[(\overline{u_{iR}^c} \gamma^\mu e_{jL}) \tilde{V}_{2\mu}^+ + (\overline{u_{iR}^c} \gamma^\mu \nu_{jL}) \tilde{V}_{2\mu}^- \right] \\
& + g_{3L}^{ij}\left[-\sqrt{2}(\overline{d_{iL}^c} e_{jL}) S_3^+ - (\overline{u_{iL}^c} e_{jL} + \overline{d_{iL}^c} \nu_{jL}) S_3^0 + \sqrt{2}(\overline{u_{iL}^c} \nu_{jL}) S_3^- \right] \, ,
\end{aligned} \tag{19}$$

$$\begin{aligned}
\mathcal{L}_{F=0} = & \left[h_{2L}^{ij}(\overline{u_{iR}} e_{jL}) + h_{2R}^{ij}(\overline{u_{iL}} e_{jR}) \right] S_2^+ \\
& + \left[h_{2L}^{ij}(\overline{u_{iR}} \nu_{jL}) - h_{2R}^{ij}(\overline{d_{iL}} e_{jR}) \right] S_2^- \\
& + \tilde{h}_{2L}^{ij}\left[(\overline{d_{iR}} e_{jL}) \tilde{S}_2^+ + (\overline{d_{iR}} \nu_{jL}) \tilde{S}_2^- \right] \\
& + \left[h_{1L}^{ij}(\overline{u_{iL}} \gamma^\mu \nu_{jL} + \overline{d_{iL}} \gamma^\mu e_{jL}) + h_{1R}^{ij}(\overline{d_{iR}} \gamma^\mu e_{jR}) \right] V_{1\mu}^0 \\
& + h_{3L}^{ij}\left[\sqrt{2}(\overline{u_{iL}} \gamma^\mu e_{jL}) V_{3\mu}^+ + (\overline{u_{iL}} \gamma^\mu \nu_{jL} - \overline{d_{iL}} \gamma^\mu e_{jL}) V_{3\mu}^0 + \sqrt{2}(\overline{d_{iL}} \gamma^\mu \nu_{jL}) V_{3\mu}^- \right]
\end{aligned} \tag{20}$$

Here, the scalar and vector leptoquark fields are denoted by S and V, their subscripts indicating the dimension of their $SU(2)_L$ representation, and the superscripts indicating the sign of the weak-isospin of each component. We allow for generation non-diagonal couplings with the indices i and j indicating the quark and lepton generation numbers, respectively. The subscript L or R on the coupling constants indicate the chirality of the lepton involved in the interaction. For simplicity, color indices have been suppressed. The leptoquarks $S_1, \vec{S}_3, V_2, \tilde{V}_2$ carry fermion number $F = 3B + L = -2$, while the leptoquarks $S_2, \tilde{S}_2, V_1, \vec{V}_3$ have $F = 0$. The interactions that affect neutrino oscillation are those with $(ij) = (12)$ or (13).

Table 1. Constraints on the leptoquark couplings with all the leptoquark masses set to 300 GeV. To obtain the bounds for a different leptoquark mass M_{LQ}, simply rescale these numbers with the factor $(M_{LQ}/300\ \text{GeV})^2$.

LQ	C_{LQ}	$\delta\lambda^2_{LQ}$	upper bound from $\lvert\xi\rvert \le \xi_0$
S_1	$+3$	$\lvert g^{12}_{1L}\rvert^2 - \lvert g^{13}_{1L}\rvert^2$	0.01
$\vec{S}_3$	$+9$	$\lvert g^{12}_{3L}\rvert^2 - \lvert g^{13}_{3L}\rvert^2$	0.003
S_2	-3	$\lvert h^{12}_{2L}\rvert^2 - \lvert h^{13}_{2L}\rvert^2$	0.01
$\tilde{S}_2$	-3	$\lvert \tilde{h}^{12}_{2L}\rvert^2 - \lvert \tilde{h}^{13}_{2L}\rvert^2$	0.01
V_2	$+6$	$\lvert g^{12}_{2L}\rvert^2 - \lvert g^{13}_{2L}\rvert^2$	0.005
$\tilde{V}_2$	$+6$	$\lvert \tilde{g}^{12}_{2L}\rvert^2 - \lvert \tilde{g}^{13}_{2L}\rvert^2$	0.005
V_1	-6	$\lvert h^{12}_{1L}\rvert^2 - \lvert h^{13}_{1L}\rvert^2$	0.005
$\vec{V}_3$	-18	$\lvert h^{12}_{3L}\rvert^2 - \lvert h^{13}_{3L}\rvert^2$	0.002

It is straightforward to calculate the effective potentials due to the exchange of these leptoquarks, as well as the effective values of ξ.[7] Assuming a common mass for leptoquarks in the same $SU(2)_L$ weak-isospin multiplet, the effective ξ due to the exchange of any particular type of leptoquark can be written in the form

$$\xi_{LQ} = C_{LQ}\,\frac{\delta\lambda^2_{LQ}/M^2_{LQ}}{g^2/M^2_W} = \frac{C_{LQ}}{4\sqrt{2}G_F}\left(\frac{\delta\lambda^2_{LQ}}{M^2_{LQ}}\right). \tag{21}$$

Here, C_{LQ} is a constant prefactor, and $\delta\lambda^2_{LQ}$ represents

$$\delta\lambda^2_{LQ} = \lvert\lambda^{12}_{LQ}\rvert^2 - \lvert\lambda^{13}_{LQ}\rvert^2\,, \tag{22}$$

where λ^{ij}_{LQ} is a generic coupling constant. The values of C_{LQ} and $\delta\lambda^2_{LQ}$ for the different types of leptoquark are listed in Table 1. The constraint $\lvert\xi_{LQ}\rvert \le \xi_0$ translates into:

$$M_{LQ} \ge \sqrt{\frac{\lvert C_{LQ}\rvert\lvert\delta\lambda^2_{LQ}\rvert}{4\sqrt{2}G_F\,\xi_0}} \approx \sqrt{\lvert C_{LQ}\rvert\lvert\delta\lambda^2_{LQ}\rvert}\times(1700\,\text{GeV})\,. \tag{23}$$

Alternatively, one can fix the leptoquark mass and obtain upper bounds on the leptoquark couplings:

$$\lvert\delta\lambda^2_{LQ}\rvert \le \left(\frac{4\sqrt{2}G_F\,\xi_0}{C_{LQ}}\right)M^2_{LQ} \approx \frac{0.03}{C_{LQ}}\left(\frac{M_{LQ}}{300\,\text{GeV}}\right)^2. \tag{24}$$

The values when $M_{LQ} = 300\,\text{GeV}$ are listed in the rightmost column of Table 1. Thought it is often stated that generation non-diagonal couplings of leptoquarks are strongly constrained by the absence of flavor changing neutral currents, it is only the products of the $(ij) = (12)$ and (13) couplings with other couplings that are constrained.[12] The limits on the individual couplings can be improved considerably. The current leptoquark mass bounds from direct searches at the Tevatron, LEP, and HERA are in the 200∼300 GeV range assuming generation diagonal couplings set equal to $\sqrt{4\pi\alpha}$. At the LHC, leptoquarks, if they exist, can be expected to be pair-produced copiously through gluon-gluon fusion. The expected sensitivity is up to about 1.5 TeV.[13] Depending on the value assumed for $\delta\lambda_{LQ}^2$, the bound from Eq. (23) can be competitive.

Acknowledgments

We would like to thank Drs. Andrew Akeroyd, Mayumi Aoki, Masafumi Kurachi, Robert Shrock, and Hiroaki Sugiyama for helpful discussions. This research was supported in part by the U.S. Department of Energy, grant DE–FG05–92ER40709, Task A (Kao, Pronin, and Takeuchi).

References

1. M. Honda, N. Okamura, and T. Takeuchi, arXiv:hep-ph/0603268.
2. NUMI Technical Design Handbook, available at http://www-numi.fnal.gov/numwork/tdh/tdh_index.html
3. Y. Itow *et al.*, arXiv:hep-ex/0106019; updated version available at http://neutrino.kek.jp/jhfnu/.
4. C. T. Hill, Phys. Lett. B **345**, 483 (1995); G. Buchalla, G. Burdman, C. T. Hill, and D. Kominis, Phys. Rev. D **53**, 5185 (1996).
5. T. Appelquist, M. Piai and R. Shrock, Phys. Rev. D **69**, 015002 (2004).
6. J. Dorenbosch *et al.* [CHARM Collaboration], Phys. Lett. B **180**, 303 (1986); P. Vilain *et al.* [CHARM-II Collaboration], Phys. Lett. B **320**, 203 (1994).
7. M. Honda, Y. Kao, N. Okamura, A. Pronin, and T. Takeuchi, in preparation.
8. R. S. Chivukula and J. Terning, Phys. Lett. B **385**, 209 (1996); W. Loinaz and T. Takeuchi, Phys. Rev. D **60**, 015005 (1999).
9. D. Acosta *et al.* [CDF Collaboration], Phys. Rev. Lett. **95**, 131801 (2005).
10. W. M. Yao *et al.* [Particle Data Group], J. Phys. G **33** (2006) 1.
11. W. Buchmüller, R. Rückl and D. Wyler, Phys. Lett. B **191**, 442 (1987); M. Tanabashi, p.412 of Ref. 10.
12. S. Davidson, D. C. Bailey and B. A. Campbell, Z. Phys. C **61**, 613 (1994); M. Leurer, Phys. Rev. D **49**, 333 (1994); M. Leurer, Phys. Rev. D **50**, 536 (1994); E. Gabrielli, Phys. Rev. D **62**, 055009 (2000).
13. ATLAS detector and physics performance. Technical design report. Vol. 2, CERN-LHCC-99-15, ATLAS-TDR-15; CMS physics : Technical Design Report v.2 : Physics performance, CERN-LHCC-2006-021, CMS-TDR-008-2; V. A. Mitsou, N. C. Benekos, I. Panagoulias and T. D. Papadopoulou, Czech. J. Phys. **55**, B659 (2005).

Study of the Change from Walking to Non-Walking Behavior in a Vectorial Gauge Theory as a Function of N_f

M. KURACHI* and R. SHROCK†

C.N. Yang Institute for Theoretical Physics,
State University of New York,
Stony Brook, New York 11794, USA
**speaker;E-mail: masafumi.kurachi@sunysb.edu*
†E-mail: robert.shrock@sunysb.edu

Based on recent works,[1,2] we present the results of calculations for several physical quantities (meson masses, the S parameter, etc.) in a vectorial gauge theory, as a function of the number of fermions, N_f. Solutions of the Schwinger-Dyson and the Bethe-Salpeter equations with the improved ladder approximation are used for the calculations. We focus on how the values of physical quantities change as one moves from the QCD-like (non-walking) to walking regimes.

1. Introduction

We consider a $(3+1)$-dimensional vectorial gauge theory (at zero temperature and chemical potential) with the gauge group $\mathrm{SU}(N_c)$ and N_f massless fermions transforming according to the fundamental representation of this group. For $N_c = 3$, if one took $N_f = 2$, this would be an approximation to actual QCD. We restrict here to the range $N_f < (11/2)N_c$ for which the theory is asymptotically free. An analysis using the two-loop beta function and Schwinger-Dyson equation leads to the inference that for N_f in this range, the theory includes two phases: (i) for $0 \le N_f \le N_{f,cr}$ a phase with confinement and spontaneous chiral symmetry breaking (SχSB); and (ii) for $N_{f,cr} \le N_f \le (11/2)N_c$ a phase with no spontaneous chiral symmetry breaking (plausibly a non-Abelian Coulomb phase). We shall refer to $N_{f,cr}$, the critical value of N_f, as the boundary between these two phases.

For N_f slightly less than $N_{f,cr}$, the theory exhibits an approximate infrared (IR) fixed point with resultant walking behavior. That is, as the energy scale μ decreases from large values, $\alpha = g^2/(4\pi)$ (g being the $\mathrm{SU}(N_c)$ gauge coupling) grows to be $\mathrm{O}(1)$ at a scale Λ, but increases only rather slowly as μ decreases below this scale, so that there is an extended interval in energy below Λ where α is large, but slowly varying. Associated with this slowly running behavior, the resultant dynamically generated fermion mass, Σ, is much smaller than Λ. In addition to its intrinsic field-theoretic interest, this walking behavior has played an important role in theories of dynamical electroweak symmetry breaking.[3–9] As N_f approaches $N_{f,cr}$ from below, quantities with dimensions of mass vanish continuously; i.e., the chiral phase transition separating phases (i) and (ii) is continuous. Recently, meson masses and other quantities such as the generalized pseudoscalar decay constant f_P and the S parameter[10] were calculated in the walking limit of an SU(N_c) gauge theory.[11,12]

It is of interest to investigate how meson masses and other quantities change as one decreases N_f below $N_{f,cr}$, moving away from the boundary, as a function of N_f, between phases (i) and (ii), deeper into the confined phase. For this purpose, in Ref. 1, as in Refs. 11 and 12, we use the Schwinger-Dyson (SD) equation to compute the dynamical fermion mass Σ (generalized constituent quark mass) and then insert this into the Bethe-Salpeter (BS) equation to obtain the masses of the low-lying mesons and other quantities. We restrict to an interval of N_f values for which the theory has an infrared fixed point. For definiteness, we take $N_c = 3$; however, N_c enters only indirectly, via the dependence of the value of the infrared fixed point α_* on N_c. Hence, our findings may also be applied in a straightforward way, with appropriate changes in the value of α_*, to an SU(N_c) gauge theory with a different value of N_c.

In order to study meson masses and other quantities as one moves away from the boundary between phases (i) and (ii), it is first necessary to know as accurately as possible where this boundary lies, as a function of N_f, i.e., to know the value of $N_{f,cr}$. For sufficiently large N_f, the beta function (calculated to the maximal scheme-independent order, namely two loops) has an IR fixed point at

$$\alpha_* = \frac{-4\pi(11N_c - 2N_f)}{34N_c^2 - 13N_cN_f + 3N_c^{-1}N_f} . \tag{1}$$

Requiring that α_* be sufficiently large as to yield spontaneous symmetry breaking in the context of an approximate solution to the SD equation for a fermion yields the condition that $N_f < N_{f,cr}$, where[9]

$$N_{f,cr} = \frac{2N_c(50N_c^2 - 33)}{5(5N_c^2 - 3)} . \tag{2}$$

For $N_c = 3$ this gives $N_{f,cr} \simeq 11.9$. These estimates are only rough, in view

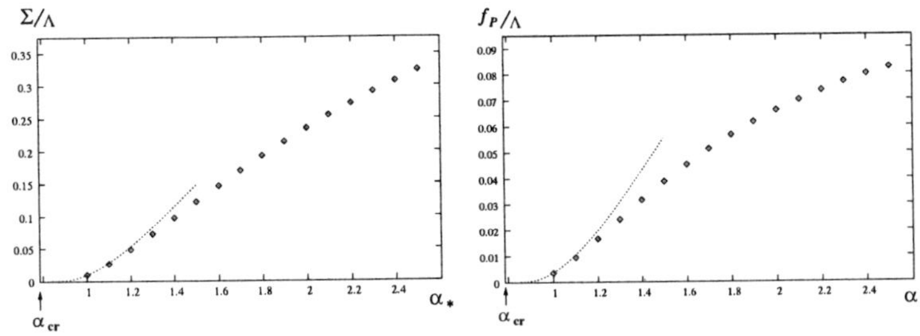

Fig. 1. Numerical solutions for Σ (left panel) and f_P (right panel), for several values of α_* (indicated by $\diamond$). For comparison, we show eq. (3) with $c = 4.0$ (left panel), $c = 1.5$ (right panel) from a fit to the results for $0.89 \leq \alpha_* \leq 1.0$.

of the strongly coupled nature of the physics. Effects of higher-order gluon exchanges and instantons have been studied in Ref. 13.

In our analysis, what we actually vary is the value of the approximate IR fixed point α_*, which depends parametrically on N_f. Thus, although our SD and BS equations are semi-perturbative, the analysis is self-consistent in the sense that our α_{cr} really is the value at which, in our approximation, one passes from the confinement phase with SχSB to the chirally symmetric phase, and our values of α_* do span the interval over which there is a crossover from walking to QCD-like (i.e., non-walking) behavior.

2. Schwinger-Dyson Equation

We first use the Schwinger-Dyson equation for the fermion propagator to calculate the dynamically generated mass Σ of this fermion. In Fig. 1 (left panel) we show the solution for the dynamical fermion mass Σ as a function of α_*. A fit to the numerical solution in the walking region $0.89 \leq \alpha_* \leq 1.0$[11] found agreement with the functional form

$$\Sigma = c\Lambda \exp\left[-\pi\left(\frac{\alpha_*}{\alpha_{cr}} - 1\right)^{-1/2}\right], \tag{3}$$

with $c = 4.0$ (see also Refs. 6 and 9). Our calculations for larger α_* show the expected shift away from walking behavior. This shift is evident in Fig. 1 for α_* larger than about 1.2. Our calculation of Σ, shown in Fig. 1, shows that Σ/Λ increases substantially, by about a factor of 30, from a value of about 0.01 at $\alpha_* = 1.0$ to 0.32 at $\alpha_* = 2.5$, much closer to the value of O(1) for this ratio in QCD.

Another quantity of interest is the pseudoscalar decay constant f_P, the N_f-flavor generalization of the pion decay constant, f_π. In Fig. 1 (right

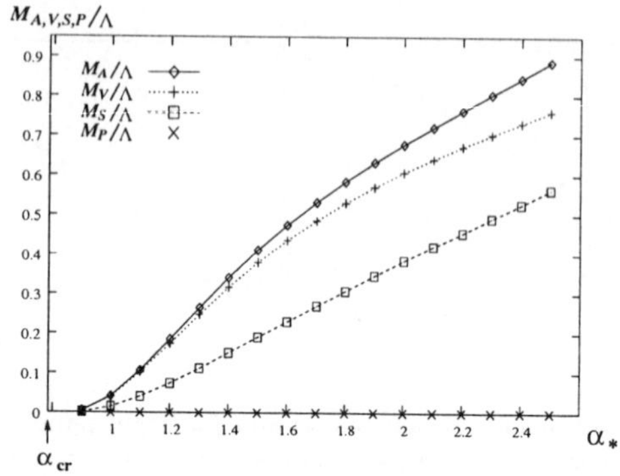

Fig. 2. Values of meson masses divided by Λ calculated from the Schwinger-Dyson and Bethe-Salpeter equations in the range $0.9 \leq \alpha_* \leq 2.5$.

panel) we show our results for f_P calculated by substituting our solution for $\Sigma(k^2)$ into the Pagels-Stokar formula. In the walking limit, f_P has been shown to satisfy a relation similar to eq. (3), i.e., it is exponentially smaller than the scale Λ. We display, as the dotted curve, the fit from Ref. 11 for the walking interval $0.89 \leq \alpha_* \leq 1.0$, given by eq. (3) with $c = 1.5$. Our results show the change from this walking type of behavior as α_* increases above this range; specifically, as α_* increases from 1.0 to 2.5, f_P/Λ increases substantially, from about 3×10^{-3} to about 0.08. This is similar to the factor by which we found that Σ/Λ increased as α_* increased through this interval.

3. Calculation of Meson Masses

We next present the results of the numerical calculations for meson masses, obtained by solving the homogeneous BS equation.[1] (As in Ref. 11, we have checked and confirmed that the flavor-adjoint pseudoscalar meson mass is zero to within the numerical accuracy of our calculation.) In Fig. 2, we show the values of meson masses divided by Λ calculated from the SD and BS equations. In Fig. 3 we plot the values of $M_{A,V,S}/f_P$ (left panel) and $M_{A,S}/M_V$ (right panel). Here, the subscripts V, A and S represent vector, axial-vector and scalar, respectively.

Our calculations yield a number of interesting results. We summarize these for the changes in these meson masses as α_* increases from 0.9 to 2.5. The ratios of the meson masses divided by Λ increase dramatically, by factors of order 10^2, approaching values of order unity at $\alpha_* = 2.5$. This amounts to the removal of the exponential suppression of these masses

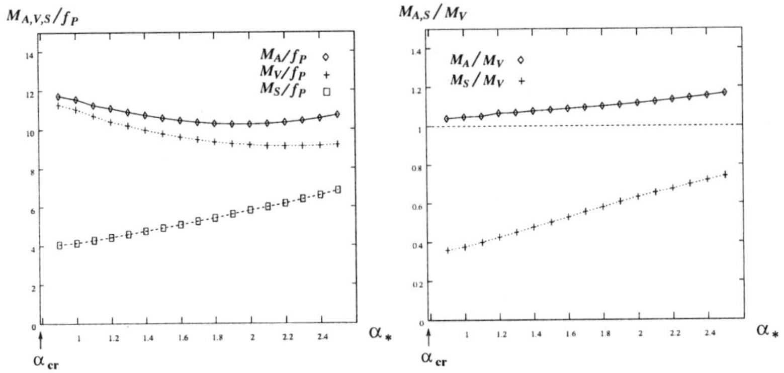

Fig. 3. Values of meson masses divided by f_P (left panel) and $M_{A,S}/M_V$ (right panel) calculated from the Schwinger-Dyson and Bethe-Salpeter equations.

which had described the walking limit near $N_{f,cr}$, as one moves away from this limit into the interior of the confined phase. For example, M_S/f_P increases monotonically from about 4 to 7, thereby approaching to within about 35 % of the value 10.7 in QCD for M_{a_0}/f_π, while M_V/f_P decreases from about 11 to 9, rather close to the value 8.5 for M_ρ/f_π and M_ω/f_π in QCD. The ratios M_A/M_V and M_S/M_V, which were found in Ref. 11 to have values close to 1.0 and 0.36, respectively, in the walking limit, both increase in the interval of α_* that we study, reaching about 1.2 and 0.74, respectively, at $\alpha_* = 2.5$. For comparison, these ratios are approximately 1.6 and 1.3 in QCD.

4. Calculation of the S parameter

In this section we present the results of our calculations of $\hat{S}$[2] in the crossover region of the theory between the walking limit at $\alpha_* \searrow \alpha_{cr}$ and larger values of α_* that move toward the QCD-like regime. Here, $\hat{S}$ represents the contribution to the S parameter from one fermion isodoublet. (Studies of S in the walking limit include.[12,14]) We calculate $\hat{S}$ via the relation $\hat{S} = 4\pi(\Pi'_{VV}(0) - \Pi'_{AA}(0))$, where $\Pi_{VV}(q^2)$ and $\Pi_{AA}(q^2)$ are the vector and the axial-vector current-current correlation functions. These correlators are computed by solving the SD equation and the inhomogeneous BS equation.[2]

In Fig. 4, as a function of α_*, we plot the value of $\hat{S}_n$, the value of $\hat{S}$ normalized by its value at $\alpha_* = 1.8$, namely, 0.47. This figure shows that $\hat{S}_n$, and hence also $\hat{S}$, decreases by about 40 % as α_* is reduced from 1.8

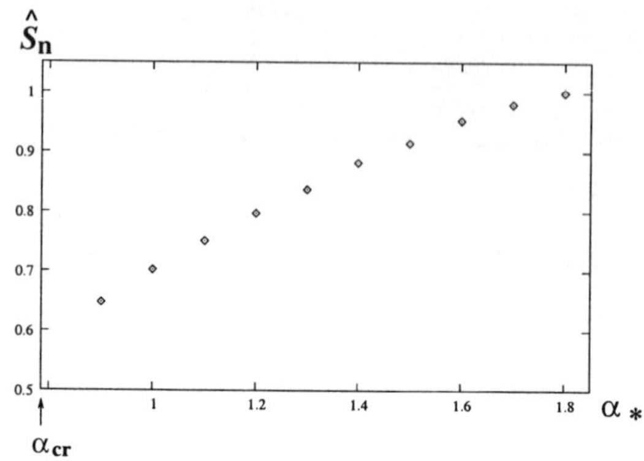

Fig. 4. Plot of $\hat{S}_n$ for several values of α_* in the range of $0.9 \leq \alpha_* \leq 1.8$. As indicated by the subscript n, the values are normalized by the value of $\hat{S}$ at $\alpha_* = 1.8$, i.e., 0.47.

to 0.9, or equivalently as N_f is increased from 10.3 to 11.6. Reinserting the factor of the number of fermion isodoublets, $N_D = N_f/2$, to get S itself, we obtain a decrease by about 30 % in S, since N_D only increases by about 10 % over this range. Thus, our calculation shows that for this range of values, S decreases significantly as one moves from the QCD-like to the walking regimes. We recall that the (improved) ladder approximation to the BS equation can overestimate S in QCD by as much as 30 %.[15] Hence, in addition to the demonstrated decreasing trend of $\hat{S}$ and S as α_* decreases from 1.8 to 0.9, one may, separately, comment that the absolute magnitude of these quantities could be about 30 % smaller than the values yielded by our ladder approximation. Our results thus strengthen the evidence for the reduction of $\hat{S}$ in a walking, as opposed to QCD-like, gauge theory, and are relevant to assessing the impact of the S-parameter constraint on technicolor theories.

5. Summary

In summary, using numerical solutions of the Schwinger-Dyson and Bethe-Salpeter equations, we have calculated several physical quantities, including f_P, meson masses, and the S parameter, as a function of the approximate infrared fixed point, α_*, or equivalently, the number of massless fermions, N_f, in a vectorial, confining SU(N) gauge theory. Our results show the crossover between walking and non-walking behavior in a gauge theory, and demonstrate that $\hat{S}$ and also S decrease significantly as α_* decreases in this range.

Acknowledgments

M.K. thanks Profs. M. Harada and K. Yamawaki for the collaborations on the related Refs. 11 and 12. This research was partially supported by the grant NSF-PHY-03-54776.

References

1. M. Kurachi and R. Shrock, JHEP **0612**, 034 (2006).
2. M. Kurachi and R. Shrock, Phys. Rev. D **74**, 056003 (2006).
3. B. Holdom, Phys. Lett. B **150**, 301 (1985).
4. K. Yamawaki, M. Bando, and K. Matumoto, Phys. Rev. Lett. **56**, 1335 (1986).
5. T. Appelquist, D. Karabali, and L. C. R. Wijewardhana, Phys. Rev. Lett. **57**, 957 (1986); T. Appelquist and L. C. R. Wijewardhana, Phys. Rev. D **35**, 774 (1987); Phys. Rev. D **36**, 568 (1987).
6. T. Appelquist, J. Terning, and L. C. R. Wijewardhana, Phys. Rev. Lett. **77**, 1214 (1996).
7. V. Miransky and K. Yamawaki, Phys. Rev. D **55**, 5051 (1997); *ibid.* **56**, E 3768 (1997).
8. R. S. Chivukula, Phys. Rev. D **55**, 5238 (1997).
9. T. Appelquist, A. Ratnaweera, J. Terning, and L. C. R. Wijewardhana, Phys. Rev. D **58**, 105017 (1998).
10. M. E. Peskin and T. Takeuchi, Phys. Rev. Lett. **65**, 964 (1990); Phys. Rev. D **46**, 381 (1992).
11. M. Harada, M. Kurachi, and K. Yamawaki, Phys. Rev. D **68**, 076001 (2003).
12. M. Harada, M. Kurachi, and K. Yamawaki, Prog. Theor. Phys. **115**, 765 (2006).
13. T. Appelquist, K. Lane, and U. Mahanta, Phys. Rev. Lett. **61**, 1553 (1988); T. Appelquist et al., Phys. Rev. D **43**, 646 (1991); T. Appelquist and S. Selipsky, Phys. Lett. B **400**, 364 (1997).
14. T. Appelquist and G. Triantaphyllou, Phys. Lett. B **278**, 345 (1992); R. Sundrum and S. Hsu, Nucl. Phys. B **391**, 127 (1993); T. Appelquist and F. Sannino, Phys. Rev. D **59**, 067702 (1999); S. Ignjatovic, L. C. R. Wijewardhana, and T. Takeuchi, Phys. Rev. D **61**, 056006 (2000).
15. M. Harada, M. Kurachi, and K. Yamawaki, Phys. Rev. D **70**, 033009 (2000).

DARK MATTER FROM TECHNICOLOR THEORIES

Christoforos Kouvaris
Niels Bohr Institute

I will discuss the possibility of having a component of dark matter from new strong coupled technicolor theories that have not been excluded by the Electroweak Precision Measurements. New technicolor models can provide naturally neutral objects that can be stable and therefore candidates for dark matter. I will also address the issue of the detection of these particles on earth based experiments.

Techni-Orientifold beyond the Standard model

Deog Ki Hong

Department of Physics, Pusan National University,
Busan 609-735, Korea
E-mail: dkhong@pusan.ac.kr

We study a technicolor model with techniquarks in second-rank tensor representaion for physics beyond the standard model. The model has nice features such as light composite Higgs and small oblique corrections, while not suffering from the falavor-changing neutral-current problem. We then construct the holographic model of technicolor and estimate the Peskin-Takeuchi S parameter for various technicolor theories.

Keywords: Technicolor; Oblique corrections; AdS/CFT.

1. Introduction

At the Large Hadron Collider (LHC) we certainly will find out the origin of electroweak symmetry breaking, which is the only remaining piece of the standard model not yet verified experimentally. Uncovering the origin of the electroweak symmetry is quite important, since it will not only provide the missing ingredient of the standard model but also unfold the physics beyond the standard model, for which we already have several hints from the experiments like the neutrino oscillation or the dark matter.

There are several proposals on how the electroweak symmetry is broken. The oldest one is to break the symmetry dynamically by a new interaction, called technicolor (TC),[1] which becomes strong at the electroweak scale and breaks the electroweak symmetry by forming a condensation of techniquarks. Dynamical breaking of electroweak symmetry is a very interesting idea and is the only mechanism of symmetry breaking observed in nature. However, not much progress has been made in this direction, because the very nature of strong interaction makes it difficult to calculate reliably the technicolor corrections to the electroweak processes, which are precisely measured. Recently, however, the development in string theory allows us to calculate such quantities reliably to be compared with the electroweak precision data.[2]

2. Techni-Orientifold

When the chiral symmetry of quantum chromodynamics (QCD) is dynamically broken by the condensate of quarks,

$$|\langle \bar{q}_L q_R + \text{h.c.}\rangle| = v_{\text{QCD}}^3\,, \tag{1}$$

the electroweak symmetry is also broken, since the left-handed and the right-handed quarks transform differently under the electroweak symmetry. However, because the QCD scale is much smaller than the electroweak scale, $v_{\text{QCD}} \ll v_{\text{EW}}$, the quark condensate is not big enough to account for the mass of electroweak gauge bosons.

It was noted some time ago by Marciano[3] that if exotic quarks in a higher representation exist, QCD alone might explain the electroweak symmetry breaking, since the scale, v_R, of the condensate of exotic quarks gets enhanced due to a large color charge carried by them:

$$\frac{v_R}{v_{\text{QCD}}} \simeq \exp\left[\frac{2\pi}{7\alpha_s(\mu_3)}\left(\frac{C_2(R)}{C_2(3)} - 1\right)\right]\,, \tag{2}$$

where $C_2(R)$ is the eigenvalue of the quadratic Casimir operator in a representation R and μ_3 is the dynamical mass scale of triplet quarks. Unfortunately, this interesting possibility has been ruled out, because the exotic quarks are seen neither yet in cosmic rays nor in the collider experiments.

Fermions in a higher representation was revived recently in models of technicolor theory, as one can build viable models with a minimal number of techniquarks in a higher representation. One necessary condition for the technicolor models to operate is that they should have a quasi-infrared (IR) fixed point.[4–6] Near the fixed point the technicolor coupling is almost constant and thus the bilinear operator of techniquarks has a large anomalous dimension, $\gamma_m \simeq 1$, suppressing dangerous flavor-changing four-Fermi interactions generated by extended technicolor interactions, while having reasonable mass spectra for ordinary quarks.

In models proposed in Refs. 7 and 8 2nd-rank tensor techniquarks are introduced. Fermions in such a representation naturally arise in string theory with orientifold compactification.[9] Furthermore, in the large (techni)-color limit such model is shown to be equivalent to the bosonic sector of $\mathcal{N} = 1$ pure super Yang-Mills theory,[10] which allows to make a reliable estimate on the Higgs mass.[8] The Higgs mass and its chiral partner, η_T', mass were found to be in the large N limit

$$m_H = m_{\eta'_T} = M \sim 170 - 500\ GeV\ . \tag{3}$$

where M is related to the gaugino condenstate as

$$M = \frac{2\,\alpha}{3}\left[\frac{3\,\langle\widetilde{q}q\rangle}{32\pi^2\,N}\right]^{\frac{1}{3}} = \frac{2\hat{\alpha}}{3}\Lambda\ , \tag{4}$$

with $\langle\widetilde{q}q\rangle = 3N\Lambda^3$ and $\hat{\alpha} = 1 \sim 3$. However, quantities, sensitive to flavors like the Peskin-Takeuchi parameters,[11] are difficult to estimate by using the planar equivalence, since it holds strictly for a Yang-Mills theory with a single Dirac fermion of 2nd rank tensor representation. For this the recent development on the string/gauge duality has been used.[2]

3. AdS/CFT and Holographic dual of Technicolor

One of the recent breakthrough in string theory is to establish the correspondence between string theory and a conformal field theory, called AdS/CFT correspondence. The original correspondence is between 10D Type IIB superstring in $AdS_5 \times S^5$ with N units of RR flux and 4D $\mathcal{N}$=4 U(N) super-Yang-Mills.[12] Several gravity duals have been found since then. But, it is quite nontrivial to find gravity duals of gauge theories. Instead one may take a bottom-up approach to engineer the gravity dual for gauge theories. The bottom-up approach was quite successful in describing the low energy dynamics of QCD.[13]

For the boundary CFT, we choose a strongly coupled $SU(N)_{TC}$ with N_{TF} techniquarks of 2nd rank tensor with $SU(N_{TF})_L \times SU(N_{TF})_R$ chiral symmetry. The boundary operators we are interested in are

$$\mathcal{O}(x) = \bar{q}_\alpha q^\beta,\ \bar{q}\gamma^\mu(1 \pm \gamma_5)t^a\,q\,. \tag{5}$$

According to the AdS/CFT correspondence, for each boundary operators we have bulk fields, $\phi(x,z) = X_\alpha^\beta, A_L^{aM}, A_R^{aM}$. Then the gravity dual of the boundary gauge theory is described as a 5D gauge theory of above bulk fields in an anti-de Sitter background, whose action becomes at low energy

$$S_{5D}^{\rm eff} = \int d^5x\sqrt{g}\,{\rm Tr}\left[|DX|^2 - m_5^2\,|X|^2 - \frac{1}{2g_5^2}\left(F_L^2 + F_R^2\right)\right] \tag{6}$$

where $D_\mu X = \partial_\mu X - iA_{L\mu}\,X + iX\,A_{R\mu}$, and $F_{L(R)}$ are the field strength tensor of $A_{L(R)}$. The bulk mass is related to the scaling dimension,Δ, of the boundary operator as $m_5^2 = \Delta\,(\Delta - 4)$. The action is defined in a slice of an AdS space, $\epsilon \leq z \leq z_m$, with metric

$$ds^2 = \frac{1}{z^2}\left(-dz^2 + \eta^{\mu\nu}dx_\mu dx_\nu\right)\ . \tag{7}$$

The most useful result of the AdS/CFT correspondence in calculating the oblique corrections is that the generating functional for the correlation functions of an operator $O(x)$ is given by the classical action of the corresponding bulk field, ϕ, satisfying the equation of motion,

$$W[\phi_0(x)] \equiv -i \ln \left\langle \exp\left[-i\int_x \phi_0\, \mathcal{O}\right]\right\rangle = S_{5D}^{\rm eff}[\phi(\epsilon, x)]\,. \tag{8}$$

Therefore the generating functional for the techniquark condensate $\bar{q}\,q$ is

$$W[S] = \int_x \left(-\frac{1}{4z^3}\,\partial_z X_0\, X_0^\dagger + h.c.\right)_{z=\epsilon} \tag{9}$$

where the classical solution of the bulk field X is given as

$$X_0(z) = c_1\, z^{4-\Delta} + c_2\, z^{\Delta}\,, \tag{10}$$

Identifying the source for the condensate to be

$$S(x) = \frac{2}{z^{4-\Delta}}\, X_0(x,z)\bigg|_{z=\epsilon}\,, \tag{11}$$

we find

$$\langle \bar{q}_L\, q_R\rangle = i\frac{\delta}{\delta S}\left\langle \exp\left[-i\int_x (\bar{q}_L S q_R + h.c.)\right]\right\rangle_{S=0} = \frac{\Delta}{4}c_2\,. \tag{12}$$

For the case of walking technicolor, $\gamma_m \simeq 1$ and thus the solution becomes

$$X_0(z) = \frac{M}{2}\, z^2 + \frac{\langle \bar{q}\, q\rangle}{4}\, z^2 \ln(z/\epsilon)\,. \tag{13}$$

4. Constraint from Electroweak Precision Data

Since the corrections to the electroweak interaction due to new physics involve massive particles, they appear mostly through the loop corrections to the vacuum polarization,

$$\int_x e^{-iq\cdot x}\, \langle J_X^\mu(x) J_Y^\nu(0)\rangle = i g_T^{\mu\nu} \Pi_{XY}(q^2) + q^\mu q^\nu \Pi_L$$

$$\Pi_{XY}(q^2) = \Pi_{XY}^{\rm sm} + \Pi_{XY}^{\rm new}\,, \tag{14}$$

where X, Y denotes the $SU(2)\times U(1)$ generators. At low energy the corrections due to the new physics can be expanded in powers of momenta. Peskin-Takeuchi introduced following parameters for the corrections:

$$S = 16\pi \left[\Pi_{33}^{\prime\rm new}(0) - \Pi_{3Q}^{\prime\rm new}(0)\right] \tag{15}$$

$$T = \frac{4\pi}{s^2 c^2 M_Z^2}\left[\Pi_{11}^{\rm new}(0) - \Pi_{33}^{\rm new}(0)\right] \tag{16}$$

$$U = 16\pi \left[\Pi_{11}^{\prime\rm new}(0) - \Pi_{33}^{\prime\rm new}(0)\right] \tag{17}$$

Then any physical observables can be expressed as a sum of the standard model value plus the corrections from the new physics,

$$x = x_{\rm sm}(m_t, m_H) + a_x S + b_x T + c_x U \,. \tag{18}$$

Experimentally, the Peskin-Takeuchi parameters are measured to be $S = 0.07 \pm 0.10$, $U \approx 0 \approx T$.[14]

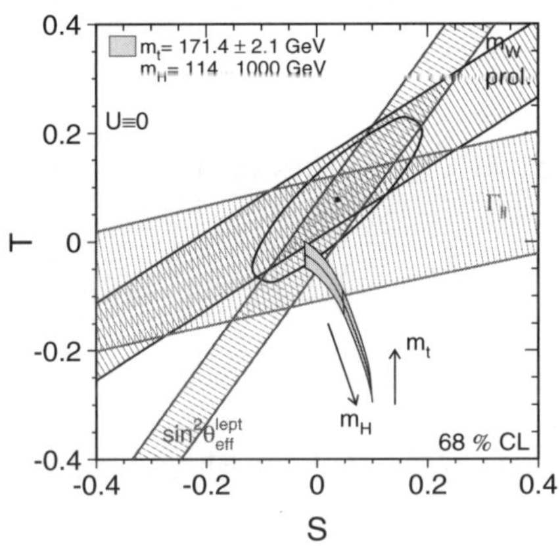

Fig. 1. Electroweak precision tests.

Since the Peskin-Takeuchi S parameter is given as

$$S = -4\pi \frac{\mathrm{d}}{\mathrm{d}q^2} \left[\Pi_V(q^2) - \Pi_A(q^2)\right]_{q^2=0} , \tag{19}$$

we need to know the vector and axial vector current correlators. We introduce the 4D Fourier transform of the gauge fields as, with $V(q,\epsilon) = A(q,\epsilon) = 1$ and $\partial_z V(q, z_m) = \partial_z A(q, z_m) = 0$ in a gauge $A_z = 0 = V_z$,

$$\begin{aligned} V^\mu(q,z) &= \left(g^{\mu\nu} - \frac{q^\mu q^\nu}{q^2}\right) \tilde{V}_\nu(q) V(q,z) \,, \\ A^\mu(q,z) &= \left(g^{\mu\nu} - \frac{q^\mu q^\nu}{q^2}\right) \tilde{A}_\nu(q) A(q,z) + \frac{q^\mu q^\nu}{q^2} \tilde{A}_\nu(0) A(0,z) \,, \end{aligned} \tag{20}$$

where $V(q,z)$ and $A(q,z)$ satisfy

$$\left[z\partial_z\left(\frac{1}{z}\partial_z\right)+q^2\right]V(q,z)=0\,,$$
$$\left[z\partial_z\left(\frac{1}{z}\partial_z\right)+q^2-\frac{g_5^2X_0^2}{z^2}\right]A(q,z)=0\,. \tag{21}$$

The correlation functions become by the AdS/CFT correspondence

$$\Pi_V(-q^2)=\left.\frac{\partial_z V(q,z)}{g_5^2\,z}\right|_{z=\epsilon}\,,\ \Pi_A(-q^2)=\left.\frac{\partial_z A(q,z)}{g_5^2\,z}\right|_{z=\epsilon}\,. \tag{22}$$

By solving Eq. (21) we find the S parameter

$$S=\frac{4\pi}{g_5^2}\int_\epsilon^{z_m}\frac{\mathrm{d}z'}{z'}\left[1-e^{2\int_\epsilon^{z'}\mathrm{d}\omega\,T^{(0)}(\omega)}\right]\,, \tag{23}$$

where $T^{(0)}(z)\equiv\partial_z\ln A(0,z)$ satisfies with $T^{(0)}(z_m)=0$ and $T^{(0)}(z)/z\big|_\epsilon=-g_5^2F_T^2$

$$z\partial_z\left(\frac{1}{z}T^{(0)}\right)+\left(T^{(0)}\right)^2=\frac{g_5^2X_0^2}{z^2}\,. \tag{24}$$

Our results for the S parameter depend only on two dimensionless parameters ϵ/z_m and F_Tz_m. For the former, we take $1/300=\Lambda_{TC}/\Lambda_{ETC}$, while the latter must be estimated by other means. We first note that our results agree well with the data (S=0.33) for the case of QCD-like technicolor, where we take $m_\rho/F_T\simeq 8.85$ or $F_T\simeq 0.29/z_m$. If we use the ladder approximation, we obtain $F_Tz_m\simeq 0.86/g_5$ as a prediction. If the chiral perturbation theory is valid up to the scale m_ρ, we have $4\pi F_T\simeq m_\rho$ or $F_T\simeq 0.19/z_m$. The corresponding S parameters for Techni-orientifold (S) and walking technicolor (F) are listed in the Table 1. We also note that by analyzing Eq. (24) one can show rigorously that the S parameter is always positive and is reduced at least about 10-20% by walking.

F_Tz_m	N=2,S	N=2,F	N=3,S	N=3,F	N=4,S	N=4,F
$1/4\pi$	0.031	0.031	0.031	0.031	0.031	0.031
$0.86/g_5$	0.086	0.057	0.17	0.086	0.29	0.12
0.19	0.15	0.14	0.17	0.15	0.17	0.16
0.29	0.28	0.22	0.34	0.26	0.37	0.31

Table 1. The S parameters of technicolor models.

5. Conclusions and Outlook

Physics beyond the standard model is soon to be discovered at LHC. One possibility is that it is a strongly coupled gauge theory based on techni-orientfold. The AdS/CFT correspondence allows us to estimate the oblique corrections of technicolor to the electroweak physics, removing the major hurdle for technicolor theories. We construct a holographic dual model of technicolor theories to show that the S parameter is always positive and reduced by walking. For the technicolor in the conformal window the ladder approximation is reliable and we predict the technipion decay constant $F_T = 0.86\, g_5^{-1}$. Then the S parameters for walking TC and techni-orientfold are found to be quite small, compatible with the electroweak precision data.

To conclude, we find the dynamical electroweak symmetry breaking with techniquarks of 2nd rank tensor is quite promising, as it does not suffer from the flavor-changing neutral-current problem and has small oblique corrections.

References

1. S. Weinberg, Phys. Rev. D **13**, 974 (1976); D **19**, 1277 (1979); L. Susskind, Phys. Rev. D **20**, 2619 (1979).
2. D. K. Hong and H. U. Yee, Phys. Rev. D **74**, 015011 (2006).
3. W. J. Marciano, Phys. Rev. D **21**, 2425 (1980).
4. B. Holdom, "Techniodor," Phys. Lett. B **150**, 301 (1985).
5. T. W. Appelquist, D. Karabali and L. C. R. Wijewardhana, "Chiral Hierarchies And The Flavor Changing Neutral Current Problem In Technicolor," Phys. Rev. Lett. **57**, 957 (1986).
6. K. Yamawaki, M. Bando and K. i. Matumoto, "Scale Invariant Technicolor Model And A Technidilaton," Phys. Rev. Lett. **56**, 1335 (1986).
7. F. Sannino and K. Tuominen, "Techniorientifold," Phys. Rev. D **71**, 051901 (2005).
8. D. K. Hong, S. D. H. Hsu and F. Sannino, "Composite Higgs from higher representations," Phys. Lett. B **597**, 89 (2004).
9. A. Dabholkar and J. Park, Nucl. Phys. B **477**, 701 (1996).
10. A. Armoni, M. Shifman and G. Veneziano, Phys. Rev. Lett. **91**, 191601 (2003).
11. M. E. Peskin and T. Takeuchi, Phys. Rev. Lett. **65**, 964 (1990).
12. J. M. Maldacena, Adv. Theor. Math. Phys. **2**, 231 (1998) [Int. J. Theor. Phys. **38**, 1113 (1999)]; S. S. Gubser, I. R. Klebanov and A. M. Polyakov, Phys. Lett. B **428**, 105 (1998); E. Witten, Adv. Theor. Math. Phys. **2**, 253 (1998).
13. J. Erlich, E. Katz, D. T. Son and M. A. Stephanov, Phys. Rev. Lett. **95**, 261602 (2005); L. Da Rold and A. Pomarol, Nucl. Phys. B **721**, 79 (2005).
14. http://lepewwg.web.cern.ch/LEPEWWG/plots/

One-Loop Corrections to the S and T Parameters in a Three Site Higgsless Model*

Shinya Matsuzaki

Department of Physics, Nagoya University
Nagoya, 464-8602, Japan
E-mail: synya@eken.phys.nagoya-u.ac.jp

R. Sekhar Chivukula and Elizabeth H. Simmons

Department of Physics and Astronomy, Michigan State University,
East Lansing, MI 48824, USA
E-mail: sekhar@msu.edu, esimmons@msu.edu

Masaharu Tanabashi

Department of Physics, Tohoku University
Sendai 980-8578, Japan
E-mail: tanabash@tuhep.phys.tohoku.ac.jp

In this talk, we discuss one-loop chiral logarithmic corrections to the S and T parameters in a highly deconstructed Higgsless model with only three sites. In addition to the electroweak gauge bosons, this model contains a single extra triplet of vector states (which we denote $\rho^{\pm}$ and ρ^{0}), rather than an infinite tower of "KK" modes. Calculations are done in 't Hooft-Feynman gauge, including the ghost, unphysical Goldstone-boson, and appropriate "pinch" contributions required to obtain gauge-invariant results for the one-loop self-energy functions. We show that the chiral-logarithmic corrections naturally separate into two parts, a model-independent part arising from scaling below the ρ mass, which has the same form as the large Higgs-mass dependence of the S or T parameter in the standard model, and a second model-dependent contribution arising from scaling between the ρ mass and the cutoff of the model.

1. The Three-Site Model

The three-site model[1,2] can be considered as a highly deconstructed version of Higgsless models,[3] as illustrated in Fig. 1 using "moose notation".[4] The

*Talk was given by S.M. based on the work done in Ref. 29.

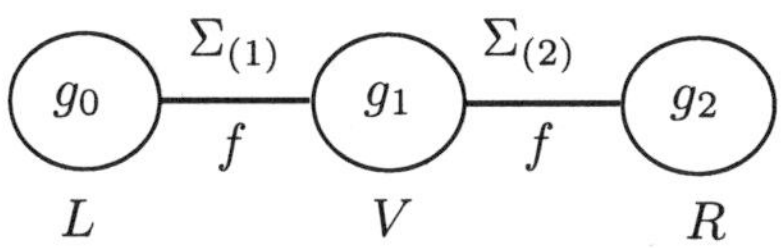

Fig. 1. The three-site Higgsless model analyzed in this talk is illustrated using "moose notation".[4] The model incorporates an $SU(2)_L \times SU(2)_V \times U(1)_B$ gauge group with couplings g_0, g_1, and g_2 respectively, and two nonlinear $(SU(2) \times SU(2))/SU(2)$ sigma models in which the global symmetry groups in adjacent sigma models are identified with the corresponding factors of the gauge group.

leading order lagrangian in the model is given by

$$\mathcal{L}_{(2)} = \frac{f^2}{4} \sum_{i=1}^{2} \mathrm{tr}[D_\mu \Sigma_{(i)}^\dagger D^\mu \Sigma_{(i)}] - \frac{1}{2g_0^2}\mathrm{tr}[L_{\mu\nu}]^2 - \frac{1}{2g_1^2}\mathrm{tr}[V_{\mu\nu}]^2 - \frac{1}{2g_2^2}\mathrm{tr}[R_{\mu\nu}]^2 , \tag{1}$$

where $L_{\mu\nu}$, $V_{\mu\nu}$, and $R_{\mu\nu}$ are the matrix field-strengths of the three gauge groups, $R_\mu = B_\mu \frac{\sigma_3}{2}$, and the covariant derivatives acting on $\Sigma_{(i)}$ are defined as $D_\mu \Sigma_{(1)} = \partial_\mu \Sigma_{(1)} - iL_\mu \Sigma_{(1)} + i\Sigma_{(1)} V_\mu$ and $D_\mu \Sigma_{(2)} = \partial_\mu \Sigma_{(2)} - iV_\mu \Sigma_{(2)} + i\Sigma_{(2)} R_\mu$. The 2×2 unitary matrix fields $\Sigma_{(1)}$ and $\Sigma_{(2)}$ may be parametrized by the Nambu-Goldstone boson fields $\pi_{(1)}$ and $\pi_{(2)}$ as $\Sigma_{(i)} = e^{2i\pi_{(i)}/f}$, for $i = 1, 2$, with the decay constant[a] f.

This model approximates the standard model in the limit[1] $x = g_{0,2}/g_1 \ll 1$, in which case we expect a massless photon, light W and Z bosons, and a heavy set of bosons $\rho^\pm$ and ρ^0 with $M_W \ll M_\rho$. Numerically, then, $g_{0,2}$ are approximately equal to the standard model $SU(2)_W$ and $U(1)_Y$ couplings, and we therefore define an angle θ such that $s = \sin\theta$, $c = \cos\theta$, and $g_0^2 \approx (4\pi\alpha)/s^2 = e^2/s^2$, $g_2^2 \approx (4\pi\alpha)/c^2 = e^2/c^2$, $s/c = g_2/g_0$, where α is the fine-structure constant and e the charge of the electron.

1.1. *Fermion Couplings and αS at Tree-Level*

In general, the standard model fermions may be delocalized along the three-site moose in the sense that their weak couplings arise from both sites 0 and 1,[5,6] $\mathcal{L}_f = \vec{J}_L^\mu \cdot ((1 - x_1)L_\mu + x_1 V_\mu) + J_Y^\mu B_\mu$, where $\vec{J}_L^\mu$ and J_Y^μ are the fermionic weak and hypercharge currents, respectively, and $0 \le x_1 \ll 1$ is a measure of the amount of fermion delocalization. This expression

[a]For simplicity, here we take the same decay constant f for both links.

is not separately gauge invariant under $SU(2)_0$ and $SU(2)_1$. Rather, the fermions should be viewed as being charged under $SU(2)_0$, and the terms proportional to x_1 should be interpreted as arising from the operator of the form

$$\mathcal{L}'_f = -x_1 \cdot \bar{\psi}_L (i\not{D}\Sigma_{(1)}\Sigma^\dagger_{(1)})\psi_L \ , \tag{2}$$

in unitary gauge. We are now interested only in the light fermions (*i.e.* all standard model fermions except for the top-quark), and therefore ignore the couplings giving rise to fermion masses [b].

Some degree of fermion delocalization is desirable for phenomenological reasons. Diagonalizing the gauge-boson mass matrix and computing the relevant tree-level four-fermion processes, one may compute the value of the S-parameter at tree-level, with the result[5,6] $\alpha S^{tree} = 4s^2(M_W^2/M_\rho^2)\left[1 - 2x_1(M_W^2/M_\rho^2)\right]$. Current experimental bounds on αS are $\mathcal{O}(10^{-3})$.[7] Since exchange of the ρ meson is necessary to maintain the unitarity of longitudinally polarized W-boson scattering, we must require that $M_\rho \le \mathcal{O}(1\,\text{TeV})$ – leading, for localized fermions with $x_1 = 0$, to a value of αS^{tree} which is too large. For the three-site model to be viable, therefore, the value of the fermion delocalization parameter must be chosen to reduce the value of αS^{tree}.[6,8–12]

1.2. *Duality and The Size of Radiative Electroweak Corrections*

By duality,[13] tree-level computations in the 5-dimensional theory represent the leading terms in a large-N expansion[14] of the strongly-coupled dual gauge theory akin to "walking technicolor".[15–20] The mass of the W boson, M_W, is proportional to a weak gauge-coupling, g_{ew}, (which is fixed in the large-N approximation) times the f-constant for the electroweak chiral symmetry breaking of the strongly-coupled theory. Therefore, we expect[14,21] $M_W^2/M_\rho^2 = \mathcal{O}\left(g_{ew}^2 N/(4\pi)^2\right)$, which is the expected behavior of the S-parameter in the large-N limit.[22,23]

We now specify the limit in which we will perform our analysis. Examining the mass matrices of the gauge bosons extracted from the lagrangian (1), in the small $x = g_0/g_1$ limit, we find $M_W^2/M_\rho^2 \approx x^2/4$ so that the tree-level value of αS vanishes if $x_1 = x^2/2$. In what follows, therefore, we will assume that $x_1 = \mathcal{O}(x^2)$. Overall, then, we work in the limit, $1 \gg |\alpha S^{tree}| = \mathcal{O}\left(g_{ew}^2 N/(4\pi)^2\right) \gg |\alpha S^{one-loop}| = \mathcal{O}\left(g_{ew}^2/(4\pi)^2\right) > 0$, which is

[b]These are discussed in detail in Ref. 1.

manifestly consistent with the large-N approximation. Once we choose the value of x_1 to make the size of αS^{tree} consistent with the phenomenological bound of $\mathcal{O}(10^{-3})$, one-loop electroweak corrections $\alpha S^{one-loop}$ become potentially relevant.

Note also that $\alpha T^{tree} \approx 0$ in these models, independent of the degree of fermion delocalization.[6,24] The one-loop corrections to αT are therefore of interest [c].

In practice, in calculating corrections to the gauge-boson self-energy functions (Π_{IJ})[22,25] defined as $\langle T I_\mu(p) J_\nu(-p)\rangle = i g_{\mu\nu}\Pi_{IJ}(p^2) + \cdots$, for $I, J = W, Z, A$, we work in the leading-log approximation, and to order α; we neglect corrections $\mathcal{O}(\alpha x^2 M_W^2)$ or $\mathcal{O}(\alpha x^2 p^2)$, but keep those of order $\mathcal{O}(\alpha x^2 M_\rho^2)$.

2. Calculations of Radiative Electroweak Corrections in Three-Site Model

2.1. *Gauge boson self-energies*

Expanding the non-linear sigma model fields $\Sigma_{(1,2)}$ in powers of the Goldstone boson fields $\pi_{(1,2)}$ as $\Sigma_{(1,2)} = 1 + 2i\pi_{(1,2)}/f + \cdots$, we calculate the gauge-boson self-energy amplitudes $\Pi_{ZZ,WW,ZA,AA}$ rewriting the non-linear interaction terms in Eqn.(1) in terms of the mass-eigenstate fields [d]. The amplitudes are evaluated in the 't Hooft-Feynman gauge by using dimensional regularization and renormalized at the cutoff scale Λ of the effective theory. The relevant one-loop diagrams are shown in Fig. 2 [e].

2.2. *Pinch Contributions in the Three-Site Model*

It should be noted that results obtained solely by calculating the gauge boson self-energy amplitudes are not gauge-invariant: As is well-known, a gauge-invariant result is obtained only after inclusion of the appropriate pieces (the so-called "pinch contributions") of one-loop vertex corrections and box diagrams.[26,27] In 't Hooft-Feynman gauge the only such contributions arise from diagrams containing triple-vector-boson vertices, as illustrated for the electroweak and ρ gauge bosons in Fig. 3. [f]

[c]Those arising from the extended fermion sector have been shown to place strong lower bounds on the masses of the KK fermions.[1] Those arising from the gauge sector are considered in this talk.

[d]For explicit expressions of these interaction terms, see Ref. 29.

[e]The expressions of the corresponding amplitudes are shown in Ref. 29.

[f]The explicit calculation of the pinch part in the three-site model is described in Ref. 29.

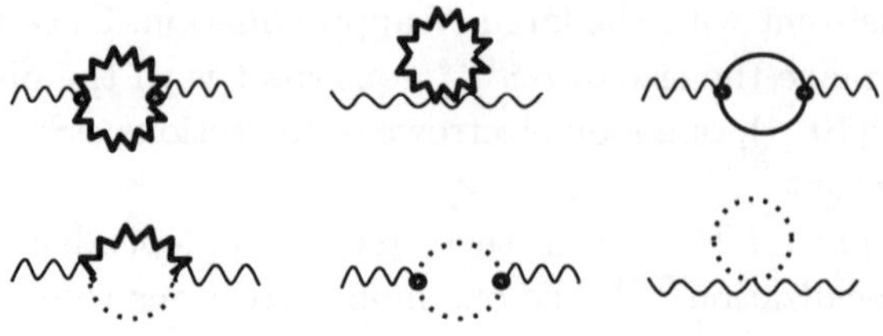

Fig. 2. One-loop diagrams for the self-energies $\Pi_{WW,ZZ,ZA,AA}$ in the three-site model. The fields flowing in loops correspond to the gauge fields A, W, Z, ρ (denoted as wavy line), the ghost fields $C_{A,W/Z,\rho}$ (straight line) and the scalar fields $\pi_{W/Z,\rho}$ (dotted line), where the $\pi_{W/Z,\rho}$ and $C_{A,W/Z,\rho}$ fields, respectively, denote the 't Hooft-Feynman gauge unphysical Goldstone bosons and the ghost fields corresponding to the electroweak and ρ bosons.

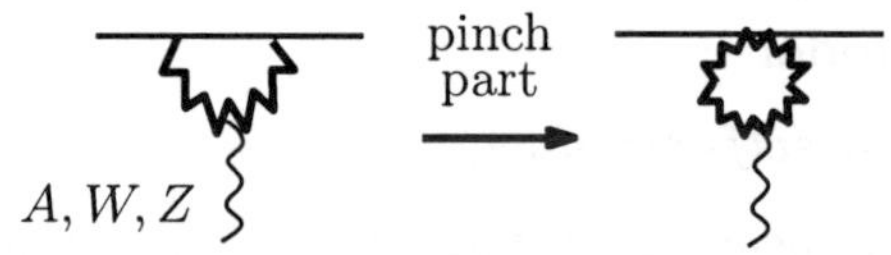

Fig. 3. Vertex correction diagrams constructed from the SM and ρ gauge boson loops which contribute to the pinch part of the self-energies[26,27] for the W/Z and A bosons. The external fermion lines are arbitrary, and may represent fermions charged under any combination of $SU(2)_0$ or $SU(2)_1$.

2.3. *Fermion Delocalization Contributions*

Consider the fermion delocalization operator of Eqn. (2). As illustrated in Fig. 4, the couplings in the delocalization operator give rise to vertex-corrections to neutral- and charged-current four-fermion processes [g]. Note that these vertex corrections are *universal* – *i.e.* proportional to the charges of the external fermions. As in the case of the "pinch" contributions, their effects may be incorporated into corrections of the gauge-boson self-energy functions.

3. Precision Electroweak Corrections – αS and αT –

Including the standard model and three-site "pinch corrections", and the corrections arising from the fermion delocalization operator, we obtain the

[g] The corrections coming from these diagrams are evaluated in Ref. 29.

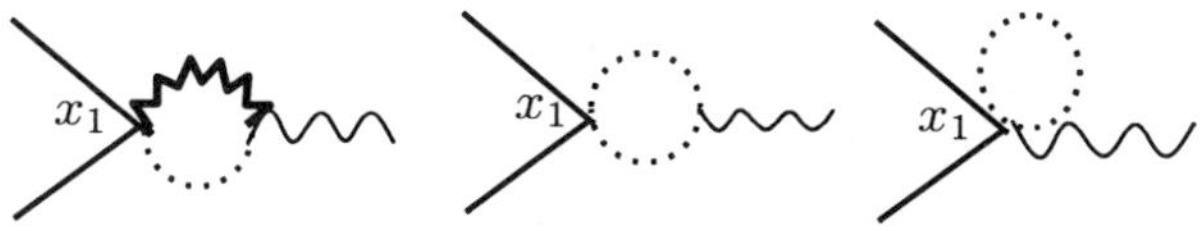

Fig. 4. One-loop vertex contributions to neutral/charged-current processes arising from the delocalization operator $\mathcal{L}'_f$. The delocalization operator is proportional to $x_1 = \mathcal{O}(x^2)$, and therefore only the contributions proportional to M_ρ^2, which are illustrated here, contribute to this order.

total neutral and charged gauge boson self-energies[h] ($\bar{\Pi}$), which should be associated with the S and T parameters.[22] As a result, the leading corrections to the S and T parameters evaluated in the limit $M_\rho^2 \gg M_W^2$ are found to be

$$\alpha S = \left[\frac{4s^2 M_W^2}{M_\rho^2}\left(1 - \frac{x_1 M_\rho^2}{2M_W^2}\right)\right]_{\mu=\Lambda} + \frac{\alpha}{12\pi}\log\frac{M_\rho^2}{M_W^2} - \frac{41\alpha}{24\pi}\log\frac{\Lambda^2}{M_\rho^2} + \frac{3\alpha}{8\pi}\left(\frac{x_1 M_\rho^2}{2M_W^2}\right)\log\frac{\Lambda^2}{M_\rho^2}, \tag{3}$$

$$\alpha T = -\frac{3\alpha}{16\pi c^2}\log\frac{M_\rho^2}{M_W^2} - \frac{3\alpha}{32\pi c^2}\log\frac{\Lambda^2}{M_\rho^2}. \tag{4}$$

Note that the $\log(M_\rho^2/M_W^2)$ terms for both αS and αT arise from "scaling" between M_W and M_ρ – and have coefficients precisely equal to the leading-log contributions from a heavy Higgs boson.[22] This allows us to match our calculation to the phenomenological extractions of S and T which depend on a reference standard model Higgs-boson mass.

The dependence on the renormalization scale (here taken to be the cut-off Λ of the effective theory) is cancelled by the scale-dependence of the appropriate counterterms[28i].

References

1. R. Sekhar Chivukula, B. Coleppa, S. Di Chiara, E. H. Simmons, H. J. He, M. Kurachi and M. Tanabashi, Phys. Rev. D **74**, 075011 (2006).
2. R. Casalbuoni, S. De Curtis, D. Dominici and R. Gatto, Phys. Lett. B **155**, 95 (1985).

[h]The explicit expressions for the total self-energies ($\bar{\Pi}$), evaluated up to $\mathcal{O}(\alpha x^2 p^2)$ or $\mathcal{O}(\alpha x^2 M_W^2)$, are shown in Ref. 29.

[i]The tree-level expression and such appropriate counterterms are understood to be evaluated at scale Λ.[30]

3. C. Csaki, C. Grojean, H. Murayama, L. Pilo and J. Terning, Phys. Rev. D **69**, 055006 (2004).
4. H. Georgi, *Nucl. Phys.* **B266** (1986) 274.
5. L. Anichini, R. Casalbuoni and S. De Curtis, Phys. Lett. B **348**, 521 (1995).
6. R. S. Chivukula, E. H. Simmons, H. J. He, M. Kurachi and M. Tanabashi, Phys. Rev. D **71**, 115001 (2005).
7. R. Barbieri, A. Pomarol, R. Rattazzi and A. Strumia, Nucl. Phys. B **703**, 127 (2004).
8. G. Cacciapaglia, C. Csaki, C. Grojean and J. Terning, Phys. Rev. D **71** (2005) 035015.
9. G. Cacciapaglia, C. Csaki, C. Grojean, M. Reece and J. Terning, *Phys. Rev. D* **72**, (2005) 095018.
10. R. Foadi, S. Gopalakrishna and C. Schmidt, Phys. Lett. B **606** (2005) 157.
11. R. Foadi and C. Schmidt, Phys. Rev. D **73** (2006) 075011.
12. R. Sekhar Chivukula, E. H. Simmons, H. J. He, M. Kurachi and M. Tanabashi, Phys. Rev. D **72**, 015008 (2005).
13. J. M. Maldacena, *Adv. Theor. Math. Phys.* **2** (1998) 231–252.
14. G. 't Hooft, Nucl. Phys. B **72**, 461 (1974).
15. B. Holdom, *Phys. Rev.* **D24** (1981) 1441.
16. B. Holdom, *Phys. Lett.* **B150** (1985) 301.
17. K. Yamawaki, M. Bando, and K.-i. Matumoto, *Phys. Rev. Lett.* **56** (1986) 1335.
18. T. W. Appelquist, D. Karabali, and L. C. R. Wijewardhana, *Phys. Rev. Lett.* **57** (1986) 957.
19. T. Appelquist and L. C. R. Wijewardhana, *Phys. Rev.* **D35** (1987) 774.
20. T. Appelquist and L. C. R. Wijewardhana, *Phys. Rev.* **D36** (1987) 568.
21. R. S. Chivukula, M. J. Dugan and M. Golden, Phys. Rev. D **47**, 2930 (1993).
22. M. E. Peskin and T. Takeuchi, *Phys. Rev.* **D46** (1992) 381–409.
23. G. Burdman and Y. Nomura, Phys. Rev. D **69**, 115013 (2004).
24. R. Sekhar Chivukula, E. H. Simmons, H. J. He, M. Kurachi and M. Tanabashi, Phys. Rev. D **71** (2005) 035007.
25. G. Altarelli and R. Barbieri, *Phys. Lett.* **B253** (1991) 161–167.
26. G. Degrassi and A. Sirlin, Phys. Rev. D **46**, 3104 (1992).
27. G. Degrassi and A. Sirlin, Nucl. Phys. B **383**, 73 (1992).
28. M. Perelstein, *JHEP* **10** (2004) 010.
29. S. Matsuzaki, R. S. Chivukula, E. H. Simmons and M. Tanabashi, arXiv:hep-ph/0607191.
30. R. Sekhar Chivukula, Shinya Matsuzaki, Elizabeth H. Simmons, and Masaharu Tanabashi, in preparation.

CONSTRAINED ELECTROWEAK CHIRAL LAGRANGIAN

Qi-Shu Yan

NCTS (Theory Division), 101, Section 2 Kuang Fu Road, Hsinchu, Taiwan
** E-mail: yanqs@phys.nthu.edu.tw*

We update the uncertainty analysis on S parameter of the electroweak chiral Lagrangian (EWCL) by including the LEP-II W pair production data. We find that experimental data still allow positive $S^{EXP}(1\text{TeV})$.

1. Introduction

Technicolor models are one of the natural candidates beyond the standard model.[1] But it is widely said that the Technicolor models are ruled out by electroweak precision data. The claim[2] is established in two logical steps: 1) by extrapolating the electroweak precision data from $\mu = m_Z$ to 1 TeV (1 TeV is argued as the scale of compositeness) by perturbation calculation with only 6 quadratic operators, of the EWCL it was found that the value of the parameter S is negative. In our fit, when triple gauge coupling (TGC) effects are not included, S is determined as

$$S^{EXP}(1\text{TeV}) = -0.17 \pm 0.10\,, \tag{1}$$

2) by using the unsubtracted dispersion relations with the assumption of custodial symmetry and vector meson dominance and the low energy hadronic QCD data of ρ and a_1 mesons as input, and by scaling up the value of S to 1 TeV, it was found that the value of S parameter is positive,[2] which is given as

$$S^{TH}(1\text{TeV}) = 0.3\frac{N_{TF}N_{TC}}{6}\,. \tag{2}$$

The discrepancy between S^{EXP} and S^{TH} is at least 3σ, which is interpreted as an evidence that Technicolor models are ruled out by precision data.

In this article we summarize our study on the uncertainty analysis in $S^{EXP}(1\text{TeV})$.[3] Part of results was reported in.[4] Our result show that the uncertainty induced by TGC measurement dominates the error bar of

S^{EXP}(1TeV) and experimental data still allows positive S^{EXP}(1TeV) in the parameter space. Therefore we argue that it is premature to claim that the electroweak precision data has ruled out Technicolor models.

2. Our knowledge on the chiral coefficients

We follow the standard analysis of chiral Lagrangian method[5,6] and include 14 operators up to mass dimension four in the EWCL.[7] Our study extends the RGE analysis of Bagger *et. al.*,[8] who have considered the effects of 6 out of the 14 operators.

The 14 gauge invariant operators constructed in the EWCL are supposed to describe EWSB models defined at 1TeV in a model independent way, either strong or weak interaction models. Below we describe how to determine 14 chiral coefficients of the EWCL in our analysis at $\mu = m_Z$. Three of six two-point chiral coefficients g, g', and v, are determined by the following inputs $1/\alpha_{\rm em}(m_{\rm Z}) = 128.74$, $m_{\rm Z} = 91.18$ GeV, and $G_F = 1.16637\times10^{-5}$ GeV^{-2}. They are assumed to be free from error bars. The next three two-point-function chiral coefficients α_1, α_0, and α_8 are extracted from Z-pole data, precise W mass measurement, and top quark mass measurement (Here we use data of PDG2004). We perform the analysis with three best measured quantities $m_{\rm W} = 80.425 \pm 0.038$ GeV, $\sin^2\theta_W^{\rm eff} = 0.23147 \pm 0.00017$ and the leptonic decay width of Z, $\Gamma_\ell = 83.984 \pm 0.086$ MeV for the S, T and U fit. We take $m_{\rm t} = 175$ GeV in our S, T, and U fit.

The central values with 1σ errors of the S, T, and U parameters are found as

$$\begin{array}{l} S(m_Z) = (-0.06 \pm 0.11) \\ T(m_Z) = (-0.08 \pm 0.14) \\ U(m_Z) = (+0.17 \pm 0.15) \end{array} \rho_{co.} = \begin{pmatrix} 1 & & \\ .9 & 1 & \\ -.4 & -.6 & 1 \end{pmatrix} \qquad (3)$$

which roughly agrees with.[8]

The relations among two-point chiral coefficients α_1, α_0, and α_8 of EWCL with the STU parameters are found to be

$$\begin{aligned} \alpha_1(\mu) &= -\tfrac{1}{16\pi}\, S(\mu), \\ \alpha_0(\mu) &= \tfrac{1}{2}\,\alpha(\mu)_{\rm EM}\, T(\mu), \\ \alpha_8(\mu) &= -\tfrac{1}{16\pi}\, U(\mu)\,. \end{aligned} \qquad (4)$$

From Eqs. (3-4), $\alpha_1(m_Z)$, $\alpha_0(m_Z)$, and $\alpha_8(m_Z)$ are determined as

$$\begin{aligned} \alpha_1(m_Z) &= (+0.13 \pm 0.21) \times 10^{-2} \\ \alpha_0(m_Z) &= (-0.03 \pm 0.05) \times 10^{-2} \\ \alpha_8(m_Z) &= (-0.35 \pm 0.29) \times 10^{-2} \end{aligned} \qquad (5)$$

Three three-point chiral coefficients α_2, α_3 and α_9 are extracted from the LEP-II measurements via the process $e^+e^- \to W^+W^-$.[9] The experimental observables of anomalous TGC[10] between δk_γ, δk_Z, δg_1^Z, and three-point chiral coefficients, α_2, α_3, α_9 , can be simplified as

$$\begin{aligned}\delta k_\gamma &= (\alpha_2 + \alpha_3 + \alpha_9) g^2 \,, \\ \delta k_Z &= (\alpha_3 + \alpha_9) g^2 - \alpha_2 g'^2 \,, \\ \delta g_1^Z &= \alpha_3 (g^2 + g'^2) \,. \end{aligned} \tag{6}$$

There is no experimental data relaxing the custodial symmetry except L3 collaboration[11] from where we take $\delta k_Z = -0.076 \pm 0.064$ as one of the inputs. Other inputs $\delta k_\gamma = -0.027 \pm 0.045$ and $\delta g_1^Z = -0.016 \pm 0.022$ are taken from LEP Electroweak working group.[12,13] We found TGC errors are quite large as reported in D0 collaboration[14] at Tevatron. Because of this fact, we use LEP data in our analysis.

We also relax the custodial $SU(2)$ gauge symmetry as it is natural in the framework of the EWCL to have a non-vanishing α_9 if the underlying dynamics break this symmetry explicitly.[15] By assuming these data are extracted independently, we can obtain three-point chiral coefficients as

$$\begin{aligned}\alpha_2(m_Z) &= (+0.09 \pm 0.14) \\ \alpha_3(m_Z) &= (-0.03 \pm 0.04) \rho_{co.} = \begin{pmatrix} 1 & & \\ 0 & 1 & \\ -.7 & -.3 & 1 \end{pmatrix}. \\ \alpha_9(m_Z) &= (-0.12 \pm 0.12) \end{aligned} \tag{7}$$

We observe that $\alpha_3(m_Z)$ is more tightly constrained than $\alpha_2(m_Z)$ and $\alpha_9(m_Z)$. In our numerical analysis, we consider the two-parameter fit data from L3 collaboration and this combined data as two scenarios to show the effects of TGC to S^{EXP}.

There is no experimental data to bound 5 four-point chiral coefficients, which usually are assumed to be of order one. Partial wave unitary bounds of longitudinal vector boson scattering processes can be used to put bounds on the magnitude of those chiral coefficients. We use the following five $J = 0$ channels to bound 5 chiral coefficients (Λ is arbitrary, which should correspond to the UV cutoff of the EWCL), α_4, α_5, α_6, α_7, and α_{10}:

$$\begin{aligned}|4\alpha_4 + 2\alpha_5| &< 3\pi \tfrac{v^4}{\Lambda^4}, \\ |3\alpha_4 + 4\alpha_5| &< 3\pi \tfrac{v^4}{\Lambda^4}, \\ |\alpha_4 + \alpha_6 + 3(\alpha_5 + \alpha_7)| &< 3\pi \tfrac{v^4}{\Lambda^4}, \\ |2(\alpha_4 + \alpha_6) + \alpha_5 + \alpha_7| &< 3\pi \tfrac{v^4}{\Lambda^4}, \\ |\alpha_4 + \alpha_5 + 2(\alpha_6 + \alpha_7 + \alpha_{10})| &< \tfrac{6\pi}{5} \tfrac{v^4}{\Lambda^4}, \end{aligned} \tag{8}$$

where bounds are obtained from $W_L^+W_L^+ \to W_L^+W_L^+$, $W_L^+W_L^- \to W_L^+W_L^-$, $W_L^+W_L^- \to Z_LZ_L$, $W_L^+Z_L \to W_L^+Z_L$, and $Z_LZ_L \to Z_LZ_L$, respectively.

Quartic gauge couplings will not contribute to S parameter, but do affect T parameter. For completeness, we list these theorectical bounds here.

3. Uncertainty in $S^{EXP}(\Lambda)$ parameter

In the framework of effective field theory method,[16] all chiral coefficients of the EWCL depend on renormalization scale μ. $S(\Lambda)$, $T(\Lambda)$, and $U(\Lambda)$ are values of parameters S, T, and U at the matching scale Λ, where the EWCL matches with fundamental theories, Technicolor models, extra dimension models, Higgsless models, etc. In the perturbation method, theoretically, the $S(m_Z)$, $T(m_Z)$, and $U(m_Z)$ can be connected with the $S(\Lambda)$, $T(\Lambda)$, and $U(\Lambda)$ by improved renormalization group equations.[3]

The effect of TGC measurement is depicted on $S(\Lambda) - T(\Lambda)$ plane as shown in Fig. 1(a) and Fig. 1(b). We highlight some features of these two figures. **(1)** In Fig. 1(a), in absence of TGC contribution (dashed-line contours), $S(\Lambda)$ becomes more negative as Λ increases with reference to the reference LEP-I fit contour at $\Lambda = m_Z$. This is roughly in agreement with the observation of Refs. 8 and 17, and PDG.[18] Two-parameter fit data with custodial symmetry condition from L3 collaboration shows that $S(\Lambda)$ is driven to the positive value region (the solid line). The central value of $S^{EXP}(1\text{TeV})$ is 0.8. Furthermore, the error bars of $S - T$ are greatly enlarged by the uncertainty of TGC measurements.

(2) In Fig. 1(b), without the custodial symmetry assumption on the experimental data, as given in Eq. (7), inclusion of TGC contribution makes the central value of $S(\Lambda)$ almost unchanged. Meanwhile, the error bars of $S - T$ are enlarged at least by a factor of 2.

Fig. 1 clearly demonstrate a fact that the TGC measurement affect the value of $S^{EXP}(\Lambda)$ significantly.

4. Discussions and Conclusion

In the unsubtracted dispersion relation,[2,5] the fact that S^{EXP} is a running parameter might not be transparent. In the RGE analysis, the quantum fluctuations of active degree of freedoms S parameter to run:

$$S^{EXP}(\Lambda) = S^{EXP}(m_Z) + \beta_S \ln \tfrac{\Lambda}{m_Z} \,. \tag{9}$$

Active degree of freedoms and new resonances can contribute to β_S function and affect the value of $S^{EXP}(\Lambda)$. We propose a naive subtracted dispersion relation without assuming custodial symmetry and vector meson

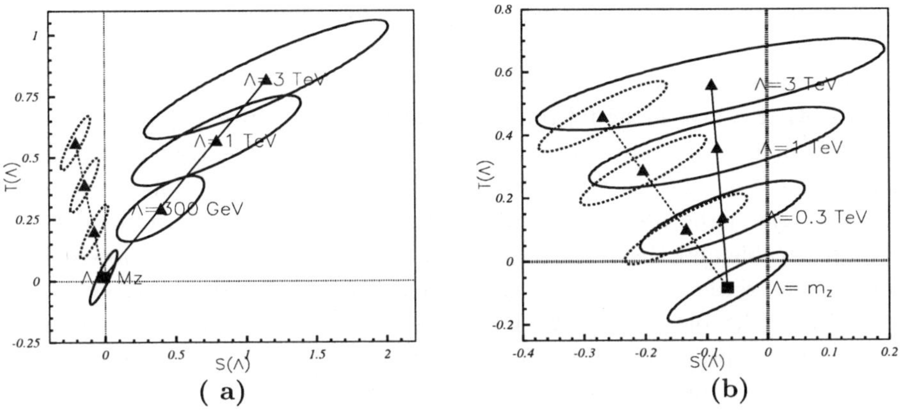

Fig. 1. *$S(\Lambda) - T(\Lambda)$ contours at $\Lambda = m_Z$, 300 GeV, 1 TeV, and 3 TeV, respectively. TGC uncertainty is included in solid line contours while not included in dashed line. Fig1(a) corresponds to the L3 two-parameter fit data. The parameter $U(m_Z)$ is taken as its best fit value, $U(m_Z) = +0.17$. Fig1(b) corresponds to the combined data. The parameter $U(m_Z)$ is taken as its best fit value, $U(m_Z) = 0.00$.*

dominance, which can read as

$$S(q^2) = S(q^2 = 0) - \frac{q^2}{3\pi} \int_{s > m_z^2}^{\infty} ds \frac{R_{3Y}(s)}{s(s - q^2)} \,. \tag{10}$$

Where $R_{3Y} = -12\pi Im\Pi'_{3Y}$, which is to count the degree of freedoms belonging to the representations of both $U_Y(1)$ and $SU_L(2)$. The $S(q^2 = 0)$ corresponds to the value determined from LEP-I Z-pole data. The second term include the contributions of active degree of freedoms and new resonances, either fermionic or bosonic. Wth Eqs. (9-10), the equivalence between the description of RGE and dispersion relation becomes obvious.[a]

Our most conservative numerical result from Fig. 1(b) can be put as

$$S^{EXP}(1\text{TeV}) = -0.08 \pm 0.20 \,. \tag{11}$$

To obtain these numerical values, we have assumed the perturbation method is valid from the energy scale m_Z to 1TeV. When data in PDG2006 is used, the central valus of $S^{EXP}(1\text{TeV})$ shifts to -0.02, which is in agreement with the observation in.[20] If there exist new resonances below or near 1TeV, the perturbation method might be invalid before $\Lambda = 1$TeV and effects of threshold and tail of new resonances might modified the value

[a]We thank Han-Qing Zheng and Kenneth Lane for comments on this point.

of S(1TeV) drastically. This might occur, for example, in the low energy Technicolor model.[21] Hence, we stress that both the center values and error bars of S(1TeV) can only be interpreted as reference values obtained in perturbation method.

Whether S and T should run in a logarithmic or power way or whether the decoupling theorem should hold in the process of extrapolating the data from m_Z scale to the matching scale is still a debatable issue. In the analysis of the minimal standard model with a Higgs,[18] the model is renormalizable and Higgs boson plays the role of regulator. Therefore only logarithmic terms are taken in the standard global fit. In the EWCL (a non-renormalizable theory), when operators beyond the $O(p^2)$ order are included, terms proportional to the power of Λ_{UV}/v enter into the radiative corrections. As the most conservative calculation, we adopted the logarithmic running. However, if power running is used, error bars of S parameters would be much larger than those shown Fig. 1.

One-loop contribution of TGC in our analysis can be attributed as part of $O(p^6)$ order effects in the standard chiral derivative power counting rule. There are analysis by including $O(p^6)$ operators in order to accommodate data of $e^+e^- \to f\bar{f}$ above Z pole, as done in Ref. 19. Even when these tree level effects of $O(p^6)$ operators are included, near TeV region, the sign of S^{EXP}(1TeV) can not change from negative to positive. Another remarkable fact is that when more operators are introduced the error bar of S^{EXP}(1TeV) becomes a few larger. But effects of these operators are smaller than those of TGC operators.

One may worry about the two-loop contributions of $O(p^2)$, which are also part of $O(p^6)$ order effects. However, due to the two loop suppression factor, they must be tiny. Therefore, uncertainty induced by the TGC dominates the error bar of S^{EXP}(1TeV). Our results show that the sign of S(1TeV) can be changed from negative to positive by TGC.

It is an open question to construct a realistic model which can have a large deviation from the prediction of the SM in TGC sector. In the Higgsless model with ideally delocalized fermions and the gauge-Higgs unification model in the warped space-time, it was found that the deviation is small.[22] The Higgsless model in warped 5D space-time might provide a solution. [b]

We show here that electroweak precision data have constrained both the oblique parameters STU significantly and the anomalous TGC considerably. But, current precision of electroweak data is not sufficient to rule

[b] We thank Kinya Oda mention this to us.

out Technicolor models, due to the large uncertainty in α_2 and α_9. Technicolor models can provide dark matter candidates.[23] Therefore, in our opinion, Technicolor models are still quite competitive and promising as a candidate of EWSB.[24]

Acknowledgments

This work is partially supported by the JSPS fellowship program and by NCTS (Hsinchu, Taiwan). We would like to thank Ulrich Parzefall for communication on the TGC measurements, Masaharu Tanabashi and Masayasu Harada for stimulating discussions.

References

1. C. T. Hill and E. H. Simmons, Phys. Repts. **381** (2003) 235 [Erratum-ibid. **390** (2004) 553].
2. B. Holdom and J. Terning, Phys. Lett. **B247**, 88 (1990); M. E. Peskin and T. Takeuchi, Phys. Rev. Lett. **65**, 964 (1990); Phys. Rev. **D46**, 381 (1992).
3. S. Dutta, K. Hagiwara, Q.S. Yan, and K. Yoshida, in prep.
4. S. Dutta, K. Hagiwara, Q.S. Yan, heh-ph/0603038.
5. J. Gasser and H. Leutwyler, Annals Phys. **158** (1984) 142.
6. M. Harada and K. Yamawaki, Phys. Rept. **381** (2003) 1.
7. A. C. Longhitano, Phys. Rev. D **22**, 1166 (1980); Nucl. Phys. B **188**, 118 (1981). T. Appelquist and G. H. Wu, Phys. Rev. D **48**, 3235 (1993).
8. J. A. Bagger *et. al.* , Phys. Rev. Lett. **84**, 1385 (2000).
9. P. Azzurri, ICHEP06, Moscow, Russia.
10. K. Hagiwara *et. al.* , Nucl. Phys. B **282**, 253 (1987).
11. P. Achard *et al.*, Phys. Lett. B **586**, 151 (2004).
12. LEPEWWG/TGC/2005-01 @ http://www.cern.ch/LEPEWWG/lepww/tgc.
13. A. Heister *et al.*, Eur. Phys. J. C **21**, 423 (2001); G. Abbiendi *et al.*, Eur. Phys. J. C **33**, 463 (2004); S. Schael *et al.* Phys. Lett. B **614**, 7 (2005).
14. V. M. Abazov *et al.*, hep-ex/0504019.
15. P. Sikivie *et al.*, Nucl. Phys. **B173**, 189 (1980).
16. H. Georgi, Annu. Rev. Nucl. Part. Sci. **43** (1993) 209.
17. M. E. Peskin and J. D. Wells, Phys. Rev. D **64**, 093003 (2001).
18. S. Eidelman, *et al.*, Phys. Lett. **B592**, 1 (2004).
19. R. Barbieri *et. al.* , Nucl. Phys. B **703**, 127 (2004).
20. D. D. Dietrich, F. Sannino and K. Tuominen, Phys. Rev. D **73**, 037701 (2006).
21. K. Lane and S. Mrenna, Phys. Rev. D **67**, 115011 (2003).
22. R. S. Chivukula, *et. al.*, Phys. Rev. D **72** (2005) 075012. Y. Sakamura and Y. Hosotani, arXiv:hep-ph/0607236.
23. S. B. Gudnason, C. Kouvaris and F. Sannino, Phys. Rev. D **74**, 095008 (2006); Phys. Rev. D **73**, 115003 (2006).
24. D. D. Dietrich and F. Sannino, arXiv:hep-ph/0611341.

TOWARD A TOP-MODE ETC*

Hidenori Fukano[1] and Koichi Yamawaki[2]

Department of Physics, Nagoya University,
Nagoya 464-8602, Japan
[1] *E-mail: fukano@eken.phys.nagoya-u.ac.jp*
[2] *E-mail: yamawaki@eken.phys.nagoya-u.ac.jp*

We attempt a modest step to build an explicit model of the extended technicolor (ETC) incorporating the top quark condensate as well as the walking/conformal technicolor realized near the conformal window associated with the Banks-Zaks infrared fixed point. It yields a low energy effective theory in the form of a certain version of the topcolor-assisted technicolor (TC2) with universal coloron. The top quark condensate is triggered by the combined effects of topcolor and ETC (and fine-tuned $U(1)_Y$) interactions near the topcolor criticality, which then becomes a main source of the mass of up-type quarks. The mass of down-type quarks and leptons comes from the walking/conformal technicolor as a dominant contribution to the electroweak symmetry breaking.

1. Introduction

The Standard model (SM) has a mysterious part, the electroweak symmetry breaking (EWSB), to give masses of quarks, leptons and W/Z-bosons. The EWSB via the elementary Higgs field in the SM has some problems. In order to solve these problems, we should build a scenario beyond the SM. One of candidates for such scenarios is Technicolor (TC).[1] TC is an attractive idea for the EWSB without the elementary Higgs, based on an analogy to the QCD with technifermion condensate, instead of the quark condensate in QCD, responsible for the mass of W/Z bosons (For a recent review see Ref. 2). In order to give mass to the SM fermions, we have to extend the TC into a larger picture which communicates the technifermion to the SM fermion. A typical one is the extended TC (ETC)[3] in which the TC group is embedded into a larger gauge group including the horizontal gauge group

*Talk was given by H.Fukano

of three families of the SM fermion.[a] This *old-type* ETC model which is based on the scale-up of the QCD has some problems; Flavor Changing Neutral Current (FCNC) problem,[3] light pseudo Nambu-Golodstone (NG) bosons and deviation from the LEP precision experiments, the so-called S parameter problem.[5]

The FCNC problem and light pseudo NG bosons have already been solved by the walking (=scale invariant/conformal) TC[6,7] with the TC gauge coupling almost constant or running very slowly, which was shown to develop a large anomalous dimension $\gamma_m \simeq 1$.[7] It was suggested[8,9] that a realistic walking/conformal gauge theory is given by the large N_f QCD, a jargon for a version of the "QCD" with N_c colors and many massless flavors $N_f(\gg N_c)$, in which the two-loop beta function possesses the Banks-Zaks infrared (IR) fixed point α_* for large N_f[b,10] such that $N_f^* < N_f < 11N_c/2$, with $N_f^* \simeq 8.01$ for $N_c = 3$. Note that $\alpha_* \searrow 0$ when $N_f \nearrow 11N_c/2$, and hence there exists a certain region $(N_f^* <)\, N_f^{\rm cr} < N_f < 11N_c/2$ ("conformal window") such that $\alpha^{\rm cr} > \alpha_* > 0$ where the chiral symmetry gets restored $\langle\bar\psi\psi\rangle = 0$: $\alpha^{\rm cr}$ is the critical coupling for the chiral symmetry breaking and may be evaluated as $\alpha^{\rm cr} = \pi/(3C_2)$ in the ladder approximation, in which case we have $N_f^{\rm cr} \simeq 4N_c$.[c,8] In the broken phase near the conformal window we have a condensate or the dynamical mass of the fermion m which is much smaller than the intrinsic scale of the theory $\Lambda(\gg m)$. In a wide region $m < \mu < \Lambda$ the coupling is walking due to the Banks-Zaks IR fixed point and the theory develops a large anomalous dimension $\gamma_m \simeq 1$ and enhanced condensate $\langle\bar\psi\psi\rangle|_\Lambda \sim \Lambda m^2$ at the scale of Λ.

As to the S parameter, it is rather difficult at this moment to draw a definite conclusion in the walking/conformal TC, since there is no reliable non-perturbative calculation of the S parameter in the walking/conformal TC which is strongly coupled and has no simple scale-up of the known dynamics like QCD. Actually, it was argued[13] that the S parameter is suppressed in walking/conformal theories with $\gamma_m \simeq 1$. Straightforward computation of S parameter was also performed, based on the ladder SD equation and Bethe-Salpeter equation, which indicates decreasing tendency

---

[a]Another possibility is a composite model where both the SM fermions and the technifermions are composite on equal footing, in which case the residual four-fermion interactions among composites play the role of the ETC-induced effective four-fermion interactions between technifermions and SM fermions.[4]

[b]There is another possibility to have the Banks-Zaks IR fixed point without large N_f by introducing higher dimensional representation.[11]

[c]In the case of $N_c = 3$, this value $N_f^{\rm cr} \simeq 4N_c = 12$ is somewhat different from the lattice value,[12] $6 < N_f^{\rm cr} < 7$.

in the walking/conformal regime.[14] Recently, a reduction of the S parameter was further claimed[15] in a version of the holographic QCD with deformation of the replacement of the anomalous dimension $\gamma_m \simeq 0$ by $\gamma_m \simeq 1$ based on AdS/CFT correspondence. Therefore, it would be worth engineering an ETC model building based on the large N_f QCD near the conformal window as a walking/conformal TC. There have been such an attempt[16] based on a generalized version of Most Attractive Channel (MAC) analysis, which showed reasonable phenomenological result. However, it does not seem to explain a large mass of the top quark (top-bottom splitting) in a way consistent with the ρ/T parameter constraint from the LEP precision experiments, the problem being more serious for the walking/conformal TC.[17]

Such a large top quark mass may be explained by the top quark condensation.[18] So, it would be natural to seek a model which accommodates top quark condensate into an explicit scheme of ETC, which, after ETC breaking, would yield a low energy effective theory something resembling the topcolor-assisted technicolor (TC2)[19,20] of a type of the *Flavor–universal TC2*[21] rather than of other variants: *Classic TC2*[19] and *Type II TC2*.[20] In such a scenario the mass of W/Z bosons mainly comes from the TC condensate with small portion from the top quark condensate, while the top quark mass comes mainly from the top quark condensate with small portion from TC condensate. In our setting of ETC framework, other up-type quarks would acquire mass mainly from the top quark condensate, through part of the ETC (horizontal gauge) interactions, with small portion from the TC condensate through ETC (sideway gauge). Down-type quarks and leptons would only acquire mass from TC condensate.

In this talk we shall experiment such a model building incorporating the top quark condensate into an explicit ETC model through sideway gauge interactions. The result roughly reproduces realistic mass hierarchy of quarks and leptons, and nontrivial CKM matrix.

2. A Top Mode ETC[22]

We use a typical one-family TC model[2] with $N_f = 8$ technifermions and with $SU(2)_{\rm TC}$ as a TC gauge group, which is a walking/conformal TC near the edge of the conformal window $N_f \sim 4N_c$[8] evaluated in the ladder approximation.[d] The simplest ETC model would be $SU(5)_{\rm ETC}$ which accom-

[d]In the SCGT06 Workshop we presented a $SU(4)_{\rm TC}$ case which is a rather complicate scenario. Here we present $SU(2)_{\rm TC}$ case which is much simpler for our purpose.

modates $SU(2)_{\rm TC}$ one-family technifermions and three families of quarks and leptons. As usual the $SU(5)_{\rm ETC}$ is expected to successively break down as $SU(5)_{\rm ETC} \to SU(4)_{\rm ETC} \to SU(3)_{\rm ETC} \to SU(2)_{\rm TC}$. Also we introduce the universal coloron-type topcolor symmetry $SU(3)_1 \times SU(3)_2$ which is expected to break down to $SU(3)_{\rm QCD}$.

Particle contents in the model is listed in Table. 1. Q, U, D, L, N, E include technifermions and 3-family quarks and leptons : $Q_L = (Q^a\,,\, q_3\,,\, q_2\,,\, q_1)_L^T\,,\, L_L = (L^a\,,\, l_3\,,\, l_2\,,\, l_1)_L^T\,,\, U_R = (U^a\,,\, t\,,\, c\,,\, u)_R^T, \cdots$, where $a = 1, 2$ is TC indices, $q_i(l_i)$ represents i-th family $SU(2)_L$-doublet quark(lepton) and $Q^a = (U^a\,,\, D^a)^T$.

Table 1. Fermion contents in a suggested model.

field	$SU(5)_{\rm ETC}$	$SU(3)_1$	$SU(3)_2$	$SU(2)_L$	$U(1)_Y$
Q_L	5	3	1	2	1/6
U_R	5	3	1	1	2/3
D_R	5	3	1	1	−1/3
L_L	5	1	1	2	−1/2
N_R	5	1	1	1	0
E_R	5	1	1	1	−1

Apart from the ETC group, this is the same as the *Flavor–universal TC2*[21] in the sense that all quarks have the same charge under the topcolor symmetry $SU(3)_1 \times SU(3)_2$, but for simplicity we do not introduce the additional (strong) $U(1)_Y$ which in the *Flavor–universal TC2* models is coupled only to the third generation quark to trigger the top condensate. Instead, the ETC gauge interaction in our case discriminates the third generation quarks from others near the criticality of the strong coloron interaction. Once the third generation is selected, top is distinguished from bottom by the usual $U(1)_Y$ interaction in the Standard Model near the criticality. [e]

For simplicity, we prepare three (electroweak singlet) effective "Higgs" fields

$$
\begin{aligned}
H_1 &\sim (\underline{5}, \underline{1}, \underline{1}) \text{ (under } SU(5)_{\rm ETC} \times SU(3)_1 \times SU(3)_2)\,, \\
H_2 &\sim (\underline{4}, \underline{1}, \underline{1}) \text{ (under } SU(4)_{\rm ETC} \times SU(3)_1 \times SU(3)_2)\,, \\
H_3 &\sim (\underline{3}, \underline{1}, \underline{1}) \text{ (under } SU(3)_{\rm ETC} \times SU(3)_1 \times SU(3)_2)\,, \\
\Phi &\sim (\underline{1}, \underline{3}, \bar{\underline{3}}) \text{ (under } SU(2)_{\rm TC} \times SU(3)_1 \times SU(3)_2)\,. \qquad (1)
\end{aligned}
$$

[e] This requires some fine tuning which may be avoided by introducing extra strong $U(1)_Y$ as in the *Flavor–universal TC2* models. Such an $U(1)_Y$ to distinguish the third generation from others may not straightforwardly be incorporated into an ETC model, however.

in order to trigger both ETC breaking : $SU(5)_{\rm ETC} \to SU(4)_{\rm ETC} \to SU(3)_{\rm ETC} \to SU(2)_{\rm TC}$ and topcolor breaking : $SU(3)_1 \times SU(3)_2 \to SU(3)_{\rm QCD}$. Here "Higgs" fields H_1 , H_2 , H_3 break $SU(5)_{\rm ETC}$, $SU(4)_{\rm ETC}$ and $SU(3)_{\rm ETC}$ successively through hierarchical condensates $\langle H_1\rangle \gg \langle H_2\rangle \gg \langle H_3\rangle$. Φ breaks $SU(3)_1 \times SU(3)_2$ through $\langle\Phi\rangle$ which we take $\langle\Phi\rangle < \langle H_3\rangle$.

Of course, instead of introducing explicit "Higgs" , the ETC breaking in this model can actually be generated dynamically through a generalized MAC analysis in the same way as in Ref. 16. As to the topcolor breaking which is strongly coupled gauge theory, we may replace the Higgs picture (due to explicit "Higgs" VEV $\langle\Phi\rangle$) by the equivalent confinement picture based on the complementarity, in which case the coloron is regarded as a composite gauge boson of Hidden Local Symmetry[23] as was briefly described in Ref. 24. We shall report detailed analyses elsewhere.[22]

After all ETC and topcolor gauge group breaking by the above symmetry breaking, we obtain the $SU(2)_{\rm TC} \times SU(3)_{\rm QCD}$ invariant four fermion interaction :

$$\mathcal{L}^{4f} = \sum_{i,j} \mathcal{L}^{i-j}_{\rm ETC} + \sum_{i,j} \mathcal{L}^{i-j}_{\rm topC} , \tag{2}$$

where i, j is TC or SM and for example $\mathcal{L}^{\rm TC-SM}_{\rm ETC}$ represents four fermion interactions between technifermions and SM-fermions via massive ETC gauge bosons. $\mathcal{L}^{\rm TC-SM}_{\rm ETC}$ and $\mathcal{L}^{\rm SM-SM}_{\rm ETC(topC)}$ are

$$\begin{aligned}\mathcal{L}^{\rm TC-SM}_{\rm ETC} = & -G_1\Big(\bar{U}_R\gamma^\mu u_R + \bar{D}_R\gamma^\mu d_R\Big)\Big(\bar{q}_{1L}\gamma_\mu Q_L\Big) \\ & -G_2\Big(\bar{U}_R\gamma^\mu c_R + \bar{D}_R\gamma^\mu s_R\Big)\Big(\bar{q}_{2L}\gamma_\mu Q_L\Big) \\ & -G_3\Big(\bar{U}_R\gamma^\mu t_R + \bar{D}_R\gamma^\mu b_R\Big)\Big(\bar{q}_{3L}\gamma_\mu Q_L\Big) ,\end{aligned} \tag{3}$$

$$\begin{aligned}\mathcal{L}^{\rm SM-SM}_{\rm ETC} = & -G_1\Big[\Big(\bar{t}_R\gamma^\mu u_R\Big)\Big(\bar{q}_{1L}\gamma_\mu q_{3L}\Big) + \frac{1}{5}\Big(\bar{u}_R\gamma^\mu u_R\Big)\Big(\bar{q}_{1L}\gamma_\mu q_{1L}\Big)\Big] \\ & -G_2\Big[\Big(\bar{t}_R\gamma^\mu c_R\Big)\Big(\bar{q}_{2L}\gamma_\mu q_{3L}\Big) + \frac{1}{4}\Big(\bar{c}_R\gamma^\mu c_R\Big)\Big(\bar{q}_{2L}\gamma_\mu q_{2L}\Big)\Big] \\ & -\frac{1}{3}G_3\Big(\bar{t}_R\gamma^\mu t_R\Big)\Big(\bar{q}_{3L}\gamma_\mu q_{3L}\Big) + [\text{down sector}, L \leftrightarrow R] ,\end{aligned} \tag{4}$$

$$\begin{aligned}\mathcal{L}^{\rm SM-SM}_{\rm topC} = & -G_1^{\rm topC}\Big[(\bar{q}_{3L}\gamma^\mu q_{3L})(\bar{t}_R\gamma_\mu t_R) + (\bar{q}_{2L}\gamma^\mu q_{2L})(\bar{c}_R\gamma_\mu c_R) \\ & +(\bar{q}_{1L}\gamma^\mu q_{1L})(\bar{u}_R\gamma_\mu u_R) + [\text{down sector}, L \leftrightarrow R]\Big] .\end{aligned} \tag{5}$$

Here, $G_i = c_i g^2_{N(\mathrm{ETC})}/(2\Lambda_i^2)\,, (i = 1,2,3; c_1 = 2\,, c_2 = 3/2\,, c_3 = 1)$, $g_{N(\mathrm{ETC})}\,(N = 6 - i)$ is the $SU(N)$-ETC gauge coupling at the scale of broken gauge boson mass Λ_i, and $G_1^{\mathrm{topC}} = h_1^2 \cos^2\theta/(2\Lambda_{\mathrm{C}}^2)$, where we have taken the coloron mass $\Lambda_{\mathrm{C}} < \Lambda_3$ and $h_{1(2)}$ is $SU(3)_{1(2)}$ gauge coupling and $h_1 \sin\theta = h_2 \cos\theta = g_{\mathrm{QCD}}$. Our hierarchy reads $\Lambda_1 \gg \Lambda_2 \gg \Lambda_3 > \Lambda_{\mathrm{C}}$.

After Fierz transformation, we obtain four fermion interactions for all SM-quarks from Eq.(4,5) as follows :

$$\begin{aligned}\mathcal{L}_{\mathrm{ETC+topC}} = &\left[\frac{1}{5}G_1 + G_1^{\mathrm{topC}}\right]\left((\bar{q}_{1L}u_R)^2 + (\bar{q}_{1L}d_R)^2\right)\\ &+\left[\frac{1}{4}G_2 + G_1^{\mathrm{topC}}\right]\left((\bar{q}_{2L}c_R)^2 + (\bar{q}_{2L}s_R)^2\right)\\ &+\left[\frac{1}{3}G_3 + G_1^{\mathrm{topC}}\right]\left((\bar{q}_{3L}t_R)^2 + (\bar{q}_{3L}b_R)^2\right). \end{aligned} \tag{6}$$

In the gap equation it is convenient to define dimensionless four fermion couplings: $g_i^{(\mathrm{ETC})} = G_i\, N_c \Lambda_{\mathrm{C}}^2/(4\pi^2) = c_i\, N_c \alpha_{N(\mathrm{ETC})}/(2\pi)\cdot(\Lambda_{\mathrm{C}}/\Lambda_i)^2$ and $g_{\mathrm{topC}} = G_{\mathrm{topC}}\, N_c\Lambda_{\mathrm{C}}^2/(4\pi^2)$, where $\alpha_{N(\mathrm{ETC})} = g^2_{N(\mathrm{ETC})}/(4\pi)$. Then the dimensionless four fermion couplings for the SM-quarks read:

$$g_{t,b} = \frac{g_3^{(\mathrm{ETC})}}{3} + g_{\mathrm{topC}} = \frac{N_c}{6\pi}\alpha_{\mathrm{TC}}(\Lambda_3)\left(\frac{\Lambda_{\mathrm{C}}}{\Lambda_3}\right)^2 + g_{\mathrm{topC}}\,, \tag{7}$$

where we have set $\alpha_{3(\mathrm{ETC})}(\Lambda_3) = \alpha_{\mathrm{TC}}(\Lambda_3)$ which is roughly given by the the BZ IR fixed point of the walking $SU(2)$ TC near criticality: $\alpha_{\mathrm{TC}}(\Lambda_3) \simeq \alpha_* = 2\pi/5\,(N_f = 8)$. Similarly, we have

$$g_{c,s} = \frac{g_2^{(\mathrm{ETC})}}{4} + g_{\mathrm{topC}} = \frac{3N_c}{16\pi}\alpha_{3(\mathrm{ETC})}(\Lambda_2)\left(\frac{\Lambda_{\mathrm{C}}}{\Lambda_2}\right)^2 + g_{\mathrm{topC}}\,, \tag{8}$$

$$g_{u,d} = \frac{g_1^{(\mathrm{ETC})}}{5} + g_{\mathrm{topC}} = \frac{N_c}{5\pi}\alpha_{4(\mathrm{ETC})}(\Lambda_1)\left(\frac{\Lambda_{\mathrm{C}}}{\Lambda_1}\right)^2 + g_{\mathrm{topC}}\,, \tag{9}$$

where we have used $\alpha_{4(\mathrm{ETC})}(\Lambda_2) = \alpha_{3(\mathrm{ETC})}(\Lambda_2)$ and $\alpha_{5(\mathrm{ETC})}(\Lambda_1) = \alpha_{4(\mathrm{ETC})}(\Lambda_1)$. Since the ETC gauge couplings are asymptotically free,

$$\mu\frac{\partial}{\partial\mu}\alpha_{N(\mathrm{ETC})} = -b_N\,\alpha^2_{N(\mathrm{ETC})}, \quad b_N = \frac{11N - 2N_f}{6\pi}(>0)\,, \tag{10}$$

for $N_f = 8$ and $N \geq 3$, we can arrange a hierarchy: $\frac{1}{6\pi}\alpha_{\mathrm{TC}}(\Lambda_3) = \frac{1}{6\pi}\alpha_{3(\mathrm{ETC})}(\Lambda_3) > \frac{3}{16\pi}\alpha_{3(\mathrm{ETC})}(\Lambda_2) > \frac{1}{5\pi}\alpha_{4(\mathrm{ETC})}(\Lambda_1)$, which is even amplified by the additional factors $(\Lambda_{\mathrm{C}}/\Lambda_i)^2$ for $g_i^{(\mathrm{ETC})}$, so that the condensate of the third generation quarks is favored to others.

Top Mode Conditions

Now we discuss conditions for only the top quark to condense due to the near critical four fermion interactions from topcolor and ETC with small perturbation of the SM gauge interactions, based on the critical line of the gauged NJL model[25,26] :

$$g^{\rm crit} = \frac{1}{4}\left(1+\sqrt{1-\frac{\alpha}{\alpha_{\rm c}}}\right)^2 , \quad (\alpha < \alpha_{\rm c} = \pi/3) . \tag{11}$$

The gauge coupling α for each quark has contributions from the SM gauge interactions:

$$\alpha_{t,c,u}(\mu) = \frac{4}{3}\alpha_3(\mu) + \frac{1}{9}\alpha_Y(\mu) , \tag{12}$$

$$\alpha_{b,s,d}(\mu) = \frac{4}{3}\alpha_3(\mu) - \frac{1}{18}\alpha_Y(\mu) , \tag{13}$$

where $\alpha_3(\alpha_Y)$ is $SU(3)_{\rm QCD}\,(U(1)_Y)$ gauge coupling constant. Here we set $\mu = \Lambda_{\rm C}\,(= 50\,{\rm TeV})$, in which case $\alpha_{t,c,u}(\Lambda_{\rm C}) = 0.0952 \pm 0.0009$, $\alpha_{b,s,d}(\Lambda_{\rm C}) = 0.0933 \pm 0.0009$ corresponding to $\alpha_Y(M_Z) = 0.0101684 \pm 0.0000014$, $\alpha_3(M_Z) = 0.1176 \pm 0.0020$. The corresponding critical line reads: $g^{\rm crit}_{t,c,u} = g^{\rm crit}\,(\alpha = \alpha_{t,c,u}(\Lambda_{\rm C})) = 0.9540 \pm 0.0005$, $g^{\rm crit}_{b,s,d} = g^{\rm crit}\,(\alpha = \alpha_{b,s,d}(\Lambda_{\rm C})) = 0.9549 \pm 0.0005$. In order for $\langle \bar{t}t\rangle \neq 0$ and $\langle \bar{b}b\rangle = \langle \bar{c}c\rangle = \langle \bar{s}s\rangle = \langle \bar{d}d\rangle = \langle \bar{u}u\rangle = 0$ are realized, we must tune $g_{\rm topC}$ to satisfy

$$g_t > g^{\rm crit}_t = \frac{1}{4}\left(1+\sqrt{1-\frac{\alpha_t}{\alpha_{\rm c}}}\right)^2 , \tag{14}$$

and $g_{b,c,s,d,u} < g^{\rm crit}_{b,c,s,d,u}$, where $g_{t,b,c,s,d,u}$ are given by Eqs. (7, 8, 9). Defining $g^{\rm ETC}_k = g_k - g_{\rm topC}\,(k = t,b,c,c,d,u)$, we show such conditions in the $(g^{\rm ETC}, g_{\rm topC})$ plane in Fig. 1 where $g_{t,c,u}(= g^{\rm ETC}_{t,c,u} + g_{\rm topC}) = g^{\rm crit}_{t,c,u}$ and $g_{b,s,d} = g^{\rm crit}_{b,s,d}$ are depicted by the line (a) and (b), respectively. The line (i) $g^{\rm ETC}_{t,b} = N_c/15 \cdot (\Lambda_{\rm C}/\Lambda_3)^2 = 0.0222$ is the model setting of present model for $\Lambda_3 = 150\,{\rm TeV}$. Now, the top quark condensate is possible only for $g_{\rm topC}$ sitting in the region (II) and (III) but $g_{\rm topC}$ in the region (III) implies also the bottom condensate. Thus $g_{\rm topC}$ must be in (II). The condition to forbid charm/up condensate is that $(g^{\rm ETC}_c, g_{\rm topC})$ and $(g^{\rm ETC}_u, g_{\rm topC})$ is below the line (ii) in the region (II), which is easily fullfilled: For instance, we may take $\Lambda_{\rm C} = 50\,{\rm TeV}$, $\Lambda_3 = 150\,{\rm TeV}$, $\Lambda_2 = 600\,{\rm TeV}$, $\Lambda_1 = 3000\,{\rm TeV}$, in which case we have $\alpha_{4{\rm ETC}}(\Lambda_1) = 0.225$ and $\alpha_{4{\rm ETC}}(\Lambda_2) = \alpha_{3{\rm ETC}}(\Lambda_2) = 0.488$ by Eq.(10). This implies $g^{\rm ETC}_{c,s} = 6.06 \times 10^{-4}$, $g^{\rm ETC}_{u,d} = 1.19 \times 10^{-5}$ which is much smaller than the value of (ii) in Fig. 1.

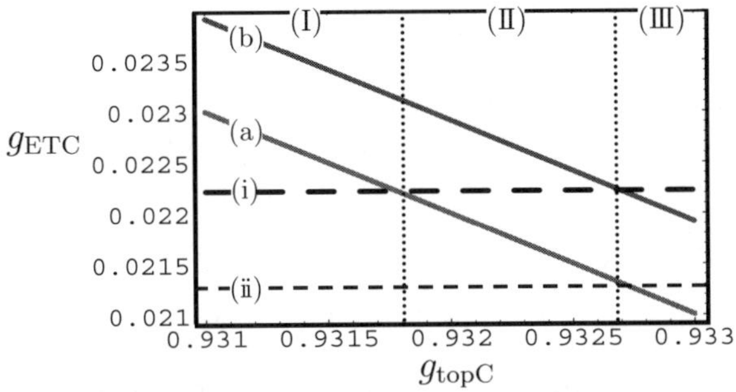

Fig. 1. Lines (a) and (b) are $g^{\rm crit}_{t,c,u}$ and $g^{\rm crit}_{b,d,s}$, respectively. The line (i) indicates our model setting $g^{\rm ETC}_{t,b}(\Lambda_3 = 150{\rm TeV}, \Lambda_{\rm C} = 50{\rm TeV}) = g_{t,b} - g_{\rm topC} = N_c/15\,(\Lambda_{\rm C}/\Lambda_3)^2$ (Eq.(7)). For $g_{\rm topC}$ in the region (II), $\langle \bar{t}t\rangle \neq 0, \langle \bar{b}b\rangle = 0$. When $g^{\rm ETC}_{c,s}(\Lambda_2) = g_{c,s} - g_{\rm topC}$ (Eq.(8)) and $g^{\rm ETC}_{u,d}(\Lambda_1) = g_{u,d} - g_{\rm topC}$ (Eq.(9)) is below (ii), and $g_{\rm topC}$ is in the region (II), we have $\langle \bar{c}c\rangle = \langle \bar{s}s\rangle = \langle \bar{u}u\rangle = \langle \bar{d}d\rangle = 0$.

The non-zero top quark mass coming from top condensate $\hat{m}_t$ ($\simeq m_t^{\rm exp} = 172 \pm 2.5{\rm GeV}$)) is related to $\Lambda_{\rm C}$ through the SD gap equation:[26]

$$g^{\rm crit}_{m_t} = \frac{1}{4}\frac{(1+\sqrt{1-\alpha_t/\alpha_c})^2 - (1-\sqrt{1-\alpha_t/\alpha_c})^2(P+Q)}{1-P+\frac{3-\sqrt{1-\alpha_t/\alpha_c}}{1+\sqrt{1-\alpha_t/\alpha_c}}Q}, \tag{15}$$

where P, Q are

$$P \equiv \frac{\Gamma(1-\sqrt{1-\alpha_t/\alpha_c})\Gamma(3/2+\frac{1}{2}\sqrt{1-\alpha_t/\alpha_c})^2}{\Gamma(1+\sqrt{1-\alpha_t/\alpha_c})\Gamma(3/2-\frac{1}{2}\sqrt{1-\alpha_t/\alpha_c})^2}\left(\frac{\hat{m}_t^2}{\Lambda_{\rm C}^2}\right)^{\sqrt{1-\alpha_t/\alpha_c}},$$
$$Q \equiv \frac{(1+\sqrt{1-\alpha_t/\alpha_c})^2}{4(1-\sqrt{1-\alpha_t/\alpha_c})}\frac{\hat{m}_t^2}{\Lambda_{\rm C}^2}. \tag{16}$$

Note that $g^{\rm crit}_{m_t}$ is reduced to the critical line $g_t^{\rm crit}$ (Eq. 11) for $\hat{m}_t \to 0$. In order to forbid bottom condensate, $g^{\rm crit}_{m_t}$ should not exceed $g^{\rm crit}_{b,s,d}$ (line (b) in Fig. 1), which implies $\Lambda_{\rm C} > 33\,{\rm TeV}$ for $\hat{m}_t \simeq 170\,{\rm GeV}$. Here we take $\Lambda_{\rm C} = 50\,{\rm TeV}$. In order to have a realistic top mass $\hat{m}_t/\Lambda_{\rm C} \simeq (170{\rm GeV})/50{\rm TeV} \sim 3\times 10^{-3}$, we would need a fine-tuning of $g_{\rm topC}$ of $\mathcal{O}(10^{-4})$ to arrange $g^{\rm crit}_{m_t} - g_t^{\rm crit} \sim P \sim (\hat{m}_t/\Lambda_{\rm C})^2 = \mathcal{O}(10^{-5})$.

Estimation of mass in the present model

In our model, the TC theory is walking/conformal below Λ_3 and the technipion (π_T) decay constant: f_{π_T} (by Pagels-Stokar formula[27]) and technifermion condensation: $\langle\bar{Q}Q\rangle$ are given by

$$f_{\pi_T}^2 = \frac{N}{4\pi^2}\int_0^\infty dx x \frac{\Sigma^2(x) - \frac{x}{4}\frac{d}{dx}\Sigma^2(x)}{(x+\Sigma^2(x))^2} = \frac{N}{8\pi}\frac{3-\frac{\gamma_m}{2}}{(3-\gamma_m)^2}\frac{m_{\rm TC}^2}{\sin(\frac{\pi}{3-\gamma_m})}\,, \quad (17)$$

$$\langle\bar{Q}Q\rangle_{\Lambda_3} = -\frac{N}{4\pi^2}\int_0^{\Lambda_3^2} dx \frac{x\Sigma(x)}{x+\Sigma(x)^2} = -\frac{N}{4\pi}\frac{2}{\gamma_m}\left(\frac{\Lambda_3}{m}\right)^{\gamma_m} m_{\rm TC}^3\,, \quad (18)$$

with $m_{\rm TC}$ being a dynamical mass of the technifermion $\Sigma(x = m_{\rm TC}^2) = m_{\rm TC}$, where

$$\Sigma(x) \sim m_{\rm TC}\left(\frac{x}{m_{\rm TC}^2}\right)^{\gamma_m/2-1}. \quad (19)$$

Thus in the case of walking dynamics $\gamma_m \simeq 1$, Eq.(17,18) lead to a relation between $\langle\bar{Q}Q\rangle_{\Lambda_3}$ and f_{π_T} : $\langle\bar{Q}Q\rangle_{\Lambda_3}/f_{\pi_T}^2 \sim -\frac{32}{5}\Lambda_3$. Similarly the top-pion(π_t) decay constant f_{π_t} is

$$f_{\pi_t}^2 = \frac{3}{8\pi^2}\hat{m}_t^2 \ln\frac{\Lambda_{\rm C}^2}{\hat{m}_t^2}\,. \quad (20)$$

If we choose $\Lambda_{\rm C} = 50\,{\rm TeV}$, we have $f_{\pi_t} = 78\,{\rm GeV}$ for $\hat{m}_t = 168\,{\rm GeV}$ and $f_{\pi_T} = 117\,{\rm GeV}$ such that $(246\,{\rm GeV})^2 = 4f_{\pi_T}^2 + f_{\pi_t}^2$. Hence $m_b = G_3\langle\bar{Q}Q\rangle_{\Lambda_3} \simeq 4\,{\rm GeV}$ for $\Lambda_3 = 150\,{\rm TeV}$ and $m_t = \hat{m}_t + G_3\langle\bar{Q}Q\rangle_{\Lambda_3} \simeq 172\,{\rm GeV}$. We obtain other quark masses ($u_i = u, c$, $d_i = d, s$)

$$m_{u_i} = G_i\Big[\langle\bar{Q}Q\rangle_{\Lambda_3} + \langle\bar{t}t\rangle_{\Lambda_{\rm C}}\Big]\,, \quad m_{d_i} = G_i\langle\bar{Q}Q\rangle_{\Lambda_3}\,, \quad (21)$$

where for simplicity we have assumed $\gamma_m \simeq 0$ for $\Lambda_3 < p < \Lambda_1$, while we estimated $\langle\bar{t}t\rangle_{\Lambda_{\rm C}} \simeq -3\hat{m}_t\Lambda_{\rm C}^2/4\pi^2$ for constant top mass function ($\gamma_m \simeq 2$). If we take $\Lambda_1 = 3000\,{\rm TeV}, \Lambda_2 = 600\,{\rm TeV}$, then $\alpha_{5\rm ETC}(\Lambda_1) = \alpha_{4\rm ETC}(\Lambda_1) = 0.225$, $\alpha_{4\rm ETC}(\Lambda_2) = \alpha_{3\rm ETC}(\Lambda_2) = 0.488$, $\alpha_{3\rm ETC}(\Lambda_3) = \alpha_{\rm TC}(\Lambda_3) \simeq 2\pi/5$, and hence we obtain

$$\begin{aligned} m_t \simeq 172\,{\rm GeV}\,, \; m_c \simeq 0.6\,{\rm GeV}\,, \; m_u \simeq 10\,{\rm MeV}\,, \\ m_b \simeq 4\,{\rm GeV}\,, \; m_s \simeq 160\,{\rm MeV}\,, \; m_d \simeq 4\,{\rm MeV}\,, \end{aligned} \quad (22)$$

in rough agreement with the reality except for the charm mass. The charm mass prediction should be improved if we take into account the non-zero anomalous dimension of technifermion condensate for $\Lambda_3 < p < \Lambda_2$.

3. Summary

We have experimented a simple idea to combine the topquark condensate into the explicit ETC model. It resulted in a reasonable quark mass hierarchy, although fine-tuning of order $\mathcal{O}(10^{-4})$ is needed for the topcolor-induced four fermion coupling. We can further expect the followings:

(1) *Quark–lepton mass splitting*
After ETC and topC breaking, $SU(2)_{\rm TC} \times SU(3)_{\rm QCD}$ remains and $SU(2)_{\rm TC}$ is near critical, so techniquarks should form their condensation before technileptons condensate due to $SU(3)_{\rm QCD}$ effects.[28] Thus it is likely to be able to explain a mass splitting between quarks and leptons.

(2) *Quark mixing angles*
Apart from the topcolor, the ETC breaking in the present model is essentially the same as Ref. 16 and hence the mass matrix of down-type fermion is essentially the same as Ref. 16, while the up-type mass matrix comes from both TC and top condensates in contrast to Ref. 16. We then naturally expect a nontrivial CKM matrix.

Details of the analysis including the phenomenology relevant to the LHC will be given in the forthcoming paper.[22]

References

1. S. Weinberg, Phys. Rev. D **13**, 974 (1976); Phys. Rev. D **19**, 1277 (1979); L. Susskind, Phys. Rev. D **20**, 2619 (1979).
2. C. T. Hill and E. H. Simmons, Phys. Rept. **381**, 235 (2003) [Erratum-ibid. **390**, 553 (2004)].
3. S. Dimopoulos and L. Susskind, Nucl. Phys. B **155**, 237 (1979); E. Eichten and K. D. Lane, Phys. Lett. B **90**, 125 (1980).
4. K. Yamawaki and T. Yokota, Phys. Lett. B **113** (1982), 293; Nucl. Phys. B **223** (1983), 144.
5. M. E. Peskin and T. Takeuchi, Phys. Rev. Lett. **65**, 964 (1990); B. Holdom and J. Terning, Phys. Lett. B **247**, 88 (1990); M. Golden and L. Randall, Nucl. Phys. B **361**, 3 (1991).
6. B. Holdom, Phys. Lett. B **150**, 301 (1985).
7. K. Yamawaki, M. Bando and K. Matumoto, Phys. Rev. Lett. **56**, 1335 (1986); T. Akiba and T. Yanagida, Phys. Lett. B **169**, 432 (1986); T. W. Appelquist, D. Karabali and L. C. R. Wijewardhana, Phys. Rev. Lett. **57**, 957 (1986).
8. T. Appelquist, J. Terning and L. C. R. Wijewardhana, Phys. Rev. Lett. **77**, 1214 (1996); T. Appelquist, A. Ratnaweera, J. Terning and L. C. R. Wijewardhana, Phys. Rev. D **58**, 105017 (1998).

9. V. A. Miransky and K. Yamawaki, Phys. Rev. D **55**, 5051 (1997) [Erratum-ibid. D **56**, 3768 (1997)].
10. T. Banks and A. Zaks, Nucl. Phys. B **196**, 189 (1982).
11. D. K. Hong, S. D. H. Hsu and F. Sannino, Phys. Lett. B **597**, 89 (2004).
12. Y. Iwasaki, K. Kanaya, S. Kaya, S. Sakai and T. Yoshie, Phys. Rev. D **69**, 014507 (2004).
13. T. Appelquist and F. Sannino, Phys. Rev. D **59**, 067702 (1999).
14. M. Harada, M. Kurachi and K. Yamawaki, Prog. Theor. Phys. **115**, 765 (2006).
15. D. K. Hong and H. U. Yee, Phys. Rev. D **74**, 015011 (2006); M. Piai, arXiv:hep-ph/0608241; K. Haba, S. Matsuzaki and K. Yamawaki, in preparation.
16. T. Appelquist and J. Terning, Phys. Rev. D **50**, 2116 (1994); T. Appelquist, M. Piai and R. Shrock, Phys. Rev. D **69**, 015002 (2004); R. Shrock's talk in SCGT06.
17. R. S. Chivukula, Phys. Rev. Lett. **61**, 2657 (1988).
18. V. A. Miransky, M. Tanabashi and K. Yamawaki, Phys. Lett. B **221**, 177 (1989); Mod. Phys. Lett. A **4**, 1043 (1989); Y. Nambu, Enrico Fermi Institute Report No.89-08,1989; W. A. Bardeen, C. T. Hill and M. Lindner, Phys. Rev. D **41**, 1647 (1990).
19. C. T. Hill, Phys. Lett. B **345**, 483 (1995).
20. C. T. Hill, Phys. Lett. B **266**, 419 (1991); G. Buchalla, G. Burdman, C. T. Hill and D. Kominis, Phys. Rev. D **53**, 5185 (1996).
21. K. D. Lane and E. Eichten, Phys. Lett. B **352**, 382 (1995); R. S. Chivukula, A. G. Cohen and E. H. Simmons, Phys. Lett. B **380**, 92 (1996); K. D. Lane, Phys. Lett. B **433**, 96 (1998); M. B. Popovic and E. H. Simmons, Phys. Rev. D **58**, 095007 (1998).
22. H. Fukano and K. Yamawaki, in preparation.
23. M. Bando, T. Kugo, S. Uehara, K. Yamawaki and T. Yanagida, Phys. Rev. Lett. **54** (1985), 1215; M. Bando, T. Kugo and K. Yamawaki, Nucl. Phys. B **259** (1985), 493; Phys. Rept. **164** (1988), 217; M. Harada and K. Yamawaki, Phys. Rept. **381**, 1 (2003).
24. M. Hashimoto, M. Tanabashi and K. Yamawaki, *Prepaed for TMU - Yale Symposium on Dynamics of Gauge Fields: An External Activity of APCTP, Tokyo, Japan, 13-15 Dec 1999.*
25. K.-I. Kondo, H. Mino, and K. Yamawaki, Phys. Rev. D **39**, 2430 (1989); K. Yamawaki, in *Proc. Johns Hopkins Workshop on Current Problems in Particle Theory 12, Baltimore, June 8-10, 1988*, edited by G. Domokos and S. Kovesi-Domokos (World Scientific Pub. Co., Singapore 1988); T. Appelquist, M. Soldate, T. Takeuchi, and L. C. R. Wijewardhana, *ibid.*
26. T. Nonoyama, T. B. Suzuki and K. Yamawaki, Prog. Theor. Phys. **81**, 1238 (1989).
27. H. Pagels and S. Stokar, Phys. Rev. D **20**, 2947 (1979).
28. T. Appelquist, J. Terning and L. C. R. Wijewardhana, Phys. Rev. Lett. **79**, 2767 (1997).

RADIATIVE ELECTROWEAK SYMMETRY BREAKING IN TeV-SCALE STRING MODELS

N. KITAZAWA

Department of Physics, Tokyo Metropolitan University,
Hachioji, Tokyo 192-0397, Japan
E-mail: kitazawa@phys.metro-u.ac.jp

We examine the possibility of necessary and inevitable radiative electroweak symmetry breaking by one-loop radiative corrections in a class of string models which are the string realization of "brane world" scenario. As an example, we consider a simple quasi-realistic model based on a D3-brane at non-supersymmetric $\mathbf{C}^3/\mathbf{Z}_6$ orbifold singularity, in which the electroweak Higgs doublet fields are identified with the massless bosonic modes of the open string on that D3-brane. We calculate the one-loop correction to the Higgs potential, and find that the its vacuum expectation value can be realized in this specific model.

Keywords: String phenomenology; Electroweak symmetry breaking.

1. Introduction and Motivations

The electroweak symmetry breaking in the standard model is not necessary and inevitable. Though technicolor theories realize necessary and inevitable electroweak symmetry breaking, like the chiral symmetry breaking in QCD, it is not easy to construct simple viable models which is consistent with the experimental data.[1,2] The radiative electroweak symmetry breaking scenario in viable models with TeV-scale supersymmetry is also a candidate of the necessary and inevitable electroweak symmetry breaking, but it requires a certain level of fine tuning among parameters in models,[3] which reduces the necessity and inevitability. There are many scenarios utilizing extra space dimensions, but it seems very difficult to construct theoretically well-defined concrete models with necessary and inevitable electroweak symmetry breaking.

Although much efforts have been made towards clarify the mechanism of the electroweak symmetry breaking in the framework of the quantum field theory, there are few efforts in the framework of the string theory, in spite of its potential as the unified theory including gravity. The radiative electroweak symmetry breaking in string models using D-branes is first examined by Antoniadis, Benakli and Quirós.[4] They have analyzed a concrete string model with no supersymmetry, and have shown that some tree-level massless scalar fields (open string states) can follow non-trivial potential at one-loop level, and obtain non-zero vacuum expectation values. The scale of the vacuum expectation value is the radius of some compactified space, which is difficult to determine in the framework of perturbative (even non-perturbative) string theory.

The author has pointed out the possibility of the radiative electroweak symmetry breaking in which the scale of the vacuum expectation value is determined by the string scale (string tension).[5] This necessary and inevitable electroweak symmetry breaking can happen in a class of non-supersymmetric TeV-scale string models which is constructed by "bottom-up approach".[6]

In the following, the model of Ref.5 is very briefly introduced, and discuss possibility and problems towards constructing realistic models.

2. A Quasi-Realistic Model

The model is based on ten-dimensional type IIB superstring theory. The six space dimensions are considered to be compactified to some orbifold space which is the toroidal compactification with some identifications of points by some discrete transformations. Consider D3-branes whose all the three space directions are not compactified. In six-dimensional compactified space, the places of such D3-branes are specified by a point. In case that D3-branes are at a orbifold singular point (fixed point by some discrete transformation), it is shown that gauge-interacting chiral fermions can be realized in the four-dimensional world-volume of the D3-branes, and it is possible to construct models of particle physics (see Ref.6 and references therein). Depending on the nature of the singularity, the world on D3-branes becomes supersymmetric or non-supersymmetric.

Since many properties of the world on D3-brane are determined locally by the nature of the singularity, we may only specify the nature of the singularity without concretely specify the six-dimensional compactified space as the first step. Consider $\mathbf{C}^3/\mathbf{Z}_6$ orbifold singularity, where $\mathbf{C}$ denotes two dimensional space parameterized by one complex coordinate and $\mathbf{Z}_6$ denotes a special discrete rotational transformation which does not preserve the supersymmetry on the D3-branes.[5] Only $\mathbf{Z}_6$ invariant states are survived on the D3-branes. The original massless open string states on a stack of N D3-branes are described by the four-dimensional $\mathcal{N} = 4$ supersymmetric U(N) Yang-Mills theory. The $\mathbf{Z}_6$ projection can break original U(N) gauge symmetry by setting non-trivial $\mathbf{Z}_6$ transformation property on the Chan-Paton factor of the open string on the D3-branes.

By considering a stack of $N = 6$ D3-brane at a $\mathbf{C}^3/\mathbf{Z}_6$ singularity with some specific $\mathbf{Z}_6$ transformation properties of Chan-Paton indices, we have U(3)×U(2)×U(1) gauge symmetry with the following massless particle contents on the D3-branes (see Ref.5 for detail).

field	U(3)	U(2)	U(1)
$q_L \times 3$	3	2*	0
$\bar{u}_L \times 3$	3*	1	1
$H \times 3$	1	2	−1

Namely, we have three generations of left-handed quark doublets and right-handed up-type quarks and three Higgs doublet fields with the definition of the hypercharge

$$Q_Y \equiv -\left(\frac{Q_3}{3} + \frac{Q_2}{2} + Q_1\right), \tag{1}$$

where Q_n ($n = 1, 2, 3$) is the U(1) charges of U(n) on D3-brane. It is clear that this system has chiral anomalies and more chiral fermions should be introduced to become a consistent system. In the language of the string theory, we have to introduce some D7-branes to cancel Ramond-Ramond tadpoles. Introduction of D7-branes gives further massless particles and the chiral anomaly is cancelled out.

Three massless Higgs fields follow the potential

$$V = \frac{g^2}{4} \sum_{r,s=1,2,3} \left((H_r^\dagger H_s)(H_s^\dagger H_r) + (H_r^\dagger H_r)(H_s^\dagger H_s)\right) \tag{2}$$

which has no flat directions (g denotes a gauge coupling). Therefore, if one-loop radiative correction gives negative mass squared, Higgs fields necessary have vacuum expectation values. These Higgs fields also have the following Yukawa couplings.

$$\mathcal{L}_Y = -g \sum_{r,s,t=1,2,3} \epsilon_{rst} \bar{u}_R^r q_L^s H^t + \text{h.c.} \tag{3}$$

Though these Yukawa couplings do not have rich structure to generate hierarchical quark mass spectrum, the mixing of three Higgs doublet fields may realize hierarchy. Three Higgs field is not necessary equivalent at one-loop level depending on the way of introduction of D7-branes.

3. One-loop correction to the Higgs Potential

The one-loop radiative correction to the Higgs mass can be calculated in string world-sheet theory. The two-point function of a Higgs state is divided by two parts

$$A_{\mathrm{D3-D3}} = A^{\mathrm{NS}}_{\mathrm{D3-D3}} + A^{\mathrm{R}}_{\mathrm{D3-D3}} \tag{4}$$

where D3-D3 means the contribution of open string on the D3-branes, NS means Neveu-Schwarz sector describing boson loop contributions, and R means Ramond sector describing fermion loop contributions. There is a similar contribution by the open string whose one edge is on the D3-branes and another edge is on the D7-branes.

$$A_{\mathrm{D3-D7}} = A^{\mathrm{NS}}_{\mathrm{D3-D7}} + A^{\mathrm{R}}_{\mathrm{D3-D7}} \tag{5}$$

Since boson and fermion loop give positive and negative contributions to the Higgs mass squared, respectively, if the fermionic contribution is larger than the bosonic one, the one-loop Higgs mass squared becomes negative.

The calculation is straightforward, but some comments must be given before presenting the result.

The one-loop diagram of the open string can be understood as the tree propagation of the closed string. The one-loop correction to the Higgs mass squared can be described as the summation of closed string propagation from D3-brane to D3- or D7-branes with different transverse momenta (transverse momenta, which are not conserved, mean the momenta in the direction perpendicular to the D3 or D7-branes). If there exist massless open string states which have tadpole coupling to D3-branes, they give infrared divergences in the zero transverse momentum limit. This is so called NS-NS tadpole problem. In the present model NS-NS tadpole problem exists in the contribution of untwisted closed string, and it is absent in the contribution of twisted closed string. Twisted closed staring is the string closed up to $\mathbf{Z}_6$ identifications, and untwisted closed string is the usual closed string. (There are two untwisted closed string sector and four twisted closed string sector in the present model.) The existence of NS-NS tadpole problem is understood as the signature that the assumed background geometry (or vacuum) is not the solution and some background redefinition is required.[7,8] It is difficult to solve this problem, since the formulation of the strings propagating non-trivial background is difficult. It may be possible that we can calculate physical quantity even in wrong ground state by the method of "tadpole resummation".[9] At present, we simply neglect the contributions of untwisted closed string.

There is a tachyon mode in the untwisted open string sector, which also means that the vacuum is unstable. It is expected that the model without tachyon mode can be constructed by considering special compactification or projection. Since the origin of the tachyon mode in the present model is related with the tachyon mode in type 0B theory,[10] and since the tachyon mode in type 0B theory is removed by a special projection (Sagnotti's type 0'B model[11]), the model without tachyon mode is expected to be constructed by using type 0'B model.

The result of the calculation of one-loop Higgs mass squared is

$$m^2 \sim -\frac{g^2}{16\pi^2}\frac{2}{\alpha'}\frac{36\sqrt{3}}{\pi}\int_0^\infty \frac{ds}{s}\frac{e^{-2\pi^2/s}}{1-e^{-2\pi^2/s}}e^{-s/3}, \tag{6}$$

where the contribution of the untwisted closed string sector and the physical divergence due to the Yukawa coupling with massless fermions (see Eq.(3)) are neglected. This is an order estimate by neglecting the exponentially suppressed higher order contribution in the twisted open string sectors. The one-loop contribution to the mass squared is negative, because, technically, bosonic one-loop contributions of twisted sectors are cancelled out with each other due to the non-trivial $\mathbf{Z}_6$ transformation rule of Chan-Paton indices. The order of the "electroweak scale" is numerically obtained as

$$v \sim \sqrt{-m^2/g^2} \simeq 10^{-2}\alpha'^{-1/2}, \tag{7}$$

where we used that the Higgs quartic coupling is given by the gauge coupling (see Eq.(2)). This is consistent with the standard expectation that the scale is essentially given by the string scale, $1/\sqrt{\alpha'}$, with one-loop suppression factor. This result suggests that necessary and inevitable radiative electroweak symmetry breaking is possible in this type of models with D-branes at non-supersymmetric orbifold singularities.

It would be very interesting to explorer realistic models in this direction. If this scenario would be true, we would observe very rich phenomena in future high-energy colliders, LHC for example, which are very different from that would be expected in field theory models.[12]

Acknowledgments

The author would like to thank A.Sagnotti for fruitful discussions and the kind hospitality during the stay in The Scuola Normale Superiore in Pisa. This work has been supported in part by the Selective Research Fund of Tokyo Metropolitan University.

References

1. M.E. Peskin and T. Takeuchi, *Phys. Rev. Lett* **65**, 964 (1990).
2. M.E. Peskin and T. Takeuchi, *Phys. Rev.* **D46**, 381 (1992).
3. G.L. Kane et al., *Phys. Lett.* **B551** 146 (2003).
4. I. Antoniadis, K. Benakli and M. Quirós, *Nucl. Phys.* **B583** 35 (2000).
5. N. Kitazawa, *Nucl. Phys.* **B755** 254 (2006).
6. G. Aldazabal, L.E. Ibanez, F. Quevedo and A.M. Uranga, *JHEP* **0008** 002 (2000).
7. W. Fischler and L .Susskind, *Phys. Lett.* **B171** 383 (1986).
8. W. Fischler and L .Susskind, *Phys. Lett.* **B173** 262 (1986).
9. E. Dudas, G. Pradisi, M. Nicolosi and A. Sagnotti, *Nucl. Phys.* **B708** 3 (2005).
10. A. Font and A. Hernández, *Nucl. Phys.* **B634** 51 (2002).
11. A. Sagnotti, hep-th/9509080; *Nucl. Phys. Proc. Suppl.* **56B** 332 (1997).
12. See for example, A. Cullen, M. Perelstein and M.E. Peskin, *Phys. Rev.* **D62** 055012 (2000).

ON CYCLIC UNIVERSES

P. H. FRAMPTON

Department of Physics and Astronomy,
University of North Carolina,
Chapel Hill, NC 27599-3255, USA.
E-mail: frampton@physics.unc.edu

Dark energy in a brane world reconciles an infinitely cyclic cosmology with the second law of thermodynamics. At turnaround one causal patch with no matter and vanishing entropy is retained for the contraction.

1. Introduction

An old question in theoretical cosmology is whether an infinitely oscillatory universe which avoids an initial singularity can be consistently constructed. As realized by Friedmann[1] and especially by Tolman[2,3] one principal obstacle is the second law of thermodynamics which dictates that the entropy increases from cycle to cycle. If the cycles thereby become longer, extrapolation into the past will lead back to an initial singularity again, thus removing the motivation to consider an oscillatory universe in the first place. This has led to the abandonment of the oscillatory universe by many workers.

Some work has been started to exploit the dark energy in allowing cyclicity possibly without the need for inflation in.[4–8] Another new ingredient is the use of branes and a fourth spatial dimension as in Refs. 9–12 which examined consequences for cosmology. The Big Rip and replacement of dark energy by modified gravity were explored in Refs. 13 and 14.

If the dark energy has a super-negative equation of state, $\omega_\Lambda = p_\Lambda/\rho_\Lambda < -1$, it leads to a Big Rip[15] at a finite future time where there exist extraordinary conditions with regard to density and causality as one approaches the Rip. In the present article we explore whether these exceptional conditions can assist in providing an infinitely cyclic model. We consider a model[16] where, as we approach the Rip, expansion stops due to a brane contribution just short of the Big Rip and there is a turnaround at time $t = t_T$ when the scale factor is deflated to a very tiny fraction (f) of itself and only one causal patch is retained, while the other $1/f^3$ patches contract independently to separate universes. Turnaround takes place an extremely short time ($< 10^{-27}s$) before the Big Rip would have occurred, at a time when the universe is fractionated into many independent causal patches.[14] We discuss contraction which occurs with a very much smaller universe than in expansion and with almost vanishing entropy because it is assumed empty of dust, matter and black holes.

A bounce takes place a short time before a would-be Big Bang. After the bounce, entropy is injected by inflation,[17] where is assumed that an inflaton field is excited. Inflation is thus be a part of the present model which is one distinction from the work of Refs. 5–8. For cyclicity of the entropy, $S(t) = S(t + \tau)$ to be consistent with thermodynamics it is necessary that the deflationary decrease by f^3 compensate the entropy increase acquired during expansion including the increase during inflation.

A possible shortcoming of the proposal could have been the persistence of spacetime singularities in cyclic cosmologies,[18] but to our understanding for the model we outline this problem is avoided, provided that the time average of the Hubble parameter during expansion is equal in magnitude and opposite in sign to its average during contraction.

This model gives renewed hope for the infinitely oscillatory universe saught in Refs. 1–3.

2. Expansion

Let the period of the Universe be designated by τ and the bounce take place at $t = 0$ and turnaround at $t = t_T$. Thus the expansion phase is for times $0 < t < t_T$ and the contraction phase corresponds to times $t_T < t < \tau$. We employ the following Friedmann equation for the *expansion* period $0 < t < t_T$:

$$\left(\frac{\dot{a}(t)}{a(t)}\right)^2 = \frac{8\pi G}{3}\left[\left(\frac{(\rho_\Lambda)_0}{a(t)^{3(\omega_\Lambda+1)}} + \frac{(\rho_m)_0}{a(t)^3} + \frac{(\rho_r)_0}{a(t)^4}\right) - \frac{\rho_{total}(t)^2}{\rho_c}\right] \tag{1}$$

where the scale factor is normalized to $a(t_0) = 1$ at the present time $t = t_0 \simeq 14Gy$. To explain the notation, $(\rho_i)_0$ denotes the value of the density ρ_i at time $t = t_0$. The first two terms are the dark energy and total matter (dark plus luminous) satisfying

$$\Omega_\Lambda = \frac{8\pi G(\rho_\Lambda)_0}{3H_0^2} = 0.72 \quad \text{and} \quad \Omega_m = \frac{8\pi G(\rho_m)_0}{3H_0^2} = 0.28 \tag{2}$$

where $H_0 = \dot{a}(t_0)/a(t_0)$. The third term in the Friedmann equation is the radiation density which is now $\Omega_r = 1.3 \times 10^{-4}$. The final term $\sim \rho_{total}(t)^2$

is derivable from a brane set-up;[9,10,12] we use a negative sign arising from negative brane tension (a negative sign can arise also from a second timelike dimension but that gives difficulties with closed timelike paths). $\rho_{total} = \Sigma_{i=\Lambda,m,r}\rho_i$. As the turnaround is approached, the only significant terms in Eq.(1) are the first (where $\omega_\Lambda < -1$) and the last. As the bounce is approached, the only important terms in Eq.(1) are the third and the last. (We shall later argue that the second term, for matter, is absent during contraction.) In particular, the final term of Eq. (1), $\sim \rho_{total}(t)^2$, arising from the brane set-up is insignificant for almost the entire cycle but becomes dominant as one approaches $t \rightarrow t_T$ for the turnaround and again for $t \rightarrow \tau$ approaching the bounce.

3. Turnaround

Let us assume for algebraic simplicity $\omega_\Lambda = -4/3 =$ constant. This value is already almost excluded by WMAP3[19] but to begin we are aiming only at consistency of infinite cyclicity. More realistic values may be discussed elsewhere. With the value $\omega_\Lambda = -4/3$ we learn from[13] that the time to the Big Rip is $(t_{rip} - t_0) = 11\text{Gy}(-\omega_\Lambda - 1)^{-1} = 33\text{Gy}$ which is, as we shall discuss, within 10^{-27} second or less, when turnaround occurs at $t = t_T$. So if we adopt $t_0 = 14Gy$ then $t_T = t_0 + (t_{rip} - t_0) \sim (14+33)Gy = 47Gy$. From the analysis in[13–15] the time when a system becomes gravitationally unbound corresponds approximately to the time when the dark energy density matches the mean density of the bound system. For an object like the Earth or a hydrogen atom water density ρ_{H_2O} is a practical unit.

With this in mind, for the simple case of $\omega = -4/3$ we see from Eq.(1) that the dark energy density grows proportional to the scale factor $\rho_\Lambda(t) \propto a(t)$ and so given that the dark energy at present is $\rho_\Lambda \sim 10^{-29} g/cm^3$ it follows that $\rho_\Lambda(t_{H_2O}) = \rho_{H_2O}$ when $a(t_{H_2O}) \sim 10^{29}$. We can estimate the time t_{H_2O} by taking on the RHS of the Friedmann equation only dark energy $\left(\frac{\dot{a}}{a}\right)^2 = H_0^2\Omega_\Lambda a^{-\beta}$ with $\beta = 3(1+\omega)$. When we specialize to $\omega = -4/3$ it follows that

$$\frac{a(t_{H_2O})}{(a(t_0) = 1)} = \left(\frac{(t_{rip} - t_0)}{(t_{rip} - t_{H_2O})}\right)^2 \tag{3}$$

so that $(t_{rip} - t_{H_2O}) = 33Gy \times 10^{-14.5} \simeq 10^{3.5}s \sim 1$ hour. [The value is sensitive to ω] It is instructive to consider approach to the Rip a more general critical density $\rho_c = \eta\rho_{H_2O}$ and to compute the time $(t_{rip} - t_\eta)$ such that $\rho_\Lambda(t_\eta) = \rho_c = \eta\rho_{H_2O}$. We then find, using $a(t_\eta) = 10^{29}\eta$, that

$$(t_{rip} - t_\eta) = (t_{rip} - t_0)10^{-14.5}\eta^{-1} \simeq \eta^{-1}\text{hours} \tag{4}$$

which is the required result. We shall see $\eta > 10^{31}$ so the time in (4) is $< 10^{-27}s$.

To discuss the turnaround analytically we keep only the first and last terms, the only significant ones, on the RHS of Eq.(1) which becomes for the special case $\omega = 4/3$

$$\left(\frac{\dot{a}}{a}\right)^2 = \alpha_1 a - \alpha_2 a^2 \tag{5}$$

in which

$$\alpha_1 = \frac{8\pi G}{3}(\rho_\Lambda)_0 \qquad \alpha_2 = \frac{8\pi G}{3}\frac{(\rho_\Lambda)_0^2}{\rho_c} \tag{6}$$

Writing $a = z^2$ and $z = (\alpha_1/\alpha_2)^{1/2} sin\theta$ gives

$$dt = \frac{2\sqrt{\alpha_2}}{\alpha_1}\frac{d\theta}{sin^2\theta} = \frac{2\sqrt{\alpha_2}}{\alpha_1}d(-cot\theta) \tag{7}$$

Integration then gives for the scale factor

$$a(t) = \left(\frac{\alpha_1}{\alpha_2}\right) sin^2\theta = \frac{\rho_c}{(\rho_\Lambda)_0}\left[\frac{1}{1+\left(\frac{t_T - t}{C}\right)^2}\right] \tag{8}$$

where $C = -(3/2\pi G\rho_c)^{1/2}$. At turnaround $t = t_T$, $a(t_T) = [\rho_C/(\rho_\lambda)_0] = (a(t))_{max}$. At the present time $t = t_0$, $a(t_0) = 1$ and $sin^2\theta_0 = [(\rho_\Lambda)_0/\rho_C] \ll 1$, increasing during subsequent expansion to $\theta_T = \pi/4$.

A key ingredient in our model is that at turnaround $t = t_T$ our universe deflates dramatically with efffective scale factor $a(t_T)$ shrinking before contraction to $\hat{a}(t_T) = fa(t_T)$ where $f < 10^{-28}$. This jettisoning of almost all, a fraction $(1 - f)$, of the accumulated entropy is permitted by the exceptional causal structure of the universe. We shall see later that the parameter η at turnaround lies in the range $\eta = 10^{31}$ to $\eta = 10^{87}$ which implies a dark energy density at turnaround (Planckian density of $\rho_\Lambda \sim 10^{104}\rho_{H_2O}$ can be avoided) such that, according to the Big Rip analysis of,[13,14] all known, and yet unknown smaller, bound systems have become unbound and the constituents causally disconnected. Recall that the density of a hydrogen atom is approximately ρ_{H_2O} and we are reaching a dark energy density of from 31 to 87 orders of magnitude higher.

According to these estimates, at $t = t_T$ the universe has already fragmented into an astronomical number $(1/f^3)$ of causal patches, each of which independently contracts as a separate universe leading to an infinite multiverse. The entropy at $t = t_T$ is thus divided between these new contracting universes and our universe retains only a fraction f^3. Since our model universe has cycled an infinite number of times, the number of parallel universes is infinite.

4. Contraction and Bounce

The contraction phase for our universe occurs for the period $t_T < t < \tau$. The scale factor for the contraction phase will be denoted by $\hat{a}(t)$ while we use always the same linear time t subject to the periodicity $t + \tau \equiv t$. At the turnaround we retain a fraction f^3 of the entropy with $\hat{a}(t_T) = fa(t_T)$ and for the contraction phase the Friedmann equation is

$$\left(\frac{\dot{\hat{a}}(t)}{\hat{a}(t)}\right)^2 = \frac{8\pi G}{3}\left[\left(\frac{(\hat{\rho}_\Lambda)_0}{\hat{a}(t)^{3(\omega_\Lambda+1)}} + \frac{(\hat{\rho}_r)_0}{\hat{a}(t)^4}\right) - \frac{\hat{\rho}_{total}(t)^2}{\hat{\rho}_c}\right] \tag{9}$$

where we have defined

$$\hat{\rho}_i(t) = \frac{(\rho_i)_0 f^{3(\omega_i+1)}}{\hat{a}(t)^{3(\omega_i+1)}} = \frac{(\hat{\rho}_i)_0}{\hat{a}(t)^{3(\omega_i+1)}} \tag{10}$$

but in contrast to Eq.(1) we have set $\hat{\rho}_m = 0$ because our hypothesis is that the causal patch retained in the model contains only dark energy and radiation but no matter including no black holes. This is necessary because during a contracting phase dust or matter would clump, even more readily than during expansion, and inevitably interfere with cyclicity. Perhaps more importantly, presence of dust or matter would require that our universe go in reverse through several phase transitions (recombination, QCD and electroweak to name a few) which would violate the second law of thermodynamics. We thus require that *our universe comes back empty!* Any tiny entropy associated with radiation is constant during adiabatic contraction.

The contraction of our universe will proceed from one of the $1/f^3$ causal patches following Eq.(9) until the radiation balances the brane tension at the bounce.

At the bounce, the contraction scale is given, using $\rho_c = \eta\rho_{H_2O}$, from Eq. (1) as

$$a(\tau)^4 = \left(\frac{(\rho_r)_0}{\eta\rho_{H_2O}}\right) \tag{11}$$

Now the model's bounce at $t = \tau$ must be before the electroweak transition at $t_{EW} = 10^{-10}s$ when $a(t_{EW}) = 10^{-15}$, and after the Planck scale when $a(t_{Planck}) = 10^{-32}$ in order to accommodate the well established weak transition and to avoid uncertainties associated with quantum gravity. With this in mind, here are three illustrative values for the bounce temperature T_B: at a GUT scale $T_B = 10^{17} GeV, a(t_B) = 10^{-30}$, at an intermediate scale $T_B = 10^{10} GeV, a(t_B) = 10^{-23}$, and at a weak scale $T_B = 10^3 GeV, a(t_B) = 10^{-16}$. From Eq.(11) and Eq.(4) for these three cases one finds, respectively, $\eta = 10^{87}$ and $(t_{rip} - t_T) = 10^{-87} hr$, $\eta = 10^{59}$ and $(t_{rip} - t_T) = 10^{-59} hr$, and $\eta = 10^{31}$ and $(t_{rip} - t_T) = 10^{-31} hr$.

Immediately after the bounce, we assume that an inflaton field is excited and there is conventional inflation with enhancement $\mathrm{E} = a(\tau + \delta)/\hat{a}(\tau)$. Successful inflation requires $E > 10^{28}$. Consistency requires therefore $f < E^{-1}$ to allow for the entropy accrued during expansion after inflation. The fraction of entropy jettisoned from our universe at deflation is thus extremely close to one, being less than one and more than $(1 - 10^{-28})^3$.

5. Entropy

The present entropy associated with radiation $S \sim 10^{88}$ is much less than the holographic bound $S \sim 10^{123}$. In[20] it is suggested that at least some of this difference may come from supermassive black holes. The entropy contribution from the baryons is smaller than S_γ by some ten orders of magnitude, so like that of the dark matter, is negligible.

What is the entropy of the dark energy? If it is perfectly homogeneous and non-interacting it has zero temperature and entropy. Finally, the 4th term in Eq.(1) corresponding to the brane term is neglible, as we have already estimated. The conclusion is that at present $S_{total}(t_0) \sim 10^{88}$.

Our main point is that in order for entropy to be cyclic, the entropy which was enhanced by a huge factor $E^3 > 10^{84}$ at inflation must be reduced dramatically at some point during the cycle so that $S(t) = S(t+\tau)$ becomes possible. Since it increases during expansion and contraction, the only logical possibility is the decrease at turnaround as accomplished by our causal patch idea. The second law of thermodynamics continues to obtain for other causal patches, each with practically vanishing entropy at turnaround, but these are permanently removed from our universe contracting instead into separate universes.

For contraction $t_T < t < \tau$ we are assuming the universe during contraction is empty of matter until the bounce so its entropy is vanishingly small. Immediately after the bounce inflation arises from an inflaton field, assumed to be excited. We find the counterpoise of inflation at the bounce and deflation at turnaround an appealing aspect of the present model.

Acknowledgments

This work was supported in part by the U.S. Department of Energy under Grant No. DE-FG02-06ER41418.

References

1. A. Friedmann, Z. Phys. **10,** 377 (1922).
2. R.C. Tolman, Phys. Rev. **38,** 1758 (1931).
3. R.C. Tolman, Relativity, Thermodynamics and Cosmology. Oxford University Press (1934).
4. J. Khoury, B.A. Ovrut, P.J. Steinhardt and N. Turok. Phys. Rev. **D64,** 123522 (2001). `hep-th/0103238`.
5. P.J. Steinhardt and N. Turok, Science **296,** 1436 (2002).
6. P.J. Steinhardt and N. Turok, Phys. Rev. **D65,** 126003 (2002). `hep-th/0111098`.
7. L.A. Boyle, P.J. Steinhardt and N. Turok, Phys Rev. **D70,** 023504 (2004). `hep-th/0403026`.
8. P.J. Steinhardt and N. Turok. `astro-ph/0605173`.
9. L. Randall and R. Sundrum, Phys. Rev. Lett. **83,** 3370 (1999). `hep-ph/9905221`; *ibid.* **83,** 4690 (1999). `hep-th/9906064`.
10. C. Csaki, M. Graesser, L. Randall and J. Terning, Phys. Rev. **D62,** 045015 (2000). `hep-ph/9911406`.
11. P. Binétruy, C. Deffayet and D. Langlois, Nucl. Phys. **B565,** 269 (2000). `hep-th/9905012`.
12. M.G. Brown, K. Freese and W.H. Kinney. `astro-ph/0405353`.
13. P.H. Frampton and T. Takahashi, Phys. Lett. **B557,** 135 (2003). `astro-ph/0211544`.
14. P.H. Frampton and T. Takahashi, Astropart. Phys. **22,** 307 (2004). `astro-ph/0405333`.
15. R.R. Caldwell, Phys. Lett. **B545,** 23 (2002). `astro-ph/9908168`; R.R. Caldwell, M. Kamionkowski and N.N. Weinberg, Phys. Rev. Lett. **91,** 071301 (2003). `astro-ph/0302506`.
16. L. Baum and P.H. Frampton. `hep-th/0610213`.
17. A.H. Guth, Phys. Rev. **D23,** 347 (1981).
18. A. Borde, A.H. Guth and A. Vilenkin, Phys. Rev. Lett. **90,** 151301 (2003). `gr-qc/0110012`.
19. D.N. Spergel, *et al,* Wilkinson Microwave Anisotropy Probe (WMAP) Three Year Results: Implications for Cosmology. (March 17, 2006). `astro-ph/0603449`.
   L. Page, *et al,* Wilkinson Microwave Anisotropy Probe (WMAP) Three Year Results: Polarization reults. (March 17, 2006). `astro-ph/0603450`.
   G. Hinshaw, *et al,* Wilkinson Microwave Anisotropy Probe (WMAP) Three Year Results: Temperature results. (March 17, 2006). `astro-ph/0603451`. `http://map.gsfc.nasa.gov/m_mm/pub_papers/threeyear.html`.
20. T.W.Kephart and Y.J.Ng, JCAP, **11** 011 (2003). `gr-qc/0204081`.

CLASSICAL SOLUTIONS OF FIELD EQUATIONS IN BRANE WORLDS MODELS

D. Karasik, C. Sahabandu, P. Suranyi and L.C.R. Wijewardhana
Department of Physics, University of Cincinnati, Cincinnati, Ohio, U.S.A.

In this contribution we review our recent efforts at finding analytic solutions to gravity field equations in Arkani-Hamed, Dimopolous and Dvali(ADD) and Randall Sundrum(RS) brane theories. Our solutions are obtained using an expansion method. These solutions represent objects confined to the brane. The size of the object, μ, is assumed to be smaller than the typical length scale L of the brane world model and the expansion parameter ϵ is the ratio of μ to L. The scale L is the compactification radius of the extra dimensions for ADD models and the ADS curvature length of the bulk for RS models respectively.

There has been a lot of interest in finding black hole solutions in brane world models. It is quite important to find such solutions and analyze their properties. If the real world happens to be described by one of the low scale gravity models like the ones proposed by Arkani-Hamed, Dimopolous and Dvali(ADD)[1] or by Randall and Sundrum[2] to solve the hierarchy problem, such solutions would describe physically accessible particle configurations at the next generation of particle physics accelerators. Therefore it is important to search for such solutions and work out their physical properties. These solutions would be useful even when the low scale gravity models are not phenomenologically relevant. Finding such solutions would help us address some important theoretical issues in black hole physics and semi classical gravity like the nature of the conjectured black-hole black string transition. At the moment the only known analytic black hole solution in realistic brane world models is the uniform black string solution found by Hawking Chamblin and Real[4] in RS models. Therefore it is important to develop approximation methods to look for black hole solutions in brane models.

We have found analytic black hole solutions of Einstein's field equations in ADD as well as RS brane world theories using an expansion procedure.

In Arkan-Hamed, Dimopolous and Dvali(ADD)[1] models the black hole is assumed to be localized on he brane and smaller than the size of the extra dimensions. In Randall Sundrum models,[23] it is localized on the TeV brane and smaller than the bulk ADS curvature length. The expansion parameter ϵ is the ratio between the size of the black hole μ and the compactification size of the extra dimension in the ADD model. In Randall Sundrum models ϵ is the ratio of μ to the bulk ADS curvature length.

First we review the method of solution in the ADD model. For simplicity we assume that there is only one extra dimension compactified on a S^1. Let us say that the size of the extra dimension L is much greater than the five dimensional horizon size of an object of mass M confined to the brane. First consider the near horizon region. In the limit when L is taken to be large with M fixed, the metric tensor near the horizon of the object can be described by the spherically symmetric five dimensional Schwarzschild Tangherlini[5] metric. We use this configuration as the starting point or the zeroth order in ϵ term of our expansion of the metric tensor in the near horizon region. Then we generate the near horizon expansion by setting up and solving the Einsteins equations order by order in ϵ in this region. This is in fact a $\frac{1}{L}$ expansion. Now consider the asymptotic region. In the limit when the mass of the object M goes to zero with L kept fixed the metric of the space becomes flat. This flat metric is the zeroth order approximation to the metric tensor in the region far away from the black hole. The Newtonian correction is the next order contribution to the metric in this approximation. To determine the metric in this asymptotic region we use an expansion in powers of M with L fixed. The linear in M term of the expansion is the Newton potential. Then we systematically generate the higher order terms. Once we have developed these two expansions we compare them term by term in the intermediate overlap region where both expansions are valid, to fix the expansion coefficients. This procedure is simple but cumbersome. It is well described in the appendix of the paper.[8] This method was also used by Harmark,[6] and Gorbanos and Kol[7] to generate the epsilon expansion to the leading order. In our paper we have generated the higher order corrections as well.[8]

In our analysis we use the following coordinates. ρ and r are the 4 and 3 dimensional radii. $\rho sin(\psi) = r$ and $\rho cos(\psi) = \frac{L}{\pi} sin\frac{\pi w}{L}$ where the periodic bulk coordinate w lies between $-\frac{L}{\pi}$ and $\frac{L}{\pi}$. In addition to these coordinates we also have the usual angular coordinates θ and ϕ defined on the brane. The first order asymptotic solution $h_{\mu,\nu}$ defined by $g_{\mu,\nu} = \eta_{\mu,\nu} + \epsilon h_{\mu,\nu}$

satisfies the Einsteins equation

$$h_{tt,rr} + \frac{2}{r}h_{tt,r} + h_{tt,ww} = 0\,.$$

The solution has the form

$$h_{tt}(r,w) = \sum_{n=0}^{\infty} c_n exp(\frac{-2\pi nr}{L})cos(\frac{-2\pi nw)}{L})\,.$$

Now the coeficients c_n are fixed by matching with the appropriate term in the expansion of the zeroth order near solution which is the five dimensional Schwarzschild Tangherlini metric. The explicit parametrization of the near solution could be found in refernce.[8] As described in the appendix of this paper the Einstein's equations in this region are solved in terms of a single function H satisfying a wave equation, and a gauge function F. Then the wave equation is solved and the constants of integration are determined by matching to the first order asymptotic solution. The gauge function is fixed by requiring that the surface gravity on the horizon is constant independent of ψ.

The following conditions ensure that the solution we found represents a five dimensional black hole confined in a space with one periodic or compact dimension.
(1)The surface gravity on the horizon surface $g_{tt} = 0$ is constant.
(2) This surface is a killing horizon.
(3)We have made sure that the solution is periodic in the bulk coordinate w.
(4) The solution is Z_2 symmetric$g_{\mu,\nu}(-w) = g_{\mu,\nu}(w)$, except for $g_{i,w}(-w) = -g_{i,w}(w)$, when i is not equal to w.
(5) The solution is not a black string. That is there is no cylindrical singularity.
We again refer the reader to the reference[8] for details and for an explicit expression of the metric tensor.

This black hole solution was evaluated up to the next to leading order in the ϵ expansion. The leading non trivial term is at the order $\frac{\mu}{L}^2$ and only the even powers contribute to the expansion. It has the following properties. The black hole is prolate rather than oblate. It grows into the compact space when the mass is increased. Its entropy becomes comparable to that of a uniform string when the black hole grows and nearly fills the bulk space. We would like to caution the reader that this is the region in the domain of ϵ where the expansion is least trustworthy. This is consistent with the picture proposed by Gregory and Laflamme in their seminal work,[10][11] where it is conjectured that uniform black strings become unstable to decay into black holes when their size becomes comparable to the size of the compactified extra dimension. A discussion of this is phenomena within the context of the ϵ expansion is found in Ref 8.

We have also applied a similar expansion method to find small black hole solutions is Randall Sundrum theories. There the bulk space is ADS. We have to solve the Einstein's equations subject to the Israel junction conditions[12] at the brane. The calculation is much more complicated than in the case of the ADD model. We have done this only to the first order in ϵ. The details could be found in the appendix of the references.[9] We find the following results. When we require that the asymptotic metric has only integer powers of M in its post Newtonian expansion the surface gravity on the surface $g_{tt} = 0$ is not constant and has a ψ dependence. The expected black hole horizon surface is singular. Note that the intersection of this surface with the brane has constant surface gravity. Therefore to an observor confined to the brane the horizon looks regular. We have also noted that if fractional powers of M are allowed at higher orders in the post Newtonian expansion then it is possible to get a solution with a regular horizon surface. The physical interpretation of such solutions is not very clear. Please note that such fractional powers in the post Netonian expansion are also found

This work is funded in part by the U.S. Department of Energy under the grant DE-FG-02-84ER40153.

References

1. N.Arkani-Hamed, S.Dimopolous and G.Dvali, P.L.B.429.263(1998).
2. L. Randall and R. Sundrum, Phys. Rev. Lett. 83:3370-3373, 1999.
3. L. Randall and R. Sundrum, Phys. Rev. Lett. 83:4690-4693, 1999.
4. A. Chamblin, S.W. Hawking, H.S. Reall, Phys.Rev.D61:065007,2000.
5. F.R.Tangherlini, Nuo Cimento, 27,636(1963).
6. T.Harmark, P.R.D.69:104015(2004) E-print HEP-TH/1310253.
7. Gorbanos and B.Kol, JHEP.0406:053:200, eprint HEP-TH/0406002.
8. D.Karasik, C.Sahabandu, P.Suranyi, L.C.R.Wijewardhana, P.R.D.710240024(2005).
9. D.Karasik, C.Sahabandu,P.Suranyi and L.C.R.Wijewardhana, Phys.Rev.D. 69,064022(2004) eprint gr-qc/0309276 and P.R.D.70,064007(2004) eprint gr-qc/0404015.
10. R. Gregory and R. Laflamme, Phys.Rev.Lett.70:2837-2840,1993.
11. R. Gregory and R. Laflamme, Waterloo 1993, General relativity and relativistic astrophysics* 190-19.
12. W.Israel, Nuovo Cimento **B44**,1(1966);**B 48**,463(1966).

RENORMALIZATION AND CAUSALITY VIOLATIONS IN CLASSICAL GRAVITY COUPLED WITH QUANTUM MATTER

DAMIANO ANSELMI

Dipartimento di Fisica "Enrico Fermi", Università di Pisa,
Largo Pontecorvo 3, I-56127 Pisa, Italy,
and INFN, Sezione di Pisa, Pisa, Italy
E-mail: damiano.anselmi@df.unipi.it

It is shown that classical gravity coupled with quantum fields can be renormalized with a finite number of independent couplings, plus field redefinitions, without introducing higher-derivative kinetic terms in the gravitational sector, but adding vertices that couple the matter stress-tensor with the Ricci tensor. The theory predicts the violation of causality at high energies.

Keywords: Gravity; renormalization; quantum field theory in curved space.

The renormalization of quantum field theory in curved space has been widely studied.[1] Treating the metric tensor as a c-number and neglecting its quantum fluctuations, classical gravity coupled with quantized fields can be used as a low-energy effective field theory, to include the radiative corrections to the Einstein field equations generated by the matter fields circulating in the loops. Moreover, it provides an interesting arena and a laboratory to test ideas about renormalizability beyond power counting. Although there exist persuasive reasons to believe that gravity must be quantized, definitive theoretical arguments and experimental proofs are still missing. Thus it is meaningful to study the physical consequences of the assumption that classical gravity coupled with quantized fields is a fundamental interaction, valid at arbitrarily high energies, instead of just an effective one.

When matter is embedded in a curved background, renormalization generates the gravitational counterterms $R_{\mu\nu}R^{\mu\nu}$ and R^2.[2] Such counterterms can be renormalized in two ways. One possibility is to add the same terms to the lagrangian, if they are not already present, multiplied by independent couplings, and reabsorb the divergences into those couplings. The resulting theory is higher-derivative. Expanding the metric tensor around flat space, the lagrangian contains higher-derivative kinetic terms, which are responsible for instabilities at the classical level and violations of unitarity at the quantum level. In particular, higher-derivative quantum gravity is renormalizable, but not unitary:[3] the propagator falls off sufficiently rapidly at high energies to ensure power-counting renormalizability, but propagates ghosts. An alternative way to remove counterterms, that applies only when they have an appropriate form, is to use field redefinitions. Applied to $R_{\mu\nu}R^{\mu\nu}$ and R^2, field redefinitions can convert the undesirable higher-derivative kinetic terms into new type of vertices that couple gravity to matter.

In classical gravity coupled with quantum fields the second method of subtraction can be consistently implemented to all orders in the perturbative expansion. Since the terms $R_{\mu\nu}R^{\mu\nu}$ and R^2 are proportional to the vacuum field equations, a metric redefinition $g'_{\mu\nu} = g_{\mu\nu} + \delta g_{\mu\nu}$ can obviously reabsorb them into the Einstein term to the first order in $\delta g_{\mu\nu}$. In the presence of matter the redefinition generates vertices that couple the matter stress-tensor to the Ricci tensor. Such a field redefinition can be promoted to all orders by a suitable map $\mathcal{M}$. This is possible because $R_{\mu\nu}R^{\mu\nu}$ and R^2 are not only proportional, but also *quadratically* proportional to the Einstein vacuum field equations. If gravity is classical, the map $\mathcal{M}$ preserves the renormalizability of the theory. However, while the map $\mathcal{M}$ eliminates the higher-derivative kinetic terms, and therefore the instabilities, it produces classical violations of causality at high energies.

We consider partially quantum, partially classical field theories. Let φ_c denote the classical fields, with action $S_c[\varphi_c]$, and φ the quantized fields, with classical action $S[\varphi, \varphi_c]$, embedded in the external φ_c-background. Call $\Gamma[\Phi, \varphi_c]$ the generating functional of one-particle irreducible diagrams obtained quantizing the fields φ in the φ_c-background, where $\Phi = \langle\varphi\rangle$. Then the total action $S_{\text{tot}}[\varphi_c, \varphi_q]$ of the partially classical, partially quantum theory is defined as

$$S_{\text{tot}}[\varphi_c, \varphi_q] = S_c[\varphi_c] + \operatorname{Re}\Gamma[\Phi, \varphi_c], \tag{1}$$

where $\varphi_q = \Phi$ and Φ is real if the fields φ are real bosonic, while Φ is the conjugate of $\overline{\Phi} = \langle\overline{\varphi}\rangle$ if the fields φ are complex or fermionic. For example, for classical gravity coupled with quantum matter, φ_c is the metric tensor $g_{\mu\nu}$ and S_c is the Einstein action, so

$$S_{\text{tot}}[g, \varphi_q] = \frac{1}{2\kappa^2}\int \mathrm{d}^4x\sqrt{-g}\,[R(g) - 2\Lambda] + \operatorname{Re}\Gamma[\varphi_q, g]. \tag{2}$$

The field equations are obtained functionally variating the action with respect to $g_{\mu\nu}$ and φ_q. For example, a simple way to solve the matter field equations is to set $\varphi_q = 0$ (or φ_q equal to its expectation value, if there is a spontaneous symmetry breaking). Then the gravitational field equations $\delta S_{\text{tot}}[g,0]/\delta g_{\mu\nu} = 0$, i.e.

$$R_{\mu\nu} - \frac{1}{2}g_{\mu\nu}R + g_{\mu\nu}\Lambda = -\kappa^2 \,\text{Re}\,\langle T_{\mu\nu}\rangle \,, \qquad \langle T_{\mu\nu}\rangle = \frac{2}{\sqrt{-g}}\frac{\delta\Gamma[\varphi_q, g]}{\delta g^{\mu\nu}} \tag{3}$$

describe how the spacetime geometry is affected by the quantized matter fields circulating in the loops.

Another approach to the semi-classical theory, due to Schwinger and Keldysh,[4] is to replace $\text{Re}\,\langle T_{\mu\nu}\rangle$ in (3) with the "in-in" expectation value of the stress tensor, which is both real and causal. Functional methods for the calculation of in-in expectation values have been developed.[5,6] It is important to observe that the renormalization structure of the theory does not depend on the interpretation of the right-hand side of the Einstein equations in (3). In particular, the counterterms $R_{\mu\nu}R^{\mu\nu}$ and R^2 are identical in the Schwinger-Kleydish approach.[6] The causality violations discussed here are an effect due to the renormalization of $R_{\mu\nu}R^{\mu\nu}$ and R^2 by means of field redefinitions of the metric tensor, so they are independent of the generalization of (3) to quantum field theory.

Consider an action S depending on the fields ϕ and modify it into

$$S'[\phi] = S[\phi] + S_i F_{ij} S_j, \tag{4}$$

where F_{ij} is symmetric and can contain derivative operators, $S_i \equiv \delta S/\delta\phi_i$ are the S-field equations, the index i stands also for the spacetime point and summation over repeated indices, including the integration over spacetime points, is understood. There exists a field redefinition

$$\phi'_i = \phi_i + \Delta_{ij} S_j, \tag{5}$$

with Δ_{ij} symmetric, such that, perturbatively in F and to all orders in powers of F,

$$S'[\phi] = S[\phi']. \tag{6}$$

Indeed, after a Taylor expansion, it is immediate to see that this equality is verified if

$$\Delta_{ij} = F_{ij} - \Delta_{k_1 i}\Delta_{k_2 j}\sum_{n=2}^{\infty}\frac{1}{n!}S_{k_1 k_2 k_3\cdots k_n}\prod_{l=3}^{n}(\Delta_{k_l m_l}S_{m_l})\,, \tag{7}$$

where $S_{k_1\cdots k_n} \equiv \delta^n S/(\delta\phi_{k_1}\cdots\delta\phi_{k_n})$ and for $n=2$ the product is meant to be unity. Equation (7) can be solved recursively for Δ in powers of F. The first terms of the solution are

$$\Delta_{ij} = F_{ij} - \frac{1}{2}F_{ik_1}S_{k_1k_2}F_{k_2j} + \frac{1}{2}F_{ik_1}S_{k_1k_2}F_{k_2k_3}S_{k_3k_4}F_{k_4j}$$
$$-\frac{1}{3!}F_{ik_1}S_{k_1k_2k_3}F_{k_3k_4}S_{k_4}F_{k_2j} + \mathcal{O}(F^4).$$

For example, take an ordinary free field theory

$$S[\phi] = \frac{1}{2}\phi_i S_{ij}\phi_j.$$

Then $S_{k_1\cdots k_n} = 0$ for every $n > 2$, while $S_{k_1k_2}$ is field-independent and quadratic in the derivatives. The modified action

$$S'[\phi] = \frac{1}{2}\phi_i\left(S_{ij} + 2S_{ik}F_{km}S_{mj}\right)\phi_j$$

describes a higher-derivative theory. Equation (7) simplifies to

$$\Delta_{ij} = F_{ij} - \frac{1}{2}\Delta_{k_1i}\Delta_{k_2j}S_{k_1k_2}.$$

Its solution reads, in matrix and vector form,

$$\Delta = \left(\sqrt{1+2FS}-1\right)S^{-1}, \qquad \phi' = \sqrt{1+2FS}\phi.$$

The map is not just a change of variables, since it changes the degrees of freedom of the theory.

In the interacting case, we use the free-field limit results to write

$$\Delta = \left(\sqrt{1+2FS}-1\right)S^{-1} + \mathcal{O}(\phi), \qquad \phi' = \sqrt{1+2FS}\phi + \mathcal{O}(\phi^2), \quad (8)$$

where F and S are the matrices F_{ij} and S_{ij} calculated at $\phi = 0$. Thus in the acausal theory every ϕ'-leg gets multiplied by $1/\sqrt{1+2FS}$.

With a source term the map gives

$$S_{\rm HD}[\phi, J] \equiv S'[\phi] + \phi_k J_k = S[\phi'] + \phi'_k J'_k(J) \equiv S_{\rm AC}[\phi', J],$$

where

$$J'(J) = \frac{1}{\sqrt{1+2FS}}J. \quad (9)$$

The action $S_{\rm HD}[\phi, J]$ describes a higher-derivative theory, while the action $S_{\rm AC}[\phi', J]$ describes, in general, an acausal theory. To see this more clearly, take for example $F_{ij} = \alpha^2\delta_{ij}/2$, $S_{ij} = -\Box\delta_{ij}$ then

$$J'(x) = \frac{1}{\sqrt{1-\alpha^2\Box}}J = \int \mathrm{d}^n x'\ \mathcal{C}_n(x-x')J(x'), \quad (10)$$

where n is the spacetime dimension and

$$\mathcal{C}_n(x) = \int \frac{\mathrm{d}^n p}{(2\pi)^n} \frac{\mathrm{e}^{-ip\cdot x}}{\sqrt{1+\alpha^2 p^2}}. \tag{11}$$

The Fourier transform (11) has to be defined with an appropriate prescription. The degrees of freedom that are responsible for the instabilities in the higher-derivative model are suppressed demanding that the prescription be regular in the limit $\alpha \to 0$. For example,

$$\mathcal{C}_n^{\mathrm{F}}(x) = \int \frac{\mathrm{d}^n p}{(2\pi)^n} \frac{\mathrm{e}^{-ip\cdot x}}{\sqrt{1+\alpha^2 p^2 + i\varepsilon}}. \tag{12}$$

Observe that $\mathcal{C}_n^{\mathrm{F}}(x)$ is complex, but the definition (1) of the action takes care of this.

Although the map is perturbative in F, formula (12) allows us to study some effects of the resummation of derivatives. The function $\mathcal{C}_n^{\mathrm{F}}(x)$ does not vanish outside the past light cone, so causality is violated. In some cases causal prescriptions for $\mathcal{C}_n(x)$ exist, but when the radiative corrections are taken into account, α^2 runs and its logarithmic dependence in general spoils the causal prescriptions. The violation of causality is the price paid for the elimination of instabilities.

The function $\mathcal{C}_n^{\mathrm{F}}(x)$ tends to zero or rapidly oscillates for $|x^2| \gg |\alpha^2|$, so the causality violations can be experimentally tested only at distances of the order of

$$\Delta x \sim |\alpha| \tag{13}$$

and become physically unobservable at distances much larger than this bound.

The map $\mathcal{M}$ can be used to convert a higher-derivative theory of classical gravity coupled with quantum matter into an acausal theory, preserving the renormalizability. Consider the higher-derivative theory

$$S_{\mathrm{HD}}[\overline{g},\varphi,\lambda,a,b,\kappa] = \frac{1}{2\kappa^2}\int \mathrm{d}^4x\sqrt{-\overline{g}}\left(\overline{R} + a\overline{R}_{\mu\nu}\overline{R}^{\mu\nu} + b\overline{R}^2\right) + S_m[\varphi,\overline{g},\lambda], \tag{14}$$

where $\overline{R}_{\mu\nu}$, $\overline{R}$ are the Ricci tensor and scalar curvature of the metric $\overline{g}$. Here S_m is the power-counting renormalizable matter action embedded in curved background. For simplicity assume that S_m does not contain masses and super-renormalizable parameters and use the dimensional-regularization technique. Then no cosmological constant is generated by renormalization.

Obviously $S_{\rm HD}$ is renormalizable, but physically unsatisfactory due to the higher-derivative kinetic terms in the gravitational sector. However, the theorem just proved ensures that there exists a map $\overline{g} = \overline{G}(g, a, b)$ such that

$$\int \mathrm{d}^4x\sqrt{-\overline{g}}\left[\overline{R}(\overline{g}) + a\overline{R}_{\mu\nu}\overline{R}^{\mu\nu}(\overline{g}) + b\overline{R}^2(\overline{g})\right] = \int \mathrm{d}^4x\sqrt{-g}R(g)$$

and reabsorbs the higher-derivative gravitational kinetic terms into the Einstein term. Applied to (14), the map generates new vertices that couple matter with gravity, and defines a new action $S_{\rm AC}$ such that

$$S_{\rm HD}[\overline{G}(g, a, b), \varphi, \lambda, a, b, \kappa] = S_{\rm AC}[g, \varphi, \lambda, a, b, \kappa]. \tag{15}$$

The action $S_{\rm AC}$ has the form

$$S_{\rm AC}[g, \varphi, \lambda, \lambda', \kappa] = \frac{1}{2\kappa^2}\int \mathrm{d}^4x\sqrt{-g}R + S_m[\varphi, g, \lambda] + \Delta S_m[\varphi, g, \lambda, \lambda'], \tag{16}$$

where

$$\Delta S_m = \int \mathrm{d}^4x\sqrt{-g}\,\left[-\frac{a}{2}T_m^{\mu\nu}R_{\mu\nu} + \frac{1}{4}(a + 2b)RT_m\right] + \mathcal{O}(a^2, b^2, ab), \tag{17}$$

where $T_m^{\mu\nu} = -(2/\sqrt{-g})(\delta S_m/\delta g_{\mu\nu})$ is the stress-tensor of the uncorrected matter sector and T_m denotes its trace. Precisely, ΔS_m does not contain any kinetic contributions and is made of vertices that are either proportional to $T_m^{\mu\nu}$ and (covariant derivatives of) the Ricci tensor, or quadratically proportional to (covariant derivatives of) the Ricci tensor.

Formula (15) is the relation between the classical actions. Analogous relations hold for the bare and renormalized actions, when the matter fields are quantized, and the generating functionals Γ:

$$\Gamma_{\rm AC}[g, \Phi, \lambda, a, b, \kappa] = \Gamma_{\rm HD}[\overline{G}(g, a, b), \Phi, \lambda, a, b, \kappa]. \tag{18}$$

The total actions $S_{\rm tot\ AC}$ and $S_{\rm tot\ HD}$ of (1) follow from their definitions. Observe that the resummed map $\overline{g} = \overline{G}(g, a, b)$ is complex, in general, due to the prescription (12). The acausal action $S_{\rm tot\ AC}$ is defined taking the real part of $\Gamma_{\rm AC}$ with g and Φ real, *after* applying the map $\mathcal{M}$ to $\Gamma_{\rm HD}$, which is not the same as applying the map $\mathcal{M}$ to $S_{\rm tot\ HD}$.

The map $\mathcal{M}$ preserves the renormalizability of the theory. Indeed, the function $\overline{g} = \overline{G}(g, a, b)$ is finite and does not depend on the quantum fields, so $\Gamma_{\rm AC}$ is convergent because $\Gamma_{\rm HD}$ is.

Finally, the map is not just a change of variables, but changes the physics, since it eliminates the unwanted degrees of freedom at the price of introducing violations of causality at small distances. In the expansion around flat space, $g_{\mu\nu} = \eta_{\mu\nu} + 2\kappa\phi_{\mu\nu}$, $\eta_{\mu\nu}$ =diag$(1, -1, -, 1-1)$, the traceless part $\widetilde{\phi}_{\mu\nu}$ of $\phi_{\mu\nu}$ is mapped as

$$\widetilde{\phi}_{\mu\nu} = \frac{1}{\sqrt{1-a\Box}}\widetilde{\phi}'_{\mu\nu}. \tag{19}$$

The other components of $\phi_{\mu\nu}$ are multiplied by $1/\sqrt{1-a\Box}$ or $\sqrt{1-b'\Box}$, where $b' = -2(a+3b)$. Thus, the causality violations due to the map $\mathcal{M}$ become detectable at distances of the order of $\sqrt{|a|}$, $\sqrt{|b'|}$ or smaller.

The arguments can be generalized straightforwardly to include masses, super-renormalizable couplings and the cosmological constant. The map $\mathcal{M}$ does not extend to quantum gravity. Nevertheless, the techniques just described might have some impact also on the task of quantizing gravity. More details are reported in Ref. 7. Generalizations are worked out in Ref. 8.

References

1. B. DeWitt, *Dynamical theory of groups and fields*, Gordon And Beach Science Publishers, New York, 1965;
N.D. Birrel and P.C.W. Davies, *Quantum fields in curved space*, Cambridge University Press, Cambridge, 1982.
2. G. 't Hooft and M. Veltman, One-loop divergences in the theory of gravitation, Ann. Inst. Poincarè, 20 (1974) 69.
3. K.S. Stelle, Renormalization of higher derivative quantum gravity, Phys. Rev. D 16 (1977) 953;
E.S. Fradkin and A.A. Tseytlin, Renormalizable asymptotically free quantum theory of gravity, Nucl. Phys. B 201 (1982) 469.
4. J. Schwinger, Brownian motion of a quantum oscillator, J. Math. Phys. 2 (1961) 407;
L.V. Keldysh, Diagram technique for nonequilibrium processes, Sov. Phys. JETP 20 (1965) 1018.
5. R.D. Jordan, Effective field equations for expectation values, Phys. Rev. D. 33 (1986) 444.
6. L.H. Ford and R.P. Woodard, Stress tensor correlators in the Schwinger-Keldysh formalism, Class. Quant. Grav. 22 (2005) 1637 and arXiv:gr-qc/0411003.
7. D. Anselmi, Renormalization and causality violations in classical gravity coupled with quantum matter, JHEP 01 (2007) 062 and hep-th/0605205.
8. D. Anselmi and M. Halat, Renormalizable acausal theories of classical gravity coupled with interacting quantum fields, hep-th/0611131, to appear in Class. Quantum Grav.

GAUGE-HIGGS UNIFICATION AND LHC/ILC

Yutaka Hosotani*

Department of Physics, Osaka University,
Toyonaka, Osaka 560-0043, Japan
**E-mail: hosotani@phys.sci.osaka-u.ac.jp*

In the gauge-Higgs unification scenario the 4D Higgs field is identified with the zero mode of the extra-dimensional component of gauge potentials. The mass of the Higgs particle in the unification in the Randall-Sundrum warped spacetime is predicted to be in the range 100 GeV - 300 GeV. The WWZ gauge couplings remains almost universal as in the standard model, but substantial deviation results for the Higgs couplings. The WWH and ZZH couplings are suppressed by a factor $\cos\theta_H$ from the values in the standard model, where θ_H is the Yang-Mills AB phase along the fifth dimension. These can be tested at LHC and ILC.

Keywords: Gauge-Higgs unification; Hosotani mechanism.

1. Origin of the Higgs boson

There is one particle missing in the standard model of electroweak interactions. It is the Higgs boson. The Higgs boson must exist, either as an elementary particle or as a composite particle. The electroweak unification is possible, only if there is something which breaks $SU(2)_L \times U(1)_Y$ symmetry to $U(1)_{EM}$ symmetry. In the standard model the Higgs boson, whose potential is such that the electroweak symmetry is spontaneously broken, gives masses to W and Z bosons. It also gives quarks and leptons masses through Yukawa couplings.

The standard model seems economical, but it hides dirty secret. Physics ought to be based on simple principles, but there seems no good principle for the Higgs sector. As a result the standard model is afflicted with many arbitrary parameters. There have been many proposals. Technicolor theory views the Higgs boson as a composite state resulting from strong technicolor interactions. Supersymmetry (SUSY) is a leading candidate beyond the standard model which cures the gauge hierarchy problem. However, the situation concerning a large number of arbitrary parameters becomes worse in the minimal supersymmetric standard model. There are other proposals such as the little Higgs theory and the Higgsless theory as well.

In this article I would like to argue that the Higgs field is "clean". The Higgs field is a part of gauge fields in higher dimensions, the Higgs sector being controlled by the gauge principle. The difference between the Higgs particle and gauge bosons originates from the structure of the extra-dimensional space. The scenario is called the gauge-Higgs unification.

The gauge-Higgs unification scenario can be tested at LHC and ILC.[1-4] It predicts that the mass of the Higgs particle is around 100 GeV - 300 GeV, exactly in the energy region where LHC can explore. Further the couplings of the Higgs particle to the W and Z bosons, and also to quarks and leptons are substantially reduced compared with those in the standard model. Thus the Higgs experiments at LHC may uncover the origin of the Higgs particle, and disclose the existence of extra dimensions.

2. Old gauge-Higgs unification

The idea of the gauge-Higgs unification is very old.[5-7] In the Kaluza-Klein theory the gravity in five dimensional spacetime of topology $M^4 \times S^1$ unifies the four-dimensional gravity with the electromagnetism. The part of the metric, $g_{\mu 5}$ $(\mu = 0, 1, 2, 3)$, contains the 4D vector potential A_μ in the electromagnetism. In the gauge-Higgs unification one considers gauge theory, instead of gravity, in higher dimensional spacetime. Extra-dimensional components, A_{y_j}, of gauge potentials transform as 4D scalars under 4D Lorentz transformations. The 4D Higgs field is identified with a low energy mode of A_{y_j}. The Higgs field becomes a part of gauge fields.

This scenario was proposed by Fairlie and by Forgacs and Manton in 1979. They tried to achieve unification by restricting configurations of gauge fields in extra dimensions with symmetry ansatz. In Ref. 7 Manton considered gauge theory with gauge group $\mathcal{G}$ defined on $M^4 \times S^2$. It is assumed that only spherically symmetric configurations are allowed and gauge fields have non-vanishing flux (field strengths) on S^2. Further it is demanded that the gauge group $\mathcal{G}$ breaks down to $SU(2)_L \times U(1)_Y$ by non-vanishing flux. There appears a Higgs doublet as a low energy mode of A_{y_j}. Quite amazingly the Higgs doublet turns out to have a negative mass squared so that the symmetry further breaks down to $U(1)_{EM}$.

There are two parameters; the radius R of S^2 and the gauge coupling g_6 in the six-dimensional spacetime. These two parameters are fixed by the Fermi constant and the four-dimensional $SU(2)_L$ gauge coupling g. m_W, m_Z, and m_H are determined as functions of g_6 and R. The Weinberg angle

Table 1. Spectrum in the gauge-Higgs unification model[7] by Manton.

$\mathcal{G}$	$\sin^2\theta_W$	m_W	m_Z	m_H
$SU(3)$	3/4	44 GeV	88 GeV	88 GeV
$O(5)$	1/2	54 GeV	76 GeV	76 GeV
G_2	1/4	76 GeV	88 GeV	88 GeV

θ_W is determined by the gauge group only. There are three gauge groups which satisfy the above requirements. The result is summarized in Table 1.

The unification is achieved and the Higgs mass is predicted, though numerical values are not realistic. There are generic problems in this scheme. First, the mass m_Z is $\sim 1/R$. In other words, it necessarily predicts a too small Kaluza-Klein scale $1/R$. Secondly, and more importantly, there is no justification for the ansatz of non-vanishing flux. The restriction to spherically symmetric configurations is not justified either.

3. New gauge-Higgs unification

There is a better way of achieving gauge-Higgs unification. The key is to consider gauge theory in a non-simply connected spacetime. It utilizes the Hosotani mechanism.[8–11]

3.1. *Yang-Mills AB phase* θ_H

When the space is not simply connected, a configuration of vanishing field strengths does not necessarily mean trivial. The phenomenon is called the Aharonov-Bohm (AB) effect in quantum mechanics. Consider $SU(N)$ gauge theory on $M^4 \times S^1$ with coordinates (x^μ, y), and impose periodic boundary conditions $A_M(x, y+2\pi R) = A_M(x, y)$. A configuration $A_y(x, y) = \text{constant}$ gives $F_{MN} = 0$, but gives

$$W \equiv P\exp\left\{ig\int_0^{2\pi R} dy\, A_y\right\} = U\begin{pmatrix} e^{i\theta_1} & & \\ & \ddots & \\ & & e^{i\theta_N}\end{pmatrix}U^\dagger \qquad (1)$$

where $U^\dagger = U^{-1}$ and $\sum_{j=1}^{N}\theta_J = 0\ (mod\ 2\pi)$. θ_j's are Yang-Mills AB phases in the theory, denoted collectively as θ_H. They cannot be eliminated by gauge transformations preserving the boundary conditions.

Classical vacua are degenerate. Yang-Mills AB phases θ_H label flat directions of the classical potential. The degeneracy is lifted at the quantum level. The mass spectrum $\{m_n\}$ of various fields depends on θ_H. The effective potential $V_{\rm eff}(\theta_H)$ is given at the one loop level by

$$V_{\rm eff}(\theta_H) = \sum \mp \frac{i}{2} \int \frac{d^4p}{(2\pi)^4} \sum_n \ln \{ -p^2 + m_n^2(\theta_H)\} \ . \tag{2}$$

The value of θ_H is determined by the location of the global minimum of $V_{\rm eff}(\theta_H)$.

3.2. *Dynamical gauge symmetry breaking*

Once the matter content is specified, the effective potential is determined and so is the value of θ_H in the true vacuum. Suppose that all fields are periodic so that the boundary conditions are $SU(N)$ symmetric. If $\theta_H \neq 0$, the symmetry breaks down to a subgroup of $SU(N)$ in general. In other words we have dynamical gauge symmetry breaking.

Take $SU(3)$ as an example. In a pure gauge theory the global minima are located at $\theta_1 = \theta_2 = \theta_3 = 0, \frac{2}{3}\pi, \frac{4}{3}\pi$. The $SU(3)$ symmetry is unbroken. Add periodic fermions in the fundamental representation. Then the global minimum is given by $\theta_1 = \theta_2 = \theta_3 = 0$, the $SU(3)$ symmetry remaining unbroken. If one has, instead, a periodic fermion in the addjoint representation, then the global minima are found at $(\theta_1, \theta_2, \theta_3) = (0, \frac{2}{3}\pi, -\frac{2}{3}\pi)$ and its permutations. The $SU(3)$ symmetry breaks down to $U(1) \times U(1)$. These results are tabulated in Table 2. Dynamical gauge symmetry breaking occurs quite naturally. It involves no fine tuning.[12]

Instead of periodic boundary conditions, one might impose more general twisted boundary conditions. For instance, one can impose $A_M(x, y + 2\pi R) = \Omega A_M(x,y)\Omega^\dagger$ ($\Omega \in SU(N)$). It can be shown that on $M^4 \times S^1$ physics does not depend on the choice of Ω, thanks to dynamics of Yang-Mills AB phases θ_H. On orbifolds such as $M^4 \times (S^1/Z_2)$ and the Randall-Sundrum warped spacetime there appear a finite number of inequivalent

Table 2. Dynamical gauge symmetry breaking in $SU(3)$ theory on $M^4 \times S^1$. $N^F_{\rm fund}$ and $N^F_{\rm add}$ denote the number of periodic fermions in the fundamental and addjoint representations, respectively.

$(N^F_{\rm fund}, N^F_{\rm add})$	global minima $(\theta_1^{\rm min}, \theta_2^{\rm min}, \theta_3^{\rm min})$	residual symmetry
$(0, 0)$	$(0,0,0), (\pm\frac{2}{3}\pi, \pm\frac{2}{3}\pi, \pm\frac{2}{3}\pi)$	$SU(3)$
$(n, 0)$	$(0,0,0)$	$SU(3)$
$(0, n)$	$(0, +\frac{2}{3}\pi, -\frac{2}{3}\pi)$ + permutations	$U(1) \times U(1)$
$(1, 1)$	$(0, \pi, \pi)$ + permutations	$SU(2) \times U(1)$

sets of boundary conditions.[9,13,14]

3.3. *Finiteness of $V_{\rm eff}(\theta_H)$ and the Higgs mass*

A mode of four-dimensional fluctuations of Yang-Mills AB phase θ_H is identified with the 4D Higgs field in an appropriate setup. Hence $V_{\rm eff}(\theta_H)$ is directly related to the effective potential for the 4D Higgs field $\varphi_H(x)$.

One significant feature is that the θ_H-dependent part of $V_{\rm eff}(\theta_H)$ is finite. The mass squared of the Higgs boson, m_H^2, is essentially the curvature of $V_{\rm eff}(\theta_H)$ at its global minimum, implying the finiteness of m_H^2.[15]

The finiteness of $V_{\rm eff}(\theta_H)$ at the one loop level has been shown explicitly in various models.[8,9] A general proof goes as follows.[12,16]

First of all large gauge invariance in theory guarantees that θ_H is related to $\theta_H + 2\pi$ by a large gauge transformation which preserves the boundary conditions. It implies that $V_{\rm eff}(\theta_H + 2\pi) = V_{\rm eff}(\theta_H)$ to all order in perturbation theory. $V_{\rm eff}(\theta_H)$ can be expanded in a Fourier series; $V_{\rm eff}(\theta_H) = \sum_n a_n^V e^{in\theta_H}$.

The one loop effective potential is given by (2). In flat space S^1 $m_n(\theta_H) = (n + \ell\theta_H/2\pi + \alpha)m_{KK}$. Here the Kaluza-Klein mass scale $m_{KK} = 1/R$ and ℓ is an integer. α is a constant determined by the boundary condition of each field. It follows that $V_{\rm eff}^{(k)}(\theta_H) = \partial^k V_{\rm eff}(\theta_H)/\partial\theta_H^k$ becomes finite for sufficiently large k almost everywhere in θ_H. $V_{\rm eff}^{(k)}(\theta_H)$ can develop infrared divergence at a discrete set of values of θ_H where $m_n(\theta_H)$ vanishes, namely at a set of points of measure zero. Hence $n^k a_n^V$ $(n \neq 0)$ becomes finite, implying the finiteness of $V_{\rm eff}(\theta_H)$ at the one loop level. The argument remains valid in the Randall-Sundrum warped spacetime as $m_n(\theta_H) \sim (n + \ell\theta_H/2\pi + \alpha)m_{KK}$ for $|n| \gg 1$.

The finiteness seems to hold beyond one loop. It has been shown that m_H^2in QED in $M^4 \times S^1$ is finite at the two loop level after renormalization in M^5.[17] There is nonperturbative lattice simulation indicating the finiteness as well.[18]

4. Electroweak interactions

To apply gauge-Higgs unification scenario to electroweak interactions, several features have to be taken into account.[19-29] First, the electroweak symmetry is $SU(2)_L \times U(1)_Y$, which breaks down to $U(1)_{EM}$. The Higgs field φ_H is an $SU(2)_L$ doublet. In the gauge-Higgs unification the Higgs field is a part of gauge fields, or must belong to the adjoint representation of the gauge group $\mathcal{G}$. This means that $\mathcal{G}$ must be larger than $SU(2)_L \times U(1)_Y$, as Fairlie, Forgacs, and Manton originally pointed out.[5–7]

Second, fermion content is chiral. This is highly nontrivial in higher dimensional gauge theory, as a spinor in higher dimensions always contains both right- and left-handed components in four dimensions. The left-right asymmetry in fermion modes at low energies can be induced from nontrivial topology of extra-dimensional space and non-vanishing flux of gauge fields in extra dimensions. There is another, simpler and more powerful, way to have chiral fermions. If the extra-dimensional space is an orbifold, appropriate boundary conditions naturally give rise to chiral fermion content.[19,20]

Let us illustrate how the orbifold structure fits in the gauge-Higgs unification, by taking gauge theory on $M^4 \times (S^1/Z_2)$. The orbifold $M^4 \times (S^1/Z_2)$ is obtained from $M^4 \times S^1$ by identifying $(x^\mu, -y)$ and (x^μ, y). There appear two fixed points (branes) at $y = 0$ and $y = \pi R$. We define gauge theory on a covering space of $M^4 \times (S^1/Z_2)$, namely for $-\infty < y < +\infty$, and impose restrictions such that physics is the same at $(x^\mu, y), (x^\mu, y + 2\pi R)$ and $(x^\mu, -y)$. The single-valuedness of physics does not necessarily mean that vector potentials A_M are single-valued. In gauge theory they may be twisted by global gauge transformation. More explicitly

$$\begin{pmatrix} A_\mu \\ A_y \end{pmatrix} (x, y_j - y) = P_j \begin{pmatrix} A_\mu \\ -A_y \end{pmatrix} (x, y_j + y) P_j^{-1} \quad (j = 0, 1) \tag{3}$$

where $y_0 = 0$ and $y_1 = \pi R$. Here P_j is an element of the gauge group $\mathcal{G}$ satisfying ${P_j}^2 = 1$. When $P_j \not\propto 1$, the gauge symmetry is partially broken by the boundary conditions. The physical symmetry, in general, can be different from the residual symmetry given by (P_0, P_1). It can be either reduced or enhanced by dynamics of θ_H.[10]

To see how an $SU(2)$ doublet Higgs field emerges, take $\mathcal{G} = SU(3)$ and $P_0 = P_1 = \text{diag}(-1, -1, 1)$. Then, the orbifold boundary condition (3) implies that $SU(2) \times U(1)$ part of the four-dimensional components A_μ are even under parity at $y = 0, \pi R$, which contains zero modes corresponding to $SU(2) \times U(1)$ gauge fields in four dimensions. On the other hand the extra-dimensional component A_y has zero modes in the off-diagonal part;

$$SU(3): \ A_y = \begin{pmatrix} & & \phi^+ \\ & & \phi^0 \\ \phi^{+*} & \phi^{0*} & \end{pmatrix} \quad , \quad \Phi = \begin{pmatrix} \phi^+ \\ \phi^0 \end{pmatrix} . \tag{4}$$

The zero mode Φ becomes an $SU(2)$ doublet Higgs field. Take $\mathcal{G} = SO(5)$ and $P_0 = P_1 = \text{diag}(-1, -1, -1, -1, 1)$ as another example. In this case the $SO(5)$ symmetry breaks down to $SO(4) \simeq SU(2)_L \times SU(2)_R$. Zero modes of A_y are

$$SO(5): \ A_y = \begin{pmatrix} & & & & \phi_1 \\ & & & & \phi_2 \\ & & & & \phi_3 \\ & & & & \phi_4 \\ -\phi_1 & -\phi_2 & -\phi_3 & -\phi_4 & \end{pmatrix} \ , \quad \Phi = \begin{pmatrix} \phi_1 + i\phi_2 \\ \phi_4 - i\phi_3 \end{pmatrix} \ . \tag{5}$$

Φ is an $SU(2)_L$ doublet. Φ is related to the Yang-Mills AB phases by (1).

Chiral fermions naturally emerge. Take $\mathcal{G} = SU(3)$ with P_j in (4). Fermions in the fundamental representation of $SU(3)$ obey the boundary condition $\psi(x, y_j - y) = P_j \gamma^5 \psi(x, y_j + y)$ so that ψ is decomposed as

$$\psi = \begin{pmatrix} \nu_L & \tilde{\nu}_R \\ e_L & \tilde{e}_R \\ \tilde{e}_L & e_R \end{pmatrix} \ . \tag{6}$$

ν_L, e_L and e_R have zero modes, whereas $\tilde{\nu}_R$, $\tilde{e}_R$ and $\tilde{e}_L$ do not. Fermion content at low energies is chiral as desired.

In the gauge-Higgs unification scenario the Higgs boson is massless at the tree level. Its mass is generated by radiative corrections. The mass of the Higgs boson is determined by the curvature of the effective potential $V_{\rm eff}(\theta_H)$ at the minimum. In Fig. 1 $V_{\rm eff}(\theta_1, \theta_2)$ is displayed in the $U(3) \times U(3)$ model of ref. 25.

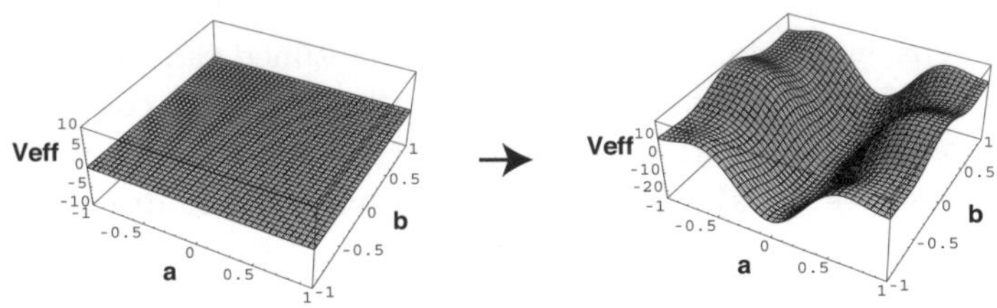

Fig. 1. The effective potential $V_{\rm eff}(\theta_1, \theta_2)$ in the $U(3) \times U(3)$ model in ref. 25 which has two θ_H's, $\theta_1 = \pi a$ and $\theta_2 = \pi b$. $V_{\rm eff} = 0$ at the classical level (in the left figure), but becomes nontrivial at the one loop level (in the right figure).

5. Difficulties in flat spacetime

The gauge-Higgs unification scenario in flat spacetime is afflicted with a few intrinsic difficulties. The electroweak symmetry is spontaneously broken by θ_H. Non-vanishing θ_H gives rise to non-vanishing masses for W and Z

bosons. m_W, for instance, is typically given by

$$m_W \sim \frac{\theta_H}{2\pi} \times \frac{1}{R} \sim \frac{\theta_H}{2\pi} \times m_{KK} \quad . \tag{7}$$

Here R is the size of the extra-dimensions. Secondly, the effective potential $V_{\rm eff}(\theta_H)$ is generated at the one-loop level, and therefore is $O(\alpha_W)$ where $\alpha_W = g_W^2/4\pi$ is the $SU(2)_L$ coupling. The Higgs mass m_H^2 becomes $O(\alpha_W)$ as well. Evaluation of $V_{\rm eff}$ shows that

$$m_H \sim \sqrt{\alpha_W} \times \frac{1}{R} \sim \sqrt{\alpha_W}\, \frac{2\pi}{\theta_H}\, m_W \quad . \tag{8}$$

The relations (7) and (8) are generic predictions from the gauge-Higgs unification in flat spacetime. Once the value of θ_H is given, m_{KK} and m_H are predicted. The value of θ_H is determined from the location of the global minimum of $V_{\rm eff}(\theta_H)$. It depends on the matter content in the theory. Given standard matter content of quarks and leptons with a minimal set of additional matter, the global minimum of $V_{\rm eff}(\theta_H)$ is typically located either at $\theta_H = 0$ or at $\theta_H = (.2 \sim .8)\pi$, as confirmed in various models. In the former case the electroweak symmetry remains unbroken. What we want is the latter. In this case $m_{KK} \sim 10 m_W$ and $m_H \sim 10\,\text{GeV}$. One has too low m_{KK} and too small m_H.

There are two ways to circumvent these difficulties. One way is to arrange the matter content such that small θ_H is obtained. This is possible as discussed by many authors, but requires either many additional fields in higher dimensional representations in $\mathcal{G}$, or fine-tuned cancellations among contributions from various fields.[23,24,26] Another way is to consider warped (curved) spacetime in extra-dimensions.[1-4,30-36] Astonishingly the warped spacetime resolves the above problems quite naturally as discussed below.

6. $SO(5) \times U(1)$ unification in warped spacetime

An attractive model is obtained by considering gauge theory in the Randall-Sundrum (RS) warped spacetime[37-39] whose metric is given by

$$ds^2 = e^{-2k\sigma(y)} \eta_{\mu\nu} dx^\mu dx^\nu + dy^2 \tag{9}$$

where $\sigma(y + 2\pi R) = \sigma(y) = \sigma(-y)$ and $\sigma(y) = k|y|$ for $|y| \le \pi R$. The topology of the spacetime is the same as $M^4 \times (S^1/Z_2)$. The spacetime is an orbifold, with fixed points (branes) at $y = 0$ and πR. It has a negative cosmological constant $\Lambda = -k^2$ in the bulk five-dimensional spacetime. The RS spacetime is an anti-de Sitter space sandwiched by the Planck brane at $y = 0$ and the TeV brane at $y = \pi R$. At low energies the spacetime resembles four-dimensional Minkowski spacetime.

We consider $SO(5) \times U(1)_{B-L}$ gauge theory[31] with gauge couplings g_A and g_B defined in the five-dimensional spacetime (9). We suppose that the structure of the spacetime is determined by physics at the Planck scale and therefore $k = O(M_{\rm Pl})$. With the warp factor $e^{-k\pi R}$ the electroweak scale m_W is naturally generated from the Planck scale.

The orbifold boundary conditions for the $SO(5)$ and $U(1)_{B-L}$ gauge fields, A_M and B_M, are given by $P_0 = P_1 = \mathrm{diag}(-1,-1,-1,-1,1)$ and $P_0 = P_1 = 1$ in (3), respectively. With this parity assignment the bulk $SO(5) \times U(1)_{B-L}$ symmetry breaks down to $SO(4) \times U(1)_{B-L} = SU(2)_L \times SU(2)_R \times U(1)_{B-L}$ on the branes. We further break the symmetry on the Planck brane by imposing the Dirichlet condition on $A_\mu^{1_R}$, $A_\mu^{2_R}$, and $A_\mu^{\prime 3_R}$ which are even under parity. Here $A_\mu^{a_R}$ $(a = 1,2,3)$ are $SU(2)_R$ gauge fields and

$$A_\mu^{\prime 3_R} = \frac{g_A A_\mu^{3_R} - g_B B_\mu}{\sqrt{g_A^2 + g_B^2}} \quad , \quad A_\mu^Y = \frac{g_B A_\mu^{3_R} + g_A B_\mu}{\sqrt{g_A^2 + g_B^2}} \quad . \tag{10}$$

A_μ^Y obeys the Neumann condition on both branes. As a result the residual symmetry is $SU(2)_L \times U(1)_Y$. The change of the boundary conditions from Neumann to Dirichlet for $A_\mu^{1_R}$, $A_\mu^{2_R}$, and $A_\mu^{\prime 3_R}$ is induced by additional dynamics on the Planck brane, and is consistent with the large gauge invariance.[3,4,40]

6.1. *Mass spectrum*

There is one mass scale in the theory, namely $k = O(M_{\rm Pl})$, and a few dimensionless parameters $k\pi R$, $g_A/\sqrt{\pi R}$ and $g_B/\sqrt{\pi R}$. The Kaluza-Klein mass scale in the RS spacetime is

$$m_{KK} = \frac{\pi k}{e^{k\pi R} - 1} \sim \begin{cases} \pi k e^{-k\pi R} & \text{for } e^{k\pi R} \gg 1 \ , \\ 1/R & \text{for } k \to 0 \ . \end{cases} \tag{11}$$

For $\theta_H \neq 0$, m_W and m_Z are given by

$$\begin{aligned} m_W &\sim \sqrt{\frac{k}{\pi R}}\, e^{-k\pi R} \sin\theta_H \\ m_Z &\sim \frac{m_W}{\cos\theta_W} \quad , \quad \sin\theta_W = \frac{g_Y}{\sqrt{g_A^2 + g_Y^2}} = \frac{g_B}{\sqrt{g_A^2 + 2g_B^2}} \quad . \end{aligned} \tag{12}$$

In a generic situation one has $\sin\theta_H = O(1)$. It follows from the relation for m_W that the dimensionless parameter $k\pi R \sim 37$ for $k = O(M_{\rm Pl})$.

Further (11) and (12) imply that

$$m_{KK} \sim \frac{\pi}{\sin\theta_H}\sqrt{k\pi R}\, m_W \quad . \tag{13}$$

For moderate values $\theta_H = (0.2 \sim 0.5)\pi$, the Kaluza-Klein scale turns out $m_{KK} = 2.6\,\text{TeV} \sim 1.5\,\text{TeV}$, which is large enough to be consistent with the current experimental limit. One of the problems in the gauge-Higgs unification scenario in flat spacetime mentioned earlier is solved. In the Randall-Sundrum spacetime there appears an enhancement factor $\sqrt{k\pi R}$.

The mass scale of low energy modes becomes much smaller than the Kaluza-Klein mass scale in the warped spacetime. This can be most clearly seen by examining the mass spectrum as a function of θ_H with various values of $k\pi R$. See Fig. 2. m_W/m_{KK} has weak dependence on θ_H for $k\pi R = 37$ and is much smaller than 1. In the flat spacetime m_W/m_{KK} becomes $O(0.1)$ for $0.1\pi < \theta_H < 0.9\pi$.

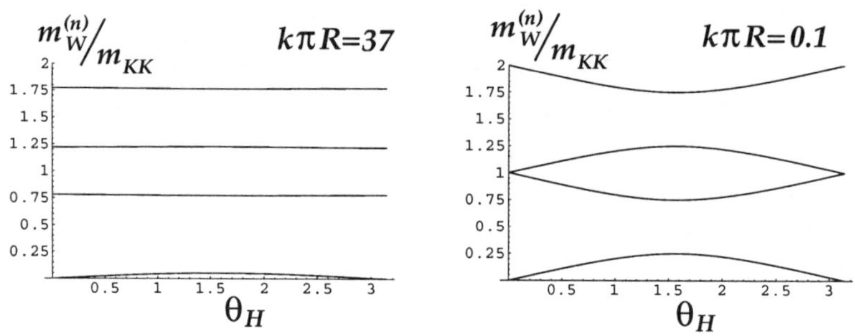

Fig. 2. $m_W^{(n)}/m_{KK}$ ($n = 0, 1, 2, 3$) is plotted for $k\pi R = 37$ and 0.1, where the former corresponds to the realistic case, whereas the latter is close to the flat spacetime limit ($k\pi R = 0$). $m_W = m_W^{(0)}$.

6.2. *Higgs mass and self-couplings*

The Higgs mass and self-couplings are generated by quantum effects, or by radiative corrections. The 4D Higgs field $\varphi_H(x)$ corresponds to four-dimensional fluctuations of θ_H. In the $SO(5) \times U(1)_{B-L}$ model

$$A_y^{\hat{4}}(x,y) = \frac{2\sqrt{2}\,k\,e^{2ky}}{g_A(z_\pi^2 - 1)}\left\{\theta_H + \frac{g_A}{2}\sqrt{\frac{z_\pi^2-1}{k}}\,\varphi_H(x)\right\} + \cdots \tag{14}$$

where $z_\pi = e^{k\pi R}$. Thus, the Higgs mass m_H, for instance, is evaluated from the curvature of $V_{\rm eff}(\theta_H)$ at the minimum. Notice that θ_H and $\varphi_H(x)$

appear in the effective potential in the combination of

$$\begin{cases} \theta_H + \dfrac{\pi g}{\sqrt{2} m_{KK}} \sqrt{\dfrac{k\pi R}{2}} \, \varphi_H(x) & \text{for } e^{k\pi R} \gg 1 \, , \\ \theta_H + \dfrac{\pi g}{\sqrt{2} m_{KK}} \, \varphi_H(x) & \text{for } k \to 0 \, , \end{cases} \tag{15}$$

where the 4D $SU(2)_L$ coupling g is given by $g = g_A/\sqrt{\pi R}$. We observe that $k\pi R/2 \sim 18$ gives various quantities in the warped space an enhancement factor compared with those in flat space.

On general ground the effective potential at one loop is estimated as

$$V_{\rm eff}(\theta_H) \sim \frac{3}{128\pi^6} \, m_{KK}^4 \, f(\theta_H) \tag{16}$$

where $f(\theta_H) = O(1)$ in minimal models. The mass m_H and the quartic coupling λ (in $\lambda \varphi_H^4/4!$) are evaluated as

$$m_H \sim \sqrt{\frac{3\alpha_W}{32\pi} f^{(2)}(\theta_H)} \, \frac{k\pi R}{2} \, \frac{m_W}{\sin\theta_H} \, , \quad \lambda \sim \frac{3\alpha_W^2}{32} f^{(4)}(\theta_H) \left(\frac{k\pi R}{2} \right)^2 \, , \tag{17}$$

where $\alpha_W = g^2/4\pi$. There is ambiguity in $f^{(2)}, f^{(4)}$ which somewhat depend on detailed content of the model. Inserting typical values $f^{(2)}, f^{(4)} \sim 4$ and $\theta_H = (0.1 \sim 0.5)\pi$, one finds that $m_H = (90 \sim 290)\,\text{GeV}$ and $\lambda \sim 0.1$. Although the precise form of $f(\theta_H)$ depends on details of the model, the feature of the enhancement by the factor $k\pi R/2$ in the RS spacetime is general. The problem of too small m_H in flat spacetime has been solved.

6.3. *WWZ coupling*

When $\theta_H = 0$, the electroweak symmetry $SU(2)_L \times U(1)_Y$ remains unbroken. The $SU(2)_L$ gauge coupling in four dimensions is given by $g = g_A/\sqrt{\pi R}$. All couplings associated with W and Z are determined by the $SU(2)_L \times U(1)_Y$ gauge principle. When $\theta_H \neq 0$, things are not so simple in the gauge-Higgs unification scenario.

With $\theta_H \neq 0$, $SU(2)_L \times U(1)_Y$ breaks down to $U(1)_{EM}$. In the standard model the W boson resides in the $SU(2)_L$ group. In the $SO(5) \times U(1)_{B-L}$ gauge-Higgs unification model, $\theta_H \neq 0$ mixes various components of $SU(2)_L$, $SU(2)_R$ and $SO(5)/SO(4)$. It also mixes various Kaluza-Klein excited states. The eigenstate W and its wave function are determined by complete diagonalization. This poses an interesting question whether or not the WWZ coupling g_{WWZ}, for instance, remains universal as in the standard model. There is no guarantee for that.

This is an important issue as the LEP2 data on the W pair production rate agrees with the WWZ coupling in the standard model within an error of a few percents. In Table 3 the ratio of g_{WWZ} in the gauge-Higgs unification to that in the standard model is tabulated for various θ_H and $k\pi R$. One sees that for the realistic case $k\pi R \sim 35$, deviation from the standard model is tiny for any values of θ_H, whereas in the flat spacetime limit ($k\pi R = 0$) substantial deviation appears for moderate values of θ_H.

Table 3. The ratio of g_{WWZ} in the gauge-Higgs unification to that in the standard model for $\theta_H = \pi/10, \pi/4, \pi/2$ and $k\pi R = 35, 3.5, 0.35$.

	$\theta_H = \frac{1}{10}\pi$	$\frac{1}{4}\pi$	$\frac{1}{2}\pi$
$k\pi R = 35$	0.9999987	0.999964	0.99985
3.5	0.9999078	0.996993	0.98460
0.35	0.9994990	0.979458	0.83378

The WWZ coupling remains almost universal in the warped space. The gauge-Higgs unification scenario in the warped space is consistent with the LEP2 data, whereas the scenario in flat space conflicts with the data unless θ_H is sufficiently small.

6.4. *WWH and ZZH couplings*

There emerges significant deviation from the standard model in various couplings of the Higgs boson. Unlike 4D gauge bosons the 4D Higgs boson is mostly localized near the TeV brane so that the behavior of wave functions of various fields on the TeV brane becomes relevant for their couplings to the Higgs boson.

Robust prediction is obtained for the WWH and ZZH couplings

$$\lambda_{WWH} \; H \, W^{\mu \dagger} W_\mu + \frac{1}{2} \lambda_{ZZH} \; H \, Z^\mu Z_\mu \; . \tag{18}$$

The detailed matter content affects the effective potential $V_{\rm eff}(\theta_H)$, but the couplings λ_{WWH} and λ_{ZZH} are determined independent of such details once θ_H is given. One finds that

$$\lambda_{WWH} \simeq g m_W \cdot p_{\rm H} |\cos\theta_H| \quad , \quad \lambda_{ZZH} \simeq \frac{g m_Z}{\cos\theta_W} \cdot p_{\rm H} |\cos\theta_H| \tag{19}$$

where $p_{\rm H} \equiv {\rm sgn}(\tan\theta_H)$. Compared with the values in the standard model, both couplings are suppressed by a factor $\cos\theta_H$. This result can be used to experimentally test the gauge-Higgs unification scenario.

6.5. *Yukawa coupling*

Couplings of the Higgs boson to quarks and leptons, Yukawa couplings, are also subject to nontrivial θ_H-dependent suppression. The Lagrangian for fermions is given by[38,39]

$$i\overline{\psi}\Gamma^a e_a{}^M \Big\{ \partial_M + \frac{1}{8}\omega_{bcM}[\Gamma^b, \Gamma^c] - i g_A A_M - i\frac{g_B}{2} \mathcal{Q}_{\rm B-L} B_M \Big\} \psi$$
$$- c\, k\, \epsilon(y)\, \overline{\psi}\psi + \text{brane interactions} . \tag{20}$$

$\mathcal{Q}_{\rm B-L}$ is a charge of $U(1)_{B-L}$. The kink mass term $ck\epsilon(y)$ naturally arises in the Randall-Sundrum spacetime where a dimensionless parameter c for each fermion multiplet plays a crucial role for determining its wave function. There can be "brane interactions" between ψ and additional brane fermion fields defined on one of the branes.

The Higgs coupling to ψ is contained in the gauge interaction involving A_y. Non-vanishing θ_H ($\langle A_y \rangle \neq 0$) induces a finite fermion mass. Although the gauge interaction is universal, the resulting 4D mass and Yukawa interaction depend on the wave function in the fifth dimension, or on c and the brane interactions. This gives flavor-dependent masses and Yukawa couplings. In the absence of brane interactions, $c = \pm\frac{1}{2}$ gives a fermion a mass of $O(m_W)$. Light fermions $(e, \mu, \tau, u, d, s, c, b)$ corresponds to $c = (0.6 \sim 0.8)$, whereas a heavy fermion (t) to $c \sim 0.4$. The large hierarchy in the fermion mass spectrum is explained by plain distribution in the parameter c.

In the minimal standard model the Yukawa coupling is proportional to the mass of a fermion. In the gauge-Higgs unification scenario this relation is modified. In the absence of brane interactions the Yukawa coupling in the gauge-Higgs unification in the RS spacetime is suppressed by a factor $\cos\theta_H$ or $\cos\frac{1}{2}\theta_H$ compared with the value in the standard model. To realize the observed spectrum of quarks and leptons, however, one needs to include brane interactions, which in turn affects the relationship between the mass and Yukawa coupling. Although the relationship depends on details of the model, it is expected that it deviates from that in the standard model.

6.6. *Gauge couplings of fermions*

Couplings of quarks and leptons to W and Z also suffer from modification, but the amount of deviation from the standard model turns out tiny. The μ-e universality in weak interactions played an important role in the development of the theory. In the modern language it says that all left-handed leptons and quarks have the same coupling to the W boson. It is dictated by the $SU(2)_L$ gauge invariance in four dimensions. In the gauge-Higgs unification, however, the universality is not guaranteed at $\theta_H \neq 0$. As explained earlier, non-vanishing θ_H mixes various components in the gauge group and various levels in the Kaluza-Klein tower. This mixing for fermions depends on, say, the kink mass parameter c, and therefore is not universal.

For $c > 0.6$ wave functions are mostly localized near the Planck brane at $y = 0$ so that the 4D gauge coupling to W becomes almost universal for any values of θ_H. Define $r_\mu(\theta_H) = g_\mu^W(\theta_H)/g_e^W(\theta_H) - 1$ where g_e^W and g_μ^W are the gauge (W) couplings of e and μ, respectively. One finds typically that $r_\mu \sim -10^{-8}$ for $\theta_H = 0.5\pi$. For τ, $r_\tau \sim -2 \times 10^{-6}$. These numbers are well within the experimental limit, being very hard to test in the near future. For top quarks, the deviation becomes bigger ($r_t(0.5\pi) \sim -2 \times 10^{-2}$), but is difficult to measure accurately.

7. Flat v.s. Warped

Why do we need the warped spacetime rather than flat spacetime? The Randall-Sundrum warped spacetime was originally introduced to naturally explain the hierarchy between the Planck scale and weak scale. When applied to the gauge-Higgs unification, there are more benefits.

See Table 4. Both Higgs mass and Kaluza-Klein mass scale turn out too small in flat space for moderate values of θ_H. The ρ parameter deviates from 1 even at the tree level and the WWZ coupling deviates from the value in the standard model in falt space. All these problems are resolved in the Randall-Sundrum warped space. Besides the observed fermion spectrum can be explained without any fine tuning of the parameters.

All of them indicate that having the Randall-Sundrum warped spacetime as background is not just an accident, but have a deeper reason. In this regard the holographic interpretation of the model in the AdS/CFT correspondence is very suggestive as explored by many authors.

8. Conclusion

The prospect of the gauge-Higgs unification in the warped spacetime is bright. The Higgs field is identified with the Yang-Mills AB phase in the extra dimension. It gives definitive prediction in the Higgs couplings, which can be tested at LHC and ILC. The model has not been completed yet. The most urgent task includes to pin down additional brane interactions for fermions so that the observed quark-lepton mass spectrum and the CKM

Table 4. Comparison of the gauge-Higgs unification in the $SO(5) \times U(1)$ model in the flat spacetime $M^4 \times (S^1/Z_2)$ and in the Randall-Sundrum warped spacetime. $\theta_H = (0.1 \sim 0.5)\pi$. $k\pi R = 37$ for the RS spacetime. The estimate of m_H has ambiguity in $f^{(2)}(\theta_H)$ in (17).

Background spacetime	$M^4 \times (S^1/Z_2)$	Randall-Sundrum
Higgs mass m_H	$3 \sim 16$ GeV	$100 \sim 300$ GeV
KK mass scale m_{KK}	$0.3 \sim 1.1$ TeV	$1.5 \sim 5.0$ TeV
$\sin\theta_W$, ρ	deviation from SM	OK
WWZ coupling	deviation from SM	OK (almost universal)
WWH coupling	—	suppressed by $\cos\theta_H$
ZZH coupling	—	suppressed by $\cos\theta_H$
Quark-lepton spectrum	fine tuning necessary	natural hierarchy
Yukawa couplings	—	generally suppressed

and MNS mixing matrices are reproduced.

Acknowledgments

This work was supported in part by Scientific Grants from the Ministry of Education and Science, Grant No. 17540257, Grant No. 13135215 and Grant No. 18204024. The author would like to thank the Aspen Center for Physics for its hospitality where a part of this work was performed.

References

1. Y. Hosotani and M. Mabe, *Phys. Lett.* B**615** (2005) 257.
2. Y. Hosotani, S. Noda, Y. Sakamura and S. Shimasaki, *Phys. Rev.* D**73** (2006) 096006.
3. Y. Sakamura and Y. Hosotani, *Phys. Lett.* B**645** (2007) 442.
4. Y. Hosotani and Y. Sakamura, hep-ph/0703212.
5. D.B. Fairlie, *Phys. Lett.* B**82** (1979) 97; *J. Phys.* G**5** (1979) L55.
6. P. Forgacs and N. Manton, *Comm. Math. Phys.* **72** (1980) 15.
7. N. Manton, *Nucl. Phys.* B**158** (1979) 141.
8. Y. Hosotani, *Phys. Lett.* B**126** (1983) 309.
9. Y. Hosotani, *Ann. Phys. (N.Y.)* **190** (1989) 233.
10. N. Haba, M. Harada, Y. Hosotani and Y. Kawamura, *Nucl. Phys.* B**657** (2003) 169; *Erratum, ibid.* B**669** (2003) 381.
11. A. Hebecker and J. March-Russell, *Nucl. Phys.* B**625** (2002) 128.
L. Hall, H. Murayama, and Y. Nomura, *Nucl. Phys.* B**645** (2002) 85.
12. Y. Hosotani, in the Proceedings of *"Dynamical Symmetry Breaking"*, ed.

M. Harada and K. Yamawaki (Nagoya University, 2004), p. 17. (hep-ph/0504272).
13. N. Haba, Y. Hosotani and Y. Kawamura, *Prog. Theoret. Phys.* **111** (2004) 265.
14. Y. Hosotani, in *"Strong Coupling Gauge Theories and Effective Field Theories"*, ed. M. Harada, Y. Kikukawa and K. Yamawaki (World Scientific, 2003), p. 234. (hep-ph/0303066).
15. H. Hatanaka, T. Inami and C.S. Lim, *Mod. Phys. Lett.* A**13** (1998) 2601.
16. Y. Hosotani, hep-ph/0607064.
17. N. Maru and T. Yamashita, NPB**754** (2006) 127.
18. N. Irges and F. Knechtli, *Nucl. Phys.* B**719** (2005) 121; hep-lat/0604006; hep-lat/0609045.
19. A. Pomarol and M. Quiros, *Phys. Lett.* B**438** (1998) 255.
20. I. Antoniadis, K. Benakli and M. Quiros, *New. J. Phys.***3** (2001) 20.
21. C. Csaki, C. Grojean and H. Murayama, *Phys. Rev.* D**67** (2003) 085012.
22. L.J. Hall, Y. Nomura and D. Smith, *Nucl. Phys.* B**639** (2002) 307;
G. Burdman and Y. Nomura, *Nucl. Phys.* B**656** (2003) 3;
C.A. Scrucca, M. Serone, L. Silvestrini and A. Wulzer, *JHEP* **0402** (2004) 49.
23. N. Haba, Y. Hosotani, Y. Kawamura and T. Yamashita, *Phys. Rev.* D**70** (2004) 015010.
24. N. Haba, K. Takenaga, and T. Yamashita, *Phys. Lett.* B**615** (2005) 247.
25. Y. Hosotani, S. Noda and K. Takenaga, *Phys. Lett.* B**607** (2005) 276.
26. G. Cacciapaglia, C. Csaki and S.C. Park, *JHEP* **0603** (2006) 099.
27. G. Panico, M. Serone and A. Wulzer, *Nucl. Phys.* B**739** (2006) 186.
28. B. Grzadkowski and J. Wudka, PRL**97** (2006) 211602.
29. C.S. Lim and N. Maru, hep-ph/0703017.
30. R. Contino, Y. Nomura and A. Pomarol, *Nucl. Phys.* B**671** (2003) 148.
31. K. Agashe, R. Contino and A. Pomarol, *Nucl. Phys.* B**719** (2005) 165.
32. K. Oda and A. Weiler, *Phys. Lett.* B**606** (2005) 408.
33. M. Carena, E. Ponton, J. Santiago and C.E.M. Wagner, *Nucl. Phys.* B**759** (2006) 202; hep-ph/0701055.
34. T. Gherghetta, hep-ph/0601213.
35. A. Falkowski, *Phys. Rev.* D**75** (2007) 025017.
36. R. Contino, T. Kramer, M. Son and R. Sundrum, hep-ph/0612180.
37. L. Randall and R. Sundrum, *Phys. Rev. Lett.* **83** (1999) 3370.
38. T. Gherghetta and A. Pomarol, *Nucl. Phys.* B**586** (2000) 141.
39. S. Chang, J. Hisano, H. Nakano, N. Okada and M. Yamaguchi, *Phys. Rev.* D**62** (2000) 084025.
40. N. Sakai and N. Uekusa, hep-th/0604121.

STRUCTURE OF S AND T PARAMETERS IN GAUGE-HIGGS UNIFICATION

C. S. Lim* and Nobuhito Maru†

Department of Physics, Kobe University, Kobe 657-8501, Japan

We investigate the divergence structure of one-loop corrections to S and T parameteres in gauge-Higgs unification. We show that these parameters are finite in five dimensions, but divergent in more than six dimensions. Remarkably, a particular linear combination of S and T parameters becomes finite in six dimension case, which is indicated from the operator analysis in a model independent way.

1. Introduction

Solving the gauge hierarchy problem motivates us to go to beyond the Standard Model (SM). Gauge-Higgs unification is one of the attractive approach to solve the gauge hierarchy problem without supersymmetry. In this scenario, Higgs is identified with zero mode of the extra component of the gauge field in higher dimensional gauge theories and the gauge symmetry breaking occurs dynamically through Wilson line phase dynamics. One of the remarkable features is that Higgs mass become finite thanks to the higher dimensional local gauge invariance. Furthermore, many applications of gauge-Higgs unification to the real world had been carried out in various aspects.

Here we would like to ask the following question; Is there any other finite (predictive) physical quantity such as Higgs mass? Noting that the gauge-Higgs sector is controlled by the higher dimensional local gauge invariance, S and T parameters

$$S = -16\pi^2 \Pi'_{3Y}(0), \tag{1}$$

$$T = \frac{4\pi^2}{M_W^2 \sin^2\theta_W} \left(\Pi_{11}(0) - \Pi_{33}(0)\right), \tag{2}$$

*e-mail: lim@kobe-u.ac.jp

†e-mail: maru@people.kobe-u.ac.jp, Speaker. This talk is based on.[1]

where $\Pi_{ij}(p^2)$ is the $g_{\mu\nu}$ part of the two-point function of currents and $\Pi'_{ij} \equiv \frac{d^2}{dp^2}\Pi_{ij}(p^2)$ and θ_W denotes the Weinberg angle, are one of the good candidates since these parameters are given as the coefficients of dimension six operators composed of the gauge fields and Higgs fields. In SM, S and T parameters are finite since SM is a renormalizable theory and these parameters are coefficients of dimension six higher dimensional operators. On the other hand, we consider here a nonrenormalizable theory, which implies that S and T parameters are in general divergent even if they are given by the coefficients of the nonrenormalizable operators. However, we know the fact that Higgs mass is finite, which is realized thanks to the higher dimensional gauge symmetry. Since S and T parameters can be also controlled by the higher dimensional gauge symmetry, we can expect that these parameters also become finite.

In this talk, we discusss the divergence structure of one-loop corrections to S and T parameters in the minimal $SU(3)$ gauge-Higgs unification on an orbifold S^1/Z_2 with a triplet fermion. We show that these parameters are finite in 5D case, but divergent in more than 6D case. The remarkable result is that in 6D case, one-loop corrections to S and T parameters themselves are certainly divergent, but a particular combination of them becomes finite. Its relative ratio agrees with that derived from operator analysis in a model independent way. This is the crucial difference from the universal extra dimension (UED) scenario.

2. Model

We introduce here a minimal model of 5D SU(3) gauge-Higgs unification on an orbifold S^1/Z_2, whose Lagrangian is given by

$$\mathcal{L} = -\frac{1}{4}F_{MN}F^{MN} + i\bar{\Psi}\not{D}\Psi \tag{3}$$

where $\Gamma^M = (\gamma^\mu, i\gamma^5)$,

$$F_{MN} = \partial_M A_N - \partial_N A_M - ig[A_M, A_N] \; (M, N = 0, 1, 2, 3, 5), \tag{4}$$

$$\not{D} = \Gamma^M(\partial_M - igA_M), \tag{5}$$

$$\Psi = (\psi_1, \psi_2, \psi_3)^T. \tag{6}$$

The periodic boundary conditions for S^1 and Z_2 parities are imposed as follows,

$$A_\mu = \begin{pmatrix} (+,+) & (+,+) & (-,-) \\ (+,+) & (+,+) & (-,-) \\ (-,-) & (-,-) & (+,+) \end{pmatrix}, A_5 = \begin{pmatrix} (-,-) & (-,-) & (+,+) \\ (-,-) & (-,-) & (+,+) \\ (+,+) & (+,+) & (-,-) \end{pmatrix}, \tag{7}$$

$$\Psi = \begin{pmatrix} \psi_{1L}(+,+) \; \psi_{1R}(-,-) \\ \psi_{2L}(+,+) \; \psi_{2R}(-,-) \\ \psi_{3L}(-,-) \; \psi_{3R}(+,+) \end{pmatrix}, \tag{8}$$

where $(+,+)$ means that Z_2 parities are even at $y = 0$ and $y = \pi R$, for instance. y is the fifth coordinate and R is the compactification radius. $\psi_{1L} \equiv \frac{1}{2}(1 - \gamma_5)\Psi$, etc. One can see that $SU(3)$ is broken to $SU(2) \times U(1)$ by these boundary conditions.

Expanding in terms of Kaluza-Klein (K-K) modes and integrating out the fifth coordinate, we obtain a 4D effective Langrangian for a fermion

$$\begin{aligned}
\mathcal{L}_{\text{fermion}} &= \sum_{n=1}^{\infty} \Big\{ (\bar{\psi}_1^{(n)}, \bar{\tilde{\psi}}_2^{(n)}, \bar{\tilde{\psi}}_3^{(n)}) \\
&\times \begin{pmatrix} i\gamma^\mu\partial_\mu - m_n & 0 & 0 \\ 0 & i\gamma^\mu\partial_\mu - (m_n + m + gh) & 0 \\ 0 & 0 & i\gamma^\mu\partial_\mu - (m_n - m - gh) \end{pmatrix} \begin{pmatrix} \psi_1^{(n)} \\ \tilde{\psi}_2^{(n)} \\ \tilde{\psi}_3^{(n)} \end{pmatrix} \\
&+ \frac{g}{2}(\bar{\psi}_1^{(n)}, \bar{\tilde{\psi}}_2^{(n)}, \bar{\tilde{\psi}}_3^{(n)}) \begin{pmatrix} W_3^\mu + \frac{\sqrt{3}B^\mu}{3} & W^{+\mu} & W^{+\mu} \\ W^{-\mu} & -\frac{W_3^\mu}{2} - \frac{\sqrt{3}B^\mu}{6} & -\frac{W_3^\mu}{2} + \frac{\sqrt{3}B^\mu}{2} \\ W^{-\mu} & -\frac{W_3^\mu}{2} + \frac{\sqrt{3}B^\mu}{2} & -\frac{W_3^\mu}{2} - \frac{\sqrt{3}B^\mu}{6} \end{pmatrix} \gamma_\mu \begin{pmatrix} \psi_1^{(n)} \\ \tilde{\psi}_2^{(n)} \\ \tilde{\psi}_3^{(n)} \end{pmatrix} \Big\} \\
&+ i\bar{t}_L\gamma^\mu\partial_\mu t_L + \bar{b}(i\gamma^\mu\partial_\mu - m - gh)b \\
&+ \frac{g}{\sqrt{2}}(\bar{t}\gamma_\mu L b W^{+\mu} + \bar{b}\gamma_\mu L t W^{-\mu}) + \frac{g}{2}(\bar{t}\gamma_\mu L t - \bar{b}\gamma_\mu L b)W_3^\mu \\
&+ \frac{\sqrt{3}g}{6}(\bar{t}\gamma_\mu L t + \bar{b}\gamma_\mu L b - 2\bar{b}\gamma_\mu R b)B^\mu
\end{aligned} \tag{9}$$

where the mass matrix for the non-zero K-K modes are diagonalized by use of the mass eigenstates $\tilde{\psi}_2^{(n)}$, $\tilde{\psi}_3^{(n)}$:

$$\begin{pmatrix} \psi_1^{(n)} \\ \tilde{\psi}_2^{(n)} \\ \tilde{\psi}_3^{(n)} \end{pmatrix} = U \begin{pmatrix} \psi_1^{(n)} \\ \psi_2^{(n)} \\ \psi_3^{(n)} \end{pmatrix} \quad U = \frac{1}{\sqrt{2}} \begin{pmatrix} \sqrt{2} & 0 & 0 \\ 0 & 1 & -1 \\ 0 & 1 & 1 \end{pmatrix} \tag{10}$$

and $L \equiv \frac{1}{2}(1-\gamma_5)$, $W_\mu^{1,2,3}$, B_μ are the $SU(2), U(1)$ gauge fields, respectively and $W_\mu^\pm \equiv (W_\mu^1 \pm iW_\mu^2)/\sqrt{2}$. $m_n = \frac{n}{R}$ is the compactification scale. $m = g\langle A_5 \rangle$ is a bottom mass, where we consider Ψ to be a third generation quark. Dirac particles are constructed as

$$\psi_{1,2,3}^{(n)} = \psi_{1,2,3R}^{(n)} + \psi_{1,2,3,L}^{(n)} \; (n > 0) \tag{11}$$

$$b = \psi_{2L}^{(0)} + \psi_{3R}^{(0)}, \tag{12}$$

and the remaining state is a Weyl spinor

$$t_L = \psi_{1L}^{(0)}. \tag{13}$$

We realized that zero mode part for t and b quarks are exactly the same as those in the SM with

$$m_t = 0, \quad m_b = m. \tag{14}$$

Thus, we can just use the result in the SM with (14). Note that the mass splitting occurs between the $SU(2)$ doublet component and singlet component. This pattern of mass splitting has a periodicity with respect to m, which is a remarkable feature of gauge-Higgs unification.

3. Calculation of S and T parameters in 5D case

In this section, we calculate one-loop corrections to T-parameter, which is obtained from the mass difference between the neutral W-boson and the charged W-bosons $\Delta M^2 \equiv \delta\Pi_{33}(0) - \delta\Pi_{11}(0)$. The result is given by

$$\begin{aligned}\Delta M^2 &= i\frac{g^2}{16}\frac{2^{D/2}}{D}\sum_{n=-\infty}^{\infty}\int\frac{d^Dk}{(2\pi)^D} \\ &\times\left[\frac{(2-D)k^2 + D(m_n^2 - m^2)}{[k^2-(m_n-m)^2][k^2-(m_n+m)^2]} - 4\frac{(2-D)k^2 + D(m_n^2+m^2)}{[k^2-m_n^2][k^2-(m_n+m)^2]}\right].\end{aligned} \tag{15}$$

Let us evaluate T-parameter in 5D by carrying out the dimensional regularization for 4D momentum integral before taking the mode sum in ordet to keep 4D gauge invariance and expanding the non-zero mode part of (15) in m/m_n, that is, we consider the case where the compactification scale is larger than the bottom mass. It is straightforward to check that the pole terms in $D \to 4$ limit are exactly cancelled and the finite value can be calculated from the log terms.

$$\Delta M^2_{(n\neq 0)} = -\frac{g^2}{40\pi^2}\sum_{n=1}^{\infty}\frac{m^4}{m_n^2} = -\frac{g^2}{240}(mR)^2 m^2, \tag{16}$$

where $\sum_{n=1}^{\infty} n^{-2} = \zeta(2) = \pi^2/6$ is used. The fact that the leading order term is proportional to m^4 corresponds to four Higgs vacuum expectation values (VEV) insertions in dimension six operator contributing to T-parameter $(\phi^\dagger D_\mu \phi)(\phi^\dagger D^\mu \phi)$($\phi$: Higgs doublet). The dependence of m_n^{-2} tells us that non-zero K-K modes effects are decoupling.

This finite value can be also obtained by taking the mode sum before 4D momentum integration. If we take the mode sum explicitly in (15), we find

$$\Delta M^2 = -\frac{g^2}{16}\frac{2^{D/2}}{D}L^{2-D}\int_0^1 dt \int \frac{d^D\rho}{(2\pi)^D}$$
$$\times\left[-\frac{D}{\rho}\left(\frac{\sinh\rho}{(\cosh\rho-1)}-1\right)-\frac{D}{2\rho}\left(\frac{\sinh\rho}{(\cosh\rho-\cos\alpha)}-1\right)\right.$$
$$-\frac{D}{2}\frac{\left(1+(D-2)\frac{\alpha^2}{\rho^2}\right)}{\sqrt{\rho^2+4t(1-t)\alpha^2}}\left(\frac{\sinh\sqrt{\rho^2+4t(1-t)\alpha^2}}{\cosh\sqrt{\rho^2+4t(1-t)\alpha^2}-\cos[(2t-1)\alpha]}-1\right)$$
$$+\frac{D}{2}\frac{\left(4+(D-2)\frac{\alpha^2}{\rho^2}\right)}{\sqrt{\rho^2+t(1-t)\alpha^2}}\left(\frac{\sinh\sqrt{\rho^2+t(1-t)\alpha^2}}{\cosh\sqrt{\rho^2+t(1-t)\alpha^2}-\cos[t\alpha]}-1\right)$$
$$\left.-\frac{3D}{2\rho}-\frac{\rho^2+D\alpha^2}{2[\rho^2+4t(1-t)\alpha^2]^{3/2}}+\frac{4\rho^2+D\alpha^2}{2[\rho^2+t(1-t)\alpha^2]^{3/2}}\right] \tag{17}$$

where $L \equiv 2\pi R, \rho \equiv Lk$ where k is an Euclidean momentum and $\alpha \equiv Lm$ (i.e. Aharanov-Bohm phase.).

By performing the dimensional regularization for 4D momentum integral, we find

$$\Delta M^2 \simeq -\frac{g^2}{(8\pi)^2}\left(m^2+\frac{4\pi^2}{15}(mR)^2m^2\right). \tag{18}$$

The m^2 is known to be coincide with zero mode contribution. The remaining m^4 term also agrees with the finite result of non-zero K-K mode contributions (16), which was calculated by performing the dimensional regularization for 4D momentum before taking the mode sum.

Similarly, one-loop corrections to S-parameter, which is obtained from the kinetic mixing term for $U(1)$ gauge bosons, can be calculated as

$$\Pi'_{3Y}(0) = i\frac{\sqrt{3}g^2}{144}2^{D/2}\int\frac{d^Dk}{(2\pi)^D}\sum_{n=-\infty}^{\infty}\left[\frac{2}{(k^2-m_n^2)^2}+\frac{1}{[k^2-(m_n+m)^2]^2}\right.$$
$$-18\int_0^1 dt t(1-t)\left\{\frac{1}{[k^2-(m_n+(2t-1)m)^2-4t(1-t)m^2]^2}\right.$$
$$\left.\left.+2\frac{(2t-1)[m_n+(2t-1)m]m+4t(1-t)m^2}{[k^2-(m_n+(2t-1)m)^2-4t(1-t)m^2]^3}\right\}\right]. \tag{19}$$

The finite value is found in a similar way.

$$\Pi'_{3Y}(0) = -\frac{23\sqrt{3}g^2}{360}\frac{1}{(2\pi)^2}\sum_{n=1}^{\infty}\left(\frac{m}{m_n}\right)^2 = -\frac{23\sqrt{3}g^2}{8640}(mR)^2. \tag{20}$$

m^2 dependence is consistent with the dimension six operator representing S-parameter $(\phi^\dagger W_\mu^a \frac{\tau^a}{2}\phi)B^\mu(\phi^\dagger\phi)$. m_n^{-2} dependence is also consistent with the decoupling nature of K-K particles.

4. $D > 5$ case

In this section, we would like to clarify whether these parameters are finite or not in the case higher than five dimensions. S and T parameters are given by the coeffcients of dimension six operators such as $(\phi^\dagger W_\mu \phi)B^\mu(\phi^\dagger\phi)$ for S-parameter and $(\phi^\dagger D_\mu\phi)(\phi^\dagger D^\mu\phi)$ for T-parameter. Naively, the corresponding operators in the gauge-Higgs unification can be regarded as the operators where Higgs doublet ϕ is replaced with A_i (i: extra space component index). Since A_i transform as $A_i \to A_i + \text{const}$ by the higher dimensional local gauge symmetry, it seems that the local operators for S and T parameters are forbidden as in the case of Higgs mass. Therefore, we are tend to conclude that S and T parameters in gauge-Higgs unification become finite, but this argument is too naive, and not correct.

The point is that the gauge invariant local operators for S and T parameters are allowed by a single gauge invariant operator $\mathrm{Tr}[(D_L F_{MN})(D^L F^{MN})]$. Therefore, there is no physical reason for S and T parameters to be finite.

$$\begin{aligned}
&\mathrm{Tr}[(D_L F_{MN})(D^L F^{MN})] \\
&\quad \supset \frac{1}{2}(8m^4)(W_\mu^3)^2 + (2m^4)W_\mu^+ W^{-\mu} + 2\sqrt{3}m^2 p^2 g_{\mu\nu} W^{3\mu} B^\nu \\
&\quad + 2\sqrt{3}m^2(p^2 g_{\mu\nu} - p_\mu p_\nu)W^{3\mu}B^\nu. \qquad (21)
\end{aligned}$$

What a remarkable thing is that we can predict some combination of S and T parameters although these parameters themselves are divergent. We can read off the ratio of them as

$$C\mathrm{Tr}[(D_L F_{MN})(D^L F^{MN})] \to \begin{cases} \Delta M^2 = 6Cm^4 \\ \Pi'_{3Y} = 4\sqrt{3}Cm^2, \end{cases} \qquad (22)$$

where C is an undetermined overall constant. Thus, we can expect the combination $\Pi'_{3Y} - \frac{2\sqrt{3}}{3m^2}\Delta M^2$ to be finite even in more than five dimensions.

In fact, we can show that $\Pi'_{3Y} - \frac{2\sqrt{3}}{3m^2}\Delta M^2$ is finite in 6D case because

$$\Delta M^2 = \frac{g^2}{40\sqrt{2}\pi^2}\sum_{n=1}^{\infty}\left[-\frac{m^4}{m_n} + \frac{1}{12}\frac{m^6}{m_n^3}\right], \qquad (23)$$

$$\Pi'_{3Y}(0) = \frac{\sqrt{3}g^2}{60\sqrt{2}\pi^2}\sum_{n=1}^{\infty}\left[-\frac{m^2}{m_n} + \frac{3}{14}\frac{m^4}{m_n^3}\right] \qquad (24)$$

where the first term indicates logatithmic divergence. Combining these results (23) and (24), we obtain the finite result

$$\Pi'_{3Y} - \frac{2\sqrt{3}}{3m^2}\Delta M^2 = \frac{11\sqrt{6}g^2}{10080\pi^2}m^4 R^3 \zeta(3). \tag{25}$$

5. Conclusions

In this talk, we have discussed the divergence structure of one-loop corrections to S and T parameters in gauge-Higgs unification. Taking a minimal $SU(3)$ gauge-Higgs model with a triplet fermion, we have calculated S and T parameters at one-loop order. In five dimensions, we have shown that one-loop corrections to S and T parameters are finite and evaluated their finite values explicitly. In more than six dimensions, S and T parameters are divergent as in the UED scenario. However, a particular combination of S and T parameters is shown to be finite in six dimension case, whose relative ratio was found to agree with that derived from the operator analysis in a model independent way. This is the crucial difference from the UED scenario.

Acknowledgments

The speaker (N.M.) would like to thank the organizers for providing him with an opportunity to present this talk in the conference. The work of the authors was supported in part by the Grant-in-Aid for Scientific Research of the Ministry of Education, Science and Culture, No.18204024.

References

1. C.S. Lim and Nobuhito Maru, "Calculable One-loop Corrections to S and T Parameters in Gauge-Higgs Unification", in preparation and related references therein.

Large gauge hierarchy in gauge-Higgs unification

KAZUNORI TAKENAGA*
Department of Physics, Tohoku University,
Sendai 980-8578, JAPAN
**E-mail: takenaga@tuhep.phys.tohoku.ac.jp*

We study a five dimensional nonsupersymmetric $SU(3)$ gauge theory compactified on $M^4 \times S^1/Z_2$. The gauge hierarchy is discussed in the scenario of the gauge-Higgs unification. We present two models in which the large gauge hierarchy is realized, that is, the weak scale is naturally is obtained from an unique large scale such as a GUT and the Planck scale. We also study the Higgs mass in each model.

Keywords: Gauge symmetry; extra dimensions; boundary conditions.

1. Introduction

Higher dimensional gauge theory has been paid much attention as a new approach to overcome the hierarchy problem in the standard model without introducing supersymmetry. In particular, the gauge-Higgs unification is a very attractive idea.[1–3] The higher dimensional gauge symmetry plays a role to suppress the ultraviolet effect on the Higgs mass. The Higgs self interaction is understood as part of the original five dimensional gauge coupling, so that the mass and the interaction can be predicted in the gauge-Higgs unification. The gauge-Higgs unification has been studied extensively.[4]

In the gauge-Higgs unification, the Higgs field corresponds to the Wilson line phase, which is nonlocal quantity. The Higgs potential is generated at the one-loop level after the compactification. Because of the nonlocality, the Higgs potential never suffers from the ultraviolet effect,[5] which is the genuine local effect, and the Higgs mass calculated from the potential is finite as well. In other words, the Higgs potential and the mass are calculable in the gauge-Higgs unification. This is a remarkable feature rarely happens in the usual quantum field theory. It is understood that the feature entirely comes from shift symmetry manifest through the Wilson line phase, which is a remnant of the higher dimensional gauge symmetry appeared in four dimensions. The Higgs mass does not depend on the cutoff at all, so that two tremendously separated energy scales can be stable in the gauge-Higgs unification.

We study a five dimensional nonsupersymmetric $SU(3)$ gauge theory, where one of spatial coordinates compactified on an orbifold S^1/Z_2. We find two models (model I, II) which realize the large gauge hierarchy.[6]

2. Gauge-Higgs unification

As the simplest example of the gauge-Higgs unification, we study a non-supersymmetric $SU(3)$ gauge theory on $M^4 \times S^1/Z_2$, where M^4 is the four dimensional Minkowski space-time and S^1/Z_2 is an orbifold which has two fixed points, $y = 0, \pi R$.

We impose the twisted boundary condition of the field for the S^1 direction and at the fixed points by using the gauge degrees of freedom,

$$A_{\hat{\mu}}(x, y + 2\pi R) = U A_{\hat{\mu}}(x, y) U^\dagger, \tag{1}$$

$$\begin{pmatrix} A_\mu \\ A_y \end{pmatrix} (x, y_i - y) = P_i \begin{pmatrix} A_\mu \\ -A_y \end{pmatrix} (x, y_i + y), P_i^\dagger, \quad (i = 0, 1) \tag{2}$$

where $U^\dagger = U^{-1}, P_i^\dagger = P_i = P_i^{-1}$ and $y_0 = 0, y_1 = \pi R$ and $\hat{\mu}$ stands for $\hat{\mu} = (\mu, y)$. The minus sign for A_y is needed to preserve the invariance of the Lagrangian under these transformations. A transformation $\pi R + y \xrightarrow{P_1} \pi R - y$ must be the same as $\pi R + y \xrightarrow{P_0} -(\pi R + y) \xrightarrow{U} \pi R - y$, so that we obtain $U = P_1 P_0$. Here we choose $P_0 = P_1 = \text{diag.}(-1, -1, 1)$.

The gauge symmetry at low energies consists of the zero modes for $A_\mu^{(0)a=1,2,3,8}$. We see that the orbifolding boundary condition P_i breaks the original gauge symmetry $SU(3)$ down to $SU(2) \times U(1)$ at the fixed points.[7] On the other hand, we observe that the zero mode for $A_y^{(0)}$ transforms as an $SU(2)$ doublet, so that we identify the Higgs field as

$$\Phi \equiv \sqrt{2\pi R}\, \frac{1}{\sqrt{2}} \begin{pmatrix} A_y^{(0)4} - iA_y^{(0)5} \\ A_y^{(0)6} - iA_y^{(0)7} \end{pmatrix}. \tag{3}$$

The VEV of the Higgs field is parametrized, by using the $SU(2) \times U(1)$ gauge degrees of freedom, as

$$\langle A_y^{(0)} \rangle \equiv \frac{a}{g_4 R} \frac{\lambda^6}{2} = A_y^{(0)6} \frac{\lambda^6}{2}, \tag{4}$$

where a is a dimensionless parameter. In order to determine a, one usually valuates the effective potential for a.[2] The gauge symmetry breaking depends on the values of a_0. It has been known that the matter content is crucial for the correct gauge symmetry breaking $SU(2) \times U(1) \to U(1)_{em}$.

3. Large gauge hierarchy in the gauge-Higgs unification

If the Higgs acquires the VEV, the W-boson becomes massive whose mass is given by $M_W = a_0/2R$. This relation defines an important ratio,

$$\frac{M_W}{M_c} = \pi a_0, \tag{5}$$

where $M_c \equiv (2\pi R)^{-1}$. Once the values of a_0 is determined as the minimum of the effective potential, the compactification scale M_c is fixed through equation (5). In the usual scenario of the gauge-Higgs unification, the VEV is of order of $O(10^{-2})$ for appropriate choice of the flavor set[8] and this yields $M_c \sim$ a few TeV. In order to realize the large gauge hierarchy such as $M_c \sim M_{GUT}, M_{Planck}$, one needs the very small values of a_0.

For the very small values of a, the effective potential can be expanded as

$$\bar{V}_{eff}(a) = -\frac{\zeta(3)}{2}C^{(2)}(\pi a)^2 + \frac{(\pi a)^4}{24}\left[C^{(3)}\left(-\ln(\pi a) + \frac{25}{12}\right) + C^{(4)}(\ln 2)\right] + \cdots, \tag{6}$$

where $V_{eff}(a) \equiv C\bar{V}_{eff}(a)$ with $C \equiv \Gamma(\frac{5}{2})/\pi^{\frac{5}{2}}(2\pi R)^5$, and the coefficient $C^{(i)} (i = 2, 3, 4)$ is defined by

$$C^{(2)} \equiv 24N_{adj}^{(+)} + 4N_{fd}^{(+)} + \frac{9d}{2}N_{adj}^{(-)s} + \frac{3}{2}N_{fd}^{(-)s} - \left(18 + 6dN_{adj}^{(+)s} + 2N_{fd}^{(+)s} + 18N_{adj}^{(-)} + 3N_{fd}^{(-)}\right), \tag{7}$$

$$C^{(3)} \equiv 72N_{adj}^{(+)} + 4N_{fd}^{(+)} - \left(54 + 18dN_{adj}^{(+)s} + 2N_{fd}^{(+)s}\right), \tag{8}$$

$$C^{(4)} \equiv 48 + 16dN_{adj}^{(+)s} + 18dN_{adj}^{(-)s} + 2N_{fd}^{(-)s} - \left(64N_{adj}^{(+)} + 4N_{fd}^{(-)} + 72N_{adj}^{(-)}\right). \tag{9}$$

We emphasize that each coefficient in the effective potential is given by the discrete values, that is, the flavor number of the massless bulk matter. This is the very curious feature of the Higgs potential, which is hardly seen in the usual quantum field theory, and is a key point to discuss the large gauge hierarchy in the gauge-Higgs unification.

3.1. *Model I*

We impose a condition $C^{(2)} = 0$ in order to obtain the hierarchically small VEV. One should note that the condition is not the fine tuning of the parameter usually done in the quantum field theory. This condition is fulfilled by the choice of the flavor set. By minimizing the Higgs potential, we obtain that

$$M_W = M_c \exp\left(-\frac{C^{(4)}}{\left|C^{(3)}\right|}\ \ln 2 + \frac{11}{6}\right). \tag{10}$$

We see that the large gauge hierarchy $M_c \sim M_{GUT}, M_{Planck}$ is realized for the large ratio, $C^{(4)}/\left|C^{(3)}\right| \gg 1$. The magnitude of the ratio for the values of $p = 11, 19$, where p is defined by $M_c \equiv 10^p$ GeV, is $C^{(4)}/\left|C^{(3)}\right| \simeq 32.54\ (p = 11), 59.12\ (p = 19)$. The large gauge hierarchy is realized if we have $C^{(2)} = 0$ and the large ratio $C^{(4)}/\left|C^{(3)}\right|$ at the same time.

Let us present a few examples of the flavor set in the model I. We choose $(k, m) = (1, 0)$ as a demonstration. Then, we find that

$$(N_{adj}^{(+)}, dN_{adj}^{(+)s}) = (1,1), (2,5), \cdots,\ (N_{fd}^{(+)}, N_{fd}^{(+)s}) = (0,3), (1,5), \cdots.$$

For $(k, m, p) = (1, 0, 19)$,

$$(N_{adj}^{(-)}, dN_{adj}^{(-)s}) = (0,29), (1,33), \cdots,\ (N_{fd}^{(-)}, N_{fd}^{(-)s}) = (42,1), (43,3), \cdots.$$

For $(k, m, p) = (1, 0, 11)$,

$$(N_{adj}^{(-)}, dN_{adj}^{(-)s}) = (0,16), (1,20), \cdots,\ (N_{fd}^{(-)}, N_{fd}^{(-)s}) = (22,1), (23,3), \cdots.$$

We observe that the flavor numbers $dN_{adj}^{(-)s}, N_{fd}^{(-)}$ are of order $O(10)$. One has to take care about the reliability of perturbation theory for such the large number of flavor because an expansion parameter in the present case may be given by $(g_4^2/4\pi^2)N_{flavor}$, and it must be $(g_4^2/4\pi^2)N_{flavor} \ll 1$ for reliable perturbative expansion.

Now, let us study the Higgs mass in the model I. The Higgs mass squared is obtained by the second derivative of the effective potential evaluated at the minimum of the potential (6),

$$m_H^2 = \frac{3g_4^2}{16\pi^2}\ M_W^2\ \left(-\frac{C^{(3)}}{6}\right)\left(= \frac{3g_4^2}{16\pi^2}\ M_W^2\ k < M_W^2\right). \tag{11}$$

The choice $k = 1$ is the most desirable one for the large gauge hierarchy, so that the Higgs mass is lighter than M_W, which is the same result in the original Coleman-Weinberg's paper.[9] Therefore, one concludes that the large gauge hierarchy and the sufficiently heavy Higgs mass are not compatible in the model I.

3.2. *Model II*

We study another model called Model II in this subsection. We introduce massive bulk fermions[10–12] in addition to the massless bulk matter in the model I. We introduce a pair of the fields, ψ_+ and ψ_- whose parity is different to each other, $\psi_\pm(-y) = \pm\psi_\pm(y)$. Then, a parity even mass term is constructed like $M\bar{\psi}_+\psi_-$.

The contribution to the mass term from the massive fermions is given by

$$\frac{1}{2}\zeta(3)C^{(2)}(\pi a)^2 \to -\frac{1}{2}\left[\zeta(3)C^{(2)} + 8N_{pair}B^{(2)}\right](\pi a)^2 \quad \text{with}$$

$$B^{(2)} = \sum_{n=1}^{\infty}\frac{1}{n^3}\left(1+nz+\frac{n^2z^2}{3}\right)\mathrm{e}^{-nz},$$

where N_{pair} stands for the number of the pair $(\psi^{(+)}, \psi^{(-)})$ and we have defined a dimensionless parameter $z \equiv 2\pi RM = M/M_c$. We observe that the potential is suppressed by the Boltzmann-like factor e^{-nz}, reflecting the fact that the effective potential shares similarity with that in finite temperature field theory.[13]

The essential behavior of the VEV is governed by the factor $B^{(2)}$, *i.e.* $\pi a_0 \simeq \gamma B^{(2)}$ with some numerical constant γ of order 1. If we write $\pi a_0 = \mathrm{e}^{-Y}$, then, one finds, remembering equation (5), that

$$-Y = \ln(\pi a_0)\left(= \ln\left(\frac{M_W}{M_c}\right)\right) = (2-p)\ln 10 \simeq \begin{cases} -34.539 \text{ for } p = 17, \\ -20.723 \text{ for } p = 11. \end{cases} \tag{12}$$

The gauge hierarchy is controlled by the magnitude of Y, in other words, the bulk mass parameter z, and the large gauge hierarchy is achieved by $|z| \simeq 30 \sim 40$. The large gauge hierarchy is realized by the presence of the massive bulk fermion. We notice that the flavor number of the massless bulk matter is not essential for the large gauge hierarchy in the model II.

Now, let us next discuss the Higgs mass in the model II. The Higgs mass is given by

$$m_H^2 = \frac{g_4^2}{16\pi^2}M_W^2\left[-C^{(3)}\ln(\pi a_0) + \frac{4}{3}C^{(3)} + C^{(4)}\ln 2\right] = \frac{g_4^2}{16\pi^2}M_W^2\, F, \tag{13}$$

where we have defined

$$F \equiv -C^{(3)}\ln(\pi a_0) + \frac{4}{3}C^{(3)} + C^{(4)}\ln 2. \tag{14}$$

The Higgs mass depends on the logarithmic factor. We observe that the larger the gauge hierarchy is, the heavier the Higgs mass is. An important point is that the coefficient $C^{(3)}$ is not related with the realization of the large gauge hierarchy, so that it is not constrained by the requirement of the large gauge hierarchy at all.

In order to demonstrate the size of the Higgs mass in the model II, let us choose $(k, l, m) = (-4, -1, -1)$. Then, the flavor set is given by

$$(N_{adj}^{(-)}, dN_{adj}^{(-)s}) = (1,3),\ (2,7), \cdots,\ (N_{fd}^{(-)}, N_{fd}^{(-)s}) = (3,1),\ (4,3), \cdots,$$
$$(N_{fd}^{(+)}, N_{fd}^{(+)s}) = (2,1),\ (3,3), \cdots,\ (N_{adj}^{(+)}, dN_{adj}^{(+)s}) = (1,0),\ (2,4), \cdots.$$

And the Higgs mass in GeV unit is calculated as

$$m_H \simeq \begin{cases} 119.5 \text{ for } p = 17, \\ 92.6 \text{ for } p = 11, \end{cases}$$

where we have used $g_4^2 \simeq 0.42$. Here, we note that in the usual scenario of the gauge-Higgs unification, one requires $g_4 \sim O(1)$ in order to have the heavy enough Higgs mass.[8,14] The large gauge hierarchy enhances the Higgs mass sizably even for the weak coupling. We observe that for the fixed integers (k, l), the large gauge hierarchy, that is, large $\ln(\pi a_0) = -Y$ enhances the size of the Higgs mass. The larger the gauge hierarchy is, the heavier the Higgs mass tends to be.

4. Conclusion and discussion

We have studied the five dimensional nonsupersymmetric $SU(3)$ model compactified on $M^4 \times S^1/Z_2$, which is the simplest model to realize the scenario of the gauge-Higgs unification. We have discussed whether the large gauge hierarchy is realized in the scenario or not. The Higgs potential is generated at the one-loop level and is obtained in a finite form, reflecting the nonlocal nature that the Higgs field is the Wilson line phase in the gauge-Higgs unification. The Higgs potential is calculable and accordingly, the Higgs mass, too. We have found two models (model I, II), in which the large gauge hierarchy is realized. The condition $C^{(2)} = 0$ is crucial for our discussions.

In connection with the condition, it may be worth mentioning that there are examples, in which the loop correction is exhausted at the one-loop level (without supersymmetry). They are the coefficient of the axial anomaly[15] and the Chern-Simons coupling.[16] As for the latter case, a simple reason for the two (higher) loop correction not to be generated comes from the invariance of the action under the large gauge transformation. Since the shift symmetry of the Higgs potential can be regarded as the invariance under the large gauge transformation, one may be able to prove that there is no two (higher)-loop correction to the mass term of the Higgs potential. In order to confirm it, one needs more studies of the higher loop corrections to the Higgs potential (mass) in the gauge-Higgs unification.[17]

References

1. N. S. Manton, *Nucl. Phys.* **B158** 141 (1979), D. B. Fairlie, *Phys. Lett.* **82B** (1979) 97, N. V. Krasnikov, *Phys. Lett.* **273B** (1991) 246.
2. Y. Hosotani, *Phys. Lett.* **126B** (1983) 309, *Ann. Phys. (N.Y.)* **190**, 233 (1989).
3. H. Hatanaka, T. Inami and C.S. Lim, *Mod. Phys. Lett.* **A13** (1998) 2601, G. R. Dvali, S. Randjbar-Daemi and R. Tabbash, *Phys. Rev.* **D65** (2002) 064021, N. Arkani-Hamed, A. G. Cohen and H. Georgi, *Phys. Lett.* **B513** (2001) 232, I. Antiniadis, K. Benakli and M. Quiros, New J. Phys. **3**, (2001),20.
4. K. Takenaga, *Phys. Rev.* **D64** (2001) 066001, *Phys. Rev.* **D66** (2002) 085009, Y. Hosotani, S. Noda and K. Takenaga, *Phys. Rev.* **D69** (2004) 125014, *Phys. Lett.* **B607** (2005) 276, N. Haba, K. Takenaga and T. Yamashita, *Phys. Rev.* **D71** (2005) 025006, N. Maru and K. Takenaga, *Phys. Rev.* **D72** (2005) 046003, *Phys. Rev.* **D74** (2006) 015017, K. Takenaga, *Phys. Lett.* **425B** (1998) 114, *Phys. Rev.* **D58** (1998) 026004.
5. A. Masiero, C. A. Scrucca, M. Serone and L. Silvestrini, *Phys. Rev. Lett.* **87** 251601 (2001).
6. M. Sakamoto and K. Takenaga, hep-th/0609067. To appear in *Physical Review* **D**.
7. M. Kubo, C. S. Lim and H. Yamashita, *Mod. Phys. Lett.* **A17** (2002) 2249, L. J. Hall, Y. Nomura and D. R. Smith, *Nucl. Phys.* **B639** (2002) 307, G. Burdman and Y. Nomura, *Nucl. Phys.* **B656** (2002) 3.
8. N. Haba, K. Takenaga and T. Yamashita, *Phys. Lett.* **B615** (2005) 247.
9. S. Coleman and E. Weinberg, *Phys. Rev.* **D7** (1973) 1888.
10. K. Takenaga, *Phys. Lett.* **B570** (2003) 244.
11. N. Haba, K. Takenaga and T. Yamashita, *Phys. Lett.* **B605** (2005) 355.
12. N. Maru and K. Takenaga, *Phys. Lett.* **B637** (2006) 287.
13. L. Dolan and R. Jackiw, *Phys. Rev.* **D9** (1974) 3320.
14. N. Haba, Y. Hosotani, Y. Kawamura and T. Yamashita, *Phys. Rev.* **D70** (2004) 01510.
15. S. L. Adler and W. A. Bardeen, *Phys. Rev.* **182** (1969) 1517.
16. H. So, *Prog. Theor. Phys.* **73** (1985) 528, S. Coleman and B. Hill, *Phys. Lett.* **159B** (1985) 184. M. Sakamoto and H. Yamashita, *Phys. Lett.* **B476** (2000) 427.
17. Y. Hosotani, in the Proceedings of "Dynamical Symmetry Breaking", ed. M. Harada and K. Yamawaki (Nagoya University, 2004), P17. (hep-ph/0504272), hep-ph/0607064, Y. Hosotani, N. Maru, K. Takenaga and T. Yamashita, work in progress.

Higgs and Top quark coupled with a conformal gauge theory

Haruhiko Terao

Institute for Theoretical Physics, Kanazawa University,
Kanazawa, 920-1192, Japan
E-mail: terao@hep.s.kanazawa-u.ac.jp

We consider a dynamical scenario improving the naturalness of the standard model, in which the Higgs field has a large anomalous dimension through conformally invariant coupling with an extra gauge sector above TeV scale. The large top Yukawa coupling is generated effectively through mixing top quarks and the fermions of the extra gauge sector. We present an explicit model consistent with the Electro-Weak precision tests.

Keywords: Little hierarchy problem; conformal field theory; top quark.

1. Introduction

Unnaturalness of the standard model (SM) indicates that new interactions and particles should appear at some low energy scale M so as to remove or suppress the Higgs mass correction. If the Higgs mass is less than 200GeV, which is expected from the electro-weak (EW) precision tests,[1] then M should be of $O(1)$TeV or less. On the other hand, however, the scale of the new physics is required to be more than $5 \sim 10$TeV in order to explain the EW precision tests in general. The discrepancy of these scales is called the little hierarchy problem of the SM.[2]

The models eliminating the quadratic divergence by imposing global symmetries, namely the supersymmetric models and the little Higgs models, lead to a relatively light Higgs boson. This is because the quartic coupling of the Higgs boson is also suppressed due to the global symmetry as well. The light Higgs is also favored by the EW precision tests. Then is there no room for a heavy Higgs to be discovered at the LHC?

The Higgs with mass near the trivilaity bound, which is about 600GeV, indicates strongly coupled theory behind. The Technicolor models would be the stereotype leading to a heavy Higgs. However the dynamical EW symmetry breaking is known to induce too large corrections to the S-parameter.[3] Moreover, there must be extra contribution to the T-parameter of $\Delta T = 0.25 \pm 0.1$ in order that the heavy Higgs mass of 400-600 GeV is made consistent with the EW precision tests.

Here we note that the Higgs mass parameter may be protected from the quadratic divergence, provided that the Higgs field acquires a positive anomalous dimension. Indeed a sufficiently large anomalous dimension can be realized by introducing coupling of the Higgs with a strongly interacting conformal field theory (CFT). Such a mechanism has been also discussed by Luty and Okui[4] recently and they proposed the "conformal technicolor" scenario and it's AdS/CFT correspondence. Differently from this, we consider a scenario in which the electro-weak symmetry breaking (EWSB) is not dynamical and the Higgs behaves as a point particle beyond TeV scale. Therefore the strongly coupled sector does not induce a large correction to S. Besides, it is shown below that we may construct an explicit model leading to a heavy Higgs boson as well as a suitable amount of the T-parameter.

2. Dynamics generating a large anomalous dimension

Suppose that the anomalous dimension of the Higgs boson H, γ_H, is given to be a scale independent constant, then the correction to the Higgs mass parameter m_H^2 depends on the cutoff scale Λ as

$$\delta m_H^2 \sim \left(\frac{\Lambda}{\mu}\right)^{\epsilon} \mu^2, \tag{1}$$

where the power of divergence is given by $\epsilon = 2(1-\gamma_H)$ and μ denotes the renormalization scale. Note that this correction may be approximated as a logarithmic one and, therefore, naturalness may be improved drastically, when γ_H is close to one.

Such an anomalous dimension can be realized in the strongly coupled IR fixed point of gauge-Yukawa theories. It has been known for some time that an IR fixed point exists for the QCD like theory with an appropriate number of flavors, which is called the Banks-Zaks (BZ) fixed point.[5] At the fixed point, the fermions acquire negative anomalous dimensions through the gauge interaction. Therefore operators of $\lambda_i \bar{\psi}_i \psi^i \phi$ with a singlet scalar ϕ, where $i = 1, \cdots, N_f$ denotes the flavor index, is relevant there. Therefore it is expected that the Yukawa couplings λ_i are enhanced and approaches another fixed point λ_* towards the IR direction.

It is easy to confirm this by solving the perturabtive renormalization group (RG) equations. The beta functions of the gauge coupling $\alpha_g = g^2/(4\pi)^2$ and the Yukawa coupling $\alpha_{\lambda_i} = |\lambda_i|^2/(4\pi)^2$ for the $SU(N_c)$ gauge theory with N_f flavors of N_c representation are found to be

$$\beta[\alpha_g] = -2b_0\alpha_g^2 - 2b_1\alpha_g^3 - 2\alpha_g^2\sum_i \alpha_{\lambda_i} + \cdots \tag{2}$$

$$\beta[\alpha_{\lambda_i}] = 2\alpha_{\lambda_i}\left(\gamma_\phi + \gamma_{\bar{\psi}\psi} + 3\alpha_{\lambda_i}\right) + \cdots , \tag{3}$$

where γ_ϕ and $\gamma_{\bar{\psi}\psi}$ denote the anomalous dimensions of the fields ϕ and $\bar{\psi}\psi$ respectively and are given as $\gamma_\phi = 2N_c\sum_j \alpha_{\lambda j}$, $\gamma_{\bar{\psi}\psi} = -6C_2(N_c)\alpha_g$. The coeeficients in the gauge beta function are given as $b_0 = 11 - 2N_f/3$, $b_1 = 102 - 38N_f/3$.

However such perturbative beta functions are not applicable in order to evaluate the anomalous dimensions in the strongly coupled region. For this purpose, the so-called exact RG (ERG) equations based on the Wilson RG are very suitable. The ERG has been applied to the non-perturbative dynamics of chiral symmetry braking phenomena in the QCD like theory by truncating the infinite number of local operators in the Wilsonian effective action.[6] It has been found that the anomalous dimensions as well as the phase diagram obtained by solving the Schwinger-Dyson equations in the (improved) ladder approximation[7] are easily reproduced by the ERG method.

For the non-perturbative analysis, the effective four-fermi operators are found to play an important role. As an example, Fig. 1 shows schematically how non-perturabative anaomalous dimension $\gamma_{\bar{\psi}\psi}$ is represented in terms of the effective four-fermi vertex in the Wilson RG framework. In the ladder approximation, it is enough to deal with only the four-fermi operator given as

$$\mathcal{O}_{4-\text{fermi}} = \frac{2G_S}{\mu^2}\bar{\psi}_{Li}\psi_R^j\bar{\psi}_{Rj}\psi_L^i \tag{4}$$

among various $SU(N_f)_L \times SU(N_f)_R$ invariant operators. Then the RG equation for $g_S = G_S/4\pi^2$ is given simply by

$$\mu\frac{dg_S}{d\mu} = 2g_S - 2N_c\left(g_S + \frac{3C_2(N_c)}{N_c}\alpha_g\right)^2 . \tag{5}$$

Not only the four-fermi coupling but also the gauge and Yukawa coupling should be treated in the ERG framework. However we substitute the two-loop gauge beta function (2) as a practical approximation. The beta function of the Yukawa coupling of the ERG is obtained by replacing the anomalous dimension $\gamma_{\bar{\psi}\psi}$ in (3) with

$$\gamma_{\bar{\psi}\psi} = -6C_2(N_c)\alpha_g - 2N_c g_S. \tag{6}$$

Then it is an easy task find the IR fixed points and the anomalous dimensions γ_ϕ, $\gamma_{\bar\psi\psi}$ from these RG equations. It is noted that the approach of RG has an advangate over solving the SD equations in analyzing the dynamics of CFTs.

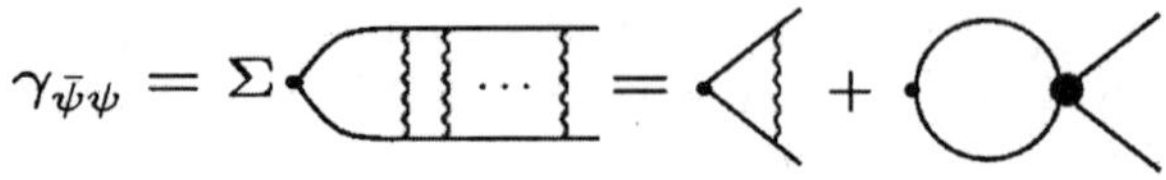

Fig. 1. The anomalous dimension $\gamma_{\bar\psi\psi}$ in terms of the effective four-fermi vertex is shown schematically.

First it is found that there is a critical value for the gauge coupling constant α_g^* of the IR fixed point, $\alpha_g^* < \alpha_g^{\rm cr} = 1/12C_2(N_c)$. For the case of $N_c = 3$, which is considered in the explicit model, the number of flavors is restricted as $16 \geq N_f \geq 12$.[8] Among them, the IR fixed point coupling, $\alpha_g^* \sim 0.06$ is very close to $\alpha_g^{\rm cr}$ in the case of $N_f = 12$. Then the anomalous dimensions are found to be $\gamma_\phi^* \simeq -\gamma_{\bar\psi\psi}^* = 0.8$. Thus we see that a large anomalous dimension close to 1 can be realized.

The scalar mass renormalized at a scale M, $m_\phi(M)$, may be also obtained by solving the RG equation and found to be

$$m_\phi^2(M) = \left[\tilde m_\phi^{*2} + \left(\frac{\Lambda}{M}\right)^\epsilon (\tilde m_\phi^2(\Lambda) - \tilde m_\phi^{*2})\right] M^2 \sim O(1-10) \times M^2, \quad (7)$$

where $\tilde m_\phi^2 = m_\phi^2/\mu^2$ stands for the dimensionless mass parameter and $\epsilon = 2(1-\gamma_\phi^*) \simeq 0.4$ and a constant $\tilde m_\phi^{*2} \simeq 4$ for $N_c = 3, N_f = 12$. Note that $(\Lambda/M)^\epsilon \sim 6.31$ even for $(\Lambda/M)^2 = 10^4$. Thus the cutoff dependence of the scalar mass is remarkably suppressed due to the anomalaous dimension.

The anomalous dimension affects the other interactions of the Higgs. First the quartic coupling is made highly irrelevant. However it is not suppressed but approaches a large fixed point coupling, since the quartic coupling is induced by the Yukawa coupling λ^*. Therefore, if we regard ϕ as the EW Higgs, then the Higgs mass must be near it's triviality bound. Meanwhile the Yukawa coupling with quarks and leptons are suppressed as $y_t(\mu) \sim (\mu/\Lambda)^{\gamma_\phi^*} y_t(\Lambda)$. Then the top Yukawa coupling is too small to generate the observed top quark mass. The model presented here avoids this problem by mass mixing between the fermions of the CFT sector with top quarks.

3. A phenomenological model

Now we describe an explicit realization of the scenario presented in the previous section briefly. First, the conformal invariance must terminate at TeV scale. We introduce another gauge symmetry $G_{\rm DSB}$ to bring about mass generation by spontaneous chiral symmetry breaking. The gauge interaction becomes strong at TeV scale and distroies the conformal invariance. In order to induce the mixing between the top quarks and the fermions in the CFT sector, we suppose that the CFT gauge symmetry is also $SU(3)_{\rm CFT}$ and that the symmetry beraking from $SU(3)_{\rm CFT} \times SU(3)'_C$ to the color gauge symmetry $SU(3)_C$ takes place through the chiral symmetry breaking. Note that the strong dynamics does not break the EW symmetry directly.

To be more explicit, let us introduce the vector-like fermions with gauge charges shown in the table:

	$SU(3)_{\rm DSB}$	$SU(3)_{\rm CFT}$	$SU(3)'_C$	$SU(2)_W$	$U(1)_Y$
ψ	**3**	**3**	**1**	**1**	any
η	**3**	**1**	**3**	**1**	any
Φ^i	**1**	**3**	**1**	**2**	1/6
ϕ^i	**1**	**3**	**1**	**1**	2/3

where $i(=1,2,3)$ denotes the species of the CFT fermions. Then the CFT sector gives an $N_c = 3, N_f = 12$ gauge theory, which we have discussed previously. Indeed, this model has an IR fixed point with a Yukawa coupling to the Higgs $\lambda^* \sim 2.64$ and the large anomalous dimension of the Higgs $\gamma_H^* \sim 0.8$ above the TeV scale, since the gauge coupling of $SU(3)_{\rm DSB}$ is small there.

The symmetry breaking of $SU(3)_{\rm CFT} \times SU(3)'_C$ to $SU(3)_C$ takes places, if the gauge interaction of $G_{\rm DSB}$ induces the diagonal vaccum expectation values (VEV) as $\langle \bar{\psi}_A \cdot \eta^a \rangle / \Lambda^2 \sim \omega \delta^a_A$, $\langle \bar{\eta}_a \cdot \psi^A \rangle / \Lambda^2 \sim \bar{\omega} \delta^A_a$, where $A = 1,2,3$ and $a = 1,2,3$ stand for indices of $SU(3)_{\rm CFT}$ and $SU(3)'_C$ respectively. The $G_{\rm DSB}$ interaction also induces the fermion condensation $\langle \bar{\psi}_A \cdot \psi^A \rangle / \Lambda^2 \sim \langle \bar{\eta}_a \cdot \eta^a \rangle / \Lambda^2 \sim M$. The mass M turns out to be the decoupling scale of the CFT sector from the EW theory and is supposed to be of TeV order.

Now we assume that the model is given at the cutoff scale Λ, which is as large as, say, 10TeV with effective four-fermi interactions such as

$$
\begin{aligned}
-\mathcal{L}_{\rm int} = {} & \lambda^* \bar{\Phi}_{iA} \phi^{iA} H + \frac{c}{\Lambda^2} \bar{Q}_{3La} \Phi^A_R (\bar{\psi}_A \cdot \eta^a) + \frac{\bar{c}}{\Lambda^2} \bar{\phi}_A u^a_{3R} (\bar{\eta}_a \cdot \psi^A) \\
& + \frac{c_\Phi}{\Lambda^2} \bar{\Phi}_A \Phi^A (\bar{\psi}_A \cdot \psi^A) + \frac{c_\phi}{\Lambda^2} \bar{\phi}_A \phi^A (\bar{\psi}_A \cdot \psi^A) + V(H), \qquad (8)
\end{aligned}
$$

where Q_{3L} and u_{3R} denote the third generation quark doublet and singlet fields respctively. Then, the effective Lagrangian obtained after the above SSB is reduced to

$$-\mathcal{L}_{\text{int}} = \lambda_*(\bar{\Phi}_L\phi_R + \bar{\Phi}_R\phi_L)H + V(H)$$
$$+M_\Phi\left(\bar{\Phi}_L + \frac{c\omega}{M_\Phi}\bar{Q}_{3L}\right)\Phi_R + M_\phi\bar{\phi}_L\left(\phi_R + \frac{\bar{c}\bar{\omega}}{M_\phi}u_{3R}\right). \quad (9)$$

These mass terms induces mixing between (Q_{3L}, u_{3R}) and (Φ_L, ϕ_R) with the mixing angles of $\tan\theta_L = c\omega/M_\Phi$ and $\tan\theta_R = \bar{c}\bar{\omega}/M_\phi$. The mixing generates the an effective top Yukawa coupling to the massless top quark fields, which we represent by $Q'_{3L} = (t, b)_L$ and $u'_{3R} = t_R$, given by

$$y_t^{\text{eff}} \sim \lambda_* \sin\theta_L \sin\theta_R. \quad (10)$$

The fixed point coupling of this model, $\lambda^* \sim 2.64$, fixes the proper mixing angle to be $\sin\theta \sim 0.6$ for the real top quark mass.

4. The Electro-Weak precision tests

Lastly we examine whether this extension of the SM model above the TeV scale can be made consisitent with the EW precision tests. We note that the oblique corrections are generated through mixing of weak doublets and weak singlets after the EWSB. It's size is given roughly by v/M, where v denotes the Higgs VEV. Therefore the parameters of T, S can be very small for a sufficiently large M. However, the decoupling scale of the CFT sector M should be relatively low in order to improve naturalness of the EW theory.

The sizeable T-parameter correction ΔT may be induced in the present model, since the weak isospin symmetry is broken explicitly. We may evaluate ΔT by calculating the one-loop corrections including the CFT fermions Φ, ϕ. Then we can fix the scale M so that this ΔT satisfies the constraint by the EWPT for the Higgs mass of $400 - 600$GeV; $\Delta T = 0.25 \pm 0.1$. It is noted that these heavy Higgs masses are close to the triviality bound values of a few TeV scale cutoff. The condition for ΔT leads to the allowed region for the scale M as $1.48\text{TeV} < M < 2.45\text{TeV}$. Similarly we may also evaluate the S-parameter and find that ΔS is much smaller than ΔT. Of course such a perturbative calculation is not so reliable, since the CFT fermions interact strongly and the higher order corrections are not negligible. These results should be taken with ambiguity of O(1). Thus, we may expect that the consistent decoupling scale exists around a few TeV.

We should mention the constraint by the amplitude of $Z \to b\bar{b}$ to this model as well. So far we have not considered the bottom Yukawa coupling, since it is not necessary to consider the mixing effect. When we add the bottom Yukawa term in the effective Lagarangian, then it is found that both of b_L and b_R are mixed with components of the weak doubles fermions Φ_L and Φ_R through EW symmetry breaking. These mixings induce deviations of the gauge couplings g_L and g_R at tree-level. However, the mixing is of order of m_b/M and, therefore $\delta g_L(R)$ is found to be less than 10^{-5}.

The SM correction for the Higgs mass parameter is now comparable with the Higgs mass itself, if the scale M of a few TeV gives the cutoff scale. However the natural size of the Higgs mass parameter $|m_H|$ is $O(1) \times M$, which is fairly larger than the Higgs mass. This parameter depends on the initial value given at the cutoff scale Λ and can be small with some fine-tuning. However fine-tuning of $O(1)$ % is necessary in order to realize the realistic EWSB.

Acknowledgments

The author is supported in part by the Grants-in-Aid for Scientific Research No. 40192653 from the Ministry of Education, Science, Sports and Culture, Japan. He also thanks T. Kobayashi, H. Nakano, M. Tanabashi and A. Tsuchiya for valuable discussions and the organizers of SCGT06.

References

1. The LEP Collaborations ALEPH, DELPHI, L3, OPAL, and the LEP Electroweak Working Group, arXiv:hep-ex/0511027.
2. R. Barbieri and A. Strumia, arXiv:hep-ph/0007265; Phys. Lett. **B462** 144 (1999).
3. M.E. Peskin and T. Takeuchi, Phys. Rev. Lett. **65**, 194 (1990); Phys. Rev. D **46**, 381 (1991).
4. M. A. Luty and T. Okui, JHEP **0609** 070 (2006).
5. T. Banks and A. Zacks, Nucl. Phys. B **196**, 189 (1982).
6. K-I. Aoki, K. Morikawa, J. Sumi, H. Terao and M. Tomoyose, Prog. Theor. Phys. **97** 479 (1997); Prog. Theor. Phys. **102** 1151 (1999); Phys. Rev. **D61** 045008 (2000).
7. For a review see *e.g.*, K. Yamawaki, arXiv:hep-ph/9603293, in Proc. 14th Symposium on Theoretical Physics, Cheju, Korea, 1995.
8. T. Appelquist, J. Terning and L. C. R. Wijewardhana, Phys. Rev. Lett. **77** 1214 (1996); V.A. Miransky and K. Yamawaki, Phys. Rev **D55** 5051 (1997); T. Appelquist, A. Ratnaweera, J. Terning and L.C.R. Wijewardhana, Phys. Rev. **D58** 105017 (1998).

PARTIALLY COMPOSITE TWO HIGGS DOUBLET MODEL

P. KO

School of Physics, KIAS
Seoul 130-722, Korea
**E-mail: pko@kias.re.kr*

We consider a possibility that electroweak symmetry breaking (EWSB) is triggered by a fundamental Higgs and a composite Higgs arising in a dynamical symmetry breaking mechanism induced by a new strong dynamics. The resulting Higgs sector is a partially composite two-Higgs doublet model with specific boundary conditions on the coupling and mass parameters originating at a compositeness scale Λ. The phenomenology of this model is discussed including the collider phenomenology at LHC and ILC.

Keywords: Dynamical electroweak symmetry breaking; composite higgs; two-higgs doublet.

1. Introduction

EWSB is the origin of the masses of chiral fermions and electroweak gauge bosons as well as CP violation in the quark sector within the standard model (SM). It is most important in particle physics to understand the origin of EWSB, and LHC will serve for this purpose. There have been many attempts to construct interesting models for EWSB beyond the SM.[1]

Dynamical symmetry breaking á la Miransky, Tanabashi, Yamawaki (MTY)[2] and Bardeen, Hill, Lindner (BHL)[3] is a particularly interesting scenario, since the heavy top mass is intimately related with a new strong dynamics that condenses the $t\bar{t}$ bilinear, and breaks the EW symmetry down to $U(1)_{\rm EM}$. Both heavy top mass and Higgs mass are generated dynamically, in anology with superconductivity of Bardeen-Cooper-Schrieffer (BCS).

However, there are basically two drawbacks in this model. First, the origin of the new strong interactions that triggers EWSB is not clear. The attractive 4-fermion interaction is simply put in by hand within the BHL model. Also, the original version of BHL with 3 families or its extension with two Higgs doublets[4] predict that the top mass should be significantly heavier than the experimental observation: $m_t = 170.9 \pm 1.1({\rm stat}) \pm 1.5({\rm sys})$

GeV.[5] However, these two drawbacks could be evaded within extra dimensional scenarios, without ruining its niceties.

If QCD is a bulk theory, then it is possible that the KK gluon exchange can induce attractive Nambu-Jona-Lasinio (NJL) type four-fermion interaction in the low energy regime, and dynamical symmetry breaking can occur in a natural way.[6] It should be emphasized that this is completely different from another popular way of symmetry breaking in extra dimension, namely symmetry breaking by boundary conditions. Therefore, in extra dimensional scenarios, electroweak symmetry can be broken by fundamental Higgs, by boundary condition or by some dynamical mechanism. Generically all three possibilities could be present altogether. In most recent studies, the gauge symmetries were broken by the nontrivial boundary conditions with or without fundamental Higgs. In this talk, I discuss another possibility, where electroweak symmetry is broken by fundamental Higgs VEV's, as well as dynamically by $t\bar{t}$ condensate.[7] This way we will find that we can avoid both drawbacks of BHL model.

2. A Model of Dynamical EWSB with a Fundamental Scalar

Our model is a simple extension of the SM. We assume there is a new strong dynamics at some high energy scale Λ, which is effectively described by the NJL type four-fermion interaction term:

$$\mathcal{L} = \mathcal{L}_{\rm SM} + G(\overline{\psi}_L t_R)(\bar{t}_R \psi_L), \tag{1}$$

where the SM Higgs doublet ϕ is included from the beginnig, unlike the BHL model. The explicit forms of the SM lagrangians can be found in Ref. 7. The Yukawa couplings for the 1st and the 2nd generations do not play any roles in our analysis, and will be ignored in the following. We don't specify the origin of this NJL type interaction, but the KK gauge boson exchange in extra dimension scenarios could be one possible origin of this new strong interaction. As a minimal extension of the SM, we assume that this new strong dynamics acts only (or dominantly) on top quark.

We can rewrite the NJL term in Eq.(2.1) in terms of an auxiliary scalar field Φ:

$$\mathcal{L} = \mathcal{L}_{\rm SM} + g_{t0}(\overline{\psi}_L t_R \tilde{\Phi} + {\rm H.c.}) - M^2 \Phi^\dagger \Phi, \tag{2}$$

where $G = g_{t0}^2/M^2$ with $M \sim \Lambda$. g_{t0} is a newly defined Yukawa coupling between the top quark and the auxiliary scalar field Φ. Φ describes the composite scalar bosons that appear when the $\langle \bar{t}t \rangle$ develops nonvanishing

VEV and breaks the electroweak symmetry. Far below the scale Λ, the Φ field will develop the kinetic term due to quantum corrections and become dynamical. The resulting low energy effective field theory will be two-Higgs doublet model, one being a fundamental Higgs ϕ and the other being a composite Higgs Φ. Thus it can be called a partially composite two-Higgs doublet (PC2HD) model.

In order to avoid too large FCNC mediated by neutral Higgs bosons, we assign a Z_2 discrete symmetry under which the lagrangian is invariant ;

$$(\Phi,\ \psi_L,\ U_R) \to +(\Phi,\ \psi_L,\ U_R), \quad (\phi, D_R) \to -(\phi, D_R). \tag{3}$$

With this Z_2 discrete symmetry, t and b couple to Φ and the SM Higgs, respectively. In consequence, our model becomes the Type-II two-Higgs doublet model as the minimal supersymmetric standard model (MSSM).

The renormalized lagrangian for the scalar fields at low energy is given by

$$\begin{aligned}\mathcal{L}_{\rm ren} &= Z_\phi (D_\mu\phi)^\dagger(D^\mu\phi) + Z_\Phi(D_\mu\Phi)^\dagger(D^\mu\Phi) - V(\sqrt{Z_\phi}\phi, \sqrt{Z_\Phi}\Phi)\\ &\quad + \sqrt{Z_\Phi} g_t(\overline{\psi}_L t_R \tilde{\Phi} + \text{h.c}) + \sqrt{Z_\phi} g_b(\overline{\psi}_L b_R \phi + \text{h.c}),\end{aligned} \tag{4}$$

with

$$\begin{aligned}V(\phi,\Phi) &= \mu_1^2\phi^\dagger\phi + \mu_2^2\Phi^\dagger\Phi + (\mu_{12}^2\phi^\dagger\Phi + \text{H.c.}) + \frac{1}{2}\lambda_1(\phi^\dagger\phi)^2 + \frac{1}{2}\lambda_2(\Phi^\dagger\Phi)^2\\ &\quad + \lambda_3(\phi^\dagger\phi)(\Phi^\dagger\Phi) + \lambda_4|\phi^\dagger\Phi|^2 + \frac{1}{2}[\lambda_5(\phi^\dagger\Phi)^2 + \text{H.c.}]\,.\end{aligned} \tag{5}$$

In the scalar potential, we have introduced a dimension-two μ_{12}^2 term that breaks the discrete symmetry softly in order to generate the nonzero mass for the CP-odd Higgs boson. Otherwise the CP-odd Higgs boson A would be an unwanted axion related with spontanesously broken global $U(1)$ Peccei-Quinn symmetry, which would be a phenomenological disaster. This μ_{12}^2 parameter will be traded with the m_A^2, the (mass)^2 parameter of the CP-odd Higgs boson, which is another free parameter of our model.

Matching the lagrangian with Eq. (2.5) at the compositeness scale Λ, we obtain the following matching conditions as $\mu \to \Lambda$:

$$\begin{aligned}&\sqrt{Z_\phi} \to 1, \quad \sqrt{Z_\Phi} \to 0, \quad Z_\phi\mu_1^2 \to m_0^2, \quad Z_\Phi\mu_2^2 \to M^2,\\ &Z_\phi\lambda_1 \to \lambda_{10}, \quad Z_\Phi^2\lambda_2 \to 0, \quad Z_\phi Z_\Phi \lambda_{i=3,4,5} \to 0.\end{aligned} \tag{6}$$

These conditions are the boundary conditions for the RG equations.

Before proceeding, we would like to compare our model with Luty's model.[4] In Luty's model, both Higgs doublets are composite, and thus the matching conditions for Z_ϕ and λ_1 become

$$\sqrt{Z_\phi} \to 0, \qquad Z_\phi \lambda_1 \to 0, \tag{7}$$

which are different from those in our model. These different matching conditions lead to very different predictions for the scalar boson spectra compared to the Luty's model. Also we have additional Yukawa coupling g_b so that we can fit both the bottom and the top quark masses without difficulty unlike the models by BHL or Luty.

3. Particle Spectra and predictions

Our model is defined in terms of three parameters: Higgs self coupling λ_{10}, the compositeness scale Λ where λ_{10} and the NJL interaction are specified, and the CP-odd Higgs boson mass m_A. Since λ_{10} is also present in the SM, our model has two more parameters compared with the SM. It is strightforward to analyze the conditions for the correct EWSB and the particle spectra. The details can be found in the original paper.[7] In the following, I highlight the main results of our model:

- We can fit both the top and the bottom masses without difficulty in our model, unlike the BHL model or the Luty model, since the bottom quark get massive due to the fundamental Higgs. The allowed region of $\tan\beta$ is rather narrow: $0.45 \lesssim \tan\beta \lesssim 1$. Therefore the W and the Z boson get their masses almost equally from the fundamental Higgs and the $t\bar{t}$ condensation in our model.
- Since $\tan\beta \lesssim 1$, there is a strong constraint from $B \to X_s\gamma$, which implies that the charged Higgs boson should be heavier than ~ 400 GeV.
- There is no CP violating mixing in the neutral Higgs boson sector, since λ_5 remains zero at all scale within our model.
- $m_H^\pm \lesssim m_A$ in our model, and the charged Higgs can be even lighter than the lightest neutral Higgs boson, when the composite scale Λ is high. See Fig. 1.
- Triple and quartic self couplings of Higgs bosons can deviate from the SM values by significant amounts.
- Higgs coupling to the top quark is enhanced in our model so that the Higgs production rate at LHC is larger than the SM values. On the other hand, the Higgs productions at the ILC through Higgs–stralhlung and the WW fusion are suppressed compared to the SM values. See Fig. 2.

4. Conclusions

In this talk, I considered a possibility that the Higgs boson produced at the future colliders is neither a fundamental scalar nor a composite scalar, but a mixed state of them. It could be a generic feature, if there exists a strong dynamics at a high scale which give rise to the dynamical electroweak symmetry breaking, in addition to the usual Higgs mechanism due to the nonvanishing VEV of a fundamental Higgs. It is interesting that this scenario could be easily realized, if we embed the SM lagrangian in a higher dimension with bulk gauge interactions. The resulting theory can accommodate the observed top mass, and give specific predictions for neutral and charged Higgs masses at a given value of Λ. Whether such scenario is realized or not in nature could be studied at LHC and ILC.

Acknowledgments

The author is grateful to B. Chung, D.W. Jung and K.Y. Lee for enjoyable collaborations, and Prof. Yamawaki for discussions and his wonderful organization of the workshop. This work is supported in part by KOSEF through CHEP at Kyungpook National University.

References

1. C. T. Hill and E. T. Simmons, *Phys. Rep.* **381**, 235 (2003) and references therein.
2. V. A. Miransky, M. Tanabashi and K. Yamawaki, *Phys. Lett. B* **221**, 177 (1989); *Mod. Phys. Lett. A* **4**, 1043 (1989).
3. W. A. Bardeen, C. T. Hill and M. Lindner, *Phys. Rev. D* **41**, 1647 (1990).
4. M. Luty, *Phys. Rev. D* **41**, 2893 (1990).
5. Tevatron Electroweak Working Group, hep-ex/0703034.
6. B. A. Dobrescu, *Phys. Lett. B* **461**, 99 (1999); H.-C. Cheng, B. A. Dobrescu and C. T. Hill, *Nucl. Phys.* **B589**, 249 (2000).
7. B. Chung, D. Jung, P. Ko and K.Y. Lee, *JHEP* **0605**, 010 (2006).
8. S.L. Glashow and S. Weinberg, Phys. Rev. D 15, 1958 (1977).
9. C. T. Hill, C. N. Leung and S. Rao, *Nucl. Phys.* **B262**, 517 (1985).

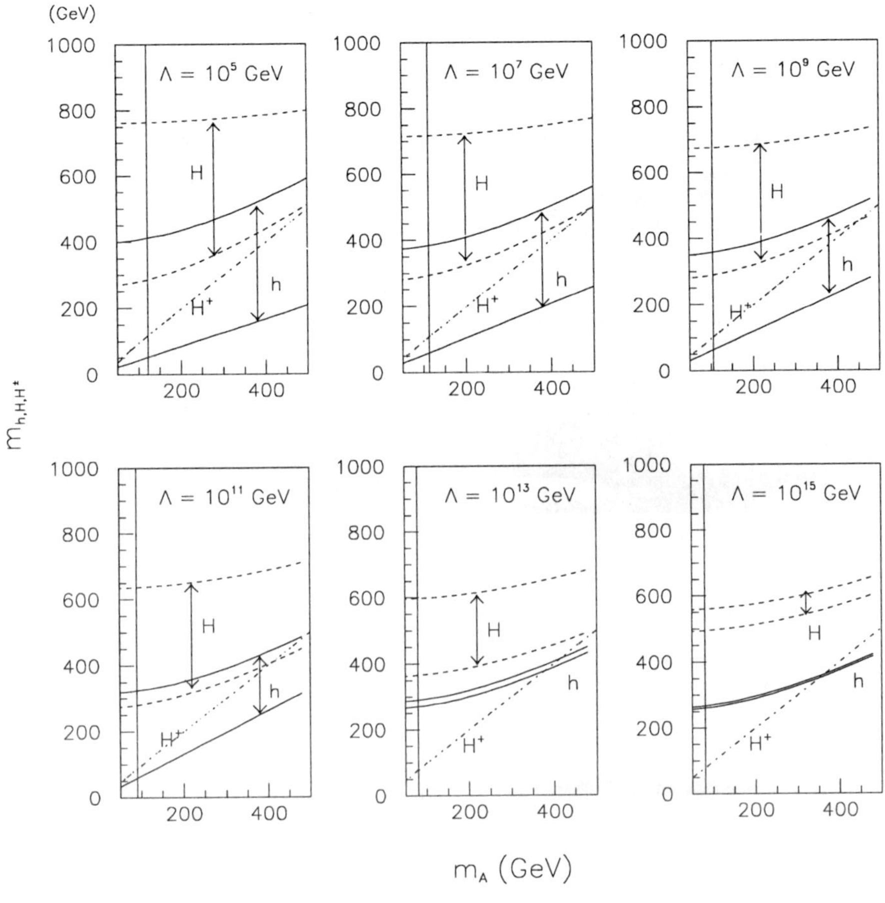

Fig. 1. Masses of neutral Higgs bosons h (inside the solid lines), H (inside the dashed lines) and the charged Higgs boson $H^{\pm}$ (dahs-dotted line) with respect to m_A.

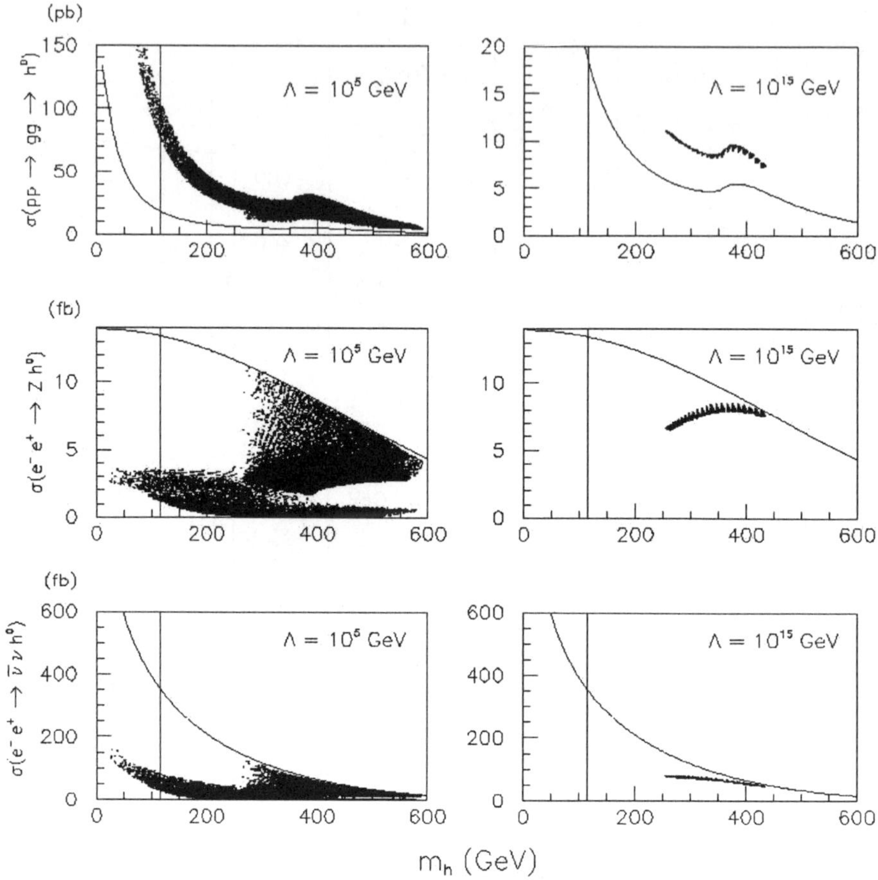

Fig. 2. Production cross section of the neutral Higgs boson at the LHC and ILC. $\sqrt{s} =$ 14 TeV for the LHC and $\sqrt{s} = 1$ TeV for the ILC are assumed. The solid curves denotes the SM predictions.

DYNAMICAL BREAKDOWN OF ABELIAN GAUGE CHIRAL SYMMETRY BY STRONG YUKAWA COUPLINGS

J. HOŠEK*

Department of Theoretical Physics, Nuclear Physics Institute ASCR
Řež near Prague, 25068, Czech Republic
**E-mail: hosek@ujf.cas.cz*
www.ujf.cas.cz

In an Abelian chirally and gauge-invariant anomaly-free model we demonstrate that strong Yukawa couplings between massless fermion fields and a massive scalar carrying the axial charge generate dynamically vastly different fermion masses, and the gauge boson mass.

Keywords: Electroweak theory; dynamical symmetry breaking; Yukawa interaction.

1. Introduction

The Higgs realization of spontaneous symmetry breakdown (SSB) in the electroweak theory is phenomenological by construction: First, there is no explanation of the 'wrong-sign' Higgs-field quadratic term. Second,[1] "who has ever heard of a fundamental theory that requires twenty-some parameters?". Moreover, the spectrum of the Yukawa couplings is sparse, irregular and extremely wide, as is the observed spectrum of the fermion masses. Moreover, no symmetry protects the scalar field mass term from quadratic renormalization. As a consequence the Higgs boson mass wants to be 'naturally'[2] heavy whereas some like it light.

Generically distinct conventional 3+1 dimensional field theory attempts at improving the drawbacks of the Higgs mechanism are not numerous,[3] and they are not cheap either: (i) Deal without elementary scalars. These options inevitably mean a strong-coupling field theory and often a great number of new fields. (ii) Deal with the elementary scalars naturally. This weak coupling option called supersymmetry does not address, however, the first two objections listed above. (iii) Deal with scalars with the ordinary sign of their mass terms.[4,5] This is the route we take. This option also yields a strong-coupling field theory with all its limitations.

Bellow we illustrate all necessary steps of the spontaneous electroweak mass generation without scalar field condensation on an Abelian prototype. Comparison with the standard Abelian Higgs mechanism can easily be made at any stage by heart.

2. Lagrangian and its symmetries

Our Abelian model is defined by its off-mass-shell perturbatively renormalizable Lagrangian

$$\mathcal{L} = \bar{\psi}_1 i \not{D} \psi_1 + \bar{\psi}_2 i \not{D} \psi_2 + \\ + (D_\mu \phi)^\dagger (D^\mu \phi) - M^2 \phi^\dagger \phi - \frac{1}{2}\lambda(\phi^\dagger \phi)^2 - \frac{1}{4} F_{\mu\nu} F^{\mu\nu} + \mathcal{L}_{\mathrm{Yukawa}}. \quad (1)$$

It contains two massless fermion fields ψ_1 and ψ_2, one massless gauge field A^μ and one massive complex scalar field $\Phi \equiv \frac{1}{\sqrt{2}}(\phi_1 + i\phi_2)$ with the Yukawa interaction (y_1,y_2 are the theoretically arbitrary real numbers)

$$\mathcal{L}_{\mathrm{Yukawa}} = y_1(\bar{\psi}_{1L}\psi_{1R}\phi + \bar{\psi}_{1R}\psi_{1L}\phi^\dagger) + y_2(\bar{\psi}_{2R}\psi_{2L}\phi + \bar{\psi}_{2L}\psi_{2R}\phi^\dagger). \quad (2)$$

Important is the Abelian axial symmetry $\psi_j \to e^{i\theta Q_j \gamma_5}\psi_j$, $\phi \to e^{-2i\theta}\phi$, with its associated axial-vector current:

$$j_A^\mu = \bar{\psi}_1 \gamma^\mu \gamma_5 \psi_1 - \bar{\psi}_2 \gamma^\mu \gamma_5 \psi_2 + 2i\left[(\partial^\mu \phi)^\dagger \phi - \phi^\dagger \partial^\mu \phi\right]. \quad (3)$$

It is the axial symmetry which prohibits the fermion and the gauge field mass terms in (1). It is this symmetry which will be broken in solutions of the Schwinger-Dyson (SD) equations for their corresponding propagators.

Once the axial $U(1)_A$ symmetry is gauged, it must not be anomalous. This important theoretical constraint can be fulfilled in many ways. Here we consider the simplest possibility, and choose the axial charges of two fermion species as $Q_1 = +1$, $Q_2 = -1$.

3. Dynamical fermion mass generation and scalar boson mass splitting

As in the SM we assume that no symmetry-breaking dynamics is due to the gauge interaction, and set $g = 0$. This interaction will be switched on perturbatively in Section 4. Unlike the SM where the scalar self-interaction is vitally important for the tree-level scalar-field condensation we believe that it is not important for the SSB here, and set $\lambda = 0$ in the following. In the full treatment to be developed this term will become necessary as

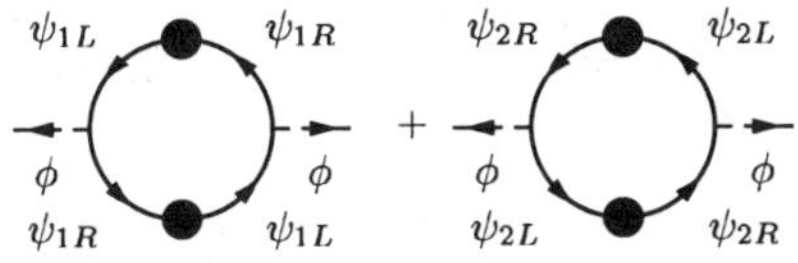

Fig. 1. Mixing in scalar sector induced by the fermion mass terms. The full blobs denote the chirality-changing part of the complete fermion propagator.

a counter-term. (In fact for clarity we omit all symmetry-preserving loop effects in the following.) The consequence of these assumptions is two-fold: First, there is no SSB in the scalar sector at tree level. Second, if SSB takes place it is a non-perturbative loop effect solely due to the Yukawa interactions (2).

We start by *assuming* that the fermion propagators $S_i^{-1}(p) = \not{p} - \Sigma_i(p^2)$ develop dynamically the chiral symmetry breaking parts $\Sigma_i(p)$. In general the Yukawa interaction and the fermion propagators yield the loop-generated scalar one-point function (tadpole). In this contribution based on[6,7] we assume that it cancels. Basic idea borrowed from models of superfluidity is[8] to arrange for the SSB in an interacting many-boson system with $\langle 0|\phi(x)|0\rangle = 0$ and $\langle 0|\phi(x)^2|0\rangle \neq 0$. The simplest possibility we can imagine is that the tadpoles due to the fermion loops pairwise cancel. This clearly happens for $y_1 = -y_2$. Since this possibility yields $\Sigma_1 = \Sigma_2$, in order to illustrate an amplification of the fermion masses we need at least two degenerate fermion pairs. In the following numerical illustration y_1 and y_2 mark the Yukawa couplings of such two independent fermion pairs and the tadpole is omitted.

With the fermion propagators the Yukawa interaction (2) gives rise to the loops depicted in Fig. 1 with both arrows of the external scalar lines going out of the graph. The resulting scalar symmetry-breaking proper self-energy Π is given by the SD equation

$$\Pi(p) = \sum_{j=1,2} 2y_j^2 \int \frac{d^4k}{(2\pi)^4} \frac{\Sigma_j(k)}{k^2 + \Sigma_j^2(k)} \frac{\Sigma_j(k-p)}{(k-p)^2 + \Sigma_j^2(k-p)}. \tag{4}$$

Thus the assumed chiral-symmetry-breaking $\Sigma_j(p)$ in the fermion sector induces the chiral-symmetry-breaking contribution $\Pi(p)$ in the boson sector. Emergence of $\Pi(p)$ modifies the full propagator of the complex scalar field. It acquires the matrix form

$$D^{-1}(p) = \begin{pmatrix} p^2 - M^2 & -\Pi(p) \\ -\Pi(p) & p^2 - M^2 \end{pmatrix} \tag{5}$$

Fig. 2. The diagrammatic representation of pole parts of proper vertex function. The pole itself is interpreted as a propagator of intermediate massless scalar particle—the Nambu-Goldstone boson—depicted by solid double line. The small empty circles are the effective vertices to the fermion and the scalar fields, respectively. Analogous expression applies to the vertex function with the scalar.

With the offdiagonal part of (5) the Yukawa interaction (2) can generate $\Sigma_j(p)$ by virtue of the SD equations

$$\begin{aligned}\Sigma_1(p) &= y_1^2 \int \frac{d^4k}{(2\pi)^4} \frac{\Sigma_1(k)}{k^2+\Sigma_1^2(k)} \frac{\Pi(k-p)}{\left[(k-p)^2+M^2\right]^2-\Pi^2(k-p)} \\ \Sigma_2(p) &= y_2^2 \int \frac{d^4k}{(2\pi)^4} \frac{\Sigma_2(k)}{k^2+\Sigma_2^2(k)} \frac{\Pi(k-p)}{\left[(k-p)^2+M^2\right]^2-\Pi^2(k-p)}\end{aligned} \tag{6}$$

We have thus arrived at three coupled homogeneous SD equations (4,6) for Π and Σ. The pole fermion and scalar boson masses are given as the solutions of the implicit equations

$$\begin{aligned} m_j^2 &= \Sigma_j^2(p^2 = m_j^2) \\ M_{1,2}^2 &= M^2 \pm \Pi(p^2 = M_{1,2}^2) \end{aligned} \tag{7}$$

The key issue is to find nontrivial solutions $\Pi(p^2)$ and $\Sigma(p^2)$ with physically interesting properties. The SD equations (4,6) clearly reveal that *if* the solutions exist they are low-momentum dominated and vanish fast in the ultraviolet (UV). At the present exploratory stage we have found them numerically for large values of the Yukawa couplings. Their properties will be briefly described in Section 5. We notice that the equations (4-7) already include the approximations described in Section 5.

4. Spontaneous breaking of the axial symmetry

Spontaneous breakdown of the chiral symmetry manifested by Π and Σ_i different from zero is accompanied by a massless pseudoscalar *composite* NG boson. It emerges as a massless pole in the proper vertex functions of the axial current. Conservation of the axial-vector current (3) guarantees this by virtue of the axial-vector Ward identities.[7] The pole parts of the vertex functions are shown in Fig. 2.

Standard analysis[9] yields the effective UV finite loop-generated couplings of the NG boson to ψ_i and ϕ as well as to the external axial-vector current. Their explicit form can be found in Ref. 7. Clearly the NG boson is a pseudoscalar excitation of the system (1) composed both of the fermion and the boson fields.

Once we switch on the gauge interaction, the gauge boson associated to the spontaneously broken axial symmetry acquires a nonzero mass. In the present model, it is given by a sum rule in terms of all Σs in the model and of Π.[7]

The symmetry-breaking proper self-energy parts Σ_i and Π of the fermion and boson propagators vanishing at large momenta induce the symmetry-breaking effective loop-generated UV-finite vertices of all participating fields. They represent the genuine predictions of the suggested nonperturbative approach. The list of all such vertices is not long. As an illustration we present the result for A^3 vertex:[7]

$$iT^{\mu\nu\rho}(p,k) = G\Big[(q^\mu k^\alpha - k^\mu q^\alpha)p^\beta \epsilon^{\nu\rho}{}_{\alpha\beta} + (p^\nu q^\alpha - q^\nu p^\alpha)k^\beta \epsilon^{\rho\mu}{}_{\alpha\beta} + (k^\rho p^\alpha - p^\rho k^\alpha)q^\beta \epsilon^{\mu\nu}{}_{\alpha\beta}\Big] . \quad (8)$$

The effective coupling constant G is given in terms of $\Sigma_i(p^2)$ and $\Pi(p^2)$.

5. Numerical results

The coupled homogeneous SD equations for Σ_i and Π are themselves derived within several common but nontrivial approximations. First, the vertices are taken as the bare ones; that is, the SD equations for all higher-point Green's functions are neglected. Second, the loop effects of the wave-function renormalizations are ignored.

Our present illustrative numerical analysis employs further approximations: (i) We switch to the Euclidean metric by the Wick rotation. By this we get rid of some poles in the propagators but not of all of them. In particular, the denominator of the scalar propagator still vanishes for some p^2. This makes some regions of parameters not accessible by our numerical procedure. (ii) We consider all proper self-energies real. (iii) The solutions are found iteratively starting from a specific Ansatz for Σ_i: $\Sigma_1(p^2) = \Sigma_2(p^2) = xM^5/(p^2+M^2)^2$ with x being a free parameter.

The main results are presented graphically. Fig. 3 shows three regions in the (y_1, y_2) plane where nonzero Σ_i were found. The region IV is *terra incognita* due to the remaining scalar propagator pole. The dashed line marks the range of $y_1 \in \langle 72, 104 \rangle$ for the fixed $y_2 = 88$ in which the fermion mass spectrum was probed. The sensitivity of the fermion masses on the Yukawa couplings is demonstrated in Fig. 4. The steep dependence of the

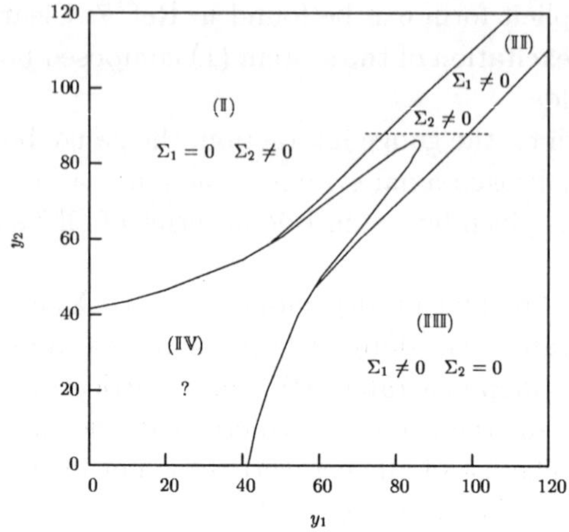

Fig. 3. The (y_1, y_2) plane with indicated areas of different behavior of the fermion self-energies.

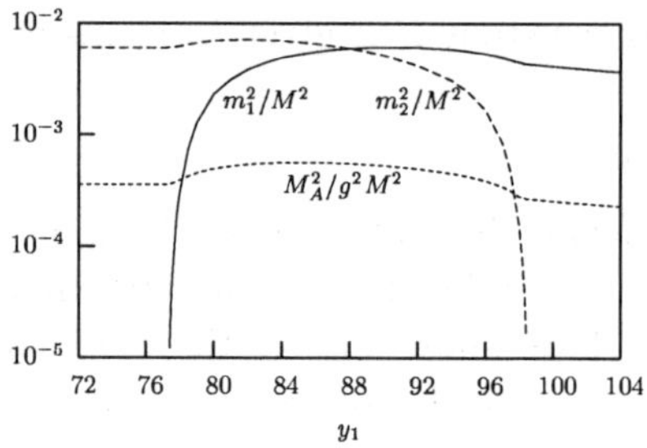

Fig. 4. The y_1-dependence of the fermion and vector boson masses $m^2_{1,2}$ and M^2_A with fixed $y_2 = 88$.

fermion masses on the Yukawa couplings close to their critical values is alluring. In order to get $m_1^2/m_2^2 = 10^{-2}$ it suffices to take $y_1/y_2 = 77.4/88$. The scalar boson mass splitting is reasonably small as expected.

6. Conclusion

Massiveness of particles is a low-momentum phenomenon. This statement is in accord with the conclusion based on the renormalization group argument that the UV stable theories can generate the masses dynamically whereas the infrared stable ones cannot.[10,11] We are thus tempted to interpret our results as an indication of the existence of the nontrivial fixed points in our model.

Within the effective field theory description of the electroweak phenomena at forthcoming energy scale we find the scalar fields welcome: Their Yukawa couplings with massless fermion fields feel flavors and by that break explicitly all huge global unwanted chiral symmetries of three electroweakly identical fermion families exactly to the sanctified $SU(2)_L \times U(1)_Y$.

Linear dependence of wildly wide, sparse and irregular fermion mass spectrum upon the Yukawa coupling constants due to the scalar field condensation is, however, a drawback. In an Abelian prototype without the tree-level scalar field condensation we have provided indications that the fermion masses can depend upon the Yukawa couplings nonanalytically. The suggested realization of the SSB seems thus 'less phenomenological'. We are confident that the corresponding $SU(2)_L \times U(1)_Y$ generalization does exist[5] thought its phenomenological viability remains to be studied.

Acknowledgments

I am grateful to Professor Tanabashi for fruitful discussions and for bringing the reference[4] to my attention, and to Tomáš Brauner and Petr Beneš for bringing me the joy from our collaboration. This work was supported by the grant GACR 202/06/0734.

References

1. T. D. Lee, *Particle Physics and Introduction to Field Theory* (Harwood Academic Publishers, New York, 1981), p.825.
2. B. Richter, *Is Naturalness Unnatural?*, SLAC-PUB-11911(2006).
3. E. Accomando *et al., Workshop on CP studies and non-standard Higgs physics*, arXiv:hep-ph/0608079.
4. K.-I. Kondo, M. Tanabashi and K. Yamawaki, in *Proceedings of 1989 Workshop on Dynamical Symmetry Breaking*, ed. T. Muta and K. Yamawaki (Nagoya University, 1989).
5. T. Brauner and J. Hošek, arXiv:hep-ph/0407339.
6. T. Brauner and J. Hošek, *Phys. Rev.* **D72**, 045007 (2005); arXiv:hep-ph/0505231.
7. P. Beneš, T. Brauner and J. Hošek, arXiv:hep-ph/0605147; to be published in Phys.Rev.D.
8. Y. Imry, in *Quantum Fluids*, eds. D. J. Amit and N. Wisser (Gordon and Breach, New York, 1970).
9. R. Jackiw and K. Johnson, *Phys. Rev.* **D8**, 2386 (1973).
10. K. Lane, *Phys. Rev.* **D10**, 1353 (1974).
11. H. Pagels, *Phys. Rev.* **D21**, 2336 (1980).

Neutron-Anti-Neutron Oscillation as a probe of B-L symmetry

R. N. Mohapatra

Department of Physics, University of Maryland
College Park, Maryland, U. S. A.
**E-mail: ab_rmohapat@physics.umd.edu*

There are various reasons to think that B-L symmetry is an intimate feature of physics beyond the standard model. I discuss how a great deal of knowledge regarding the nature of B-L can be obtained from the baryon number violating process, $N - \bar{N}$ oscillation. I argue by giving a very plausible and explicit gauge model that existence of a B-L scale far below the grand unification scale but much above the weak scale can lead to observable oscillation time for this process contrary to conventional thinking.

Keywords: Neutron-anti-neutron oscillation; baryogenesis.

1. Introduction

There are various reasons to think that B-L symmetry is an intimate feature of physics beyond the standard model. Two most compelling of them are: (i) the seesaw mechanism for understanding small neutrino masses requires the introduction of right handed neutrinos[1]i.e. in the presence of three right handed neutrinos B-L symmetry which was a global symmetry of the SM Lagrangian becomes a local symmetry; (ii) an inherent aspect of seesaw is the Majorana mass of the right handed neutrino that breaks the B-L symmetry which provides a way to understand why the seesaw scale is so much less than the Planck scale as required by observations; (iii) a third reason appears once one admits the presence of supersymmetry at the TeV scale where a common belief is that the lightest SUSY particle (LSP) ia naturally stable and is the dark matter of the universe. In the MSSM however, this stability is not guaranteed. The simplest way to have SUSY LSP naturally stable is to have B-L as a symmetry of physics beyond MSSM.[2]

If indeed B-L symmetry is present in nature, there are several questions that immediately come to mind: (a) is it a global or local symmetry ? (b) is it a broken or unbroken symmetry ? (c) if it is broken, what is the breaking scale and what is the associated physics ? In this talk I will argue that search for the process of neutron-anti-neutron oscillation[3–5] can provide a partial answers to some of these important questions: in particular the question of scale of B-L breaking and associated expanded gauge symmetry. The point is that since a $\Delta B = 2$ operator breaks B-L by two units, it is natural to associate the scale associated with the $\Delta B = 2$ operator also with the scale of B-L breaking. Since seesaw mechanism also breaks B-L symmetry, M could also be the seesaw scale. In fact, in the context of models such as those based on $SU(2)_L \times SU(2)_R \times SU(4)_c$,[6] the same mass scale M is responsible for both processes.[5] Thus the search for $N - \bar{N}$ oscillation could not only illuminate the nature of the important symmetry, B-L but could also be one way to unravel the mystery of the seesaw mechanism that is expected to be major player in the physics of neutrino mass.

2. Operator analysis of $\Delta B = 2$ processes

A simple way to get an idea about the scale probed by a physical process is to do an operator analysis i.e. assuming a particular symmetry and spectrum of the low energy theory, write effective higher dimensional operators for the process under consideration and see for what value of the mass that scales the operator, the process is observable. This argument has been a useful tool to probe physics at short distance scales e.g. proton decay, neutrino mass. This method of course has its limitations. For instance, often in these discussions, one uses the SM fermion spectrum till the new scale characterising B-L symmetry but if there is a SM non-singlet new particle with a 100 GeV mass not discovered yet, the naive scale arguments would be misleading. Similarly, the presence of unknown higher symmetries of the new theory could also invalidate these arguments. Little Higgs models provide examples of this in the context of Higgs boson physics. Let us apply this discussion to the $\Delta B = 2$ operator which would cause neutron-anti-neutron oscillation. In the standard model, the effective operator is given by:

$$\mathcal{O}_{\Delta B=2} = \frac{1}{M^5}[QQQQQQ + u^c d^c d^c u^c d^c d^c] \tag{1}$$

Thus the strength of this operator $G_{\Delta B=2} = \frac{1}{M^5}$. Note the dependence of $G_{\Delta B=2}$ on the new physics scale M. The $N - \bar{N}$ mixing mass can be deduced from this operator by the approximate formula:

$$\delta m_{N-\bar{N}} \simeq c G_{\Delta B=2} \Lambda^6_{QCD} = c \frac{\Lambda^6_{QCD}}{M^5} \tag{2}$$

where c is of order one. Taking the best current lower bound on $\tau_{N\leftrightarrow\bar{N}}$ from ILL reactor experiment[7] and is 10^8 sec. and comparable bounds from nucleon decay search experiments,[8] one can then obtain a lower limit on M to be around 300 TeV. There are proposals to improve the precision of this search by at least two orders of magnitude.[9] An observable $N-\bar{N}$ oscillation time with current reactor facilities is expected to be around $10^8 - 10^{11}$ seconds which will probe the mass scale arounf a 1000 TeV. The relevant physics question then is, can the improvement of the $N-\bar{N}$ oscillation time to the level of 10^{11} sec. throw any useful light on the seesaw scale. Since the current prejudice about the seesaw scale is that it is likely to be around $10^{11} - 10^{14}$ GeV, one would naively think that the answer to the above question is No since if M in the above equation is $10^{11} - 10^{14}$ GeV, $N-\bar{N}$ oscillation is way beyond the reach of any conceivable experiment. What I would like to show in this talk is that the above naive operator analysis arguments become invalid if there is TeV scale supersymmetry and new as yet undiscovered new particles with 100 GeV mass that do not alter SM physics; secondly I will give explicit examples of interesting supersymmetric theories, which give realistic realization of seesaw mechanism and provide examples of new TeV scale particle that lead to observable $N-\bar{N}$ oscillation.

3. Supersymmetry and enhanced $\Delta B = 2$ operator

In the presence of supersymmetry at the TeV scale, there are new particle at the TeV scale i.e. the squarks and sleptons etc. They can then enter the effective $\Delta B = 2$ operator. The leading operator the becomes:

$$\mathcal{O}_{\Delta B=2} = \frac{1}{M^3} u^c d^c \tilde{d}^c \tilde{u}^c \tilde{d}^c \tilde{d}^c \tag{3}$$

Note that the power dependence on the seesaw scale (or B-L breaking scale) has ow considerably softened. In fact naive power counting arguments then imply that if this is the leading operator one can probe the seesaw scale upto 10^8 GeV. The power dependence in fact softens even further if there are new color sextet particles at or below the TeV scale. For instance, if there is a scalar diquark sextet field of $\Delta_{u^c u^c}$ type, the leading operator becomes:

$$\mathcal{O}_{\Delta B=2} = \frac{1}{M^2} d^c d^c \Delta_{u^c u^c} \tilde{d}^c \tilde{d}^c \tag{4}$$

and the scale reach of $N-\bar{N}$ goes up to 10^{11} GeV. If on the other hand there is a field $\Delta_{u^c d^c}$, the leading order operator becomes:

$$\mathcal{O}_{\Delta B=2} = \frac{1}{M} d^c d^c \Delta_{u^c d^c} \Delta_{u^c d^c} \tag{5}$$

increasing the scale reach to the GUT scale. The question then arises whether there are plausible models where this can happen and below I give examples of such models.

4. $SU(2)_L \times SU(2)_R \times SU(4)_c$ model with light diquarks

The quarks and leptons in this model are unified and transform as ψ : $(\mathbf{2},\mathbf{1},\mathbf{4}) \oplus \psi^c : (\mathbf{1},\mathbf{2},\bar{\mathbf{4}})$ representations of $SU(2)_L \times SU(2)_R \times SU(4)_c$. For the Higgs sector, we choose, $\phi_1 : (\mathbf{2},\mathbf{2},\mathbf{1})$ and $\phi_{15} : (\mathbf{2},\mathbf{2},\mathbf{15})$ to give mass to the fermions. The $\Delta^c : (\mathbf{1},\mathbf{3},\mathbf{10}) \oplus \bar{\Delta}^c : (\mathbf{1},\mathbf{3},\overline{\mathbf{10}})$ to break the $B-L$ symmetry. The diquarks mentioned above which lead to $\Delta(B-L) = 2$ processes are contained in the $\Delta^c : (\mathbf{1},\mathbf{3},\mathbf{10})$ multiplet. We also add a $B-L$ neutral triplet $\Omega : (\mathbf{1},\mathbf{3},\mathbf{1})$ which helps to reduce the number of light diquark states. The superpotential of this model is given by:

$$W = W_Y + W_{H1} + W_{H2} + W_{H3} \tag{6}$$

where

$$W_{H1} = \lambda_1 S(\Delta^c\bar{\Delta}^c - M^2) + \lambda_C \Delta^c\bar{\Delta}^c\Omega + \mu_i \mathrm{Tr}\,(\phi_i\phi_i) \tag{7}$$

$$W_{H2} = \lambda_A \frac{(\Delta^c\bar{\Delta}^c)^2}{M_{P\ell}} + \lambda_B \frac{(\Delta^c\Delta^c)(\bar{\Delta}^c\bar{\Delta}^c)}{M_{P\ell}} \tag{8}$$

$$W_{H3} = \lambda_D \frac{\mathrm{Tr}\,(\phi_1\Delta^c\bar{\Delta}^c\phi_{15})}{M_{P\ell}}, \tag{9}$$

$$W_Y = h_1\psi\phi_1\psi^c + h_{15}\psi\phi_{15}\psi^c + f\psi^c\Delta^c\psi^c. \tag{10}$$

Note that since we do not have parity symmetry in the model, the Yukawa couplings h_1 and h_{15} are not symmetric matrices. When $\lambda_B = 0$, this superpotential has an accidental global symmetry much larger than the gauge group;[11] as a result, vacuum breaking of the $B-L$ symmetry leads to the existence of light diquark states that mediate $N \leftrightarrow \bar{N}$ oscillation and enhance the amplitude. In fact it was shown that for $\langle\Delta^c\rangle \sim \langle\bar{\Delta}^c\rangle \neq 0$ and $\langle\Omega\rangle \neq 0$ and all VEVs in the range of $10^{11} - 10^{12}$ GeV, the light states are those with quantum numbers: $\Delta_{u^cu^c}$. The symmetry argument behind is that[11] for $\lambda_B = 0$, the above superpotential is invariant under $U(10,c) \times SU(2,c)$ symmetry which breaks down to $U(9,c) \times U(1)$ when $\langle\Delta^c_{\nu^c\nu^c}\rangle = v_{BL} \neq 0$. This results in 21 complex massless states; on the other hand these vevs also breaks the gauge symmetry down from $SU(2)_R \times$

$SU(4)_c$ to $SU(3)_c \times U(1)_Y$. This allows nine of the above states to pick up masses of order gv_{BL} leaving 12 massless complex states which are the six $\Delta^c_{u^c u^c}$ plus six $\bar{\Delta}^c_{u^c u^c}$ states. Once $\lambda_B \neq 0$ and is of order $10^{-2} - 10^{-3}$, they pick up mass (call $M_{u^c u^c}$) of order of the elctroweak scale.

5. A new diagram for neutron-anti-neutron oscillation

To discuss $N \leftrightarrow \bar{N}$ oscillation, we introduce a new term in the superpotential of the form:[5]

$$W_{\Delta B=2} = \frac{1}{M_*} \epsilon^{\mu'\nu'\lambda'\sigma'} \epsilon^{\mu\nu\lambda\sigma} \Delta^c_{\mu\mu'} \Delta^c_{\nu\nu'} \Delta^c_{\lambda\lambda'} \Delta^c_{\sigma\sigma'} , \tag{11}$$

where the μ, ν etc stand for $SU(4)_c$ indices and we have suppressed the $SU(2)_R$ indices. Apriori M_* could be of order $M_{P\ell}$; however the terms in Eq.(2) are different from those in Eq. (4); so they could arise from different a high scale theory. The mass M_* is therefore a free parameter that we choose to be much less than the $M_{P\ell}$. This term does not affect the masses of the Higgs fields. When $\Delta^c_{\nu^c\nu^c}$ acquires a VEV, $\Delta B = 2$ interaction are induced from this superpotential, and $N \leftrightarrow \bar{N}$ oscillation are generated by two diagrams given in Fig. 1 and 2. The first diagram (Fig. 1) in which only diquark Higgs fields are involved was already discussed in Ref. 5 and goes like $G_{N\leftrightarrow\bar{N}} \simeq \frac{f_{11}^3 v_{BL} M_\Delta}{M^2_{u^c u^c} M^4_{d^c d^c} M_*}$, Taking $M_{u^c u^c} \sim 350$ GeV, $M_{d^c d^c} \sim \lambda' v_{BL}$ and $M_\Delta \sim v_{BL}$ as in the argument,[11] we see that this diagram scales like $v_{BL}^{-3} v_{wk}^{-2}$. In Ref. 10 a new diagram (Fig. 2) was pointed out which owes its origin to supersymmetry. We get for its contribution to $G_{\Delta B=2}$:

$$G_{N\leftrightarrow\bar{N}} \simeq \frac{g_3^2}{16\pi^2} \frac{f_{11}^3 v_{BL}}{M^2_{u^c u^c} M^2_{d^c d^c} M_{\mathrm{SUSY}} M_*} . \tag{12}$$

Using the same arguments as above, we find that this diagram scales like $v_{BL}^{-2} v_{wk}^{-3}$ which is therefore a significant enhancement over diagram in Fig.1.

In order to estimate the rate for $N \leftrightarrow \bar{N}$ oscillation, we need not only the different mass values for which we now have an order of magnitude, we also need the Yukawa coupling f_{11}. Now f_{11} is a small number since its value is associated with the lightest right-handed neutrino mass. However, in the calculation we need its value in the basis where quark masses are diagonal. We note that the $N - \bar{N}$ diagrams involve only the right-handed quarks, the rotation matrix need not be the CKM matrix. The right-handed rotations need to be large e.g. in order to involve f_{33} (which is $O(1)$), we need $(V_R^{(u,d)})_{31}$ to be large, where $V_L^{(u,d)\dagger} Y_{u,d} V_R^{(u,d)} = Y^{\mathrm{diag.}}_{u,d}$. The left-handed rotation matrices $V_L^{(u,d)}$ contribute to the CKM matrix, but right-handed rotation matrices $V_R^{(u,d)}$ are unphysical in the standard model. In

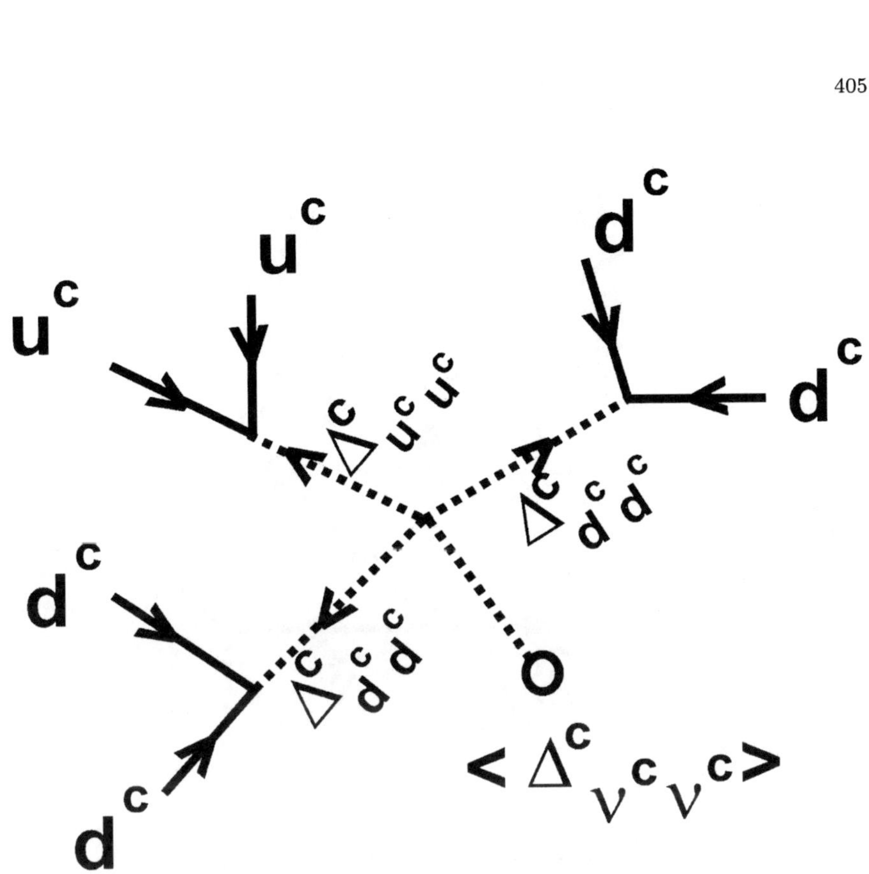

Fig. 1. The Feynman diagram responsible for $N - \bar{N}$ oscillation as discussed in Ref. 8.

this model, however, we get to see its contribution since we have a left-right gauge symmetry. Let us now estimate the time of oscillation. When we start with a f-diagonal basis (call the diagonal matrix $\hat{f}$), the Majorana coupling f_{11} in the diagonal basis of up- and down-type quark matrices can be written as $(V_R^T \hat{f} V_R)_{11} \sim (V_{31}^R)^2 \hat{f}_{33}$. Now $\hat{f}_{33}$ is $O(1)$ and V_{31}^R can be ~ 0.6, so f_{11} is about 0.4 in the diagonal basis of the quark matrices. We use $M_{\rm SUSY}$, $M_{u^c u^c} \sim 350$ GeV and $v_{BL} \sim 10^{12}$ GeV. The mass of $\tilde{\Delta}_{d^c d^c}$ i.e. $M_{d^c d^c}$ is 10^9 GeV which is obtained from the VEV of $\Omega : (\mathbf{1}, \mathbf{3}, \mathbf{1})$. We choose $M_* \sim 10^{13}$ GeV. Putting all the above the numbers together, we get

$$G_{N \leftrightarrow \bar{N}} \simeq 1 \cdot 10^{-30} \text{ GeV}^{-5}. \tag{13}$$

Along with the hadronic matrix element,[14] the $N - \bar{N}$ oscillation time is found to be about 2.5×10^{10} sec which is within the reach of possible next generation measurements. If we chose, $M_* \simeq M_{P\ell}$, we will get for $\tau_{N-\bar{N}} \sim 10^{15}$ sec. unless we choose the $M_{d^d d^c}$ to be lower (say 10^7 GeV).

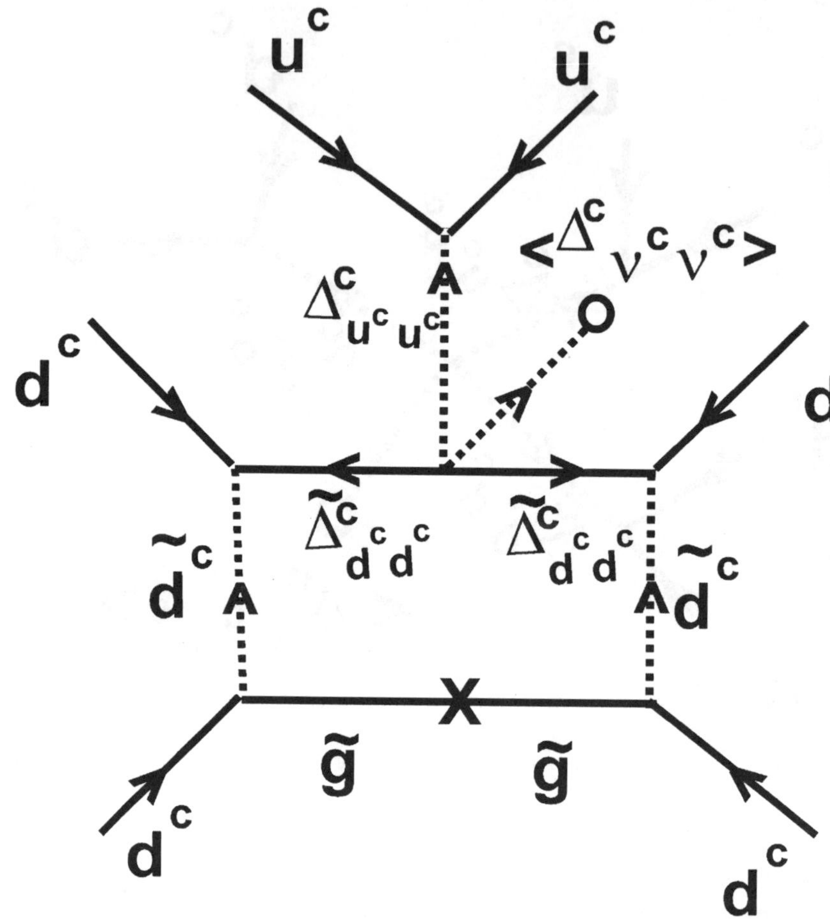

Fig. 2. The new Feynman diagram for $N - \bar{N}$ oscillation.

This is a considerable enhancement over the nonsupersymmetric model of Ref. 5 with seesaw scale of 10^{12} GeV. We also note that as noted in Ref. 5 the model is invariant under the hidden discrete symmetry under which a field $X \to e^{i\pi B_X} X$, where B_X is the baryon number of the field X. As a result, proton is absolutely stable in the model. Furthermore, since R-parity is an automatic symmetry of MSSM, this model has a naturally stable dark matter.

6. Baryogenesis and $N - \bar{N}$ oscillation

In the early 1980's when the idea of neutron-anti-neutron oscillation was first proposed in the context of unified gauge theories, it was thought that the high dimensionality of the $\Delta B \neq 0$ operator would pose major difficulty in understanding the origin of matter. The main reason for this is that the higher dimensional operators remain in thermal equilibrium until late in the evolution of the universe since the thermal decoupling temperature T_* for such interactions goes roughly like $v_{BL}\left(\frac{v_{BL}}{M_P}\right)^{1/9}$ which is in the range of temperatures where B+L violating sphaleron transitions are full thermal equilibrium. They will therefore erase any baryon asymmetry generated in the very early moments of the universe (say close to the GUT time of 10^{-30} sec. or so) in prevalent baryogenesis models. In models with observable $N - \bar{N}$ oscillation therefore, one has to search for new mechanisms for generating baryons below the weak scale. In this section, we discuss such a possibility[12] discussed in a recent unpublished work with K. S. Babu and S. Nasri. As an illustration of the way the new mechanism operates, let us assume that there is a scalar field that couples to the $\Delta B = 2$ operator i.e. $L_I = Su^c d^c d^c u^c d^c d^c / M^6$, where the scalar boson has mass of order of the weak scale and $B = 2$. This leads to baryon number violation if $< S > \neq 0$ and observable $N - \bar{N}$ transition if M is in the few hundred to 1000 GeV range. The direct decay of S in these models can lead to an adequate mechanism for baryogenesis. To discuss how this comes about, let us first note that the high dimension of L_I allows the scalar $\Delta B \neq 0$ decay to go out of equilibrium at weak scale temperatures. This clearly satisfies the out of equilibrium condition given by Sakharov conditions for origin of matter. This is the reason we require a higher dimensional operator. For direct proton decay operators such as $QQQL$, the decoupling temperature is much higher and our mechanism will not apply. To outline the rest of the details of this mechanism,[12] we consider an effective sub-TeV scale model that gives rise to the higher dimensional operator for $N \leftrightarrow \bar{N}$ oscillation. It consists of the following color sextet, $SU(2)_L$ singlet scalar bosons (X, Y, Z) with hypercharge $-\frac{4}{3}, +\frac{8}{3}, +\frac{2}{3}$ respectively that couple to quarks. These states could emerge from the supersymmetric model described in the previous section if $< \Omega >= 0$. We add to it a scalar field with mass in the 100 GeV range. One can now write down the following standard model invariant interaction Lagrangian:

$$\begin{aligned} \mathcal{L}_I &= h_{ij} X d_i^c d_j^c + f_{ij} Y u_i^c u_j^c + \\ & g_{ij} Z (u_i^c d_j^c + u_j^c d_j^c) + \lambda_1 S X^2 Y + \lambda_2 S X Z^2 \end{aligned} \tag{14}$$

The scalar field S is assumed to have $B = 2$. To see the constraints on the parameters of the theory, we note that the present limits on $\tau_{N-\bar{N}} \geq 10^8$ sec. implies that the strength $G_{N-\bar{N}}$ of the the $\Delta B = 2$ transition is $\leq 10^{-28}$ GeV^{-5}. From Fig. 3, we conclude that

$$G_{N-\bar{N}} \simeq \frac{\lambda_1 M_1 h_{11}^2 f_{11}}{M_Y^2 M_X^4} + \frac{\lambda_2 M_1 h_{11} g_{11}^2}{M_X^2 M_Z^2} \leq 10^{-28} GeV^{-5}. \tag{15}$$

For $\lambda_{1,2} \sim h \sim f \sim g \sim 10^{-3}$, we have $M_1 \sim M_{X,Y,Z} \simeq 1$ TeV. In our discussion, we will stay close to this range of parameters and see how one can understand the baryon asymmetry of the universe. The singlet field will play a key role in the generation of baryon asymmetry. We assume that $< S >\sim M_X$ but $M_{S_r} \sim 100$ GeV, where S_r is the real part of the S field after its vev is subtracted. It can then decay into final states with $B = \pm 2$. On the way to calculating the baryon asymmetry, let us first discuss the out of equilibrium condition. As the temperature of the universe falls below the masses of the X, Y, Z particles, the annihilation processes $X\bar{X} \to d^c \bar{d}^c$ (and analogous processes for Y and Z) remain in equilibrium. As a result, the number density of X, Y, Z particles gets depleted and only the S particle survives along with the usual standard model particles. One of the primary generic decay modes of S is $S \to u^c d^c d^c u^c d^c d^c$. There could be other decay modes which depend on the details of the model. Those can be made negligible by choice of parameters which do not enter our discussion of $N - \bar{N}$ and baryogenesis. For $T \geq M_S$, the decay rate of S is given by $\Gamma_S \sim \frac{T^{13}}{16\pi^9 M_X^{12}}$ where we have set the masses of X, Y and Z particle to be of same order for simplicity. This decay goes out of equilibrium around $T_* \simeq M_X \left(\frac{160\pi^9 M_X}{M_{P\ell}}\right)^{1/11}$. Here we have assumed that the coupling of the X, Y, Z particles to second and third generation quarks are of order 0.1-1. This gives $T_* \sim 0.1 - 0.2 M_X$ or in the sub-TeV range. Below this temperature the decay rate of S falls very rapidly as the temperature cools. However as soon as $T \leq M_S$, the decay rate becomes a constant whereas the expansion rate of the universe is slowing down. So at a temperature T_d, S will start to decay at

$$T_d \simeq \left(\frac{M_{P\ell} M_S^{13}}{(2\pi)^9 M_X^{12}}\right)^{1/2} \tag{16}$$

Since the corresponding epoch must be above that of big bang nucleosynthesis, this puts a constraint on the parameters of the model. For instance, for $M_S \sim 200$ GeV and $M_X \sim 3$ TeV, we get $T_d \sim$ GeV. Also this implies that the X, Y, Z masses cannot be arbitrarily high, since the heavier these particles are, the lower T_d will be. We expect this upper limit to be in the TeV range at most. It is well known that baryon asymmetry can arise only via the interference of a tree diagram with a one loop diagram. The tree diagram is clearly the one where $S \rightarrow 6q$. There are however two classes of loop diagrams that can contribute: one where the loop involves the same fields X, Y and Z. A second one involves W-exchange, which involves only standard model physics at this scale (Fig. 3). We find that the second contribution can actually dominate. It also has the advantage that it involves less number of arbitrary parameters. The baryon asymmetry is defined as follows

$$\epsilon_B \simeq \frac{n_S}{n_\gamma} \frac{\Gamma(S \rightarrow 6q) - \Gamma(S \rightarrow 6\bar{q})}{\Gamma(S)} \tag{17}$$

We find that

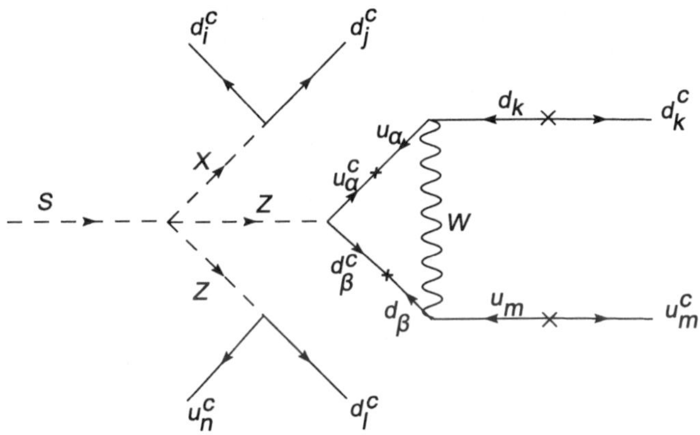

Fig. 3. One loop diagram for S decays.

$$\epsilon_B \simeq \begin{cases} 2\alpha_2 Im(V_{tb} V_{cb}^* h_{33} h_{23}^*) \frac{m_c m_b^2 m_t}{M_W^2 M_S^2}; & M_S < m_t \\ 2\alpha_2 \frac{m_s m_b m_t^2}{M_W^2 M_S^2} Im[(h_{33} h_{32}^*)(V_{tb}^* V_{cb})]; & M_S > m_t \end{cases} \tag{18}$$

Note that the trace in the above equation has an imaginary part and therefore leads to nonzero asymmetry. The magnitude of the asymmetry depends on T_d/M_S as well as the detailed profile of the various coupling matrices h, g, f and we can easily get the desired value of the baryon asymmetry by appropriately choosing them. We have checked that there is no conflict between the desired magnitude of baryon asymmetry and the present lower bound on the $N - \bar{N}$ transition time of 10^8 sec. An important point about our baryogenesis mechanism, is that as the masses of the X, Y, Z particles get larger, the amount of baryon asymmetry goes down for given M_S (due to the dilution factor T_d/M_S discussed below) as does the strength of the $\Delta B = 2$ transition giving interesting correlation between the $N - \bar{N}$ process and baryon asymmetry. In fact an adequate baryogenesis puts an upper limit on $\tau_{N-\bar{N}}$ as discussed below. Using $m_c(m_c) = 1.27\ GeV$, $m_b(m_b) = 4.25\ GeV$, $m_t = 174\ GeV$, $V_{cb} \simeq 0.04$, $M_S = 200\ GeV$ and $|h_{33}| \simeq |h_{23}|\$1$ we find $\epsilon_B \sim 10^{-8}$. There is a further dilution of the baryon asymmetry arising from the fact that $T_d \ll M_S$ since the decay of S also releases entropy into the universe. In this case the baryon asymmetry reads

$$\eta_B \simeq \epsilon_B \frac{T_d}{M_S} \tag{19}$$

In order that this dilution effect is not excessive, there must be a lower limit on the ratio T_d/M_S. From our estimate above we require that $T_d/M_S \geq 0.01$. Since the decay rate of the S boson depends inversely as a high power of $M_{X,Y}$, higher X, Y bosons would imply that $\Gamma_S \sim H$ is satisfied at a lower temperature and hence give a lower T_d/M_S. In figure 3 we plotted $M_{X,Y}$ vs M_S using $T_d \geq M_S/100$, and the constraint $G_{N\bar{N}} \leq 10^{-28}\ GeV^{-5}$. The coupling $\bar{\lambda}^4 \equiv \lambda_1 h_{11}^2 f_{11} \sim \lambda_2 h g_{11}^2$. This in turn implies that the $\tau_{N-\bar{N}}$ must have an upper limit. For instance, for choice of the coupling parameters $\lambda \sim f \sim h \sim g \sim 10^{-3}$, and $M_S \simeq 200\ GeV$ we find $\tau_{N-\bar{N}} \leq 10^{10}\ sec.$

7. Conclusion

In conclusion, we have presented a realistic quark-lepton unified model where despite the high seesaw (v_{BL}) scale (in the range of $\sim 10^{12}$ GeV), the $N - \bar{N}$ oscillation time can be around 10^{10} sec. due to the presence of a new supersymmetric graph and accidental symmetries of the Higgs potential (also connected to supersymmetry). This oscillation time is within the reach of possible future experiments. We have also found a new way to generate the baryon asymmetry of the universe for the case when $N - \bar{N}$ oscillation is observable. These results should provide a motivation to conduct a new round of search for $N - \bar{N}$ oscillation.

Acknowledgments

I would like to thank K. S. Babu, B. Dutta, Y. Mimura and S. Nasri for collaborations that led to the results reported in this talk. I like to thank Y. Kamyshkov for many discussions on the experimental prospects for $N - \bar{N}$

oscillation. I am grateful to Prof. K. Yamawaki for kind hospitality at the SCGT06 symposium in Nagoya. This work is supported by the National Science Foundation grant no. Phy-0354401.

References

1. P. Minkowski, Phys. lett. B **67**, 421 (1977); M. Gell-Mann, P. Ramond, and R. Slansky, *Supergravity* (P. van Nieuwenhuizen et al. eds.), North Holland, Amsterdam, 1979, p. 315; T. Yanagida, in *Proceedings of the Workshop on the Unified Theory and the Baryon Number in the Universe* (O. Sawada and A. Sugamoto, eds.), KEK, Tsukuba, Japan, 1979, p. 95; S. L. Glashow, *The future of elementary particle physics*, in *Proceedings of the 1979 Cargèse Summer Institute on Quarks and Leptons* (M. Lévy et al. eds.), Plenum Press, New York, 1980, p. 687; R. N. Mohapatra and G. Senjanović, Phys. Rev. Lett. **44**, 912 (1980).
2. R. N. Mohapatra, Phys. Rev. **D 34**, 3457 (1986); A. Font, L. Ibanez and F. Quevedo, Phys. Lett **B 228**, 79 (1989): S. P. Martin, Phys. Rev. **D46**, 2769 (1992).
3. V. A. Kuzmin, JETP Lett. **12**, 228 (1970).
4. S. L. Glashow, Cargese Lectures (1979).
5. R. N. Mohapatra and R. E. Marshak, Phys. Rev. Lett. **44**, 1316 (1980).
6. J. C. Pati and A. Salam, Phys. Rev. D **10**, 275 (1974).
7. M. Baldo-Ceolin *et al.*, Z. Phys. C **63**, 409 (1994).
8. M. Takita *et al.* [KAMIOKANDE Collaboration], Phys. Rev. D **34**, 902 (1986); J. Chung *et al.*, Phys. Rev. D **66**, 032004 (2002) [hep-ex/0205093].
9. Y. A. Kamyshkov, hep-ex/0211006.
10. B. Dutta, Y. Mimura and R. N. Mohapatra, Phys. Rev. Lett. **96**, 061801 (2006).
11. Z. Chacko and R. N. Mohapatra, Phys. Rev. D **59**, 055004 (1999) [hep-ph/9802388].
12. K. S. Babu, R. N. Mohapatra and S. Nasri, Phys. Rev. Lett. (2006).
13. M. Fukugita and T. Yanagida, Phys. Lett. B **174**, 45 (1986); V. A. Kuzmin, V. A. Rubakov and M. E. Shaposhnikov, Phys. Lett. B **155**, 36 (1985).
14. S. Rao and R. Shrock, Phys. Lett. B **116**, 238 (1982); J. Pasupathy, Phys. Lett. B **114**, 172 (1982); Riazuddin, Phys. Rev. D **25**, 885 (1982); S. P. Misra and U. Sarkar, Phys. Rev. D **28**, 249 (1983).

CONCLUDING REMARKS

Gerard 't Hooft

Institute for Theoretical Physics
Utrecht University
and

Spinoza Institute
Postbox 80.195
3508 TD Utrecht, the Netherlands
e-mail: g.thooft@phys.uu.nl
Internet: http://www.phys.uu.nl/~thooft

What is the origin of mass? Does this question have anything to do with 'Strongly Coupled Gauge Theory'? It is not easy to summarize all that has been said about these questions at this meeting. In this short recapitulation, I will give you a general impression of the various issues that have been considered here, and I will add my own comments.

This is not a PowerPoint talk, and there are no technical calculations or figures in this summary; for those I refer to the original contributions.

It has become en vogue to attribute particle masses to the Higgs mechanism (which actually stands short for *Brout-Englert-Higgs* Mechanism). There is some justification to that, because in many cases, mass terms in the Lagrangian have to be associated to couplings of spinor fields or vector fields to some Higgs scalar. We must keep in mind however that mass can also be due to other forms of energy, such as the complex effects due to dimensional transmutation in strongly interacting gauge fields, where no Higgs scalar was involved. Thus, in general, the Higgs may add to the masses in our world but it would not be correct to identify the Higgs with the 'origin of mass'.

This does not make the Higgs less interesting. A frequently asked question is: Is the Higgs elementary or composite? From a purely theoretical point of view, it could be either. The idea that the Higgs might be composite was well received in the physics community. After all, there are at least two precedents in our endeavors to understand Nature's laws, where a field that spontaneously breaks a symmetry turned out to be a two-fermion composite. One is the Ginzburg-Landau theory for the superconductor, where the "Higgs" is called "order parameter", and it was understood to be the product of two electron fields (the Cooper pair). Secondly, we have QCD, where, according to the Gell-Mann-Lévy σ model, chiral $SU(2) \otimes SU(2)$ symmetry is spontaneously broken by the (σ, π) system, which actually are quark-antiquark bound states.

If the Higgs is really composite, the only theory we have is to view this Higgs as being just like the σ field in QCD, but a new color force must be assumed. This color symmetry, at approximately $1000 \times$ the energy of the QCD Λ parameter, was dubbed 'technicolor'. The technicolor gauge group could be like the color group $SU(3)$, but in order to account for the couplings between the Higgs and the other particles in the Standard Model, this gauge group had to be extended to a larger symmetry, which led to the notion of 'extended technicolor'.

The competitor to the Technicolor scenario is the theory of the supersymmetric version of the Standard Model, or *MSSM*, where the first *M* stands for 'Minimal'. There is much to be said in favor of the supersymmetric option. Most of all, if the Higgs has a superpartner, then the mass of this fermionic superpartner can be protected by chiral symmetry, and hence also the Higgs mass can be kept low. Thus a natural theory with weakly coupled Higgs fields is obtained.

The step towards a supersymmetric world is a big one, and the question is whether this is really what happens above he TeV scale. In our meeting, we concentrate on different possible scenarios. Today, we find that Technicolor theories are troubled by severe restrictions from observations. Not only do we need to extend the technicolor gauge group in order to account for the Higgs Yukawa couplings, the running coupling constant also has to slow down, so as to avoid the flavor changing neutral currents. Thus, the 'Walking Technicolor' theories saw the light.

Today's experimental information appears to favor the picture that the Higgs is situated well between two important limits, one being the unitarity bound from above, which means that we are sufficiently far from the Landau ghost associated to a new strong-interaction region; the other is the stability limit from below, saying that the mass is large enough for the vacuum state to be absolutely stable. Could the relatively small, "fine tuned", Higgs mass be explained by assuming the Higgs itself to be a Goldstone boson? According to this important idea, local gauge symmetry mixes with new global symmetries, to provide the Higgs with its very special electro-weak quantum numbers. This is the "little Higgs" theory. In the "littlest Higgs" theory, unwanted Higgs amplitudes could be suppressed by a symmetry called T symmetry, where all unwanted elements in our theory are declared to be odd under T. We also saw the "Moose theory" (so named because of the shape of the diagrams being used, I think), the "Simple Little Higgs", and various other concoctions, all characterized by slightly different choices for the various symmetry groups.

The 'S parameter' is a measurable quantity that may indicate how close the walking technicolor is from 'standing still', that is, the conformal limit. An important question is whether S is positive or negative, but it was not obvious to me whether the various reports started with the same definition of S.

Many of the above models lead to the existence of more than one Higgs field, for example, two Higgs doublets, one Higgs coupled to the top sector and one to the bottom sector of the fermion generations. Could it be that there is no Higgs at all? Several contributions were given in this meeting discussing theories without Higgses. One motivation for this would be that theories without fundamental scalars would have much fewer adjustable physical parameters. Possessing adjustable physical constants is not a desirable property of fundamental theories, but it is not wrong either. 'Naturalness' is a more pressing requirement for a healthy theory. Mooselike contraptions can perhaps be used to exorcize the Higgs, and keep naturalness up to very high energies.

An interesting possibility is that the Higgs sector could actually be quite complex. Could there be a whole crowd of Higgs particles, each contributing a little bit to spontaneous breakdown of $SU(2)\times U(1)$? Quite a lot of scalars are allowed before asymptotic freedom would be endangered. If this were the case, it would be difficult to detect any Higgs particle at all at LHC. So, if no Higgs is found, this could be an explanation that will be difficult to dismiss.[a] It would be an ugly theory, however, requiring a very large number of free parameters, and it would be difficult to keep such a theory natural. It would be better to attribute large numbers of scalars to a technicolor scheme or perhaps a SUSY scheme. In these considerations, the hierarchy problem is closely related to the naturalness requirement.

The second theme in this meeting is that of strongly coupled gauge theories, in general. The prototype of the strong interaction theories is Quantum Chromodynamics, QCD. The great challenge here is whether we will ever

---

[a]There *would* be signals from such a crowd of Higgses that experimentalists can look for.

master the art of performing detailed, quantitative, precise calculations. Before addressing this question, one must first obtain a global understanding, and then a more detailed understanding, of the phenomenon of permanent quark confinement in QCD. For instance, precision calculations of the hadron mass spectrum and the hadron decay properties will not be possible if we do not understand why the quarks do not come out.

Essential ingredients are the Renormalization Group and Asymptotic Freedom. The next step towards understanding quark confinement is the so-called Abelian Projection. In this formalism, one automatically obtains color-magnetic monopole configurations. If we may assume that these monopoles Bose-condense, confinement of color charges follows naturally. This explanation works in a qualitative sense, but the question came up to what extent one must require gauge-invariance of the procedure. The answer to this question was also given: because gauge-invariance is guaranteed by the original formulation of the theory, we actually have the freedom to choose any gauge-fixing procedure that we like. We could choose, for instance, a gauge in which some field combination values vanish identically. This then may simplify a calculation. While continuing the calculation, we must insist that this gauge condition continues to be obeyed. Our calculation might not work in other gauges. In other gauges, the outcome may be less accurate or wrong. This is because we were forced to ignore field values that were actually only small in our special choice of gauge. If this is the situation, the result of the calculation will not appear to be gauge-invariant anymore. Thus, by fixing the gauge, gauge-invariance may be deliberately lost. This does not at all invalidate the result; the result would be valid in one gauge only, and a transformation towards other gauges would yield incorrect results in general.

One may however wish to produce a result that holds for more general choices of gauge. If one wants this, one must choose a calculational procedure that is correct regardless which gauge choice is made. Then the result indeed must be gauge-invariant. This is a *choice* that one can make, but it is not a requirement for doing a correct calculation.

Be this as it may, advanced methods were presented in this meeting which, as I interpret them, amount to applying the Abelian projection gauge, after which gauge transformations are performed that put the result in more general gauges. This leads to an apparently gauge-invariant result.

An entirely different approach is the application of some form of effective string theory to describe the hadronic interactions, in particular in the mesonic sector. A popular modern tool is the AdS/CFT correspondence. It appears to be quite effective, but fundamental difficulties are encountered as well. only if, in the gauge theory sector, the couplings are quite large, one can handle the corresponding anti-De Sitter space calculations in terms of effective field theories. How do we produce more realistic results?

What we have, in practice, is an expansion parameter, something like $1/(g^2N)$, where g is the gauge coupling and N is the number of colors. What we see here is a situation very similar to lattice gauge theories, where also the $1/(g^2N)$ expansion shows string-like behavior. Well, in lattice theories, we have the so-called 'improved actions", which are like the original Wilson action, but with further marginal terms added, so as to give more rapid convergence when we match the lattice expansion with the continuum theory at larger g^2N, or equivalently, at larger lattice mesh length a. Similar methods ar now proposed in the AdS/CFT approach, where it is called "perfection". Using 'perfection', one can achieve promising results for QCD. The advantage of AdS/CFT with perfection, over lattice theories with 'improved action', is that the former has more of the symmetries that one would like to see in the physical limit. The results are claimed to be promising.

Another important issue discussed at this meeting is the question: how does confinement work if one modifies the gauge couplings, the color group, the number of colors, or the number of flavors? In general, one expects various kinds on phase transitions. These things are important to know, among others, in theories with walking technicolor. Does chiral symmetry break spontaneously, or get restored? Which other symmetries survive?

It is tantalizing to realize that QCD is a theory whose predictions are as precise as those of the electro-weak theory. In QED, some effects can be checked against experiment with a precision of 10 decimal places or more. But for QCD, theory is still in a very bad situation. For the non-perturbative, low energy regime, the best we have still seems to be lattice QCD. It allows us to compute hadronic masses and decay constants with two, or at best three decimal places. Why are our predictions here so bad compared to the calculation of $g-2$ for the electron and the muon? Why can we not compute the proton—rho-meson mass ratio with a similar accuracy? What is badly needed is one or several systematic and accurate calculational approaches to QCD and related theories. One possibility would be an accurate effective string theory.

It is not obvious to me whether in fact the AdS/QFT duality will be powerful enough to do this job, since it seems to map a gauge theory onto a rather ill-understood string theory in a higher-dimensional AdS space. We do have perturbation expansions for that, but how badly do these converge? Are there alternative ways? Looking at the original Veneziano amplitude and its Koba-Nielsen generalization, one cannot avoid the impression that some effective string theory should work quite nicely. *Pure* QCD appears to be close to a bosonic string theory in just 4 space-time dimensions, with variable intercept and no massless modes. How can that be? Perhaps the complete glueball spectrum must be approximated by a 26 dimensional string with 22 compactified dimensions. The excitations in these compactified directions may well be some kind of internal excitations of the QCD vortex tubes. But, the standard bosonic string possesses massless B vector fields as well as gravitons. These do not occur in the glueball sector of QCD. Therefore, the spectrum of states must be rearranged so as to turn all photonic and gravitonic states into parts of *massive* spin-one and spin-two modes, not unlike what happens in a Brout-Englert-Higgs theory. Unfortunately, we were unable to envisage a completely acceptable BEH mechanism for spin-two gravitons. In short, although these ideas look nice in words, it has not been possible to turn them into useful tools for QCD.

Turning to the classical string that seems to be describing the QCD vortex well qualitatively, we asked the question what the 'normal modes' should be for the baryonic string (a "wrapped D-brane", according to some). Surprisingly, it was found that, at the Torricelli point, this string couples low-frequency to high frequency modes. Classical string theory here suffers from a divergence not unlike the divergence that forced Max Planck into introducing the concept of quantum mechanics. Quantizing the energy contents of the high frequency modes into integral multiples of $\hbar\omega$ removes the divergence.

We complained that lattice theory gives us far less accurate numbers than theories such as QED. Yet lattice theory is the only non-perturbative approach that, at least in principle, allows for systematic and accurate calculations. Further progress was reported on the problem of maintaining chiral symmetry on the lattice. We are now approaching accuracies that would make it worth while to consider the inclusion of quantum electrodynamical effects, which are an order of α smaller. Most of the lattice calculations are not yet accurate enough to justify these corrections, but it is important to start thinking about these effects.

Strongly coupled gauge theory is applied to study QCD at very high temperature and/or very high pressure. This topic was discussed in great detail at this meeting. For theoretical studies of the very early universe, and of the dense interior of collapsed stars, it is important to know how QCD might condense. Progress was reported on the analysis of the CQD phase diagrams for the 2 flavor and the 3 flavor case. At certain temperatures and pressures QCD might solidify into a crystalline phase: the Larkin-Ovchinnikov-Fulde-Ferrell (LOFF) phase. Its symmetries are still under speculation. Do we have cubic symmetry? Do we have something resembling a liquid crystal, where two cubes are rotated with respect to one another? In that case, an angle of 45° might be the most stable configuration, but it could also be that rotations in 3-space are coupled to rotations in color space. Only very accurate calculations can resolve which of all those configurations actually minimizes the free energy.

The other field where such calculations are relevant is that of the quark gluon plasma. Such a plasma is expected to be produced when heavy ions collide with ultra-relativistic energies. At CERN, LHC is designed to accelerate not only single protons but also heavy nuclei, so as to investigate this plasma. Although the plasma is only generated during a collision, it is expected to come close to thermal equilibrium before it expands and cools off. Strongly coupled QCD is needed to understand what happens. There appears to be no complete agreement yet concerning the phase transition from the hadronic phase towards the plasma phase. When chiral symmetry is completely restored, all particles form explicit representations of the full chiral Lie group, so that they form parity doublets, except for the massless sector. Does chiral symmetry restoration also imply a vanishing ϱ and A_1 mass? Lattice gauge theories allow for the most direct attack on such questions. Some say we should have a first order phase transition at $T \approx 170\,\text{MeV}$, others say it is not a sharp one; presumably this apparent disagreement arises because the system is not *completely* at thermal equilibrium.

There is much interest in the QCD phase diagram. Lattice QCD was used to compute the effects of strong magnetic fields on chiral symmetry. But here also, application of AdS/CFT machine was reported about. One finds several phase transitions, including a superconducting phase. How close do these approximations border to reality? Will we ever find more reliable systematic approaches that allow for precision calculations?

Strong coupling gauge theories are essential for generating technicolor models for the electro-weak sector. Here, our biggest obstacle is the hierarchy problem. How can we explain the occurrence of huge scale differences in a realistic, "natural" theory? A clever trick to avoid the 'speed trap' set by today's precision measurements, is to force the running coupling strength to slow down: "walking technicolor": if the coupling strength is only weakly scale dependent then naturally large scale differences can arise. What worries me here is, is this not a too clever trick? Is a beta function that nearly cancels out to zero not another example of unnatural fine tuning? Would the competitor, the supersymmetric version of the Standard Model, not be in a much better position to explain the data?

Many speakers expressed their eagerness for the new clues that LHC will hopefully provide for us (such as, for example, the detection of leptoquarks). In the mean time, perhaps we could also investigate a third scenario, that of very large gauge groups, with large and complex flavor groups as well. Highly complex gauge systems could mimic extra dimensions, where particles seem to be arranged in Kaluza-Klein sequences even though there are no real extra dimensions — these would make the system non-renormalizable.

Heroic attempts are made to compute the various weak mixing parameters that make up the Cabibbo-Kobayashi-Maskawa matrix. This can only conceivably be successful if the complete set of all possible models are investigated systematically, but this is easier said than done. Examples are the so-called Moose theories, or else "top color assisted technicolor". These are attempts to view the Higgs as a tightly bound $t\,\bar{t}$ bound state, but since this appears not to agree with other observations concerning the top quark, one assumes a Higgs field that strongly mixes with such a state.

Even more difficult and speculative are attempts to produce models to explain the neutrino mixing matrix, to the extent that it is known from the incomplete experimental data. Most likely we have to invoke a see-saw mechanism. Here also, new information from LHC is expected to imply crucial further restrictions on model building. Today we try meticulously to weed out the possibilities one-by-one.

Some of the questions we ask border to the more philosophical ones. Indeed, the philosophy of one's approach is far from irrelevant for our chances of success. One such question is that of the origin of chiral symmetry in the observed particle spectrum. It is not the title of this meeting, but it was the title of one of the contributions.

Can the Higgs be further "unified" with the gauge particles? To achieve this, extra dimensions would be needed, and it is unlikely that such an idea can be reconciled with renormalizability. Should the gravitational sector be quantized completely, or perhaps only partly? The gravitational field would be slightly better manageable if our world were only "partially" quantum, while gravity remains classical; this, however, would lead to other more mind-boggling difficulties, with uncertainty relations and with causality.

The latter, causality, is very high on my list of priorities for any viable theory.

Causality is also an important concept for *conformally invariant* theories, which were considered at several occasions. Conformal invariance is realized automatically as soon as the β function approaches a stable zero, be it a trivial one (where the couplings themselves vanish) or a non-trivial one. Theories can also be conformally invariant either in the UV limit or in the infrared limit. It could even be that QCD itself is conformally invariant in the infrared, but this cannot be traced back directly in the hadron spectrum. If it is true, as one participant suggested, that the QCD coupling strength reaches a fixed point in the infrared, then it is hard to regard this as more than a matter of definition: how do we define the QCD coupling in the region where it is very strong? This is known to be far from unambiguous.

In any case, conformal invariance is an important mathematical tool; it severely limits the dependence of amplitudes on the external momenta.

Finally, some miscellaneous topics were brought forward. One of these was the possibility of "cyclic cosmologies" — indeed, when speaking of wild speculations, nothing beats the cosmologists. A possible exception however is the idea that small, 'detectable', black holes might be formed in a higher-dimensional space surrounding us.

Neutron-antineutron oscillations are a speculation that is not quite as wild. We know that baryon number violating processes can occur, either within or slightly beyond the Standard Model. Neutrinos are likely to mix with anti-neutrinos, so why not neutrons with antineutrons? It would be a $\Delta B = 2$ transition, which is not favored in our most cherished models. The idea is that matter-antimatter oscillations could also be relevant for the process of baryogenesis in the early universe.

However, if neutrons mix with antineutrons, at time scales as short as 10^{10} seconds, then one would have to explain why this does not lead to explicit annihilation processes. Not one in years of Super-Kamiokande. This is why either sphaleron processes, or the transitions caused by virtual Grand-Unified gauge bosons are still our most favored explanation of baryogenesis. This can also be seen as the origin of mass.

We thank the organizers and our hosts for a lively and inspiring meeting.

LIST OF PARTICIPANTS

Anselmi, Damiano	Pisa — Italy
Aoki, Ken-ichi	Kanazawa — Japan
Bardeen, William A.	Fermilab — U.S.A.
Brodsky, Stanley J.	SLAC/ Stanford — U.S.A.
Casalbuoni, Roberto	Florence — Italy
Cheng, Hsin-Chia	California, Davis — U.S.A.
Evans, Nick J.	Southampton — U.K.
Frampton, Paul H.	UNC-Chapel Hill — U.S.A.
Fujihara, Takahiro	Hiroshima — Japan
Fujiyama, Kazuhiko	Nagoya — Japan
Fukano, Hidenori	Nagoya — Japan
Fukaya, Hidenori	Riken — Japan
Furuhashi, Yusuke	Nagoya — Japan
Haba, Kazumoto	Nagoya — Japan
Harada, Masayasu	Nagoya — Japan
Hashimoto, Michio	Nagoya — Japan
Hayakawa, Masashi	Nagoya — Japan
Hayashi, Masako	Hiroshima — Japan
Holt, Jeremy W.	SUNY Stony Brook — U.S.A.

Hong, Deog-Ki	Pusan — Korea
Hosek, Jiri	NPI — Czech
Hosotani, Yutaka	Osaka — Japan
Ichinose, Shoichi	Shizuoka — Japan
Imoto, Toshiya	Nagoya — Japan
Inagaki, Tomohiro	Hiroshima — Japan
Ishiduki, Megumi	Nagoya — Japan
Kanaya, Kazuyuki	Tsukuba — Japan
Kikukawa, Yoshio	Tokyo — Japan
Kimura, Daiji	Hiroshima — Japan
Kim, Sung-Gi	Nagoya — Japan
Kitazawa, Noriaki	Tokyo Metro. — Japan
Ko, Pyungwon	KIAS — Korea
Kohda, Masaya	Nagoya — Japan
Kondo, Kei-Ichi	Chiba — Japan
Konishi, Kenichi	Pisa — Italy
Kouvaris, Christoforos	Niels Bohr — Denmark
Kugo, Taichiro	Kyoto — Japan
Kumagai, Akira	Nagoya — Japan
Kurachi, Masafumi	SUNY Stony Brook — U.S.A.
Lee, Jong-Phil	KIAS — Korea
Maekawa, Nobuhiro	Nagoya — Japan
Maru, Nobuhito	Kobe — Japan
Matsuzaki, Akihiro	Nagoya — Japan

Matsuzaki, Shinya	Nagoya — Japan
Mohapatra, Rabindra N.	Maryland — U.S.A.
Nagao, Keiko	Nagoya — Japan
Nakamura, Satoshi	Kanazawa — Japan
Nakkagawa, Hisao	Nara — Japan
Namekawa, Yusuke	Nagoya — Japan
Nemoto, Yukio	Nagoya — Japan
Nishijima, Kazuhiko	Tokyo — Japan
Nishino, Hiroyuki	Nagoya — Japan
Noda, Shusaku	Osaka — Japan
Nojiri, Shin'ichi	Nagoya — Japan
Nonaka, Chiho	Nagoya — Japan
Okagawa, Hiroyuki	Nagoya — Japan
Ono, Akihito	Chiba — Japan
Sakurai, Kazuki	Nagoya — Japan
Sawada, Shoji	Nagoya — Japan
Schechter, Joseph M.	Syracuse — U.S.A.
Semenoff, Gordon W.	British Columbia — Canada
Shibata, Akihiro	KEK — Japan
Shrock, Robert	SUNY Stony Brook — U.S.A.
Sinclair, Donald K.	Argonne — U.S.A.
Sugimoto, Shigeki	Nagoya — Japan
Takei, Kazuaki	Nagoya — Japan
Takenaga, Kazunori	Tohoku — Japan

Takeuchi, Tatsu	Virginia Tech — U.S.A.
Tanabashi, Masaharu	Tohoku — Japan
Terao, Haruhiko	Kanazawa — Japan
't Hooft, Gerard	Utrecht — Netherlands
Tsuchiya, Akito	Kanazawa — Japan
Tsukada, Yuki	Nagoya — Japan
Uehara, Shozo	Utsunomiya — Japan
Uno, Shunpei	Nagoya — Japan
Wang, Shang-Yung	Tamkang — Taiwan
Watanabe, Noriaki	Nagoya — Japan
Wijewardhana, Rohana	Cincinnati — U.S.A.
Wu, Xiao-Hong	KIAS — Korea
Yada, Tomoki	Nagoya — Japan
Yamawaki, Koichi	Nagoya — Japan
Yan, Qi-Shu	NCTS — Taiwan
Yasuda, Junichiro	Nagoya & Tokyo — Japan
Yokoi, Naoto	Riken — Japan
Yoshikawa, Tadashi	Nagoya — Japan
Yoshimoto, Shunji	Nagoya — Japan
Zahed, Ismail	SUNY Stony Brook — U.S.A.
Zakharov, Valentine I.	INFN, Pisa — Italy